CW01429539

International Manufacturing Strategies

International Manufacturing Strategies
Context, Content and Change

Edited by

Per Lindberg
Chalmers University

Christopher A. Voss
London Business School

and

Kathryn L. Blackmon
London Business School

KLUWER ACADEMIC PUBLISHERS
BOSTON / DORDRECHT / LONDON

A C.I.P. Catalogue record for this book is available from the Library of Congress.

ISBN 0-7923-8061-4

Published by Kluwer Academic Publishers,
P.O. Box 17, 3300 AA Dordrecht, The Netherlands.

Sold and distributed in the U.S.A. and Canada
by Kluwer Academic Publishers,
101 Philip Drive, Norwell, MA 02061, U.S.A.

In all other countries, sold and distributed
by Kluwer Academic Publishers,
P.O. Box 322, 3300 AH Dordrecht, The Netherlands.

Printed on acid-free paper

All Rights Reserved
© 1998 Kluwer Academic Publishers
No part of the material protected by this copyright notice may be reproduced or
utilized in any form or by any means, electronic or mechanical,
including photocopying, recording or by any information storage and
retrieval system, without written permission from the copyright owner.

Printed in the Netherlands

TABLE OF CONTENTS

Over the past 20 years, an increasing number of factors have placed the manufacturing strategies of companies and countries in a global context. These include:

- The development and diffusion around the globe of new practices such as 'Lean Production';
- The increasing opening of domestic markets to foreign competition;
- The increasing globalisation of manufacturing by formerly domestic companies;
- The increasing role of offshore manufacturing within the host country, both as a competitor and as a source of new and improved ways of manufacturing;
- The move to global rather than local supply chains;
- The rapid political development of new and old economies, from the opening of Eastern Europe to the growth of the EC, and the development of Mercosur and NAFTA.

This study focuses on a single industry, the manufacturing of fabricated/assembled metal products, electrical/non-electrical machinery and equipment, transportation equipment, and scientific/control instruments. Within these industrial segments, companies are on the one hand being faced with new flexible technologies, lean production paradigms, time-based competition, and outsourcing; on the other, they are also being faced with strong industry and national pressures to change the way in which work is organised and managed.

In their efforts to change and improve, companies are finding that their manufacturing strategies and trajectories for improvement have been shaped in the past, and constrained in the future—by their country's history and socio-economic context, by factors such as culture, legislation, the strength or weakness of the domestic currency, and so on.

CONTENTS

This book seeks out to review and address the global manufacturing strategy area through research in the four major economic areas of the world: Europe, North America, Latin America, and Asia. The manufacturing strategies, practices, and performance of 600 manufacturing sites have been examined in twenty countries around the globe. The book sets out to review the strategies in some of the key countries and to examine in detail the relationship between national context and these strategies.

There are a number of manufacturing practices that any company must address if it is to reach world class standards of performance. These include investment in information systems and advanced manufacturing technologies, new forms of work organisation, the broad set of practices known as 'Lean Production', and linking manufacturing strategy to business strategy. However, as Gustavo Vargas points out in Chapter 13, 'best practices' can hardly claim to be neutral and therefore unaffected by the specific macroeconomic and sociocultural environment in which firms operate. Government policies, work rules, social attitudes, factor costs, and competition intensity all play a vital role in shaping what constitute 'best practices'.

Included in these chapters are examples of all of these:

- the impact of very different macroeconomic climates in two neighbours, Argentina and Brazil;
- the impact of the changing economic background, for example, the strong yen's effect on Japanese companies;
- a review of the postulated short-termism in the UK;
- the effect of distinctive cultural and trading context in countries such as Denmark; and
- the role of supranational economic trading blocs such as NAFTA on manufacturing strategies in Canada and Mexico.

A second focus is the difference in practices between Japan and the US, and other Western countries. The research builds a clear picture of these differences and some of the underlying reasons for them.

The book also explores how, in changing from traditional to leading-edge practice, there are a range of choices open to an organisation. The research has identified strong differences in the trajectories for change found in different countries and in companies within a country. The drivers and consequences of these differences are explored.

Finally, the richness of a database of 600 sites from 20 countries has led to the opportunity to build on existing research and explore new aspects of manufacturing strategy theory.

UNDERLYING RESEARCH PROJECT

This book is based on a single major research project undertaken in 20 countries around the globe, including Japan, the US, the UK, Sweden, Italy, Brazil, and Argentina. It focuses on manufacturing strategies and practices in each, and uses the research data to focus on factors specific to industrial countries or regions, and those that are common across a group of countries or the entire sample.

THE BOOK

The core of the book is a set of chapters each reviewing an individual country. An overall common approach was applied to understanding the country's socio-economic background, the distinctive results for that country from the research, and the links between the two. Most chapters are illustrated by a small case study of a company. Differences between the country chapters reflect both the individual perspectives of the authors and the differences between individual countries. In some cases, such as Japan, the observations and analysis are spread over a number of chapters. A second set of chapters compares countries or looks at country-based issues in greater depth.

The following set of chapters integrates the findings from the various countries, the different trajectories that have been followed, and the impact of external variables and socioeconomic context on these.

The final part of the book is devoted to new ideas and developments in functional areas and in manufacturing strategy that have been developed from the analysis conducted during the research.

The appendices present a description of the research methodology and the results of the research project.

ACKNOWLEDGEMENT

The Engineering and Physical Science Research Council (EPSRC) provided funding for this project at London Business School.

CONTRIBUTORS

REBECCA ARKADER, Doctoral Student, COPPEAD, Universidade Federal de Rio de Janeiro, Rio de Janeiro, Brazil.

EMILIO BARTEZZAGHI, Dipartimento di Economia e Produzione, Politecnico di Milano, Milano, Italy.

ANDREA BERT, Dipartimento di Economia e Produzione, Politecnico di Milano, Milano, Italy/ Consorzio MIP, Milan, Italy.

KATE BLACKMON, Centre for Operations Management, London Business School, London, England.

HARRY BOER, School of Management Studies, Department of Technology and Organisation, University of Twente, Enschede, Netherlands / The University of Loughborough, England.

RAFFAELLA CAGLIANO, Dipartimento di Economia e Produzione, Politecnico di Milano, Milano, Italy.

DAE-SUNG CHANG, Kyungki University, Korea.

SOONG-HWAN CHUNG, Hansung University, Korea.

DOMIEN DRAAIJER, School of Management Studies, Department of Technology and Organisation, University of Twente, Enschede, Netherlands / Philips Semiconductors, Nijmegen, The Netherlands.

JOHN ETTLIE, School of Business Administration, University of Michigan, Ann Arbor, Michigan, USA.

P. FERNANDO FLEURY, Ipiranga Professor of Manufacturing Strategy, COPPEAD, Graduate Business School, Universidade Federal de Rio de Janeiro, Rio de Janeiro, Brazil.

JAN FRICK, Department of Business Administration, Hoyskolen i Stavanger, Stavanger, Norway.

FRANK GERTSEN, Institut for Produktion, Aalborg Universitetscenter, Aalborg, Denmark.

JOAQUIM BORGES GOUVEIA, Universidade Católica Portuguesa, C.R. Porto, DEEC, FEUP, Universidade do Porto,, Porto, Portugal.

POUL H. K. HANSEN, Institut for Produktion, Aalborg Universitetscenter, Aalborg, Denmark.

ROAR HJULSTAD, Department of Business Administration, Hoyskolen i Stavanger, Stavanger, Norway.

RAIMO HYOTYLAINEN, VTT Automation, Industrial Automation, Espoo, Finland.

FRASER JOHNSON, Richard Ivey School of Business, University of Western Ontario, London Ontario, Canada.

JOHN W. KAMAUFF, Richard Ivey School of Business, University of Western Ontario, London, Ontario, Canada.

PER LINDBERG, Department of Operations Management, Chalmers University of Technology, Goteborg, Sweden.

ROBERTO LUCHI, IAE, Universidad Austral, Buenos Aires, Argentina.

MARCELO PALADINO, IAE, Universidad Austral, Buenos Aires, Argentina.

K PIETILAINEN, VTT Automation, Industrial Automation, Espoo, Finland.

EDUARDO REMOLINS, IAE, Universidad Austral, Buenos Aires, Argentina.

BOO HO RHO, Sogang University, Korea.

JENS O. RIIS, Institut for Produktion, Aalborg Universitetscenter, Aalborg, Denmark.

NORM SCHEIN, Richard Ivey School of Business, University of Western Ontario, London, Ontario, Canada.

MAGNUS SIMONS, VTT Automation, Industrial Automation, Espoo, Finland.

HONGYI SUN, Department of Business Administration, Hoyskolen i Stavanger, Stavanger, Norway.

RUI SOUSA, Universidade Católica Portuguesa, Porto, Portugal.

GIANLUCA SPINA, Dipartimento di Economia e Produzione, Politecnico di Milano, Milano, Italy.

GUSTAVO A. VARGAS, Insituto de Empresa, Madrid, Spain, and California State University, Fullerton, CA, USA.

CHRIS VOSS, Centre for Operations Management, London Business School, London, England.

PETER WARD, College of Business, Ohio State University, Columbus, Ohio, USA.

BERT WOOD, Richard Ivey School of Business, University of Western Ontario, London, Ontario, Canada.

HAJIME YAMASHINA, Kyoto University, Kyoto, Japan.

YONG-MOK YU, Dankook University, Korea.

PART I - INTRODUCTION

CHAPTER 1

INTRODUCTION: INTERNATIONAL MANUFACTURING STRATEGY:

CONTEXT, CULTURE, AND CHANGE

1.1. The global context of manufacturing practices and strategies

This book is concerned with the way in which national context has an impact on how manufacturing strategies are set, the ways in which companies and countries continually improve their manufacturing capability, and the resulting improvement in performance. It is based on a study of 600 companies in 20 countries.

Over the past decade, there has been a continual stream of innovation in manufacturing. The rate of technological innovation has led to new product options in industry and services. At the same time it has led to a reversal in the trend towards simplification and division of manufacturing work, in favour of flexible and cost efficient production of high quality products. These innovations include the development of new practices such as those embodied in Lean Production and individual techniques such as Total Productive Maintenance; and the continual development of new technologies such as flexible automation. They have commonly been seen as coming from Japan, but in reality have originated in many countries including the USA (statistical process control, materials requirements planning), Russia (group technology) and the UK (flexible manufacturing systems), as well as Japan.

The globalisation of competition and sourcing has led to increased pressures on, and opportunities for, national industries: challenging them to develop the ability to change strategies and tactics in response to new demands. Finally, social, economic and environmental demands have increasingly made traditional technological and organisational forms obsolete. Overall, the situation in the early 1990s was one of turbulence and of a need for understanding of the changes taking place in manufacturing industry.

Over the same period, manufacturing has moved from being the preserve of a few sophisticated industrial nations, to being central to economies in every quarter of the globe. As best practice, or 'world class' manufacturing, is considered to embody most of these innovations practice and technology, the strategies for the use, applicability and performance of these in different countries and regions has become of vital importance.

3

P. Lindberg et al. (eds.), International Manufacturing Strategies, 3-17.
© 1998 *Kluwer Academic Publishers. Printed in the Netherlands.*

CHAPTER 1

The development of manufacturing strategies within the context of a country or a region is of increasing concern to a wide range of parties. Multinational firms are operating across regions and are trying to develop manufacturing strategies that are tailored to the local context of their operations. Domestic firms are seeking to set manufacturing strategies for competing in markets open to global players, to tailor these to their local context and to develop effective plans for introduction of world class manufacturing. As companies in some developing countries have described it, moving from the factory of the past to the factory of the future. Governments and trading blocs are concerned with how the competitiveness of manufacturing in their areas can be promoted, and are seeking to promote change that will ensure the increasing effectiveness of their domestic manufacturing industry.

Every country and region represents a different context for manufacturing strategy. The local context will include the economic factors such as currency strength, the impact of regulations such as in the EC, the historical position of manufacturing such as in many Latin countries moving from protected manufacturing to open economies, the local investment climate, market sizes and characteristics. Its context will also include the social and cultural aspects of the country and region that impact manufacturing.

The global context within which manufacturing strategies develop is changing and the exploration and understanding of this dynamic is important. In responding to this dynamic, a company must make choices from a wide range of possible activities. The sequence of these choices over time is the trajectory of change. In this study we have sought to examine how these trajectories vary with context and country.

For at least a decade it has been argued that a systematic strategy for developing manufacturing is a necessary prerequisite for achieving good practice and performance in manufacturing. The problem that practitioners have in developing such a strategy for manufacturing is to understand what the competitive situation and market demands, what the current problems are, and what measures in manufacturing work in such a situation. This requires access to knowledge of manufacturing practice and its effect on performance. The compilation of knowledge and explanation of performance, in turn, becomes the agenda for researchers. This is the background to this book.

1.2. Assessing manufacturing practice and performance

Understanding practice and performance in manufacturing is a key managerial task and often also a key problem. Manufacturing managers must pay attention to what investments to make, what changes to support, what organisation form is appropriate, what systems to develop—in short, what kind of manufacturing practice should prevail. Given the vast range of tools, models and techniques available for the interested manager, one could assume that the process of developing a good manufacturing practice is an easy one. Unfortunately, history tells us that this is not the case. Not only are there but a few examples of excellence in manufacturing practice—let alone world class—but the task of understanding what others are doing and what best practice is all about involve a long search process and with implementation, an even longer process of rearrangement and change in operations.

In order to develop our understanding of these issues, a research study was launched by Chalmers University of Technology and London Business School. The aim of this study, the International Manufacturing Strategy Survey (IMSS), was to investigate manufacturing strategies and practices in industrialised nations around the world, and to create possibilities for comparative analysis of manufacturing strategies in the engineering industry throughout the world.

1.2.1. UNDERSTANDING THE DRIVERS FOR MANUFACTURING CHANGE

Several studies and examples have shown that manufacturing change and development is dependent on the drivers for change and thus the context in which manufacturing is undertaken. These drivers, or forces, derive not only from the competitive context of the firms, from which the needs for improvements are determined, but also from the national and regional context, in both the macro-economic sense and in the cultural sense.

1.2.2. THE STRATEGIC CONTEXT

The business in which a company operates will have an impact on the strategy for manufacturing. For example, the highly cyclical and capital-intensive nature of the paper and pulp industry puts totally different demands on inventory levels and capacity utilisation policies than the aerospace industry would. Similarly, the supply chain and co-operation between supplier and customer are different in different industries, and also vary over time. For example: ICL, the leading British computer company, redefined its structure in the early 1980s because of poor performance and a need to revitalise manufacturing. It formed a fully owned subsidiary, Design to Distribution (D2D). From being a poor internal supplier to ICL, D2D became a successful contract manufacturer in the electronics business, winning the European Quality Award in 1994. Ericsson Telecom, the leading Swedish manufacturer of public telecommunication equipment has undertaken similar structural changes in the early 1990s, leading to radically redefined supply structures and roles for component manufacturers.

A manufacturing strategy is also a function of markets and customers. Suppliers to the telecommunication operators such as Ericsson Telecom have faced a totally different customer behaviour in the wake of deregulation, increasing needs for cost efficiency and lead-time reduction. The customers in public telecommunication are, however, still quite different from the new operators in the mobile phone business. Whereas the markets and customers in public telecommunication by and large still are quite conservative and price conscious, the mobile phone business is expanding very

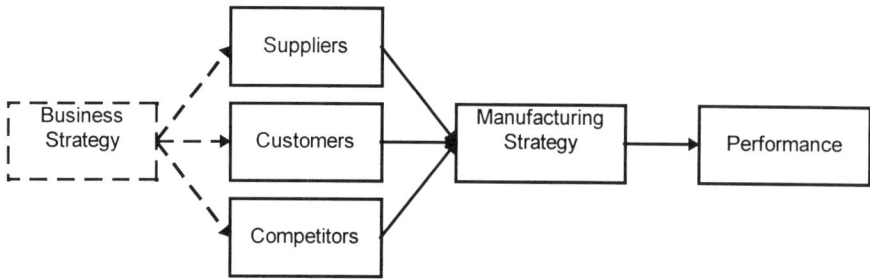

Figure 1-1. A Model of the Strategic Context of Manufacturing

rapidly, placing premium demands on technological performance and delivery performance rather than on price. A comparison between Ericsson Radio (mobile/cellular phones) and Ericsson Telecom (public telecom), will show that this results in the development strategies and manufacturing strategies in the two businesses being quite different. Thus, the impact of customers, suppliers and competitors on manufacturing is clear and obvious, with the restructuring and diffusion of Lean Production in the auto industry yet another example of strategic changes in manufacturing structure and practice. Figure 1-1 depicts a "traditional" model of these relationships.

It is generally argued that improved strategies and practices in manufacturing lead to improved performance. While this may seem as an obvious statement, it is not clear as to what strategies are effective in what contexts, and what combinations of practices that produce results. Neither is it clear whether there are certain strategies that pay off more in certain industries or companies than in other, due to the social, economic and/or cultural context of the firms. One objective of the present study was to enhance the understanding of the performance effects of different strategies and practices, and to highlight the impact of economic and cultural context.

1.2.3. THE ECONOMIC CONTEXT

The period in the early 1990s was characterised by turbulence and great economic uncertainty. The growth of the 1980s was succeeded by stagnation and in some instances depression. The political arena was characterised by the fall of the eastern block, the Gulf War and the manifestation of trade blocs (e.g., the European Union and the North American Free Trade Union), while economic policies nurtured privatisation and deregulation.

There are several noteworthy occurrences. First, in all OECD countries industrial employment decreased between 1990 to 1994 (Japan kept the same level), and with a few exceptions, industrial production fell. Since employment decreased more than production, manufacturing companies must have undertaken sharp measures to increase productivity (with a few exceptions such as Japan and Germany, where productivity decreased over the period). In addition there are marked differences between countries in the level of unemployment and interest rates, resulting in differences in costs for manufacturing investments, as well as differences in general consumption.

The economic context is of course under constant change. Table 1-1 indicates the environment under which manufacturing strategies were formed in the mid-1990s. Even if the other factors in the strategic environment are of equally great importance for manufacturing strategy as these economic circumstances, they are important as context for understanding strategic behaviour. One objective of the IMSS is therefore to analyse and discuss the impact of economic context on manufacturing, and to show that there is more to understanding manufacturing strategy than just the industrial or business logic. Even if the indicators in Table 1-1 cover a relatively short time period, the argument put forward here is that the influence of economic context on manufacturing strategy has a significant long-term impact on the way manufacturing is strategically managed.

1.2.4. THE CULTURAL CONTEXT

Not only do companies define manufacturing strategies in an economic and business environment, they need also to take into account the specific culture that is present in the country. Cultural traits, such as relationships between employees, attitude to authority, conflict resolution and motivational structures, all affect the organisational patterns of manufacturing, as well as the investment behaviour and attitude towards new practices and technologies. Thus, culture will have a profound impact on the decisions made in organisations, and thereby also on the strategies that evolve over time. This is especially true in manufacturing, since it often is labour-intensive, thereby strongly encapsulating national norms and values.

Table 1-1. Economic Indicators 1994–1995

	Trade balance (% GNP) 1995	Industrial production (1990 = index 100) 1994	Industrial employmen t 1994	Interest rates Dec. 1994	Unemploy- ment (%) 1994	GNP growth (annually) 1983–1993
USA	-2.62	105	95	7.97	7.6	2.8
Norway	—	102	94	8.02	6.0	2.7
Denmark	—	101	89	9.1	10.7	1.9
Holland	—	100	94	7.78	6.7	2.5
Canada	3.24	98	85	9.15	11.2	2.6
Finland	8.57	96	75	9.52	17.7	1.0
Great Britain	-1.56	95	86	8.44	10.2	2.2
Germany	3.04	93	98	7.5	8.8	2.8
Italy	3.55	93	—	10.27	10.8	2.2
Sweden	7.57	93	70	10.68	8.2	1.2
Portugal	—	92	78	—	5.5	2.8
Japan	3.05	91	100	4.59	2.5	3.7
Spain	N/A.	91	—	11.07	22.4	2.8

Source: OECD (1994)

In this book, we try to capture some national characteristics that may influence manufacturing strategy and practice. Trompenaars (1993) summarises different national patterns of corporate culture through two cultural dimensions; centralised-decentralised and formal-informal orientation in the organisation (see Figure 1-2). Even though the figure should be interpreted with some caution, as small companies in any culture tend to be less formal and large companies more formal, the results show very marked distinctions.

The dimensions shown in Figure 1-2 represent only a fraction of the possible cultural and social aspects that impact organisational behaviour. There are, as mentioned above, several other dimensions that may be of significance. Thus, rather than using a pre-defined set of comparative cultural measures, we refer to those dimensions and particular areas that are of significance for each specific country.

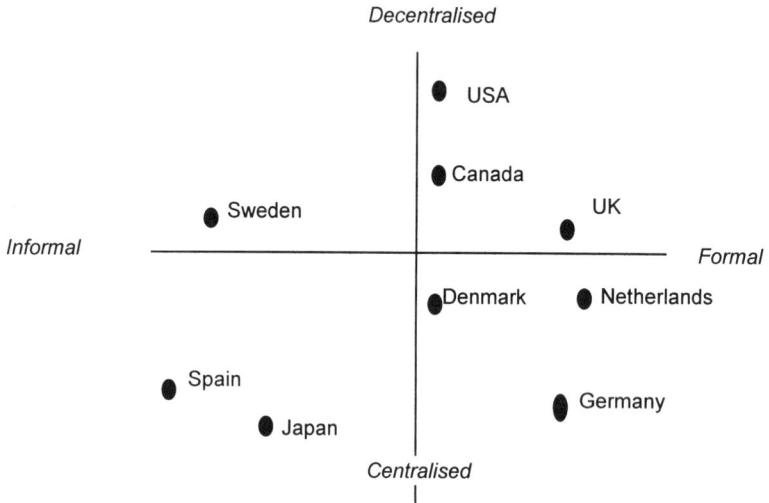

Figure 1-2. National Patterns of Corporate Climate (source; Trompenaars, 1993)

1.3. The aim and scope of this book

This book is based on the work of the research teams at Chalmers and London, and draws on the analyses done by collaborators in many of the 20 countries involved in the study. The aim and scope of this book reflect the issues discussed above and include:

• to compare manufacturing strategy and practice in 20 countries around the world;
• to examine the influence of the economic and social background of different countries on manufacturing strategies;

- to develop an understanding of the impact of trading blocs such as the EU, NAFTA and Mercosur on manufacturing companies;
- to describe the performance effects of certain manufacturing strategies and practices;
- to identify the different trajectories of change found in companies, countries and regions;
- to use the data from the study to review existing and emerging new paradigms in manufacturing;
- to analyse the impact of the socio-economic environment on the strategies of firms.

The next part of this chapter presents the theoretical background of the research and the research methodology. It then outlines some general empirical results. The final section gives an overview of the rest of this book.

1.4. Theoretical Background

Manufacturing strategy has been variously defined as:

- a plan that describes the way to produce and distribute the product (Mayer and Moore, 1983);
- the task that manufacturing must accomplish (Skinner, 1978);
- a consistent pattern of decision-making (Hayes and Wheelwright, 1984);
- a long-range plan or vision for the manufacturing function (Schroeder et al., 1986).

Skinner (1969, 1974, 1978, 1985) laid the conceptual foundations for the study of manufacturing strategy, which was followed up with empirical research in the early 1980s, and the area of manufacturing strategy became defined as a field and widely diffused beyond the United States to Japan and Europe in the early 1990s.

Leong et al. (1990) and Voss (1992) identified the two dimensions of research into manufacturing strategy as "process" research and "content" research. Content research focuses on the specifics of what was decided (e.g., Buffa, 1984; Fine and Hax, 1985; Hayes and Wheelwright, 1984; Hill, 1989; Schroeder et al., 1986; Skinner, 1979; Leong et al., 1990). On the other hand, process research focuses on how strategic decisions are addressed in an organisational setting, including its links to business and corporate strategy (e.g., Skinner, Anderson et al., 1989; Fine and Hax, 1985; Hayes and Wheelwright, 1984; Hill, 1989).

This study was designed to gather information about manufacturing strategy in an international context, in particular the relationship between business unit goals, competitive priorities and action programs and operational and business performance. In the next section of this chapter, several other global manufacturing strategy research projects are described in order to make clear the aims and goals of the present study.

1.4.1. GLOBAL MANUFACTURING STRATEGY RESEARCH

This book presents the results from our international study of the content of manufacturing strategy. The primary concern of the IMSS research project has been to investigate the content of manufacturing strategy in an international context. Until the end of the 1980s, most manufacturing strategy research was concentrated on the United States, and thus to a large extent was shaped and influenced by the special economic, social, political and cultural situation prevailing there. More recently, research in other areas, for example European views of manufacturing strategy, have begun to be heard. A number of significant international projects have emerged, including the International Manufacturing Futures Survey, the Global Manufacturing Research Group, the World Class Manufacturing Study, the Made in Europe project, and the Global Manufacturing Technology and Strategy Vision Project.

International Manufacturing Futures Survey

The International Manufacturing Futures Survey is one of the longest-running international research projects on manufacturing and competitiveness. The project was initiated in 1981 at Boston University by Professor Jeffrey G. Miller, and since 1983 has been administered in Japan by Waseda University and in Europe by INSEAD. It includes over 200 individual questions designed to assess the firm's business strategy, competitive priorities, manufacturing objectives, action programs, and performance improvement, and is answered by over 500 large and successful manufacturing businesses in these three regions.

Global Manufacturing Research Group

The Global Manufacturing Research Group (GMRG) is a 'multinational community of researchers dedicated to the study and improvement of manufacturing practices world-wide'. It was founded by Professor Clay Whybark of the University of North Carolina at Chapel Hill, and has been holding annual workshops since 1990. Data have been received from nearly 500 companies in 11 countries.

World Class Manufacturing Study

The World Class Manufacturing Study (WCM) was initiated by Professors Roger Schroeder of the University of Minnesota and James and Barbara Flynn of Iowa State University to investigate world class manufacturing, the 'set of processes designed to achieve a sustainable competitive advantage through continuous improvement of manufacturing capability'. It has since been expanded to Japan, Italy, the United Kingdom, and Germany. The project concentrates on firms in the industries of transportation, electronics, and machinery.

Made in Europe
The Made in Europe project, headed by Professor Chris Voss at London Business School, investigates the adoption of world class manufacturing projects and the resulting gains in manufacturing performance. Over 1,350 manufacturing sites in the United Kingdom, Germany, the Netherlands, and Finland have participated in the project since 1993.

Other Projects
In addition, there have been a number of one-off or narrowly focused studies. For example, Professor Roger Schmenner of Indiana University has conducted an number of international comparisons of factory productivity with international partners, including Professor Robert Collins of IMD in Lausanne, Switzerland, and Professor Boo Ho Rho of Korea. Detailed survey data has been collected from over 500 factories in the United States, Europe, and Korea to examine factors that contribute to productivity at the plant level. Another example is the Worldwide Manufacturing Competitiveness Study conducted in the automotive components industry by Andersen Consulting with British academic partners.

1.5. The IMSS Study

The data for this study were obtained through the administration of the International Manufacturing Strategy Survey (IMSS) instrument. The IMSS was administered in 20 different countries as a postal survey. Data collection was supervised by an academic partner in each of the countries and co-ordinated by Chalmers University of Technology and London Business School. Data gathering took place during 1993. The final sample included 600 surveys from companies. Table 1-2 presents the distribution of responses.

This is comparable to other international surveys of manufacturing strategy, such as Kim and Miller (1992), which used responses from 111 US firms; Vastag and Whybark (1994), which used responses from 153 firms in the US and Europe, and Ferdows and De Meyer (1988), which used responses from 222 firms in Europe.

Table 1-2. Number of Responses by Country

Country		Total	Country		Total
Argentina	ARG	41	Great Britain	GBR	36
Australia	AUL	29	Germany	GER	24
Austria	AUT	27	Italy	ITA	41
Belgium	BEL	3	Japan	JAP	27
Brazil	BRA	28	Mexico	MEX	62
Canada	CAN	23	Netherlands	NED	27
Chile	CHI	6	Norway	NOR	20
Denmark	DNK	17	Portugal	POR	41
Spain	ESP	29	Sweden	SWE	61
Finland	FIN	17	United States	USA	41
Grand Total					600

The IMSS study focused on companies in a single industry, category, ISIC 38—Fabricated metal products. Industry segments included companies in ISIC 381 (Metal products, except machinery and equipment), 382 (Machinery except electrical), 383 (Electrical machinery, apparatus, appliances and supplies), 384 (Transport equipment), and 385 (Professional and scientific and measuring and controlling equipment, not elsewhere classified, and photographic and optical goods). The distribution of respondents among these 3-digit categories was 33.6% in 381, 14.6% in 382, 21.5% in 383, 14.2% in 384, and 8.6% in 385.

Respondents included strategic business units and individual firms (inclusive of parent companies). 46.6% of the respondent were from companies, 26.1% from divisions, 26.6% from plants, and 0.7% from other types of business units.

The IMSS questionnaire included four broad categories of questions. The first category determined the profile of the company or business unit. The second category addressed the structural and infrastructural choices. The third, and largest, section asked for detailed information about manufacturing objectives and action programs. The fourth section questioned respondents about their manufacturing performance. A copy of the survey is shown in the Appendix.

1.5.1. SELECTED RESULTS

In this section, we describe the overall results in four key areas—competitive priorities, manufacturing objectives, action programs, and business unit performance. These are analysed in greater detail by country in subsequent chapters.

Competitive Priorities
Competitive priorities represent competitive abilities that a firm is seeking to acquire, sustain or improve upon, with the goal of differentiating itself relative to competitors and/or lowering its costs (Porter, 1980, 1985). Within the manufacturing context, competitive priorities have been identified as cost, quality, delivery, and flexibility (Skinner, 1979; cf. Hill, 1989). These priorities have been validated through empirical studies (e.g., Hayes and Wheelwright, 1984; Swamidass and Newell, 1987;

Table 1-3. Importance of Business Unit Goals

Priority	Description	Importance	
		Mean	S.D.
Quality	Superior product design and manufacturing quality compared with competitors	4.60	0.668
Service	Superior customer service compared with competitors	4.44	0.805
Costs	Lower manufacturing costs than competitors	4.33	0.838
Dependable delivery	More dependable deliveries than competitors	4.26	0.870
Delivery speed	Faster deliveries than competitors	4.16	0.882
Product variety	Wider product range than competitors	3.39	1.131

Ward et al., 1988). Further competitive priorities that have been identified include time (Stalk, 1988; Bower and Hout 1988), and service.

Five competitive priorities were examined: cost, quality, delivery speed, delivery dependability, and product range. For each competitive priority, the respondent was asked to indicate its degree of importance to the business unit. A five-point, self-anchoring scale was used where 1 = 'Not important' and 5 = 'Very important'. Table 1-3 presents the average scores of importance on business unit goals as reported by the respondents.

Manufacturing Objectives
Based upon these competitive priorities, a number of manufacturing objectives may be identified, including price, volume flexibility, conformance quality, performance quality, service, product flexibility, delivery dependability, and delivery speed (Vickery et al., 1992).

For each manufacturing objective, respondents were asked to indicate its degree of importance using the 5-point scale described above and whether the goal was a quantified objective for the business unit, as yes or no. Table 1-4 presents the averaged importance scores and percentage of respondents having quantified goals.

The rankings of the responses to the importance of manufacturing objectives and having quantified goals were slightly different. The most important manufacturing objectives were (1) improving conformance quality, (2) reducing unit costs, (3) reducing overhead costs, and (4) reducing materials costs. The least important manufacturing objectives were (1) increasing delivery speed, (2) improving the ability

Table 1-4. Explicit Manufacturing Objectives and Degree of Importance

Description	Quantified		Importance	
	(%)	Rank	Mean	Rank
Reduce unit cost	82.4%	1	4.28	2
Improve conformance quality	81.5%	2	4.29	1
Improve direct labour productivity	77.5%	3	3.98	5
Reduce inventories	77.4%	4	3.93	7
Reduce overhead costs	74.6%	5	4.10	3
Reduce manufacturing lead time	74.6%	6	3.96	6
Reduce materials costs	74.3%	7	4.04	4
Increase delivery reliability	67.0%	8	3.90	8
Improve supplier quality	64.4%	9	3.82	9
Reduce procurement lead time	59.6	10	3.56	12
Increase delivery speed	54.1%	11	3.52	13
Reduce new product development time	52.2%	12	3.59	11
Improve ability to make rapid volume changes	44.4%	13	3.43	14
Improve white-collar productivity	48.1%	14	3.60	10
Reduce number of suppliers	42.5%	15	2.99	16
Improve ability to make rapid design changes	38.5%	16	3.41	15

to make rapid volume changes, (3) improving the ability to make rapid design changes, and (4) reducing the number of suppliers. Fewer than 50% had adopted goals for (1) improving the ability to make rapid volume changes, (2) improving white-collar productivity, (3) reducing the number of suppliers, or (4) improving the ability to make rapid design changes.

Action Programmes
In order to support competitive priorities and manufacturing objectives, companies may implement action programs. New technologies such as CAD, CAM, FMS and CIM have been identified by Hayes and Jaikumar (1988) as examples of such action programs.

The survey also asked respondents to indicate from a list of activities the degree of use for each activity over the past two years and the relative payoff from the activity; if the activity was not currently being used, the respondent was asked to indicate whether

Table 1-5. Action Programmes

Description	Degree of use		Relative payoff	
	Mean	Rank	Mean	Rank
CAD	3.60	1	3.74	1
Health and safety programs	3.50	2	3.58	5
Implementing team approach (work groups)	3.38	3	3.65	2
Defining a manufacturing strategy	3.31	4	3.58	5
Total quality management program	3.18	5	3.38	9
Materials requirements planning (MRP)	3.16	6	3.47	8
Environmental protection programs	3.16	7	3.28	16
ISO 9000	3.14	8	3.28	16
Just-in-time manufacturing (lean production)	3.03	9	3.60	3
Kaizen (continuous improvement)	3.00	10	3.60	3
Reorganise to a 'plant within a plant'	2.92	11	3.51	7
Just-in-time (frequent deliveries to customers)	2.91	12	3.35	10
Statistical process control (SPC)	2.89	13	3.08	23
Quality policy deployment	2.78	14	3.19	20
Zero defects	2.73	15	3.28	16
CAM	2.69	16	3.30	14
MRPII	2.66	17	3.35	10
Value analysis/redesign of products	2.64	18	3.29	15
Energy conservation programs	2.58	19	2.97	27
Quality function deployment	2.56	20	3.24	19
Simultaneous engineering	2.56	21	3.13	21
Pull scheduling (Kanban)	2.55	22	3.34	12
Design for assembly/manufacturability (DFA/DFM)	2.49	23	3.34	12
Benchmarking	2.42	24	3.08	23
Single-minute exchange of dies (SMED)	2.40	25	3.07	26
Activity-based costing (ABC)	2.40	26	3.12	22
Total preventive maintenance (TPM)	2.36	27	3.08	23

it would be initiated. Five-point, self-anchoring scales were used where 1 = 'No use' and 5 = 'High use' for degree of use, and 1 = 'Low' and 5 = 'High' for relative payoff. Table 1-5 lists the average of the responses for both items.

There was a wide gap between the degree of use and the perceived payoff for several of the activity programs. By degree of use, the top four activities were CAD, health and safety programs, teamwork, and defining an manufacturing strategy, but by relative payoff they were CAD, teamwork, just-in-time, and Kaizen (continuous improvement). The bottom four activities by degree of use were benchmarking, single-minute exchange of dies (SMED), activity-based costing (ABC), and total productive maintenance (TPM), but by relative payoff they were energy conservation, single-minute exchange of dies, statistical process control (SPC), and benchmarking.

Business Unit Performance

Manufacturing can be a formidable competitive weapon (Hayes and Wheelwright, 1984; 1986; Hayes et al., 1988). It has been widely hypothesised that manufacturing strategy should be linked to operational and business performance (e.g., Hanson and Voss, 1993). Proper strategic positioning or alignment of operations capabilities may significantly impact competitive strength and business performance of an organisation (Anderson et al., 1989).

Respondents were asked to indicate the percentage by which the business unit's performance had improved or worsened from a base index of 100 for 1990. The results are shown in Table 1-6.

Table 1-6. Performance Measures

Performance Area	Mean
Conformance to specifications (manufacturing quality)	128.8
Manufacturing lead time	128.3
On-time deliveries	126.8
Inventory turnover	126.5
Delivery lead time	124.2
Speed of product development	120.2
Customer service	119.6
Procurement lead time	118.9
Equipment changeover	118.3
Product variety	117.4
Average unit manufacturing cost	114.1
Market share	111.3
Profitability	110.8

1.6. The structure of the book

1.6.1. CONTEXT AND STRATEGY

The first main section of the book addresses the context and strategies of individual countries and regions. This section is in two parts. The first examines the findings from some of the major countries in the study. Each country chapter analyses the strategies and performance of that country in the context of the economic and social context of the country. In addition, as Japan is seen as the leader in manufacturing practices, there is a detailed examination of Japanese manufacturing practices relative to the rest of the world. The overall findings from this section are analysed to develop our understanding of the impact of the socio-economic context on company and national level manufacturing strategies.

An issue that the global manufacturing company has to consider is the growth of trading blocs. The second part of this section of the book will examine the impact of three different trading blocs—the EU, NAFTA and Mercosur—on the strategies and performance of firms and examine the differences between the blocs.

1.6.2. CHANGE PATTERNS

The second section of the book is also in two parts. The first is concerned with the dynamics of change. It examines the dynamics first at the level of the individual company, and then at the level of the country. Alternative patterns of change are examined and a view of 'soft' versus 'hard' trajectories is put forward. This section then examines the impact of economic context on trajectories of change.

The second part revisits our understanding of manufacturing strategy and its implementation, drawing on the experiences of different countries. Using the study data, a new paradigm for manufacturing encompassing multi-focusedness is proposed.

References

Anderson, J.C., Cleveland, G., and Schroeder, R.G. (1989) 'Operations management: a literature review', *Journal of Operations Management*, **8**, 2, April, 133-158.

Bower, J.L., and Hour, T.M. (1988) 'Fast-cycle capability for competitive power', *Harvard Business Review*, November-December, 110-118.

Buffa, E.S. (1984) *Meeting the Competitive Challenge*, Dow Jones and Irwin, New York.

Fine, C.H., and Hax, A.C. (1985) 'Manufacturing strategy: a methodology and an illustration', *Interfaces*, **15**, 6., 28-46.

Hanson, P, and Voss, C. (1993) *Made in Britain: The True State of Britain's Manufacturing Industry*, IBM Consulting Group and London Business School, London.

Hayes, R.H. and Jaikumar, R. (1988) 'Manufacturing's crisis: new technologies, obsolete organisations', *Harvard Business Review*, September-October, 77-85.

Hayes, R.H., and Wheelwright, S.C. (1984) *Restoring Our Competitive Edge: Competing through Manufacturing*, Wiley, New York.

Hayes, R.H., Wheelwright, S.C. and Clark, K. (1988) *Dynamic Manufacturing*, The Free Press, New York.

Hill, T.J. (1989) *Manufacturing Strategy: Text and Cases*, Irwin, Homewood, IL.

Leong, G.K., Snyder, D.L. and Ward, P.T. (1990) 'Research in the process and content of manufacturing strategy', *OMEGA International Journal of Management Science*, **18**, 2, 109-122.

Porter, M.E. (1980) *Competitive Strategy*, The Free Press, New York.

Schroeder, R.G., Anderson, J. and Cleveland, G. (1986) 'The content of manufacturing strategy: an empirical study', *Journal of Operations Management*, **6**, 4, 405-415.

Skinner, W. (1979) *Manufacturing in the Corporate Strategy*, Wiley, New York.

Stalk, G. (1988) 'Time—The next source of competitive advantage', *Harvard Business Review*, July-August, 41-51.

Swamidass, P.M., and Newell, W.T. (1988) 'Manufacturing strategy, environmental uncertainty, and performance: a path-analytic model', *Management Science*, **33**, 4, 509-524.

Trompenaars, F. (1993) *Riding the Waves of Culture*, The Economist Books, London, 1993.

Vickery, S.K., Droge, C. and Markland, R.E. (1989) 'Production competence and business wtrategy: do they affect business performance?', *Decision Sciences*, **24**, 2, 435-455.

Voss, C.A. (1992) *Manufacturing Strategy: Process and Content*, Chapman and Hall.

Ward, P., Miller, J.G. and Vollmann, T. (1988) 'Mapping manufacturers' concerns and action plans', *International Journal of Operations and Production Management*, **8**, 6, 5-17.

PART II - COUNTRY STUDIES

CHAPTER 2

MANUFACTURING STRATEGY PUT IN SOCIO-ECONOMIC CONTEXT

Explaining and modelling manufacturing strategy has traditionally—and correctly—been done in the context of internationalisation (Dunning, 1988), the business (Hill, 1985) and, more specifically, the products produced and the markets served (Hayes and Wheelwright, 1984). The logic in manufacturing strategy development is to link the way in which the firm competes through its products in the market to the design of the manufacturing system—its facilities, processes and organisational arrangements.

This procedure, while highly rational from a business perspective, tends to disregard the fact that every organisation operates in a local environment, regardless of the business it primarily is engaged in. This local environment—context—is both of a social nature and economic nature. Different regions have different cultures and ethnic particularities (e.g., conservatism, conformity), skill levels, and political stability (e.g., influencing investment patterns) that have an impact on the way in which manufacturing strategy is enacted and specific actions are formed and implemented. Likewise, the economic context of different nations (e.g., interest rates, cost levels) has a significant impact on the manufacturing strategy of individual firms. Thus, we propose a perspective on manufacturing strategy that is formed and implemented not only by understanding the business logic, but also through understanding the social and economic context of the firm. In this chapter, we will develop this perspective and present an overview of some of the key social and economic features of the countries involved in the IMSS, and their impact on the manufacturing strategies in the respective country. The analysis of the contextual impact on manufacturing strategy has been done with a relatively simple model, depicted below in Figure 2-1.

In applying this model of analysis, there is no uniform form of description possible or intended. We have no ambition to present a full, comprehensive, in-depth analysis of each country, since this has not been an explicit aim of the research. Rather, we want to stimulate further work in the area, and to add to an understanding of how manufacturing strategies vary between countries.

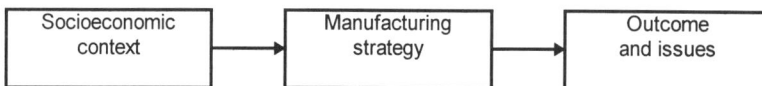

Figure 2-1. Contextual Impact on Manufacturing Strategy

21

P. Lindberg et al. (eds.), International Manufacturing Strategies, 21-43.
© 1998 *Kluwer Academic Publishers. Printed in the Netherlands.*

Since no specific *ex post* hypothesis was formulated as to what specific socio-economic dimensions would have the most significant impact on manufacturing strategy, each research team has defined the key dimensions for their respective analyses. As a moderator and point of common departure, we have also drawn upon a few key references in understanding and explaining the socio-economic context in the IMSS countries. We have used Hofstede's sociocultural dimensions (Hofstede, 1982) for discussing and describing the particularities important to manufacturing strategy development. Further, we have used other sources as a reference for economic data from each country (e.g., OECD, 1994; IMD, 1993).

It is important to emphasise that this chapter is primarily a summary and interpretation of the contextual pictures provided by the participating researchers. We believe that, while most contributors have created an overview that supports the proposition that manufacturing strategy is dependent on social and economic context *as well as* the business environment of the individual firms, the conclusions are confined to the IMSS perspective and data. Based on the contributions presented later in the book, conclusions will be made here at varying levels of abstraction. Based on the prevailing cases found in this research, a summary of countries with contextual similarities will be presented.

2.1. Argentina and Brazil: Moving out of History

South America is by no means a homogeneous continent that allows itself to be easily and rapidly described. The two countries participating in IMSS—Argentina and Brazil—are examples of relatively developed economies on this continent. Still, the countries differ substantially. Brazil still fights massive inflation and enormous imbalance in income distribution, while Argentina has fought inflation very successfully the last few years, creating a stable currency and a relatively high growth rate. However, the two countries have a similar history of political turbulence and social inequality, and more importantly, share a history of regulated economic policy, now gradually liberalised and eased to allow more competition and to develop the economy to increased standards.

2.1.1. ARGENTINA - MOVE OUT OF HISTORY

Argentina had for several years (up until 1977) a relatively closed economy with an import substitution policy. The result of this policy was a lack of competition and a concentration of production in a few large companies supplying the domestic market. In many companies, this resulted in mass production of relatively unsophisticated products. Whilst being protected by policies promoting import substitution, companies basically became non-competitive in terms of structure, organisation and processes.

Table 2-1. Socio-Economic Conditions and Manufacturing Practice in Argentina.

	Socio-economic features	Strategy	Outcome
1977	Closed economy import substitution	*Traditional Strategy* High vertical integration	
		Poor supplier relations Wide range of products and customers Make to order batch processing Planning and control centralised work force organisation reflect mass production environment.	Lack of focus
1985–91	High inflation	Uncertainty	
1991	- Convertibility plan - stability in currency - liberalisation	*Emergent strategy* Compete on international markets Cost and quality focus Restructuring of supply chains Investments in machinery & training	

Companies became highly integrated because of stability in technology and demand, whilst supplier relations with remaining suppliers became poor because of a lack of incentives for co-operation. The lack of a competitive structure led to a lack of focus in the typical Argentinean company, who produced a wide range of products and supplied a wide range of customers, resulting in non-streamlined batch production. Similarly, organisational principles were built on centralisation and reflected a mass-production environment.

1985–91 was a period of high uncertainty in Argentina, but after the convertibility plan was implemented in 1991, the currency stabilised and companies were exposed to increased international competition. This resulted in a sharper focus on cost and quality, as well as a restructuring of supply chains and increased investment in development of facilities and organisation.

Thus, the stability created in Argentina with the Convertibility Plan has in general been met by a relatively aggressive strategy from Argentine companies, with emerging competitiveness in industrial markets. Table 2-1 highlights this emergent strategy.

2.1.2. BRAZIL - MOVING FROM THE FACTORY OF THE PAST TO THE FACTORY OF THE FUTURE

Similar to those in Argentina, industrial companies in Brazil by and large have suffered from the liberalisation of the economy insofar as they have outdated strategies and need

Table 2-2. Socio-Economic Conditions and Manufacturing Practice in Brazil.

Socio-economic features	Strategy	Outcome
Protected and regulated economy	*Traditional strategy* Inadequate supply chain	
	End of month purchasing	High inventories and long lead times
	Below average investment levels	
Gradual relaxation of import restrictions in early 1990s	*Emergent strategy* High diffusion of 'lean' practices	Emerging improvements in lead-times
	'Soft' organisational development	
High inflation Demand instability	Low hardware investment levels	

modernisation. Brazil used to have a protected and heavily regulated economy. As in Argentina, the lack of industrial competition and unsophistication of traditionally industries in general created supply chains that were quite inadequate. Due to the enormous inflation rates in Brazil, companies developed an end-of-the month buying behaviour, thereby trying to cope with inflation. This in turn led to high levels of inventory and long material throughput times. According to the IMSS data, Brazilian companies have a level of investment that is below world average. Once again, it is likely that the massive inflation makes investments less attractive. The gradual relaxation of economic regulations and import restrictions in the early 1990s has put additional pressure on Brazilian manufacturers. Whilst still suffering from high inflation, a relatively low degree of foreign direct investments and demand instability, the relaxation has started a gradual modernisation of manufacturing strategies in Brazilian companies.

In the early 1990s, the diffusion of modern—*lean*—manufacturing practices has increased, and the understanding of requirements in industrial competition has also been diffused. Emerging practice in Brazil is primarily based on 'soft' measures—commonly referred to as organisational development—rather than on investments in hardware. This is a likely effect of the still uncontrolled inflation in the early 1990s, in combination with high instability and uncertainty in demand levels. Table 2.2 summarises this pattern.

2.1.3. ARGENTINA AND BRAZIL - EXAMPLES OF DEREGULATION AND EMERGENT MODERNISATION

Both economies have been regulated and relatively closed up until recently. They are both opening up to industrial competition, thereby demonstrating the movement from traditional manufacturing strategies to emerging strategies. The pattern of contextual dependence is clearly that of deregulation leading to increased exposure to industrial competition, leading to emerging redefined manufacturing structures (supply chains) and organisation forms.

2.2. Portugal and Spain: From Low-Cost to Euro-Cost Economies

Just as Argentina and Brazil have experienced dramatic changes whilst opening up their economies, Spain and Portugal may perhaps be seen as their European counterparts. These economies have been integrated into the European Union, putting pressure on stabilisation and harmonisation of policies, increasing competition but also opening up large markets for exports. While these countries do not have identical histories and current levels of development, both Portugal and Spain have, in this changing context, experienced similar changes in manufacturing strategies to cope with these new challenges.

2.2.1. PORTUGAL - CHANGE OF HISTORIC MISMATCH

Portugal used to be a low cost/low wage economy. In a European context it still is, but this history has left the companies oriented mainly towards high volumes and few variants of relatively simple goods. In spite of the high volumes and standardised products, production processes have to a large extent not been standardised to match these requirements, and production of relatively inexpensive products in large quantities has made quality performance and reputation suffer, a position that to some extent still

Table 2-3. Socio-Economic Conditions and Manufacturing Practice in Portugal

Socio-economic features	Strategy	Outcome
	Traditional strategy	
Based on high volumes/few variants		
Low wage/low cost economy	Large proposition of with non-standard production processes	Poor quality & delivery
	Emerging strategy	
EC Integration	Focus reliability in quality and delivery	
-loss of low cost edge		
-competition in regional markets		
	Changes in practices that drive quality & delivery	
	-Planning and scheduling	
	-JIT practices	
	-Capacity management	
	High percentage of regional purchasing	High raw material levels

prevails on the European market. Similarly, delivery performance levels have traditionally been low, giving Portuguese manufacturers a second-class image.

EC integration has led to a situation where the low-cost edge has to some extent became eroded, forcing Portuguese manufacturers to focus more on reliability in quality and delivery rather than cost only. This has accelerated diffusion of practices that increase quality and delivery performance, such as changes in planning and scheduling systems, Just-In-Time practices and capacity management policies.

The relatively high degree of regional purchasing of material and components still makes inventory level reduction difficult in Portuguese companies, but the overall picture suggests that Portuguese manufacturers *in general* need to change the mismatch between product/market requirements of the 1990s and the characteristics of the production processes, as shown in Table 2-3.

2.2.2. SPAIN - ISLANDS OF EXCELLENCE

Spain has gone through a remarkable revitalisation in terms of manufacturing over the last decade. From a low cost/low wage and traditional agricultural economy, Spain has developed into an industrial economy. Integration into the EC boosted the manufacturing-led growth that characterised the 1980s up until 1992. The era of growth ended with the recession beginning in 1992, exposing weaknesses in labour market policies and rising wages.

Over this period, manufacturing practice in Spain has developed to a level where diffusion of state-of-the-art manufacturing principles is on par with other developed economies. Spanish companies *on average* thus match industry practice in terms of quality efforts, logistics and just-in-time principles. This has created major improvements in quality, lead times and delivery performance, an improvement that has continued into the 1990s.

Table 2-4. Socio-Economic Conditions and Manufacturing Practice in Spain

Socio-economic features	Strategy	Outcome
EC Integration	Follow industry practice in quality, lead time & delivery Average adoption rates in most areas.	High improvement in quality, logistics, and JIT
Dependent on technology imports		Relatively low investment in flexible automation.
Strong manufacturing led growth until 1992		
1992 recession exposed weaknesses in rigid labour market and rising wages		Lowering of attractiveness to FDI
High uncertainty avoidance		
- conservatism		Lagging in organisational development

The recession has exposed weaknesses in labour markets and also pinpointed the higher wage levels, thereby lowering the attractiveness of the Spanish economy to foreign direct investment. Spanish manufacturers seem to lag other countries in a few areas such as investments in flexible automation, and organisational development. The Spanish culture is quite conservative, typified by high uncertainty avoidance. This most likely reinforces the lack of organisational development, due to a general unwillingness from management to empower personnel and lose control over how decisions are made.

Even if Spanish manufacturers seem to use state-of-the-art manufacturing practices, there are a few mismatches to consider. First, organisational development needs to go hand in hand with development of Total Quality and JIT, otherwise they will fail. Spanish manufacturers need to not only adopt tools and techniques, but also change organisational forms. Secondly, even if diffusion of practices (disregarding organisational development) seems to be at world average, this reflects a situation with 'islands of excellence' in a sea of less than average and even mediocre practice. Thus, the diffusion of best practice must accelerate and become more widespread in Spain, as the summary in Table 2-4 suggests.

2.2.3. SPAIN AND PORTUGAL - SIMILAR HISTORY BUT DIFFERENT CONTEMPORARY ISSUES

The overall picture of Spain and Portugal suggests that from a low-cost position, both economies (with Spain leading) are moving towards a position of European, relatively high-cost economies. This has forced manufacturers to move their priorities from a cost/price competitive posture, towards a stance based more on quality and reliability. In Portugal, just initiating this step, the overall issue to deal with for manufacturers in general seems to be to structurally alter historic manufacturing processes from old batch production principles to either highly efficient standardised production or to flexible make-to-order production. In Spain, an economy that started on this route before Portugal, the issue seems to be to accelerate and broaden the diffusion of best practice, so that the broad layers of Spanish manufacturers can improve their competitiveness.

2.3. Japan and the United States: Who is the Teacher and Who is the Student?

Comparing Japan and the US may seem awkward from many perspectives. They have very different historical backgrounds in terms of economic development, and also in a political and social sense. Further, the industrialisation processes of the two countries are quite different, and the process in the US by and large took place half a century before Japan.

Therefore, we will not compare, but very briefly contrast, the two economies. The two economies are the largest in the world, and there is a history of intense rivalry between companies and industries, as well as recent heated argument around trade policies, in principle based on industrial performance.

2.3.1. THE US - LEARNING THE LESSON?

Even if the US industrial and consumer market consists of several segments and buyer groups, the market as such is very large, and especially during the growth and expansion years of the early 1900s and up until the 1960s, the expanding markets were supplied by increasingly high volumes of standardised products. The US companies supplying these markets consequently developed procedures and processes for mass production. In combination with a low degree of exports, the average US company became dependent on the US market, and on mass production principles. Over time, this created a set of large and dominant companies, but they were increasingly inflexible and vulnerable to competition through other means than price and standardised products.

Table 2-5 shows the (simplified) evolution of the US industrial situation from around 1980 onwards. The general perception of US-made goods in the 1970s and early 1980s was one of questionable quality, and manufacturers were considered unreliable performers. Even if there were significant market and demand variations up until around 1980 (especially during the oil crises in the 1970s), a general characterisation is one of relatively low (however increasing) foreign competition and reasonable consumer demand levels. In the 1980s, however, the picture changes.

Table 2-5. Socio-Economic Conditions and Manufacturing Practice in the US

Socio-economic features	Strategy	Outcome
-1980	Mass production	Standard products, low quality and reliability
Strong domestic demand and relatively low foreign competition		
1980s		
Increased competition from imports	Quality (TQM)	Slow diffusion overall, but a few significant and successful early movers (Motorola, HP, Xerox)
Saturated/competitive markets	JIT (Lean production)	Significant failures to realise benefits reported
Heavy Japanese and other foreign manufacturing investment	Benchmarking	Diffusion of practice
1990s		
Continued intense international competition	Mass Customisation	Regained competitiveness
	Down-sizing	
General conditions		
Flexible labour market		Simple structural adjustments
Managerial orientation		Rapid diffusion of new concepts
Low cost of capital		Access to venture capital

Mass markets became increasingly saturated, and in combination with increased foreign competition (especially from Japan), the traditional American mass producers were put under pressure to increase quality levels, reliability and productivity.

The Japanese lesson was slowly diffused throughout the economy, especially JIT and TQM. In spite of accelerated foreign and Japanese investments, aiding the diffusion of best practice, and even if there were a few star cases in corporate turnaround, the overall diffusion was slow, and considerable problems in implementation of new practice were reported.

In the 1990s, the diffusion of industrial best practice continues (e.g., TQM and JIT), together with a number of 'home-grown' concepts such as Business Process Re-engineering and Process Management. Taken together, this process and diffusion over some 15-20 years has revitalised the US industry at the cost of millions of jobs in old-fashioned industries and companies, but with increased competitiveness and new jobs as a consequence.

The flexible US labour market has aided the structural transition of US industry. Even if many industrial jobs are lost forever in a structural transition, the willingness of individuals to move to where the jobs are and to accept available jobs keeps overall unemployment rates relatively low compared to Europe and some other regions. Another feature of the American culture is its orientation towards management and the managerial culture. The flood of new management concepts and theories, as well as the relatively short-sightedness of American management in general, speeds up the diffusion of new concepts. This is an advantage in the sense that new *concepts* have the potential of rapid spread, but it is a weakness in the sense that a thorough understanding of the application of key and fundamental concepts becomes limited, thereby hampering overall diffusion of *practice.* Thus, it is important to make a distinction between diffusion of *concepts* and *practice.*

2.3.2. JAPAN - ON THE VERGE OF A NEW ERA?

Needless to say, the Japanese development has been the subject of vast amounts of literature, research and dogma over that last two decades. Still, however, Japanese industrial practice is undergoing interesting developments, as shown in Table 2-6..

The early Japanese products exported to Western markets were known more for mass-produced poor quality goods and imitation rather than innovation. Japanese products often held low-end status and were aimed to attract relatively unsophisticated buyers. This situation prevailed to roughly the mid-1970s. Then came an era of increased diversity and variety, where strong domestic competition forced producers to increase variety and productivity, as well as compete on export markets with other Japanese companies. Up until the late 1980s, the diffusion of lean production, continuous improvement and other familiar concepts continued to be the prime vehicle for manufacturing development. As a result, inventory levels, quality costs, deliveries

Table 2-6. Socio-Economic Conditions and Manufacturing Practice in Japan

Socio-economic features	Strategy	Outcome
Historic development		
- 1975		
Rapidly growing markets	Mass Production	Simple and unsophisticated products
1975–88		
Increased diversity & variety		
Saturated, highly competitive domestic markets and competition on export markets	Lean Production, supply chain development	Capacity utilisation
		Inventory levels
	High variety production	Quality costs
	Continuous improvement	Throughput efficiency
	Strong export focus	On-time delivery
1988–1991		
Competition on new products and speed to market	Focus on rapid product development	Time to market
1992 -		
Appreciation of the yen (*endaka*)		
High cost society	Increase overseas procurement	
	Explicit national strategy on location of manufacturing;	
	- In Japan - high value added	
	- Overseas - stable, standard products	
	Manufacture inside key markets	
	National technology infrastructure	
	Company focus on cost reduction	High use of TPM
General traditional conditions		
Hierarchical & feudal background		Strong link to company
Long-term relations		
Team tradition		Focus on small improvements
Preventive culture		
Work ethic		
General emerging conditions		Dedication
Westernisation of youth		Different work ethic
Appreciation of Yen		High labour costs
Emerging unemployment		New perspective on company responsibilities

and productivity increased dramatically, combined with increased product variety and complexity.

In the late 1980s, competition evolved further, and the ability to launch new products faster than competitors became increasingly important. Focus was increasingly placed on new product development and time-to-market issues, resulting in further diffusion of integrated product development practices and in further reduction of product renewal cycle times.

In the era of the *endaka*, the strong appreciated yen, labour costs have skyrocketed in comparison to competitive countries, and the cost of manufacturing in Japan is increasingly considered to be too high. As a consequence, both overseas procurement and overseas production have increased.

The current strategy is to produce technology-intensive products with high value-added in Japan, to maximise learning from this production to improve production methods and product designs, and to develop a technological infrastructure in Japan that will enhance the capability of innovation and product development. The strategy also means production of standard products overseas—often in other Asian countries—since the aim is to establish production within emerging markets and inside competitive markets. Undoubtedly, Japanese cultural traditions have provided the foundation for the development of manufacturing practices. The linkage between employer and employee, still feudal in nature, is strong and provides the necessary dedication and loyalty to engage in lean production, TQC and continuous improvement activities. Further, a historical heritage from rice cultures has nurtured team efforts and relentless pursuit of perfection, as well as a inclination towards prevention and standardisation of methods. Similarly, frequent natural disasters as well as the aftermath of the Second World War have lent the traditional Japanese culture a sense of 'hardship', in the perception that conditions of life are hard, and hard and dedicated work is the only saviour.

To some extent, the traditional Japanese culture is changing. Japanese youth are gradually adopting Western values and lifestyle, and the increased wealth of Japan of course changes the value of work. This means that Japanese employers to an increasing extent must rely on other mechanisms to attract employees than before, and this together with the appreciation of the Yen may lead to an acceptance of unemployment and workforce reduction, a move that was almost unthinkable a few years back.

2.3.3. JAPAN AND THE US - A HISTORY OF RIVALRY AND WHO'S TEACHING WHOM

In spite of the more than obvious differences between Japan and the US, there are certain historical features in the recent history that allow interesting contrasts. First, the two nations have engaged in various conflicts over the last 60 years. The conflict during the Second World War ended in a Japanese defeat, with US occupation and a rebuilding of constitution and legal systems largely based on US principles. Immediately after the war, the US occupational forces engaged in massive education and training of Japanese industrialists in basic industrial management and in quality management. During this period, the US was the obvious teacher.

But Japan was a good and ambitious student. Applying the lessons in the Japanese context, and developing them beyond what the teacher had imagined, the Japanese

The US route

Large domestic markets	→	Early dominance of mass production. Later apparent weaknesses in lean production.	→	Leader up to 1975-80. Follower from 1980 and onwards. Weakened economy.	→	Shift from issues on off shore production to diffusion of lean production principles.
High cost	→	Early weakness in mass production. Later dominance in lean production.	→	Follower up to 1975-80. Leader from 1980 and onwards.	→	Shift from diffusion of lean production principles to issuess of off-shore production.

The Japanese route

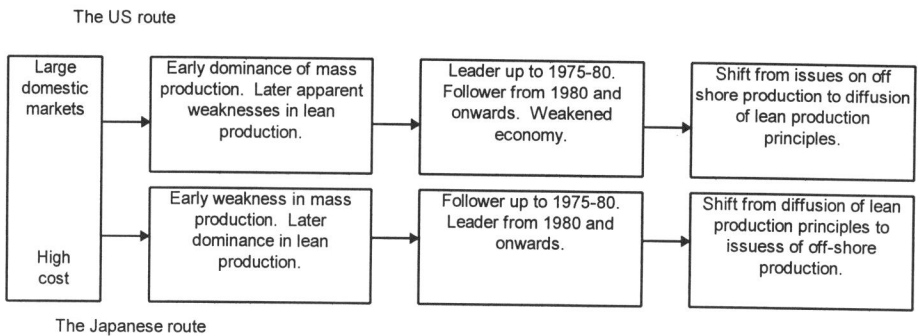

Figure 2-2. Contrasting Japanese and US Developments

perfected many manufacturing techniques to what are now world class standards and undisputed models of best industrial practice. In less than 40 years, Japan overtook the US as an industrial leader, and the teacher was made student.

Secondly, although each had a large domestic and competitive market, the two economies evolved differently. An early dominant force in mass production principles, and a leader(at least in some industries) up to the mid 1970s, the US later showed severe weaknesses in lean production, and became a follower. The Japanese, on the other hand, showed severe weaknesses in their early attempts at mass production, but from around 1980 onwards, their leadership in lean production is undisputed (Figure 2-2).

The current issues in the two countries are somewhat different. From being a strong economy with companies focusing on off-shore production, the US is now an attractive area to locate production. Even though this is due not only to competitive cost levels but also access to markets, this trend is quite different from that in Japan, where companies actively locate productive capacity overseas. Thus, while the shift—admittedly simplistic—in the US has been from a focus on outsourcing of standardised production to understanding lean production, the shift in Japan has been the opposite.

2.4. Scandinavia - Welfare States in Transition

The Scandinavian countries of Sweden, Finland, Denmark and Norway are culturally similar, and share a set of political and economic features. Culturally, the Scandinavian countries have a common history of rivalry, co-operation and brotherhood, and there are strong links between the people in the countries. The political scene has been quite similar in the countries as well, dominated by social concern and welfare policies.

Economically and industrially, however, the countries are somewhat different. The Danish and especially the Norwegian economies are sound and in balance, and the industries in these countries—especially Denmark—are dominated by small companies.

Sweden and Finland, on the other hand, have a greater dominance of large companies in the engineering industry, and also share a set of economic imbalances.

2.4.1. SWEDEN - TRADITIONS AND BELIEFS UNDER RE-EVALUATION

The Swedish economy and industry came out of the Second World War intact, and the already strong niche engineering firms became the engine of the creation of the classical welfare state, generally associated with Sweden, during the export surge after the war. The growth in the economy and welfare continued through the 1960s, and into the early 1970s. By this time, the size of the public sector increased dramatically, and the once impressive productivity growth levelled out, and showed worrying stagnation.

In the 1980s, productivity growth was still very weak, at times even negative, and signs of a too large public sector keeping unemployment at almost negligible levels were emerging. In manufacturing development, most initiatives were heavily biased towards social issues, in general ignoring or not being able to bring about revitalisation of manufacturing. The slow growth in productivity, and high wage costs, further cemented the position of Swedish manufacturers in high cost/high value niches, thus limiting volumes and possible capacity expansion. A few devaluations of the currency during the 1970s and 1980s did not significantly alter the situation, except for the fact that real incomes started to decrease in comparison to other nations. As a compensatory mechanism, wages rose accordingly, further adding to cost burdens of manufacturing and to inflation. Thus, there was a vicious circle.

In the 1980s, the capital market was deregulated, increasing the availability of credits dramatically. As a response to the slow growth of productivity, to some extent

Table 2-7. Socio-Economic Conditions and Manufacturing Practice in Sweden

Socio-economic features	Strategy	Outcome
1970s and early 1980s		
Tradition of welfare, full employment	Heavily socially oriented	Slow productivity growth
Lack of skilled work force development		
High cost economy	Niche companies	Limited volumes
Strong trading nation	High exports	Currency dependent
1980s		
De-regulation of capital market	Hardware investments	Slow productivity growth
1990s		
Focus on productivity	Lead-time reduction	Strong improvement in LT leading to strong productivity growth and reduction in inventory levels
	Decentralisation	
	Competence development	
Devaluation	Export focus	Increase in exports
	Profit maximisation	Record profits

driven by the ease of credit policies and by the lack of skilled work force, companies invested heavily in automated equipment and hardware. Unfortunately, this did not help productivity rise.

During the late 1980s there was a strong focus on the lack of productivity growth in Sweden. A national program for productivity improvement was launched in the early 1990s, just in time to parallel some significant change efforts in the manufacturing sector of the industry.

The change efforts of the early 1990s were a reaction to the dramatic downturn in production demand in this period. In combination with a significant devaluation in 1992, this brought industrial production back to record volumes by 1993, but with some 250,000 fewer jobs. The margins created by the devaluation together with the relatively strong market position of several Swedish companies made profits very high in the period. At the same time, the Swedish state and public sector was under-financed largely due to the size and generosity of the welfare systems, and to the increase of unemployment created by the productivity improvements in industry. Thus, the 1990s have been a decade of re-evaluation of old assumptions and systems in Sweden, but also a decade of strong revitalisation of manufacturing practice.

2.4.2. FINLAND - RECOVERING FROM THE FALL OF THE SOVIET UNION

Finland has a recent history that is very similar to that of Sweden. The industrialisation started later, but the growth of the manufacturing sector was high through the 1970s and 1980s. The strategy in Finnish companies has been one of high growth and high investment, thereby rapidly bringing Finnish companies to a competitive standard in

Table 2-8. Socio-Economic Conditions and Manufacturing Practice in Finland

Socio-economic features	Strategy	Outcome
Historically high growth in 1975-late 1980s	High growth / high investment strategy	High growth (average manufacturing sector - 5% annual
Lack of skilled work force	High degree of flexible automation	
Liberalisation of lending - high borrowing	Low adoption of 'soft' developments	
1990s 1990 - loss of Soviet Market 1991 - devaluation	Emergent strategy Move to market globalisation Over-borrowed, therefore limited hardware investments	Export led recovery
1992 - float currency	Move to focusing/less variety Leveraging organisation development/HR	
High unemployment		Dual economy - thriving exports, domestic/public sectors sluggish

terms of technology. In the 1980s, Finnish (as well as Swedish) companies tried to compensate for the lack of skilled work force with increased automation, leading to a proportionally very high diffusion of flexible automation in Finland, and, underpinned by liberalisation of capital markets, this led to high borrowing levels. The simultaneous reliance on technological solutions led to a low adoption of organisational renewal efforts.

In 1990, the fall of the Soviet Union was a serious blow to the Finnish economy. Dependent on the exports from the niche engineering companies, this loss of its main trade partner threw Finland into a crisis, followed by a devaluation in 1991 and floated currency in 1992. While still suffering from very high unemployment levels (reaching almost 20%), the weakened currency has led to an export-led recovery in the mid-1990s. The trend in Finland in the mid-1990s is to leverage hardware investments with developments in organisation and human resource issues.

The high unemployment rates in Finland have led to a two-track economy—the companies exporting to other regions than the former Soviet Union thrive, while other parts of the economy are suffering from sluggish demand, lowered compensations and decreased wealth.

2.4.3. NORWAY - MANUFACTURERS WITH STRONG LOCAL IDENTITY

The Norwegian economy has recovered from the turbulence of the early 1980s, and is today one of the strongest in Europe. A small country with a narrow customer base, the strategies of Norwegian manufacturers is customer orientation and specialisation on a narrow product range. Customers are typically from specialised applications such as oil/off-shore industry or defence. The sales to specific customers of specialised products, in combination with exports, lead to high levels of profitability.

An interesting feature of Norwegian manufacturers is their strong local identity and linkage. This means a close link to customers and a socially controlled situation, resulting in flat organisations and high stability. Norway's high educational standards provides an additional strong feature of Norwegian manufacturers—local linkage to specialised customers. Stability and competence facilitate niche strategies (Table 2-9).

Table 2-9. Socio-Economic Conditions and Manufacturing Practice in Norway

Socio-economic features	Strategy	Outcome
Small country	Customer orientation	High market share
Narrow customer base	Focus on service and delivery improvement	High delivery performance
Specification oriented customers (oil/offshore military)	Specialisation/narrow product range	High profitability
Export orientation	ISO 9000	
Companies have strong local linkages	Flat organisations, high stability high education	
Unionisation	Flat compensation schemes	

2.4.4. DENMARK - IS PROFIT A PRIME CONCERN?

Similar to other Scandinavian countries, Denmark has a tradition of being a welfare state. Labour costs are high, and the trading tradition and small economy of Denmark lead to an export orientation. In contrast to Sweden and Finland, Denmark is a country of small firms. The high individualism of Danes leads not only to Euro-scepticism, but also to a desire to develop and control a small business. The culture is also quite egalitarian, leading to non-hierarchical and decentralised organisations.

Located in a small economy, companies in Denmark as well as companies in other Scandinavian countries, are export-oriented. The smallness of Danish firms make them even more niche-oriented than other Scandinavian companies. Manufacturing strategy is often based on flexibility and responsiveness to customers, rather than on specialisation, leading to a customer service strategy with high prices. Large investments in machinery and equipment have, however, resulted low ROI levels. In the Danish context, it may be a greater driver to control and run a business than to maximise profit. Table 2-10 summarises the Danish position.

2.4.5. SCANDINAVIA - SIMILARITIES AND DIFFERENCES

As noted previously, the Nordic countries share a few basic features, such as being small and export-dependent economies, having a high commitment to social responsibilities and equality in distribution of income, and being relatively high cost societies.

Table 2-10. Socio-Economic Conditions and Manufacturing Practice in Denmark

Socio-economic features	Strategy	Outcome
Trading country/ small economy	Global niches, export orientation Regional sourcing Focus on quality, customer service, delivery (and high price)	
Small firms	Non-specialised, flexible production	
High R&D		
High Labour cost environment	High investment in machinery/equipment Multi-focusedness	Low ROI
Welfare state	Decentralised, non-hierarchical organisations	
High individualism		
Egalitarian society	Fewer organisational levels	
Low Power distance	Some EC scepticism	
Low uncertainty		
Low masculinity		

Denmark	Small economies	Small companies	Niche markets	Large market shares
		Strong currencies	High cost / price	
Norway	High exports			

- →

| Sweden | High costs | Large and small companies | Globalisation Productivity/cost advantage | High profits |
| Finland | High social responsibility | Weak currencies | High exports | |

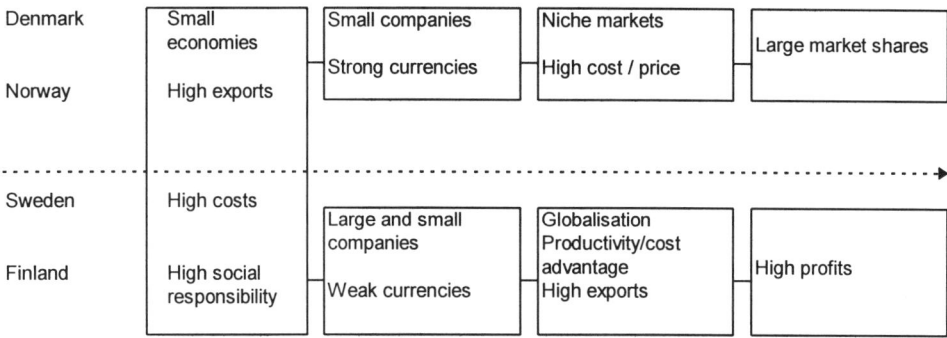

Figure 2-3. Schematic Development in the Scandinavian Countries

From this position, however, two development patterns have emerged. Danish and Norwegian economies in the early 1990s were strong compared to the Finnish and Swedish ones, resulting in relatively 'hard' currencies. Danish and Norwegian companies are predominantly small, aiming for specific niches and for high price/high cost business. The specific niche orientation leads to high market shares in the respective business.

Swedish and Finnish manufacturers seem to have developed in a somewhat different way. Building on a larger scale, these manufacturers have utilised the weak Swedish and Finnish economies over the last few years to use the regained productivity and relative cost advantage to go global and to increase business and margins on export markets. Of course, this is a strategy most likely applicable also to Norwegian and Danish companies, but the difference seems to be that Swedish and Finnish companies have been in a better position due to favourable exchange rates and to a very strong productivity improvement rate in the early 1990s. This development is depicted in Figure 2-3.

2.5. Continental Europe: Italy, Netherlands, and the UK

Within a broadly similar economic and cultural area (Europe and the European Community), a comparison of the three economies of Italy, Netherlands and the United Kingdom shows how there are substantial local economic and social differences leading to differences in manufacturing strategy.

2.5.1. ITALY - NEEDING ORGANISATIONAL DEVELOPMENT?

Italy suffers in the mid-1990s from a similar 'duality' of the economy as Sweden and Finland—export industries are flourishing, while state and public sectors are in crisis. From a situation of stable growth in the 1980s, the economy was hit severely in the deep recession in the early 1990s, resulting in a devaluation of the *lira* by 40% in 1992. Thus, the cost advantage of Italian manufacturers has improved significantly, thereby helping exports.

In addition, the Italian industries are fragmented in that small businesses are often highly competitive on international markets, and also often operate in networks based on the social and cultural norms of Italy and almost exclusively family-owned. Examples of this are the packaging industry around Bologna, the eyeglass industry and the Benetton company, which are prime examples of network organisation. The large industries, on the other hand, are often state-owned, with monolithic structures, rigid and steep hierarchies and mass volumes of products with an often dubious quality reputation. This reputation may or may not be deserved, but it is still a reputation that may inflict hesitance among potential customers. It is, however, fair to say that these larger companies are consolidating on quality and cost, thus gaining in industrial competitiveness. (See Table 2-11.)

A few issues seem to be shared among the Italian manufacturers. A prominent one is design and new product innovation, which seem to be a general Italian strategy. Whether innovation is actually realised may be debated, but the design and fashion skills of the Italian industry are indisputable.

Another issue faced by Italian manufacturers is that there seems to be a general lack of organisational development, at least in comparison to other countries. Here, tradition

Table 2-11. Socio-Economic Conditions and Manufacturing Practice in Italy

Socio-economic features	Strategy	Outcome
	Small Companies	
Small family owned, & large, (often) state-owned firms;	Flexibility, quality and responsiveness	High product variety
		Poor inventory and cost performance.
Fragmented industry	Internationalisation	
	Networks	Low profitability
Little 'managerial' culture		
	Large Companies	
Stable growth in 1980s	Consolidation of international competitiveness in quality and cost	
State crisis in 1990s; Duality of economy		
	Common Strategies	
Devaluation (40%) in 1992	Investment in product innovation Lack of investment in process, especially in organisational development Good supply chain management Lack of coherent manufacturing strategies	

and culture are the most likely explanations for the lagging position. Italian industry needs to improve on its organisational skills, especially as they have invested heavily in machinery and automation. There also seems to be a need to develop a more coherent picture of manufacturing, involving technological *and* organisational development, as well multi-focusedness in the sense that the ability to simultaneously improve on flexibility, responsiveness, quality and cost must be improved.

2.5.2. NETHERLANDS - PROGRESSIVENESS AND CONSERVATISM

Located strategically in western Europe, the Netherlands is one of the distribution hubs of Europe. This gives Dutch companies a highly industrial profile, especially as the Netherlands traditionally is a trading nation. Exports are high, mainly to central and continental Europe.

The profile of Dutch manufacturers is otherwise somewhat contradictory. The progressive character of the culture seems to be embedded in strategic ambitions to differentiate, while the conservative character seems to slow down adoption of practices to support this. For example is there a low emphasis on new product design practices, and also a relatively low progress in development of manufacturing processes (flow orientation, machinery). The Dutch thus seem to have progressive strategic ambitions, but somewhat conservative measures to reach these ambitions.

The Dutch also seem more inclined to development of manufacturing through 'soft', organisational means rather than through hardware investment. This seems to be consistent with the high individuality and low power distance of the Dutch culture (according to Hofstede (1982)).

2.5.3. THE UK - FROM STRONG DEFENCE TO OFFENCE

Up until around 1985, manufacturing in the UK was characterised by defensive strategies and very low investment levels. The traditional and conservative management of British firms led to poor labour relations and short-sighted financial control rather

Table 2-12. Socio-Economic Conditions and Manufacturing Practice in the Netherlands

Socio-economic features	Strategy	Outcome
Distribution hub of Europe	Storage and warehousing	High levels of inventory
High, regional, exports		
Progressiveness and	Close to IMSS average in	
conservatism	organisational development	
High individuality	Slow in development of	
	production processes	
Low power distance		
Average uncertainty avoidance	Mismatch in strategy	Low emphasis on product design practices
	Aim is differentiation, action	
	does not support this	

than long term investment policies and world class objectives. The British unions were also strong and the mistrust between employers and employees/unions led to a poor manufacturing climate in general, and a very poor climate in development and change terms in particular. On top of this, the high inflation in Britain in this period did create large uncertainty and unwillingness to invest in 'risky' manufacturing operations. The once proud and acknowledged manufacturing skill in Britain gradually became second class. Industries vanished and regions suffered.

In the 1980s, the rules of the game were changed, for better or for worse. The Conservative government under Mrs. Thatcher took power away from unions and deregulated labour markets, and under this period, a shakeout in manufacturing took place. This initially resulted in increased unemployment. The government policies also gradually created a labour cost position that would make Britain attractive to foreign direct investments (FDI).

In combination with an active industrial policy to promote FDI, there was a major wave of FDI into the UK, for example by Japanese auto manufacturers (e.g., Nissan and Honda) and electronics manufacturers. This aided the diffusion of best practice in Britain, gradually leading to improved competitiveness of domestically owned and managed manufacturers as well. Though painful in many ways, this process has led to improved competitiveness of the manufacturing industry in Britain, and in the late 1980s created growth and new industrial jobs.

In the 1990s, the process of diffusion of best practice has continued, and there were further major improvements in quality and reliability among British manufacturers, even if Britain as well as other countries was hit by the recession of the early 1990s.

Table 2-13. Socio-Economic Conditions and Manufacturing Practice in the UK

Socio-economic features	Strategy	Outcome
-1985	Neglect and gradual decline of manufacturing	Low investment
Volatility & uncertainty		Defensive
High inflation		
Poor labour relations		
Strong finance control		
1983–88 Thatcherism		
Taking power form unions	Shakeout in manufacturing	Major productivity improvement
Deregulation of labour market		Initial high unemployment, later growth with new jobs
Inflation control		
Major foreign direct investments	Diffusion of world class practice	
Late 1980s on		
	Adoption of JIT, TQM and organisational development ISO 9000	Major improvements in quality & reliability

Mid-sized European economies	The UK	Deregulation		FDI, productivity growth, early lean production adoption.		Wide variance - leaders and laggards.
		Low cost				
	Italy	Strong networks of family SMEs. Large public companies.		Fragmentation.		SMD - flexibility and specialisation. Large cos - cost & quality.
Large and small companies	Netherlands	History of trade. Conservative culture. Social responsibility.		Slow and careful adoption of new practices.		Mismatch of intentions and action.

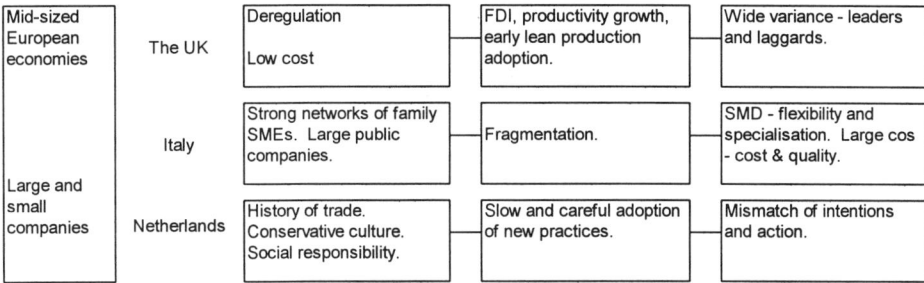

Figure 2-4. Schematic Development in Three Medium-Sized European Economies.

2.5.4. ITALY, NETHERLANDS AND THE UK - THREE PATTERNS OF DEVELOPMENT

The three are located in a broadly similar area—in both cultural and economic terms. They are all mid-sized economies with a combination of small and large companies. Still, the developments are different.

The Netherlands seems to be situated midway between Scandinavia and central Europe, with a strong social responsibility ethos. In the sectors studied in IMSS, the Dutch companies by and large are regional players, somewhat dominated by their German neighbours. There has also been a history of dependence on others, through trade and through conflicts during the wars. This may have created some cautiousness and carefulness in adoption new practice. The conservative culture seems also to result in a slow and careful adoption of new practice, leading to a possible mismatch in intentions and actions (see Figure 2-4).

The UK has through deregulation and a low cost policy attracted foreign investments and thereby speeded up diffusion of new practice. This has lead to strong productivity growth, but there are still leaders and laggards within the manufacturing community. Italy, finally, have strong networks of family-owned SMEs on the one hand, and large often publicly owned companies on the other. While the SMEs use their flexibility for responsiveness and specialisation, the large companies are trying to consolidate quality levels while keeping cost advantages.

2.6. The Contextual Dependence of Manufacturing Strategy

In this chapter, we have shown that there are linkages between the context—economic and social/cultural—of manufacturing and the manufacturing strategy content of firms in the country. The conclusions presented here are mainly drawn from the IMSS data and book contributions. Although the depicted linkages are simplified and schematic, we think that they serve as a useful starting point for a discussion on context and strategy, as well as a short introduction to the rest of the material.

While this linkage may seem to be obvious, looking back at the interdependence between context and manufacturing strategy shown here, this is generally not recognised in theories and models of manufacturing strategy, nor is it reflected in general management literature. Often practices are assumed to be applicable regardless of context, and diffusion rates and patterns are often assumed to follow simplistic models and curves. While this *in some instances* may lead to correct assumptions of reality, it may more likely be by chance than by anything else.

Looking at how the country specific contexts influence manufacturing strategy in our overview, we may first of all conclude that what we have described are very general patterns, and that there most likely are high variability *within* countries, especially where contexts vary significantly (e.g., in Brazil and the US). Obviously, each company also has its specific context, which puts our general patterns in perspective. But once again, our ambition has been to show these general patterns, rather than individual and unique cases.

One linkage that we have seen is the *geographical context*. Depending on the geographical location, the manufacturing strategy may imply different supply policies (e.g., close location of suppliers leads to sequenced JIT deliveries), and different inventory levels (e.g., the distribution hubs in the Netherlands). Such geographical preconditions are obvious in their influence on manufacturing strategy.

Another linkage is the *investment context*, that is, the cost of capital, the willingness to take risks and the availability of alternative solutions. Taking the example of Finland (and Sweden), there was a lack of skilled work force in the 1980s, and simultaneously there was a deregulation of capital markets (leading to easier borrowing) and willingness by companies to invest and expand. The consequence was high investment in flexible automation, which in turn became a burden during the recession of the early 1990s. The UK, on the other hand, has had very defensive investment policies, due to inflation, unrest on labour markets, distrust in manufacturing as sound investments, and so on. This led in the 1970s to under-investment, in turn leading to further degradation of manufacturing performance, which led to further distrust in manufacturing as an investment opportunity. What these cases show is that behaviour in manufacturing is not only dependent on the business environment in the industry, but also on the managerial context. Manufacturing strategy is a function of investments as well as of the plans and intentions of company executives.

A third linkage that we have shown is the *cultural context*. The Dutch example shows that conservatism leads to slow and careful adoption of new manufacturing practice (in the Dutch case, new product development practice), perhaps regardless of strategic intentions. The Italian case shows that individualism and family-like co-operation may lead to excellent small- and medium-sized companies, but also to hierarchical and relatively low-performing large companies. Individuality also leads to fragmentation in general. The Japanese example shows that a culture of dedication, feudalism and loyalty may create an extremely sound environment for high performance manufacturing. Finally, the inclination towards social responsibility leads to a rationality in manufacturing development that is not only economic, but also based on human values.

A fourth, and classic, linkage is the *product/market* linkage. The size of the market and the variety in consumer demand lead to a specific manufacturing process. In the US and Japan, for example, markets are large, leading to large volumes and lean (mass) production processes. In Denmark and Norway, with companies operating in small niches, volumes are small, but the demands on flexibility and responsiveness are large.

Moreover, the linkages are *dynamic*. The order-winning criteria change over time. The US/Japan case shows this clearly. Where once a well-managed mass production philosophy led to dominance (the US), it eventually became outdated in many industries, and was replaced by lean manufacturing principles. In Argentina and Brazil, deregulation and import liberalisation led to exposure to industrial competition, forcing the local manufacturers to rapidly move out of the past. In Portugal and Spain, the integration in Europe and the development of the economies has led to significant pressure on manufacturing development and a move from low cost to quality and reliability.

There seems to be a more or less global recognition of key concepts in manufacturing. Examples of this are concepts such as Just-In-Time, Total Quality, teamwork, etc. However, this may not necessarily reflect an equal distribution across countries. The application and interpretation of a given concept are contextually dependent. As Lillrank (1995) argued, concepts and practices are not easily transferred between cultures and contexts. Tools and simple techniques are easily transferred, while more complex practices cannot be transferred without specific understanding of context. This leads to a necessary aggregation and conceptualisation of practice into a level of description that is contextually independent. When this is done, the conceptualised knowledge is transferable between contexts, but the concepts and practices must be disaggregated and interpreted in the new, local, context. Understanding this, is most likely fundamental in order to understand diffusion and application of *best practice*.

References
Dunning, J.H. (1988) *Explaining International Production*, Harper Collins
Hayes, R.H. and Wheelwright, S.C. (1984) *Restoring our Competitive Edge - Competing Through Manufacturing*, Wiley & Sons
Hill, T. (1985) *Manufacturing Strategy*, McMillan
Hofstede, G. (1991) *Cultures and Organizations*, McGraw-Hill
Lillrank, P. (1995) 'The Transfer of Management Innovations from Japan', *Organisation Studies*, **16**, 6

CHAPTER 3

MANUFACTURING STRATEGY IN ARGENTINA: THE CHALLENGE OF
CHANGE

*Marcelo Paladino, Roberto Luchi, and Eduardo Remolins, IAE, Universidad Austral,
Buenos Aires, Argentina*

3.1. Introduction

Argentina has faced many structural economic changes in the recent past. Since the
beginning of the Stabilisation Program (Convertibility Plan) in 1991 and the process of
opening up the economy after decades of an Imports Substitution Process, firms now
face low exchange rates and less protection from imports. As a result, businesses are
under considerable pressure to adopt best practice in manufacturing management.

The main objective of this chapter is to analyse current practices in Argentine
manufacturing to see whether they are related to the new state of the economy, to see
what the model of manufacturing organisation used by Argentine firms in the IMSS
sample has been, and to see what the model for competitiveness will be in the future.

3.2. Environment: the Recent Macroeconomic History of Argentina

To understand the current changes in Argentina's economy, we must look back at the
country's recent macroeconomic history. Most of the Argentina industrial base was
created during the Imports Substitution Process (ISP) that lasted until 1977. This period
was marked by the adoption of the Fordist approach to production organisation. In the
developed countries, this approach was characterised by rigid production, based on
scale economies in the local market, narrow product mix, and high inventories,
simultaneously accompanied by a large and stable demand and a low interest rate.

Argentine firms adapted the Fordist approach to production to their smaller and
more closed economy. Lacking scale economies relative to the international market,
these firms grew despite their inefficiencies because of protectionist policies. The State
maintained strong interventionist policies during that period, both indirectly through
regulation and directly through investments in infrastructure and in public enterprises in
'strategic' sectors.

The main economic actors during this period were multinational corporations
(especially after the 1960s), followed by small and medium-sized domestic firms, and
public enterprises. The end of ISP saw the emergence of national holdings, whilst at the

45

P. Lindberg et al. (eds.), International Manufacturing Strategies, 45-61.
© 1998 *Kluwer Academic Publishers. Printed in the Netherlands.*

same time, public enterprises disappeared due to privatisation. As well, multinational corporations began to rationalise products globally, which affected their Argentine subsidiaries.

Since 1977, Argentina has undergone a number of failed attempts to stabilise, open, and reorganise the economy. Over the past two decades, these attempts have been followed by periods of high inflation and rising fiscal deficits. This period marked the end of the Imports Substitution Process (ISP), which had dominated the macroeconomic scenario since around 1930.

The last thirty years have seen the adoption of 'Toyotist' or lean production, with a special emphasis on flexibility, low inventories, and quick response to market demand, based on a wider product mix, multiskilled operators, and fewer organisation levels. Among other changes have been the implementation of techniques such as Just-in-Time, *Kanban*, and Total Quality Management.

We are presently seeing the convergence of production and structural economic reform in Argentina, Greater exposure to competition, a stable economy and new production processes are leading to a new kind of Argentine firm. In order to understand the current situation, it is necessary to explore the most recent changes in the business environment, which clearly affect those 'surviving firms' that are now trying to change.

3.2.1. THE LAST TEN YEARS

During President Alfonsin's government (1983-1989), poor management of public funds and erratic stabilisation programs caused great instability and, consequently, uncertainty about the future behaviour of the principal economic variables. The Consumer Price Index (CPI) rose 10,418,096.03% between January 1985 and March 1991 after the Convertibility Plan was initiated. This rise had a negative effect on economic activity and business decisions. The high rate of inflation clearly reduced incentives to produce and to be more competitive. Inflation also gave confusing signals to businessmen: relative prices became even more volatile with the growth of Price Indices. Consequently, many investments in R&D, equipment, and training were greatly reduced or even curtailed.

The management of public policies, especially a balanced budget, was ignored until the Convertibility Plan was announced in March 1991, although there were some attempts to reduce inflation, such as the Austral Plan in 1985, which was based on a price control system. Neither the Austral Plan, nor its successor the Primavera Plan in 1988, achieved sustained success. In May 1989, just as President Menem took office, Argentina experienced its first hyper-inflationary period, and the CPI reached its highest one-month growth level in history: 196.9%. Similarly in 1990, after the Bonex plan was implemented, Argentina had a commercial surplus in its balance of payments account, but this came at the expense of a large deficit in its capital account; in other words, there was a strong capital outflow (Figure 3-1).

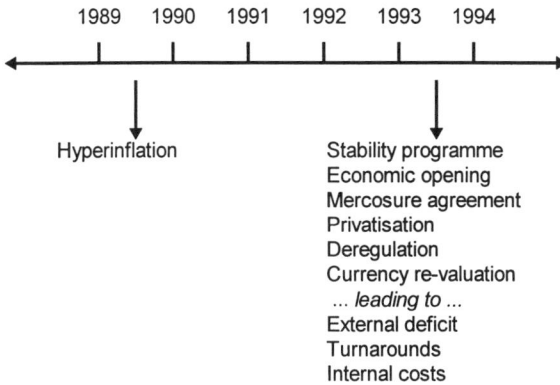

Figure 3-1. The Argentine Economic Environment

3.2.2. THE CONVERTIBILITY PLAN

On March 31, 1991, Minister Cavallo launched the so-called Convertibility Plan. This plan was based mainly on the Gold Exchange System, but the major difference was that the Convertibility Law established strict parity between the domestic currency and the US dollar. The parity rate was first 'US$1 = A10,000' (ten thousand Australs), but on January 1, 1992, the peso was substituted for the Austral, so that '$1 = US$1'; in other words, the nominal (fixed) exchange rate became one.

After Convertibility took place in April 1991, CPI monthly rates of change were rapidly reduced. The rate of inflation began to converge with international levels. Of course, this progress was slower than people wanted. At the beginning of the plan, prices of non-tradable goods (in particular services) did not fall as rapidly as the prices of tradable goods. As a result, the real exchange rate was reduced in term of non-tradable goods, which especially affected firms in those sectors included in this survey.

Once the Convertibility Plan started, manufacturing activity began to increase, and installed capacity utilisation could be filled, reaching more 'natural' levels, and, as a result, employment increased. The plan produced a 'production reactivation' almost immediately, but investment in increasing manufacturing capacity fell. Since 1992/1993, however, investment data have indicated that firms have started to invest again in capital equipment.

3.3. Manufacturing Strategy in Argentina - Evidence from the IMSS Survey

We have set out to find out whether and how manufacturing firms are adapting their business practices in order to fit the new structural conditions reported above. Firms are under pressure to survive and compete in their new environment, and this affects their behaviour.

Based on a survey of Argentine businessmen, we can identify the starting points for increasing competitiveness in Argentina as improvements in internationalisation and the financial system.[1] Internationalisation is based on market freedom, expectations of stability, wider potential for and protection of foreign investment, and so on. The financial system includes adequate regulation, a sufficient level of sophistication, and free access to capital markets. Weaknesses were also perceived in the areas of infrastructure, and science and technology.

In this section we will examine the strategies, goals and financial outcomes of firms in light of the changed economic context of Argentina. The IMSS sample from Argentina included 41 firms, with a high level of heterogeneity. In particular, there are manufacturers from five different 3-digit ISIC categories: metal products, machinery (except electrical), electrical machinery, transport equipment, and professional and scientific equipment (including photographic and optical goods).[2] Although the firms' average number of employees was 301, there was a wide range of firm sizes. The companies were market leaders, with an average share of almost 50% of the market for their principal product line. The sample seems to be representative in terms of production volume.

3.3.1. THE FOCUS OF FIRM EFFORTS

As pointed out earlier, Argentina is undergoing a prolonged period of economic stability. One result of this is that firms today face a considerable amount of accumulated under-investment. We would therefore expect to see an increasing level of investment in process equipment. The survey data supported this, indicating a high level of investment in equipment (13.3% of turnover). Firms are clearly giving investment priority to process equipment compared with their levels of investment in R&D (4.53%) and Training and Education (3.03%), which is reasonable in a turnaround context. Over the longer term, however, a lower gap between these three areas may be desirable. Research elsewhere has indicated that productivity differences between plants with similar levels of automation may be explained by differences in organisational factors.[3] This is a powerful argument for increasing the relative proportion of investment in training and education over the longer term.

The dramatic adjustments that firms are making in response to the changing economic environment are reflected in the business objectives of the firms in the survey. The major competitive goals for the businesses are reducing cost and improving quality. This is consistent with firms 'playing the global game': competitive costs and

---

[1]   See M. Paladino, A. Carrera, and R. Luchi, 'Informe Mundial de Competitividad', IAE, 1994.
[2]   The complete presentation of data and their analysis is in 'International Manufacturing Strategy Survey - Argentina: Results and Analysis'.
[3]   See J. Womack, D. Jones, and D. Roos, *The Machine that Changed the World,* Harper Perennial, 1991.

high levels of quality are the basic requirements for entering global markets or competing in domestic markets against international firms. Until recently, Argentine firms had not paid sufficient attention to international marketplaces or competition, a behaviour that results from decades of competing within a closed economy, but this is now changing as firms adjust to the new, more open, competitive conditions.

This pattern is consistent with the experience of another country that has previously gone through a similar transition: Korea.[4] Firms in a developing country first focus their efforts on the manufacturing basics (cost and quality), then once they gain a share of international markets, the focus changes to other targets such as flexibility and service to customers. Argentina is clearly in the first part of this cycle. In addition, the top ranking given to cost seems to reflect the past history of Argentina, dominated by protectionist policies

3.3.2. FIRM PERFORMANCE

Many firms did not provide full reports of their financial data in the survey but some analysis is possible nevertheless.

Return on Investment (ROI)
It is clear that ROIs are generally low (median 0.11%). This low figure could result from Argentina's history of high inflation, which has made it difficult to measure real performance. High inflation requires re-evaluating assets, revenues and stocks, and the distortion caused by this is probably reflected in the reported ROIs. It may also reflect the disturbances in information reporting due to previous levels of high inflation. The reported ROIs should thus be considered with care.

Inventory Turnover
Inventory turnover averaged 98 days, exceptionally high in today's world of lean production. However, it is not possible to tell from this static data whether this is in fact a reduction from even higher levels in previous years. This high level may also in part reflect the fragmented nature of markets.

Cost Structure
The cost structure (direct materials - 48%; direct wages - 26%; overhead - 26%) reflected a relatively high level of wages as a proportion of the total. Reducing wage costs is the core of the government Labour Reform initiative, which is still ongoing. In addition, the labour market is still very rigid, unable to react appropriately to external shocks or changing external variables. As a result, slow labour market reaction to the rapidly changing environment has caused traumatic adjustments, leading to high unemployment.

---

4   See J. Miller, A. De Meyer, and N. Jinchiro, *Benchmarking Global Manufacturing*, Richard D. Irwin, 1992, and Chapter 25 in this volume.

50   CHAPTER 3

3.3.3. CURRENT MANUFACTURING PRACTICES

Suppliers and Vertical Integration
There is a particularly interesting pattern of suppliers. Not only is the number of suppliers per firm high but, in contrast to other countries, the number is increasing. Argentine firms may be in a process of transition, trying to find the 'right team' to work with in a long-term relationship, and during the transition process this relationship might not be as close as might be desirable. The dynamics of this process are shown in Figure 3-2.

As noted earlier, the historically-closed economy has led to serious scale problems for manufacturers. 69% of value added comes from manufacturing, with just 31% from assembly, which may indicate a relatively high level of vertical integration. This may in part explain the poor relationships between firms and their suppliers, with little development of partnership or outsourcing. There are a number of explanations for this finding. First, it may reflect a lack of confidence in suppliers. Second, it could also reflect restrictions designed to promote labour stability. In recession, these may lead to keeping production inside the firms in order to avoid layoffs and subsequently bringing in previously outsourced production.

Facilities Use
Low scale, high labour costs and fragmented markets, coupled with a need to reduce costs, have led to a distinctive pattern of facilities use. As indicated above, equipment in Argentina is used to produce a wide range of products for a wide range of customers, resulting in a low level of focus. To minimise costs, the emphasis is on maximising the

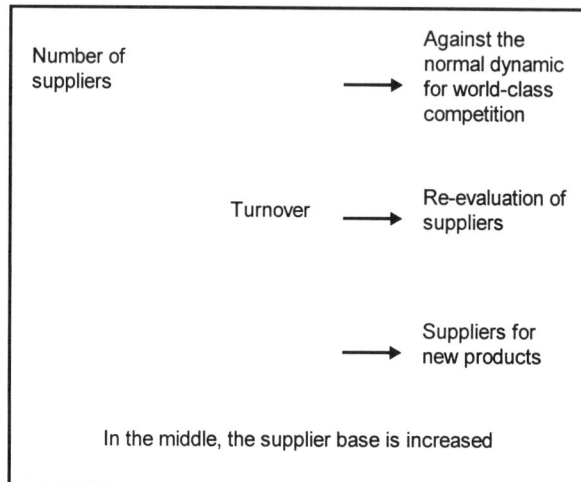

Figure 3-2. Suppliers

```
┌─────────────────────────────────────────────────────────────┐
│                                                               │
│  Facilities                        ────►   Diseconomies of    │
│  Sales and suppliers                       scale             │
│                                            History of closed  │
│                                            economies          │
│                                                               │
│  Relationship with suppliers       ────►   Low               │
│                                                               │
│                                                               │
│  Differences in orders/customers                             │
│  Types of Equipment                ────►   Lack of focus     │
│  Types of processes/products                                 │
│                                                               │
│                                                               │
│  So:      To compete on cost               Efficiency through │
│                                     ────►   using existing    │
│                                             equipment         │
│                                                               │
└─────────────────────────────────────────────────────────────┘
```

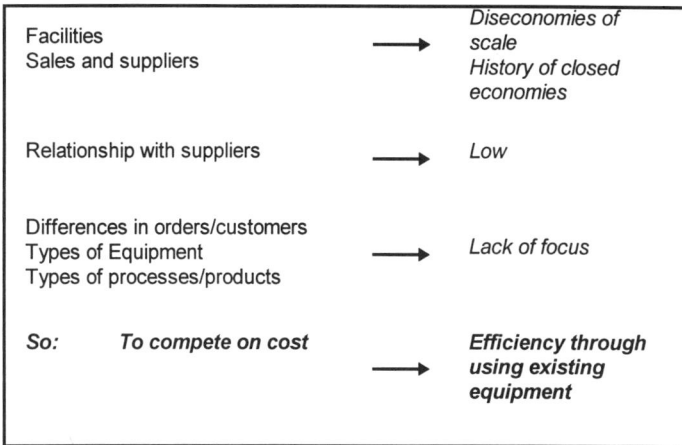

Figure 3-3. Common Practices in Manufacturing

utilisation of equipment (capacity utilisation of 76% is high). This, however, sharply reduces production flexibility. This strategy reflects the current objectives of competing on cost rather than flexibility or time to market. This pattern is described in Figure 3-3.

As internationalisation leads to increasing scale, we might expect to see increasing use of cellular manufacturing in the future to benefit from reducing the response time to market changes. It might be beneficial to trade off some loss of cost efficiency in return for gaining access to different markets.

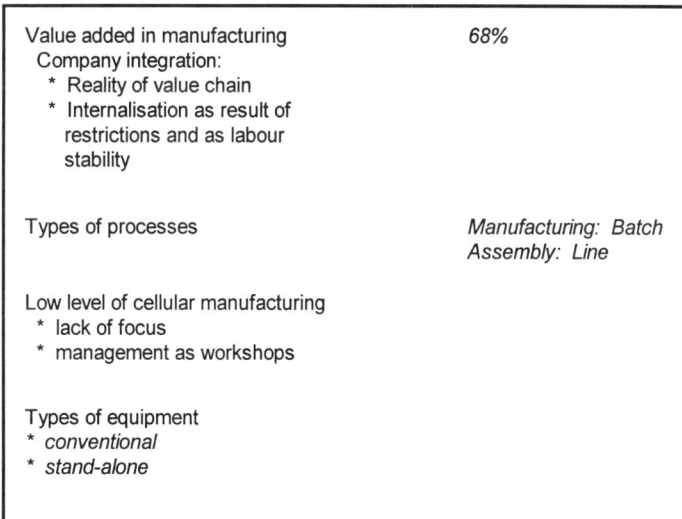

```
┌─────────────────────────────────────────────────────────────┐
│                                                               │
│  Value added in manufacturing            68%                  │
│    Company integration:                                       │
│      *  Reality of value chain                                │
│      *  Internalisation as result of                          │
│         restrictions and as labour                            │
│         stability                                             │
│                                                               │
│  Types of processes                      Manufacturing: Batch │
│                                          Assembly: Line       │
│                                                               │
│  Low level of cellular manufacturing                          │
│    *  lack of focus                                           │
│    *  management as workshops                                 │
│                                                               │
│                                                               │
│  Types of equipment                                           │
│  *  conventional                                              │
│  *  stand-alone                                               │
│                                                               │
└─────────────────────────────────────────────────────────────┘
```

Figure 3-4. Manufacturing Process Characteristics

Capacity and Inventory Planning
The capacity strategies of firms are consistent with fragmented markets, a relatively small economy, and the past economic chaos that made forecasting very unreliable, as well as with batch and job-shop strategies. 84% of firms chose to set capacity equal or greater than forecast demand (41% and 43% respectively). This is consistent with and creates flexibility to react to changes in demand. Finally, this is accompanied by high levels of inventories, in particular raw materials. Overall, the strategies of firms in this area are clearly related to the previous stages of the country's economic performance (see Figures 3-4 and 3-5).

Manufacturing Automation
As noted previously, Argentine manufacturing is dominated by make-to-order batch processes. The data on automation show that the level of computer integration is low (median of 2 and maximum of 4 on a 1-10 scale). In addition, the level of use of automation is relatively low, with conventional machines dominating production, and low use of numerically-controlled (NC) machines, robots, and FMC (see Appendix). These data suggest that in the past in Argentina there has not been a clear need to invest in automation for productivity improvement. Any changes in this area will be related to external changes in relationships with both customers and suppliers. There was unlikely to be much benefit from technology in the traditional fragmented relationships found in the past

Organisation
A consistent pattern of work organisation was found. Group work was relatively low (about 30%), particularly considering the predominance of batch manufacture. Only 31% of firms offered any incentive payments to workers, and these were mainly individual (62%) rather than group-based incentives. Based on the number of

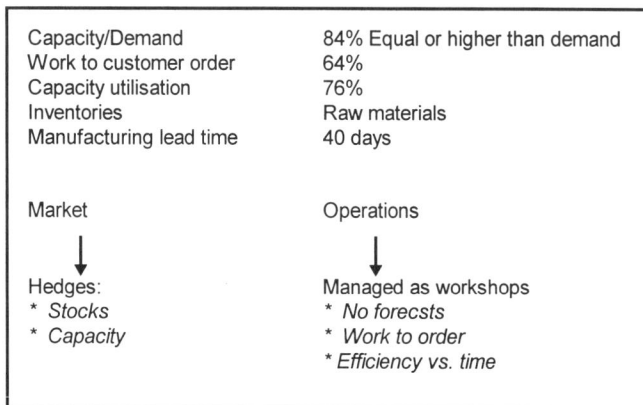

Capacity/Demand   84% Equal or higher than demand
Work to customer order   64%
Capacity utilisation   76%
Inventories   Raw materials
Manufacturing lead time   40 days

Market   Operations

↓   ↓

Hedges: Managed as workshops
* Stocks   * No forecsts
* Capacity   * Work to order
                               * Efficiency vs. time

Figure 3-5. Capacity and Inventory Planning

suggestions per worker (fewer than two per year), the data indicate relatively low levels of participation by workers in manufacturing. This is consistent with both organisational culture and the low level of incentives.

This was accompanied by low levels of workforce training, 112 hours for new workers and 33.5 hours for old ones. This is considerable lower than many other countries and gives cause for concern, as low levels of training lead to lower levels of productivity and consequently real higher levels of costs.

In summary, a workshop management style was found, characterised by a relative lack of attention to human resources in comparison to investment in new processes. Obviously, the priority in Argentina has been 'hard' investment rather than 'soft'.

Planning and Control Systems
Several characteristics of the planning and control systems were studied. First, schedules were frozen for an average of 5 weeks, although there was a high variability between firms that may reflect the lack of focus in processes and the production of many products on the same line. Second, relatively few components were delivered Just-in-Time (22%), although again with a high variability. This is consistent with the high level of stocks. Third, 22% of deliveries were late, with materials shortages, lack of machine capacity, production bottlenecks, and due date changes – all of which are characteristic of job shops – being reported as the main causes. Finally, there is a high level of centralisation of planning and control (70% is done at the planning department level), suggesting a Tayloristic approach to planning and control. These findings are all consistent with the batch production process used, the low empowerment of human resources, and poor supplier relations. They are also consistent with the reported delivery lead time of 40 days.

Maintenance and Quality
Maintenance expenditure is primarily corrective (nearly two-thirds of costs) rather than preventive. This pattern is consistent with the low levels of training, as more problems will occur with untrained workers and fewer incentives. A similar pattern is found in quality costs, where control and quality costs account for more than 70% of the total costs of quality. This may reflect a situation of unstable processes. These practices are consistent with traditional ways of manufacturing and are a considerable distance from lean manufacturing and international best practice.

Co-ordination of Manufacturing and Design
The co-ordination of design and manufacturing again reflects a traditional style of manufacturing, with co-ordination primarily through rules and standards rather than formal or informal meetings (40% and 35% respectively). A relatively low score on the degree to which there was active contribution of manufacturing to design processes (3.28 on a 1-5 scale) was also reported, which implies non-concurrent engineering rather than lean production approaches.

3.3.4. SPECIFIC MANUFACTURING GOALS AND ACTIVITIES

Examining the full list of goals (see IMSS Tables in Appendix), it can be seen that there were a wide range of goals that were considered important by firms, the highest being reducing unit costs and manufacturing lead times. The highest rankings by degree of importance were given to reducing unit costs, improving direct labour productivity, improving conformance quality, reducing overhead costs, and reducing manufacturing lead times, while the lowest ranking was for reducing the number of suppliers. These rankings are consistent with both the stated business objectives and the manufacturing practices described earlier.

Relatively few of the sample reported their improvement activities fully, so the data must be treated with care. For those who reported (see IMSS Tables in Appendix), the data indicate a very low level of use compared with other countries, but for all elements the payoff score is higher than the degree of use. This may indicate that firms are at the beginning of implementation and are envisaging high payoffs for activities that are not yet being used.

3.3.5. PERFORMANCE IMPROVEMENTS 1990-1992

The improvements in performance were highly variable, with some items such as procurement lead time and customer service having low response rates, possibly indicating a lack of knowledge of these data (see Table 3-1). Overall, the profitability increase of 7.15% is relatively low, and was possibly affected by the foreign policies in place after 1991. Some sectors, specifically industrial commodities, have seen

Table 3-1. Improvement in Performance Indicators, 1990-1992

Indicator	Better				Worse			
	N	Avg.	S.D.	Dev. Coeff.	N	Avg.	S.D.	Dev. Coeff.
Conformance to specifications	36	39.58	34.76	87.81	0	-	-	-
Average unit manufacturing cost	35	26.57	42.56	160.15	3	26.57	7.64	65.47
Inventory turnover	30	32.27	50.28	156.31	4	31.25	14.36	45.96
Speed of product development	27	35.37	45.19	147.76	2	7.50	3.54	47.14
On-time deliveries	32	31.28	43.25	138.27	3	7.33	4.62	62.98
Equipment changeover	31	24.32	36.03	148.12	4	12.75	9.14	71.71
Market share	28	26.46	59.04	223.10	3	13.33	7.64	57.28
Profitability	29	14.07	42.29	300.62	9	15.11	9.64	63.77
Customer service	28	33.39	42.27	126.60	0	-	-	-
Manufacturing lead time	31	30.84	53.00	171.85	0	-	-	-
Procurement lead time	25	26.64	44.57	167.31	0	-	-	-
Delivery lead time	32	24.84	32.22	129.67	1	10.00	-	-
Product variety	29	21.79	57.00	26.56	5	55.00	81.39	147.99

considerable falls in profitability; nevertheless, the fight for quality and service has begun in Argentine firms, although it will be affected by low profitability and uncertainty about important aspects such as product variety.

3.4. Case Study: Parana Shock Absorbers

Paraná is a leading manufacturer of shock absorbers, located in Rosario, a city to the north of Buenos Aires. In 1993, it had sales of $35 million and 300 employees. The firm supplied the market for original and replacement equipment as well as three assemblers of motor vehicles and light trucks.

Ownership of the shareholders' equity was shared between an American group (33%), who was a leader in shock absorber manufacturing with operations on a global scale, one of the main Brazilian assemblers of components (26%), and an Argentine partner (26%); with the remainder quoted over the Stock Exchange. The company operated with technology licences belonging to the American group.

In 1989, the Argentine partner requested assistance from the American group to initiate a turnaround process at the company. American executives and consultants sent to Argentina collaborated with the local management team to develop a diagnosis. After this, the process of turning around operations began, which was monitored by the American group. Two years later, in 1991, in a seminar that brought together firms related to the American group at a world-wide level, the local company was presented as a 'leading case' on how to adapt operations after a management change with minimum investment.

The association with the American group provided a number of advantages in addition to the technological links. Several operating parameters had been standardised across all group companies world-wide, which made constant 'benchmarking' possible[5] and also allowed local management to keep ahead of the industry dynamics. As an example, the impulse to change in 1989 was inspired at least partly by a similar adjustment process implemented in the company's US operations, experience that could be capitalised on locally, as Argentine workers made periodical visits to US for training in work methods.

The corporate group gave priority to the competitiveness and benchmarking criteria among its firms in shifting its production to the least expensive alternative, be it local, American or Brazilian. Thus, the local organisation, up to middle management, had to know and perform a continuous supervision of the local costs level. At present, Paraná considers its productivity to be competitive at an international level, taking into account automation and product flexibility demanded by the local market. Productivity had increased from 16.8 shock absorbers per person/day at the beginning of the turnaround

---

[5]   Some parameters used were: shock absorbers productivity per day and per worker, inventory of finished and in process products, area occupied, manufacturing lead time, wastes, development and audits of suppliers.

process to 52 in May 1993, with the target for that year set at 64 units. The number of employees had decreased to 300 from the original 700 in 1989.

Current scheme of work organisation: the cell concept
Whilst the initial phase of the process had been essentially based on a management change, when the business had become profitable again, some investment was performed.[6] The entire plant was focused around the organisation of work into cells. Initially, the cells were organised by the type of process: tube cutting, rod operations, assembly, etc.. At present, the different cells have been integrated by products, with a work scheme synchronised with the final assembly plan needs. The flow of materials and operations makes intensive use of *Kanban* signals.

Management gave priority to the philosophy of 'lean' operations and to meeting the operational parameters of the group's firms, emphasising that mode of working in spite of the strong volume increases experienced in the industry.

The reality of JIT
Paraná made daily deliveries to the automobile assemblers. Geographical distances from the plant and its customers could be managed through adequate logistic co-ordination,[7] as the company did not make direct deliveries to the assembly line. If a customer were to request JIT delivery, it would be possible only at a nearby location. As has been already mentioned, the philosophy of internal inventory reduction had arisen through the company's own initiative.

Materials management was performed through the MRP system and *Kanban* signals. The MRP explored the monthly needs and divided them per day. *Kanban* signals regulated the flow of internal materials. MRP was also used to communicate future needs to the plant's suppliers, whilst the physical deliveries were controlled with *Kanban*.

Utilisation of CEP and systems
CEP is intensively used in all characteristic variables. The objective was to evolve from shock absorbers to complete suspension systems. This would imply, for example, adding components such as springs, bearings and bushings to the shock absorber.

Technological links
The plant delivered complete shock absorber systems to the automobile assemblers. The technology was licensed from the parent company. The international group made basic investments in engineering and then tried to optimise the final product development in

---

[6]   The automation and flexibility levels required for the process were the variables that had an impact on productivity. For example, the group had plants in USA with a high automation level, a productivity index of 200 shock absorbers per worker/day and manufactured only two product references. It was estimated that the local automation level would increase over time to 85, but the number of references handled was 400.

[7]   The distances to the assemblers varied from 300 to 400 kilometres approximately.

different companies of several countries, which were able to achieve economies of scale as suppliers at a global level. The intention of the Argentine company was to participate actively in this process. The plant's technological capacity was determined by the availability of licenses, depending on the division of products among the global group and its capacity to adapt a product to the local environment. For example, Paraná made adaptations of original designs to increase the shock absorbers' durability, as demanded by the local market.

3.5. Case Study - Mercogearbox

The company supplies one local terminal, with capacity of 300,000 gear boxes per year. An investment of US $220 million was required to set it up. From the beginning, it produced purely for export. Initially, it was expected to export 90% of its production to Brazil and 10% to Germany. The company owned 238 machining centres, 3 automatic transfer lines, and a number of semiautomatic assembly lines. It had implemented advanced management concepts including cells and JIT with suppliers. In 1993, the firm had sales of US $200 million and 1,300 employees.

The experience of competing in global terms
From its beginning, the project foresaw exports to Brazil and Germany. Because the launch of the vehicle in Brazil was delayed, the firm began exporting to Germany. A sales agreement was signed with a plant owned by the German partner of Mercogearbox, which had a manufacturing capacity of 1,200 vehicles per day, and was exporting to European Union, Japan and USA.Production began on January, 1992 and shortly after, when 310 transmissions had been assembled, six rejects were found, what meant a 2% defect level. This percentage was unacceptable for the operational performance of the plant, because the customer only accepted a maximum level of 0.2%. The firm had to recover and control the 5,200 transmissions that were in the transport and warehouse system. But, it was essential to develop a new strategy for training the plant personnel that was oriented to the concept of supplier-customer and decentralised decision making. Each worker was intended to be an administrator of the process, with authority to stop a machine if it was required.

In the month of June of that year, controls were tightened. The company began to integrate customer requirements into the process. The errors of the internal test beds had been corrected.

Mercogearbox then worked with a volume of 1,200 daily boxes, supplied to Brazil and the local terminal, with a failure index of 0.05%, which was much lower than that accepted by the German plants. This positioned the competitiveness level of the company equal to that of the assemblers of other regions.

The firm considered that the unquestionable values, which were the essence of the operative unit, were to be competitive as an exporter and to meet international quality levels. For example, when the boxes for Germany had to be reworked, the top executive level of the company had communicated to the union management that it was a good chance to defend the working source. One of the union leaders had gone to Germany to receive training.

Operative unit design in order to compete in global terms
Preparation had begun in 1991. 40% of the machines were numerically-controlled, a high percentage of them were interconnected, and 41% of the machining centres used intelligent measurement systems. Much attention had been paid to the early involvement of the personnel with the technology. Intensive training had previously been carried out. Each worker from the machining centres had travelled to Germany in order to see the equipment creation from its origin, observing for example how the screws were mounted and taking notes in the assemblers' plants.

People had worked intensely to establish a suppliers-customers regimen within the plant. For example, in the case of forged parts, a major component in operations, those responsible for this material used to go to the plant frequently to contact the worker in charge of its processing. The work scheme had several levels of supplier-customer relations: external supplier and manufacturing of parts; transmission lines and test bed assembly; test beds; final customer (terminal) and subsidiaries. In Germany, 'the customer voice' had been listened to every day at 7 a.m.; the customer was not expected to phone, the plant workers contacted him on their own initiative.

The plant demanded a high level of personnel training, which included some engineering students working in the line. The process of operative excellence had been made easier by employing younger workers. The company also wanted to evolve towards a system of categories according to apprenticeship.

The objective in which they were working was the evolution to a greater quantity of autonomous cells of work for controlling quality, logistics and maintenance. The workers in the cells were under the charge of a 'team leader' and were responsible for the allocation of resources in order to fulfil the daily production plan. A constant exercise of benchmarking against operating units in other regions was performed to monitor the competitive position of the company in a global competitive environment.

Process for implementing JIT deliveries
The operating philosophy had been extended to suppliers. For example, as already mentioned, each supplier of forged parts has a person in its firm. Mercogearbox has autonomy with relation to the main assemblers in the process of supplier development and audits. There were at that time 10 pieces using JIT delivery (up to 3 deliveries per day) and the philosophy was deliveries free from inspections.

Plant scale - Optimal capacity of the business
The plant capacity had been designed to be competitive at an international level. The management thought the actual level of 1,200 boxes per day placed them in a highly

competitive position with regard to costs. The firm had the potential to increase annual production from 300,000 boxes to 500,000. The transfer system they had gave them enough flexibility to change from one model to another; they were handling eight product references.

3.6. Summary

According to the opinions of businessmen, the starting points for increasing competitiveness in Argentina were identified as: Internationalisation and Finances.[8] Internationalisation is based on market freedom, expectations of stability, wider potential for and protection of foreign investment, and so on. Finances includes adequate regulation, a sufficient level of sophistication, and free access to capital markets. Weaknesses were perceived in the areas of infrastructure, and science and technology.

We have set out to find out whether and how manufacturing firms are adapting their business practices in order to fit the new structural conditions. Firms are under pressure to survive and compete in their new environment, and this affects their behaviour.

It is evident that firm policies are an aggregation of the characteristics acquired in the period of import substitution policies and the prevailing economic conditions. There is a high degree of consistency between firm policies across many areas. There is now a new direction, and the direction of this change is reflected in the stated objectives of the companies. The speed of such change is crucial for the competitiveness of the country in future years.

The challenge for firms is to begin the difficult part of operations, developing their skills to improve the 'soft' side of manufacturing and to change the prevailing paradigm of manufacturing to one that matches the future market and economic environment. This new paradigm will be very different from the present one.

3.6.1. CONCLUSIONS

Our analysis has shown that Argentine firms are in the process of adapting to new economic conditions under the pressure of increasing competition brought about by the change from import substitution with closed markets and high inflation to an open, deregulated and stable economy in a strong process of regional integration with Brazil. We can identify the elements of the current manufacturing paradigm for Argentina as being:

---

[8]   See M. Paladino, A. Carrera, and R. Luchi, 'Informe Mundial de Competitividad', IAE, 1994.

- *Suppliers*. Contrary to the paradigm of lean manufacturing being adopted elsewhere, Argentine firms are increasing the number of their suppliers as part of redefining their supplier base in the context of structural change. This is accompanied by a lack of close supplier relationships.
- *Facilities*. General-purpose equipment predominates, and is used to manufacture a wide variety of products with the goal of maximising equipment utilisation (and hence resulting in a minimisation of flexibility). This is consistent with competing on competitive objectives of costs and quality rather than flexibility or service.
- *Capacity strategies*. Production is make-to-order, and excess capacity and high levels of inventory are being used to hedge against uncertainty. These are consistent with a small economy, lack of scale opportunities, and information chaos, which has made forecasting useless due to inflation and hyperinflation.
- *Manufacturing processes*. These have been predominantly batch manufacturing, with a high level of vertical integration, consistent with the make-to-order environment.
- *Organisation*. Low workforce incentives, low participation, little teamwork, and low levels of training show a lack of attention to human resources. This can be contrasted with a high level of investment in new equipment.
- *Planning and control*. Planning and control systems are highly centralised, predominantly Tayloristic, and reflect the low level of empowerment of human resources.
- *Maintenance and Quality*. These emphasise corrective rather than preventive approaches, consistent with low levels of training.

In summary, the Argentine manufacturing paradigm can be summarised as being predominantly a traditional batch manufacturing approach, reflecting Argentina's historical economic and industrial situation. This is consistent with the rankings of market aims and financial and inventory performance described in Section 3.3. However, it is in contrast with the goals and focus of the companies, which clearly reflect the current situation and the future in Argentina. The context of Argentine manufacture, its present position and future directions is shown in Figure 3-6.

The new challenge for companies in Argentina is to make the following changes required to compete regionally and globally:

- To develop manufacturing strategy processes, as indicated by the lack of alignment between the companies' stated goals and objectives and the actual manufacturing practices in place;
- To improve human resources management, in areas such as teamwork, training, and linking manufacturing and design;
- To emphasise quality more strongly;
- To develop supplier networks, especially managing relationships with suppliers.

Figure 3-6. Manufacturing Paradigm in Argentine Firms

Groups of variables	Criterion	Operations Paradigm	Challenges	Declared degree of use	High payoff, but low degree of use
Suppliers	Poor relations	Organisation as workshop	Manufacturing strategy	High	Design
Facilities	No focus	Traditional organisation	Improve labour relations	High	
	Different products run on same line	Investment in equipment			
Capacity	Excess capacity	Quality not so important	Quality improvement	High	MRP
Manufacturing process	Batch				
Organisation	No incentive pay	Low costs by using capacity			Pull-*Kanban*
	Low participation				
Planning & control	Centralised	Customer orders			
Maintenance and quality	Corrective, not preventive	Factory view w/o including suppliers	Supplier development		
Design/ manufacturing	Little co-ordination				Important difficulties in how to rethink operations Real change inherent in the manufacturing paradigm

Generally speaking, whilst actions have already been taken on the 'hard' aspects, there is a need to improve the 'soft' aspects and redesign organisational structures.

References

Paladino, M., Carrera, A., and Luchi, R. (1994) 'Informe Mundial de Competitividad'.

Paladino, M., Carrera, A., and Luchi, R.'International Manufacturing Strategy Survey - Argentina: Results and Analysis', IAE.

Womack, J. Jones, D. and Roos, D. (1991) *The Machine that Changed the World,* Harper Perennial.

Miller, J. De Meyer, A. and Jinchiro, N. (1992) *Benchmarking Global Manufacturing*, Richard D. Irwin.

CHAPTER 4

MANUFACTURING MODERNISATION IN BRAZIL: SCOPE AND
DIRECTION IN THE METAL PRODUCTS, MACHINERY AND EQUIPMENT
INDUSTRY

*P. Fernando Fleury and Rebecca Arkader, COPPEAD/UFRJ, Federal University of Rio
de Janeiro, Rio de Janeiro, Brazil*

4.1. The environment for Brazilian manufacturing, past and present

Among developing nations, Brazil has one of the largest and most diversified
manufacturing sectors, developed over the last four decades. Over this period, however,
and especially during the last twenty years, much has changed in the way companies
organise their manufacturing operations and compete in the marketplace. The lean
production concept of manufacturing, developed in Japan by Toyota, is rapidly
spreading to the West. The increase in global competition, coupled with the perception
that lean production is a superior form of organisation, has contributed to this rapid
diffusion (see, for example, Womack, Jones, and Roos, 1990). The challenge before
Brazilian manufacturing firms, in the face of such an international change, has been to
cope with this shift in the manufacturing practice paradigm, at the same time as
significant changes in domestic economic policies have been forcing them to adopt a
new competitive stance.

Brazil is a vast country, the eighth largest economy in the world in 1993.[1] Its
income distribution is, however, deficient. Despite a reasonable export performance in
recent years and the achievement of sizeable trade surpluses (as shown in Table 4-1
below), the country can still be considered to be a mostly closed economy, taking into
account the weight of exports in its GNP. After a period of considerable growth in the
1970s, since the early 1980s the country has experienced a critical cycle of economic
problems, which one government plan after the other has not been able to address
seriously and solve. On one hand, the poor general prospects, as well as the consistently
high cost of capital, have kept away both local and foreign private investment in
production. On the other hand, the fiscal crisis—adversely affected by low levels of tax
collection throughout the economy, as well as by an unfavourable tax system—has
resulted in insufficient investment in education, health, and social infrastructure.

---

[1]   Unless otherwise indicated, the country data in this section are taken from the 1993 IMD (Lausanne,
      Switzerland) *World Competitiveness Report.*

P. Lindberg et al. (eds.), International Manufacturing Strategies, 63-79.
© *1998 Kluwer Academic Publishers. Printed in the Netherlands.*

Table 4-1. Recent Brazilian Foreign Trade Performance (in US$ million)

	1990	1991	1992	1993
Exports	31,414	31,620	35,793	38,783
Imports	20,661	21,041	20,554	25,711
Balance of Trade	10,753	10,579	15,239	13,072

Source: Central Bank of Brazil Bulletin, March 1995

Brazil has been widely perceived as a highly protectionist and regulated economy. This stems from issues such as the participation of the State in the economy, price control policies, and antitrust and industrial property legislation. Performance in terms of social indicators has been equally unsatisfactory; for example, the country had in 1993 the highest illiteracy[2] index among all countries in the IMSS database. Workforce productivity has been among the lowest in the newly industrialised world, and productivity growth practically stagnated in the 1980s—in fact, between 1982 and 1989, it rose only 0.92%. The outlook is better, however, relative to managerial capabilities and the availability of qualified engineers.

It is against this background that economic policy in the last decades has helped shape the Brazilian manufacturing sector. At the outset of the 1990s, the situation of the Brazilian manufacturing sector was clearly a result of the more than thirty years under strongly protectionist economic policies, which had led to scant internal competition, and a strong presence of the State in the economy, both in terms of regulation and of production activities. In such an environment, managerial issues related to efficiency and competitiveness had typically been put aside. In fact, corporate performance was mostly determined by factors outside market forces. The result was that quality and productivity suffered, and that high levels of waste were hidden by price distortions brought by high inflation and lack of competition.

Accordingly, during a period of accelerated development in process and product technology in developed nations there was little effort to upgrade plant and equipment in the Brazilian manufacturing sector, or to acquire the skills that might enhance competitiveness in increasingly demanding international markets for industrialised goods.

The first three years of the 1990s have brought significant change and turbulence to the Brazilian environment. The economy has since undergone a process of redefinition of priorities and patterns of competition in the domestic market. The gradual relaxation

---

[2] Some recent data published in the press illustrate this specific shortcoming for Brazilian manufacturing firms in terms of workforce qualification. In Autolatina, the country's largest private company, holding company for Volkswagen and Ford operations in Brazil and Argentina, in the beginning of the 90's, 28,000 of the 41,000 employees had not concluded grade school and, of the former, 3,700 had not even made it to 4th grade. In Azaleia, Brazilian's largest shoe exporter, 89% of the workforce had not concluded grade school. In the 1980s, an internal survey conducted by COFAP, a large company in the auto parts industry with $500 million in sales, 16,000 employees, and a strong export record, led to the conclusion that the average word usage domain of its workforce was limited to 150 words. The issue of education and competitiveness in Brazilian industry is addressed in Fleury (1993).

of protectionist measures and the reduction of the strong position of the state in the economy have been pushing Brazilian companies toward considerable efforts to promote changes conducive to higher levels of competitiveness.

Therefore, in the face of more intense external competition and under the influence of an adverse macroeconomic environment featuring high inflation rates, instability on the demand side, and high interest rates, there has been an urgent need for Brazilian companies to promote significant changes. The decision on the part of the Brazilian government to open and deregulate the economy, on the one side, and the rapid diffusion, on a global scale, of the 'lean production' paradigm, constitute the main ingredients of the present shifting environment for Brazilian manufacturing companies. Facility modernisation and change in management practices are required for Brazilian companies who wish not only to thrive, but first of all to survive in the medium and long run.

The IMSS research has presented an opportunity not only to evaluate the scope, measure the pace, and identify results of manufacturing practices and strategy in a relevant industrial sector in Brazil, but also to compare them with those of companies in the same industries in other countries. The data gathered on the Brazilian sample of 28 companies enable us to look into the present state of Brazilian manufacturing and to probe into the scope and direction of manufacturing modernisation in Brazil under an environment such as the one described above.

In the next section we present and discuss specific results concerning issues in manufacturing that we consider to be of the greatest relevance to help position Brazilian firms in terms of the international competitive scene. The third part of the chapter illustrates the paths to manufacturing improvement in Brazilian firms with the case of a locally-owned bus manufacturing and assembly company, one of the firms in the IMSS sample. The paper concludes with a discussion of the preferential path toward manufacturing modernisation that evidence indicates is being followed by Brazilian firms, and of the competitive outlook it implies.

4.2. Manufacturing performance and strategies: highlights from the Brazilian sample of firms in the IMSS database

4.2.1 OPERATIONAL CHARACTERISTICS AND OUTCOMES

The operational characteristics of the 28 Brazilian firms in the sample seem to reflect the economic environment in which they have been embedded. The combination of a regulated and closed economy with scant competition has resulted in an industrial structure in which the leading companies tend to be large, inefficient, inward looking, and less differentiated than the average company in the sample as a whole.

The average size of Brazilian firms in the IMSS survey, measured by the number of employees, was almost twice that of the firms in the whole sample (1,476 compared with 867). With an average market-share of 40%, Brazilian firms were leaders in their main product line segments, but their return on investment, at 10%, was not as high as it might have been when compared with the average company in the whole sample. The non-Brazilian companies reported a 14% return on investment, in spite of an average 34% market-share.

The closed environment also seems to have influenced the international positioning of Brazilian companies. They were more inward looking when compared with the sample as a whole, made up mostly of firms in developed countries. In fact, Brazilian firms in the sample reported fewer international links with customers and suppliers than the average firm in the whole sample, both in terms of percent of export sales and of imported supplies.

The issue of differentiation may be examined by observing the performance achieved by companies in dimensions like innovativeness, delivery performance, and flexibility. Compared with the whole sample, Brazilian firms performed poorly on several measures related to innovativeness. They had very narrow product lines: the average number of products they offered was 180, compared with 440 for the total sample; their percentage of income derived from new products was 11%, compared with 19% for the whole sample; and their investment in R&D was 3.9% of annual sales, compared with a general average of 4.9%.

Delivery performance was one of the most striking results regarding Brazilian firms in the sample. In terms of lead time, they lagged well behind world average: 99 days compared with 52 days. In terms of late deliveries, their record was not very different from the average firm in the total sample—12% compared with 11.8%—but still worse than more competitive nations like Japan and Germany.

One possible reason for the long lead times, despite the existence of an average capacity slack, could be that the high-inflation context tended to induce an end-of-the month purchasing pattern, leading to a monthly-measured average slack and periodical production peaks. Most dependability problems arose out of materials shortages—50% in average—a proportion far above that for the total sample, which posted 34.1%. Therefore, material shortage also contributed to poor lead time performance. This highlights the issue of supply-related problems adversely affecting materials management in Brazilian firms.

With undependable deliveries due to materials shortages, Brazilian firms carried substantial levels of raw materials inventory—41.5 days, compared with 33.1 days for the sample as whole. Several reasons might have accounted for such large raw materials inventories, one of them being the reduced possibility of accurate forecasting of material needs in view of the end-of-the-month purchasing/ordering behaviour: firms ordered materials one end-of-month to meet their clients' orders one month ahead. In addition, the lack of standardisation might be leading to excess in materials specifications, as well as possible mismatches between what was in stock and what in fact was needed on the shop-floor. It may also be recalled that Brazilian firms do not yet report intense use of MRP programs to help in materials management. These could be explanations for the fact that, despite large average raw materials inventories, deliveries were still late due to materials shortage.

Average WIP inventories held by Brazilian firms, on the other hand, ranked with those for the sample as a whole—23.2 days compared with 24.3 days of production—and their average finished goods inventories were also quite low—11.4 days of production compared with 20.9 for total sample companies. This might be the result of a predominant use of the one-off type of production process, of the lower proportion of forecast orders reported by Brazilian companies (though not considerably lower than that of other countries), and of capacity policies. It seems possible to propose that the typically higher levels of raw materials inventories in Brazilian firms signal supply chain problems, be it in terms of supplier relations or of adverse economic conditions, as low average WIP and finished goods inventories would suggest they were not the result of deliberately conservative policies.

The problems in the supply channels were further evidenced by the small proportion of firms receiving materials on a just-in-time basis: 19.6%, compared with 29.3% for the sample as a whole. The reported intensity of use of external just-in-time practices by Brazilian firms was relatively low compared with that of other nations. In the specific case of Brazilian firms, this relatively low use of just-in-time in supply and delivery could be traced to several different causes, among which were the large geographical distances, poor transportation infrastructure, and instability in the economic environment, which was not conducive to co-operative behaviour among firms and came to favour an atypical purchasing behaviour. The preference for end-of-the month purchases, for instance, can be seen as a disincentive for external just-in-time practices. In addition, peculiarities in tax laws may induce firms to forego logistics efficiency and give preference to purchases from out of their own state (for example, for one specific product, value-added tax might be 17% for trade inside the state and 12% for interstate trade - the consequent price differentials might be quite relevant and discourage geographical clustering of firms and just-in-time purchases).

Further evidence of troubles in the supplier relations area was the fact that, even though Brazilian firms reported a friendly relationship with suppliers, friendlier in fact than world average, they maintained a much larger supplier base than the rest—an average of 995 suppliers, compared with 440 for the total sample.

One way to explain the strategic disadvantages of Brazilian firms is to look into some of their organisational characteristics. For example, they tended to use more of the *one-off* type of production process than firms in other countries, a result that might be credited, at least in part, to the lack of comprehensive product standardisation efforts. They were also making less intensive use of cellular layout than the overall average firm, both in terms of manufacturing and assembly operations. They were spending relatively less on process equipment than firms in the total sample—9.6% against 10% of sales revenues.

The critical issue of hardware modernisation deserves a special comment. Brazilian firms used much less advanced process technology than their counterparts in other countries. 88% of the total number of machines consisted of conventional equipment, compared with 81% for the whole sample. The average Brazilian company had adopted a very low number of FMS/FMC, robots, and machining centres, a clear indication that the country was far behind others in the use of flexible automation equipment.[3] This can be confirmed by a more precise comparison of the number of machines per thousand employees in total sample companies against that in Brazilian companies: 130 vs. 107 for conventional machines; 17.6 vs. 12.9 for NC machines; 5.0 vs. 0.8 for machining centres; and 4.4 vs. 0.3 for robots.

A few other organisational and social aspects should be further considered. Brazilian firms spent an average of 60 hours on training their new production workers, compared with 113 hours in sample IMSS firms, even though they stood somewhat above the average for the latter relative to training given to the regular workforce—43 hours compared with 35 hours. In terms of incentive payments of all kinds, Brazilian firms were quite behind, with very few companies (some 10%) adopting such a practice. This contrasts with total sample averages, in which 24% of the companies paid based on individual incentives and 22% practised group incentives. In addition, Brazilian firms scored poorly in terms of job classifications in manufacturing: they had 49 job classifications, compared with only 10.2 for the whole sample. All this must have negatively influenced the performance of Brazilian companies in terms of the number of suggestions per employee—only 1.6 per year, compared with 7.4 for the sample as a whole.

It is interesting to note that even in an adverse situation like the one just described Brazilian firms managed to score quite well in terms of direct worker absenteeism, as well as employee turnover rates. Compared with international standards, absenteeism was 2.3% against 4.8% and turnover was 3.6% compared with 8.1%.

All the above comments lead us to the conclusion that, up to 1993, the general performance of the Brazilian companies in the sample was far behind the average standard for companies in the sample as a whole. This can be summarised by the gap analysis presented in Table 4-2.[4]

From the table it can be seen that Brazilian companies were far behind on most indicators. Two main problem areas could be identified. The first of these regards time-related performance, reflected in long lead times, late deliveries, and supplier relations, which may be credited to: uncertainty in the economic environment, deficiencies in the physical infrastructure, and organisational difficulties. The other problem area was

---

[3]   The average Brazilian firm had a proportion of robots that was one twentieth that of Japanese firms' and one tenth that of total sample firms'.

[4]   Gaps discussed in this and other sections indicate in percentage terms the distance between the higher and the lower value; positive gaps indicate Brazilian firms appear to be better off concerning the indicator; the opposite is the case where the gap (shown between brackets) is negative. For example, a higher inventory turnover being the better, the 11% gap is indicated as negative; as a lower number of days of finished goods inventory is considered to be better, the 87% gap is indicated as positive.

Table 4-2. Selected Comparative Performance Indicators for Brazilian and Total Sample Companies

Performance Indicator	Total Sample	Brazil	Performance Gap
Inventory turnover	8.2	7.4	(11%)
Finished goods inventory (days)	21.3	11.4	87%
Raw materials inventory (days)	32.6	41.5	(27%)
Percent of just-in-time purchases	29.7	19.6	(52%)
Number of suppliers	437	996	(130%)
Delivery lead time (days)	52	99	(90%)
Percent of late delivered orders	11.7	11.9	(2%)
Percent of throughput efficiency	32.7	68.0	108%
Percent of preventive quality costs	22.1	26.9	22%
Percent of revenue from new products	18.9	11.2	(69%)
Product variety (number of products in line)	736	165	(346%)
Investment in R&D (% of turnover)	4.9	3.9	(26%)
Investment in training (% of turnover)	2.3	2.5	9%
Investment in equipment (% of turnover)	10.0	9.6	(4%)

product variety and development, which had not yet even become a competitive priority to firms in the sample. These shortcomings might be credited, among other factors, to the lack of competition in the economy, an inadequate level of technological capabilities and skilled resources, the lack of appropriate investment funds, and an outdated firm strategy which privileged cost over differentiation.

The outdated strategy of Brazilian firms can be confirmed by IMSS data concerning the quantified goals for improvements they reported. According to these data, the most relevant goals were reductions in inventory, manufacturing lead time, material costs, and unit costs, as well as improvement in direct labour productivity. The main concerns in terms of their operations came out in a fairly evident fashion— efficiency in the use of resources and supply chain management. On the other hand, even though they were shown to be lagging behind in terms of innovativeness, Brazilian firms still ranked two related goals—the reduction of new product development cycle and the ability to make rapid design changes—among those with the lowest priority.

4.2.2 PROGRAMS AND ACTIVITIES FOR IMPROVEMENT

Having analysed the organisational characteristics and performance of Brazilian firms, the next step is to look into their efforts to improve operational performance. In general, it could be said that they were ahead of the average firm in the total sample in terms of adoption of improvement/modernisation programs and activities. From the 27 improvement programs listed in the IMSS survey, Brazilian firms were ahead in 21 of them, measured by the percentage of firms adopting each program. The widest gaps in favour of Brazilian firms were: pull scheduling, with a 25.1% gap; value analysis, with 19.5%; quality function deployment, with 15.1%; ISO 9000, with 14.4%; and SPC, with 13.7%. Among the six programs in which Brazilian firms had a negative gap in relation to firms in other countries, only one could be considered significant, MRPII, with a 30% negative gap.

Brazilian firms also indicated an above-world average intensity of use in most of the improvement programs and activities, with high reported payoffs. The most favourable gaps in intensity of use were in *Kanban* (a 27% gap); value analysis (7.5%); quality function deployment (11%); ISO 9000 (3.0%); and SPC (3.0%). They were making relatively intensive use of most of the so-called 'Japanese' production techniques, with sizeable reported payoffs. One interesting result was the high payoffs reported on the non-Japanese ABC and CAM techniques for the Brazilian sample, compared with their only moderate ranking in terms of adoption rate and intensity of use, pointing perhaps to a gap that has been perceived by local companies.

4.2.3   PERFORMANCE IMPROVEMENT IN THE PREVIOUS 2 YEARS

As reported above, the efforts made by Brazilian firms in search of operational improvement seem to be paying off. Brazilian companies, as well as total sample companies, reported substantial performance improvements in terms of the parameters considered in the IMSS survey. Brazilian firms, in particular, have shown a remarkable improvement record, above 20% in eight out of the 13 parameters. The largest advances were in terms of quality conformance, inventory turnover, on-time delivery, customer services, and delivery lead-time. It is particularly relevant to point out that this means that the most important reported goal, inventory reduction, has been successfully addressed, and that three out of these top five performance areas were related to what has been identified above as one of the most critical problems for the Brazilian IMSS sample firms, namely supply chain management.

Despite the difficulty in comparing this kind of data due to different stages of manufacturing advancement, Brazilian firms indicated superior performance improvement relative to total sample companies, with few but important exceptions. To compensate for the difficulties in lead time/delivery performance, and in product variety, Brazilian firms would have had to show higher positive variation than the rest of the companies. Results indicated that despite reported advances in the corresponding parameters, Brazilian firms improved as much as the total sample in terms of on-time delivery; were only slightly ahead in terms of delivery lead time and procurement lead time improvement; and advanced less in manufacturing lead time. Prospects were no better in terms of new products and product variety—they were not gaining ground in terms of speed of product development and were advancing quite slower than the rest of the world in terms of product variety. There was also a considerable negative gap in terms of equipment changeover performance gains—the advance for the sample as a whole was quite larger, implying that Brazilian firms would be facing difficulties in competing through flexibility.

4.3. An overall assessment of the Brazilian sample in the IMSS database

Compared with the manufacturing performance of firms in other nations, Brazilian firms in the IMSS database have been shown to be in an overall poorer situation. In fact, concerning several important indicators, there is still plenty of ground for improvement. In particular, severe deficiencies could be identified in terms of their handling of supply-related and product development aspects. This could mean a serious shortcoming in terms of facing competition based on dimensions such as delivery and innovativeness.

A delay has also been identified in the adoption by Brazilian firms in the IMSS database of modern advanced process technology. This setback, resulting from the reduced rate of investment in new plant and equipment in recent years, due to economic difficulties on a national level, could mean Brazilian firms might be facing problems in competing in international markets.

However, Brazilian firms have been shown to be engaged in substantial efforts to improve their performance. This is shown not only by improvement results reported for the previous two years, as commented above, but also by the significant rates of adoption and intensity of use of programs and activities designed for bringing improvement in manufacturing practices and results.

Good as these prospects are, there are also reasons for concern, as improvement efforts have not yet been enough to adequately bridge the gap in relation to more competitive economies in those areas where they have been showing a poorer performance. This might be impairing their possibilities to set out to compete on differentiation, holding them instead restrained in outdated price/standardisation competitive strategies.

4.4. Marcopolo SA - an illustration of manufacturing improvement effort

Overview

Founded in August of 1949 in Caxias do Sul, an industrial city in the Southern state of Rio Grande do Sul, and originally owned by eight local partners, Marcopolo S.A. had a tough start, but began a period of significant growth in 1954. Besides the original facilities in the neighbourhood of Planalto, the company had a larger plant in Ana Rech, launched in 1981, as well as smaller units in the state of Parana and in Portugal.

In 1993, Marcopolo held 32% of the overall domestic market for bus bodies, ranking world-wide as one of the largest producers in this industry. Worth noting was the export performance of the company, which had grown from $9 million, in 1987, to a record high of $104 million, in 1992, when its products were sold to over 30 countries in four continents. Mexico was, by that time, the company's main export market. In 1992, the company had sales of $210 million and earnings of $21 million.

In 1993, Marcopolo's product line included inter-city buses and city buses, mini-buses for diverse uses, and bodies for other specialised vehicles. Several structural components, frames, and fabricated parts (representing about 30% of the company's materials needs) were manufactured in the Planalto plant and forwarded to the Ana Rech unit, 9 km away, to be assembled into new vehicles.

The company had 4,100 employees in 1993, of whom 3,400 were directly employed in production activities in two production shifts.

The problems: the factory of the past
Before 1986, Marcopolo's facilities could be described as dirty and disorderly. The company suffered from several different types of waste, which contributed to higher costs and brought several kinds of problems to production activities.

Production was organised on the basis of a functional layout and followed traditional Taylorist principles. Both production runs and purchasing orders contributed to excess inventories, as large orders were adopted based on an EOQ concept, which the company believed would minimise set-up and purchasing costs. Due to little planning and control of inventories, it was common to have excess production and purchases, resulting in waste. For example, when a new generation of buses was launched in 1983, loads of parts for the old models were thrown away because there was no use for them in the new line.

Because planning and control were deficient, management felt it was necessary to keep significant amounts of buffer parts and WIP inventories. Despite this policy, large waiting times were quite usual due to materials and parts shortages. Sometimes materials were lacking; at other times there was a surplus.

Materials management and production scheduling were complicated by the lack of adequate standardisation of parts and components for the different vehicle models, as well as in tooling. The assembly of one bus body required the use of about 50 different jigs for the various parts of the vehicle. In addition, there were around 40 different models of windows used in the buses.

All this caused considerable delays in production, extending the lead times. An order for a bus had to wait 15 to 20 days before it went into production. Average lead time was 33 days. Average waiting time was 14.5 days for inter-city buses and 14 days for city buses.

Maintenance was typically a 'fire-fighting' activity, centrally organised. According to the prevailing functional organisation, workers mostly specialised in specific tasks. Worker morale was not particularly high, and the company had to cope with significant levels of absenteeism: 4.5% in 1986.

It is unnecessary to point out that quality and productivity suffered.

The recognition of the problem: improvement programs

The turnaround at Marcopolo began in 1986, when its CEO and COO took a tour of 13 Japanese companies, where they had the opportunity to observe their facilities and operational practices. Their attention was particularly caught by the orderliness and cleanliness of the shop-floor, as well as by the organisation in manufacturing cells and the co-operation among the workers. As a result of what they saw and learnt during this trip to Japan, Marcopolo's management was convinced of the need to implement dramatic changes in their operations to achieve the desired levels of productivity.

Management set out to gradually implement a series of operations improvement and waste reduction programs, some of which were still under way in 1993. The starting point was an internal series of lectures for the workforce on the principles on which modern manufacturing practices were based. The idea was to adopt the Japanese principles, but to adapt them to their own environment so as to reach the best results.

The first initiative concerned the working environment—a program called 'Suggestions for the Improvement of the Marcopolo Environment' was initiated, and workers were urged to improve their working surroundings and conditions and to express their feelings relative to their tasks and the company. This was followed by the creation of improvement working groups, of which by 1993 there were 223, with 1,680 participants, representing 38% of the workforce—yielding an average of 5.9 problem solutions per employee. The housekeeping effort was based on the Japanese five 'S' philosophy.

A second overall program, called 'Integrated System for Co-operative Production', sought to motivate workers, to optimise the flow of materials, to avoid waste and to reduce lead times. It involved several different initiatives, aimed at the improvement of operations.

A new layout concept was adopted, based on manufacturing cells, led by an operational co-ordinator who, being himself a worker in the group, was responsible for guidance of his fellow workers in the 'mini-plants', as well as for their satisfaction in the workplace. For example, the metalworking area of the Planalto plant had well defined sections for the production of seats, windows, and so forth that went directly into the assembly of the bus bodies.

The flow of materials was totally redesigned, as the company adopted a pull system. This first meant the adoption of a Kanban card system, leading to just-in-time production that adjusted internal manufacturing to assembly needs. In 1993, 40% of the parts and components outsourced by the company were delivered on a just-in-time basis, mostly by firms located in the same geographical region. Materials supply was considered, however, one of the main production problems yet to be solved.

A TPM program was initiated in 1989. As a result, workers took on responsibility for the maintenance of the equipment they themselves operated. By 1992, however, only 10% of maintenance expenditures represented prevention costs.

There was no formal quality management program in the company. Instead, it was considered that quality should be present in everything they did, as well as in the relationship with and among workers. Practices of Japanese inspiration, like *poka-yoke* mechanisms, were introduced into the production process. There was a score system for product quality evaluation, which was compared with pre-determined and increasing quality goals. There was also a system for monitoring the satisfaction of workers with their internal and external suppliers, and a permanent attention to the need to encourage worker participation and co-operation.

In order to achieve worker motivation and higher levels of productivity, the company directed substantial funds to basic education, as well as to functional training activities. The company increased its employee training expenditures, and training programs resulted, in 1992, in almost 300,000 hours of training for existing and new employees. One of the goals in the training programs was to have workers capable of job rotation—in fact, by 1992, 70% of the workforce were considered by the company to be multiskilled workers. Another important initiative was launched in 1989, when the company set up a school for the children and family of employees, where they received technical training, earning salaries just like regular employees, so as to coach them for future employment in the company. Great emphasis was given to efforts leading to enhanced workforce qualification and motivation.

Special attention began to be given to manufacturability issues, with the implementation of CAD and CAM programs. Considerable care was taken to optimise the design of new bodies; accordingly, the latest generation of models, launched in 1992, featured only four variants, which differed only in terms of height; customisation was left to interior decoration and accessories. As a result, one jig, instead of 50, was required to manufacture the frames and assemble a body. All windows, for instance, were identical. This program resulted in reduced inventories and transportation within facilities.

It is important to note the widespread concern with waste reduction—specific goals were set and continually updated concerning materials consumption. Special care was taken to keep decreasing inventory levels through the above mentioned programmes.

In 1993, there were still few initiatives in terms of automation—the painting section was to be the first major effort, as it was considered to be the bottleneck operation.

Increased competitiveness: performance improvement in the factory of the present
Several indicators illustrate the results achieved under the improvement programs. Worker morale was greatly improved: by 1992, absenteeism had dropped 38% from previous levels; worker turnover dropped below 1%; workers had taken over responsibility for quality and maintenance on the line, as well as for shop-floor cleaning. Between 1987 and 1991, there was a 45% increase in the average compensation (in constant prices).

Soon the results achieved under the improvement programs the company implemented were reflected in operational performance. WIP inventories, for instance, were 70% lower in the improved factory compared with their previous levels. There was considerable reduction in parts waste—on the line model renewal of 1992, adequate design and production planning guaranteed that no parts were manufactured in excess and lost in the transition. Waiting times were down from 50 to 63%, depending on the job. Lead times were drastically cut, reaching 7.2 days for inter-city buses and 5.2 days for city buses by 1992.

As a result of all the improvement initiatives, production costs dropped by 30%. Productivity levels increased remarkably since the beginning of the improvement programs, and the monitoring system adopted by the company set increasingly ambitious goals for efficiency and productivity.

These efforts were rewarded with outstanding results in its industry. According to the annual ranking of the 500 largest Brazilian companies published by a major business periodical (Exame, 1993), Marcopolo stood out as the best in its sector in 1992, based on a pool of financial performance indicators by which the different companies are ranked. Sales were up 63.4%; return on equity was 22.8%. Profitability was especially noteworthy, considering the fact that half the firms in the industry posted losses in that year. In a highly competitive international market, Marcopolo had managed to have 42% of its sales from export activities in 1992, up from 16% in the previous year.

By 1993, the company's CEO had no doubts that the restructuring the company had gone through and the successful implementation of operational improvement programs based on the Japanese model were the main reasons for the excellent performance and competitive position Marcopolo had achieved.

4.5. Conclusions

World-wide competition is a multidimensional and dynamic process, in which new technologies, managerial practices, and forms of organisation shape and are shaped by the evolution of firms and markets. Nowadays, more than ever, it is difficult to focus on an exclusive key to competitive success. Low cost and good quality are still pre-requisites for competitiveness, but increasingly they have to be joined by innovativeness and flexibility in production and marketing.

Some authors[5] have pointed out that there is no predetermined sequence in the adoption of innovation leading to competitive success or failure. There are however alternative paths to modernisation, based on different factors and skills, which may lead companies and nations to unique competitive positioning vis-à-vis other players. Whatever the choice, it is important for companies and policy-makers alike to understand the requirements on which a particular path rests and the competitive implications that such a path may entail.

---

[5] See, for instance, on the issues treated in this section, Kaplinsky (1993); Humphrey, J. (1993); and Mayer-Stamer et al. (1991).

PATHS TO INDUSTRIAL MODERNIZATION

SOFTWARE

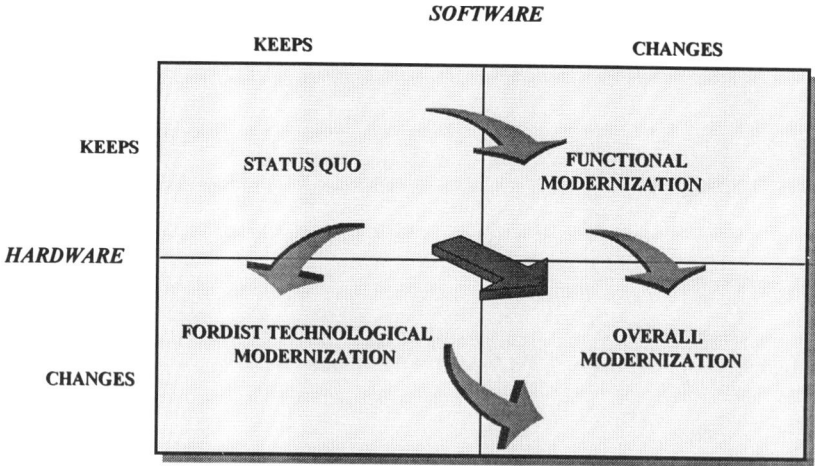

Source: Abranches et al., 1994.

Figure 4-1. Paths to Industrial Modernisation

The framework[6] above (see Figure 4-1) schematically shows the basic sequences leading from the status quo to overall modernisation, which implies in an integral change in technological foundations as well as in operational systems, managerial and production processes, and labour organisation.

In such a conceptualisation, there may be several roads to the overall modernisation stage located in the lower right quadrant, in which change involves both 'hardware' issues—the technical base of production—and 'software' issues—the set of systems, processes, and practices adopted by firms. One is the 'hard' or technological path, in which movement to overall change passes primarily by changes in hardware. The other is the 'soft' path, that moves primarily along functional modernisation, in which there is a change in the procedures, systems, and logistics used by firms in their operations. A firm may adopt a gradual path or may opt for a jump from one stage to the other; likewise, it may go along only one of the paths or along both. In overall modernisation, both the 'hard' and 'soft' sides have to be well developed and co-ordinated, and geared to the adopting firm's competitive priorities.

The previous results and discussion in this paper indicate that substantial effort is under way in Brazilian manufacturing firms toward modernisation and enhanced competitiveness. In addition, they allow us to observe that, in view of a combination of environmental circumstances and existing skills and capabilities, firms have been adopting the 'soft' path to modernisation, seeking, more specifically, in a first stage,

---

[6]   This discussion on paths of change in technology and management practices is based on Abranches, Fleury, and Amadeo (1994) and Fleury and Arkader (1995).

what has been termed as the 'lean production' or 'Japanese' model of manufacturing improvement. This path is suggested by results concerning adoption and intensity of use of programs and activities for improvement in the Brazilian sample in the IMSS data base (e.g., Kanban, SPC, group approach), as discussed above. The case study also pointed out the preference for the 'soft' path, based on observation of examples of waste reduction activities in Japanese companies.

Ultimately, some firms may already be envisaging the 'flexible specialisation' model of manufacturing development. The 'Japanese' model implies identifying inefficiencies and waste and seeking to streamline production so as to achieve better quality, lower costs, and overall dependability. The 'flexible specialisation' model goes a step further in gearing production toward competitive goals: the 'soft' changes introduced in organisational and social practices precede and condition those in hardware, and the adopting firm seeks, at the same time, to move along several competitive dimensions, namely cost, quality, dependability, flexibility, and innovativeness.

Based on information on firm modernisation projects appearing in the country's most influential business periodical, a recent study (Abranches, Fleury and Amadeo, 1994) has pointed out that the movement toward industrial modernisation in Brazilian firms have been strongly based on changes in managerial concepts and procedures. In fact, 94% of reports on modernisation referenced changes in 'soft' issues, against 6% investment in hardware modernisation, both for local and foreign firms. Of the changes in 'soft' issues, 71% had to do with organisational issues—mainly outsourcing, cellular layout, SPC, pull scheduling, and simultaneous engineering; and the rest with social issues, among which were worker participation and motivation, multiskilling, flexible scheduling, and productivity incentive payments.

The IMSS survey confirms this trend for the industry on which it focused. Data show that efficient use of resources is the main goal for Brazilian firms in the sample. They have been heavy adopters of 'Japanese style' manufacturing practices and programs. Starting from a far from favourable situation, they have been achieving remarkable results in terms of efficiency and quality improvement. They are still quite far behind, however, in terms of hardware modernisation—figures on existing automated equipment show that the average Brazilian company in the sample has invested substantially less in flexible machinery and robots.

The patterns of adoption of the 'soft' path to manufacturing modernisation have shown that organisational aspects have prevailed over social ones. In fact, Brazilian firms have been above world average in terms of the rate of adoption of, for instance, pull scheduling, SPC, QFD, plant-within-plants, work teams, and reduction in the number of hierarchical levels in production. In terms of social aspects, on the other hand, they have not advanced as much, for instance in the issues of new worker training, reduction of job classifications, initiatives for incentive payment schemes, and worker participation in decision making.

The capabilities Brazilian firms possess and the environment in which they operate may help to explain their options in terms of modernisation paths. Unlike their counterparts in more developed countries, they lack tradition and skills in engineering, which require time and adequate resources to be developed, and the economic context is not conducive to commitments in plant and equipment investment. They would have difficulties, therefore, in opting for models based on higher technological skills and strong hardware investment. The alternative approach of low costs with quality seems to fit more into existing conditions. In fact, there are indications that firms have pursued the programs and techniques identified with the 'soft' approach more in an effort to survive under changing domestic competitive circumstances than to qualify for opportunities the market might be offering.

But, considering the characteristics of competition in today's world markets, flexibility and innovativeness if pursued indefinitely may eventually lead Brazilian companies to a competitive impasse. Due to their present weaknesses, as previously highlighted, they tend to have little potential for pursuing a competitive positioning based on differentiation, having instead to opt for low price strategies. This could in fact prove quite a competitive drawback, since it has been shown that only dynamic advantages, that is, those based on continuous innovation and advanced factors, can be sustainable in the long run.

It is, however, possible to end this assessment of the paths of Brazilian companies to manufacturing modernisation on an optimistic note. The progress so far has awakened Brazilian companies to the benefits of manufacturing improvement, and the positive results indicated by companies in the IMSS survey sample on most aspects, seem to have prompted them to seek more ambitious improvement goals in the near future, in the direction of the flexible specialisation model. If the changes in the prevailing economic model are matched by success in economic stabilisation efforts, there may be renewed perspectives for investment in the Brazilian manufacturing sector. In this case, a new wave of investment in modern plant and equipment may be anticipated, adding new perspectives for Brazilian firms in their paths to overall modernisation.

References

Abranches, S., Fleury, P.F., and Amadeo, E. (1994) 'O novo contexto da competição internacional e o posicionamento do Brasil'. Working Paper No.1, Rio de Janeiro, Brazil: Finep, January (in Portuguese).

Banco Central do Brasil (1995) *Boletim Mensal*, March (in Portuguese).

Exame (1993) 'Melhores e Maiores', August.

Figueiredo, K.F., Arkader, R., and Reis, H. L. (1994) Marcopolo S. A. Carrocerias e Ônibus, case text for classroom discussion, COPPEAD/UFRJ (in Portuguese).

Fleury, P. F. (1993) 'Educação, Competitividade e o Papel do Setor Produtivo'. in Proceedings of the 17th ENANPAD (in Portuguese).

Fleury, P. F., and Arkader, R. (1995) 'Ameaças, Oportunidades e Mudanças: trajetórias de modernização industrial no Brasil', Draft research document. Rio de Janeiro, Brazil: COPPEAD/UFRJ (in Portuguese).

Humphrey, J. (1993) 'The Management of Labour and the Move Toward Leaner Production Systems in Third World Countries: the case of Brazil', Mimeo.

IMD (1994) World Competitive Report 1993, Lausanne, Switzerland.

Kaplinsky, R. (1993) 'Implementing JIT in LDCs: from theory to practice', Paper presented at the Workshop on Intra-firm Reorganization in the Third World: Institute of Development Studies, April 14-16.

Meyer-Stamer, J. et al. (1991) 'Comprehensive modernization on the shop floor: a case study of the Brazilian machinery industry'. Mimeo, Rio de Janeiro.

Reis, H.L. (1994) 'Implantação de Programas de Redução de Desperdícios na Indústria Brasileira: um estudo de casos', MBA Dissertation, COPPEAD/UFRJ (in Portuguese).

Womack, J.P., Jones D.T. and Roos D. (1990) *The Machine that Changed the World*, MIT Press.

CHAPTER 5

DENMARK: HORSE SENSE MANUFACTURING

Frank Gertsen & Jens O. Riis, Institut for Produktion, Aalborg Universitetscenter, Aalborg, Denmark.

5.1. Macroeconomic background

From 1989–1993, Denmark had the highest Gross Domestic Product (GDP) per capita and the lowest consumer price inflation among the IMSS countries after Japan, according to the Economic and Social Research Foundation (1994). However, GDP growth was moderate, in fact below the rest of the European Union during this period (Danish Industries, 1994).

Denmark is currently a successful trading state. After several years of being negative, the trade balance evened out in 1990 and reached a surplus of 34.7 billion DKK in 1993 (8.2 in 1990; 14.1 in 1991; 28.8 in 1992). Several political initiatives have succeeded in reducing consumer spending and thus also resulted in lower imports (Danish Industries, 1994), whilst years of fixed exchange rate policy have created faith in the Danish *krone.*

Lacking natural resources (except for some oil and building materials), the Danish economy has always been fairly open and heavily dependent on foreign trade. The economy grew overall at a rate of 4% from 1987–1992. As domestic sales declined due to the slow-down in consumption, growth became more dependent on export markets: exports grew from 42% to 53% of production between 1986 and 1993 (Danish Industries, 1994). The Danish economy is closely tied to the German economy, with Germany, its biggest trading partner, accounting for 25% of Danish exports, and Sweden and UK following with nearly 10% each. Imports are similarly distributed, but the value of imports did not increase nearly as much during the last 10 years.

Though the Danish total foreign debt has started to decline after peaking in 1988, total central government debt is quite high—62.4% of GDP in 1992 (Economic and Social Research Foundation, 1994)—and increased rapidly from 1983 to 1994 (Danish Industries, 1994). This growth is mainly due to the interest burden of the debt, which by itself accounted for the central government budget deficit. Public spending doubled between 1970 and 1992, and has shifted from public investment towards transfer payments and public consumption (Danish Industries, 1994), reinforcing the perception that Denmark is a 'welfare state'.

P. Lindberg et al. (eds.), International Manufacturing Strategies, 81-101.
© 1998 *Kluwer Academic Publishers. Printed in the Netherlands.*

Balance problems in the Danish economy, 1982 and 1991

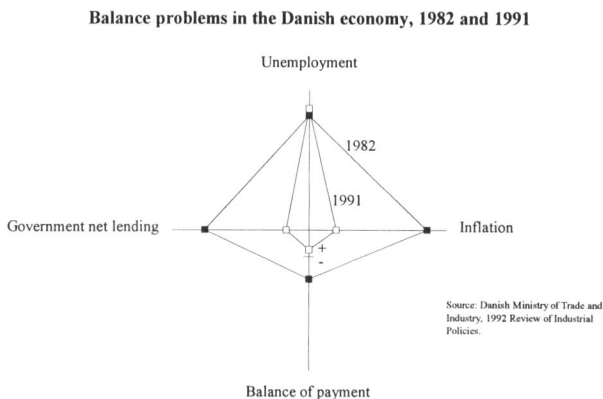

Figure 5-1. The Danish Macroeconomic Balance (Reproduced from Andersen, 1993).

5.1.1. NEVER HAVE SO FEW WORKED SO HARD TO FEED SO MANY LOOKING FOR WORK

Although economic conditions are quite healthy overall, this has come as the result of trade-offs. Unemployment has grown from 220,000 in 1987 to 350,000 in 1993. This unemployment rate of 12.4% exceeded that in Germany, UK, USA or Sweden. However, the government is now talking about *'bending the curve'* and some initiatives such as various types of sabbatical leaves have been launched. Unemployment has dropped to approximately 11% (January, 1995) and is expected to continue to fall. The macroeconomic balance problems are summarised in Figure 5-1.

5.1.2. STRUCTURE OF TRADE AND INDUSTRY

Whilst Denmark has a large service sector (public and private), in 1991 only 20% of the total work force were occupied in manufacturing and 15% in industrial manufacturing (Danish Industries, 1994). Figure 5-2 illustrates the value added (GDP at factor cost) by sector for 1993. Manufacturing, private services and construction have increased since then.

The iron and metal industries are the largest manufacturing industries in terms of production value (approximately one-third), number of employees, and investments (Danish Industries, 1994). They have also achieved one of the highest industry growth rates during the past ten years.

The structure of the manufacturing industry is characterised by some cross-industry clusters. The term agro-industrial complex is often used to characterise the cross-industry complex of production, distribution and consultants more or less related

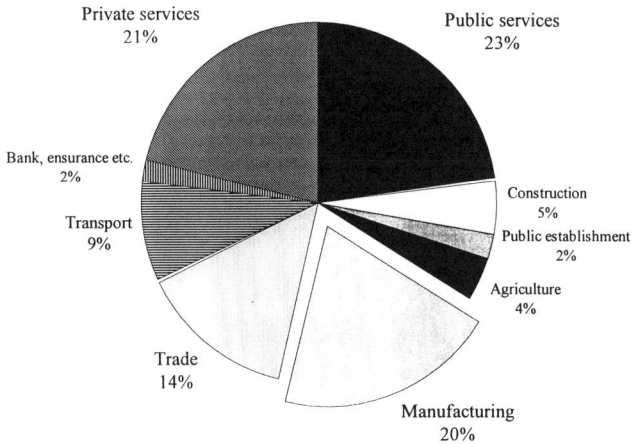

Figure 5-2. Gross Domestic Product (GDP) at Factor Cost (Production Value less Intermediary Input) by Trade (Source: Danish Industries, 1994).

to agriculture. In 1983 this complex accounted for 18% of the GDP and provided the most competitive products in terms of their percentage share of the world's industrial exports of these products (Møller, 1988).

Porter (1990) has pointed out four blocks of Danish industries that could serve as a basis for future growth:

- Food, agriculture, fish, and related process machinery (part of the agro-industrial complex). This industry has its historical roots in farming and fishing. A strong co-operative movement has played an important role in refining products and industrialisation of processes.
- Construction/housing, including building materials, electrical household machines and furniture. A high domestic standard of housing has been a platform for competitive advantage. The furniture industry is based on small companies who have uncertain access to raw materials but a reputation for good product design and quality.
- Health, for example, pharmaceuticals and technical equipment for hospitals. The pharmaceutical industry has one of the highest export rates (90% of its production) compared with the industry in other nations and it differentiates itself by being based on raw materials from animal production.
- Transport, particularly shipbuilding is based on traditions and a rather opportune natural location of the country.
- Other product areas of some substantial value are entertainment (TV, radio, etc.), mink furs, energy, environmental protection, communication and orthopaedic devices.

Employment in Manufacturing by Size of Firm

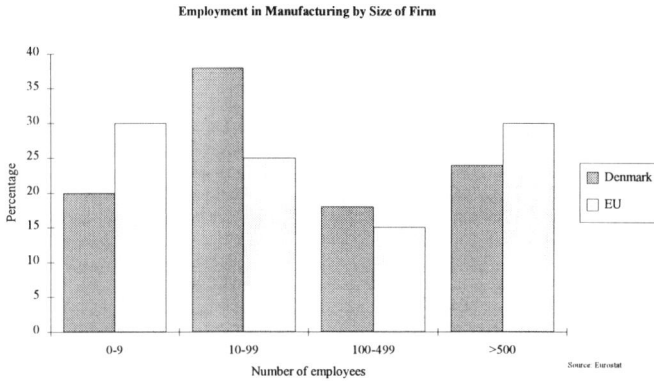

Figure 5-3. Employment by Company Size for Manufacturing Industries Compared with the Average for the European Union (Eurostat, 1992).

Compared with the European Union, more Danish people are employed by manufacturing companies that have between 10 and 500 employees, and fewer by the very small and the large companies. Figure 5-4 shows the distribution by size of companies and employment in each size category. Denmark is the only IMSS country that is not represented by any industrial companies in the *Fortune 500* (Economic and Social Research Foundation, 1994). Even though there are only a few large companies with more than 500 employees, they do account for more than 25% of the total employment in this industry, similar to the situation in most other industries.

There is an ongoing political discussion as to whether these 'industrial locomotives' are missing as drivers for the rest of the industries, or whether competitive advantage, on the contrary, is to be found in small, innovative and flexible companies, ultimately

Iron and metalwork industry, size and employment

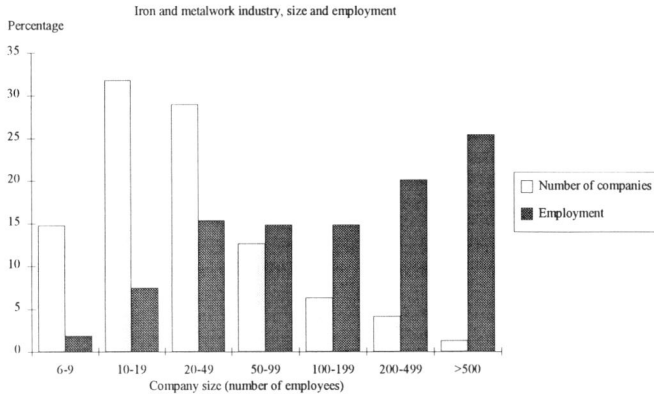

Figure 5-4. Size of Companies and Employment in the Iron and Metal Industries (ISIC 38) (Source: Danish Statistics, 1989).

tied together in some inter-organisational structure. The latter structure seems to be deeply rooted in the way the downstream agricultural industry was organised, and the way the education and apprenticeship system worked during the shift from agriculture to industry.

5.2. Sociocultural dimensions

According to Hofstede's cultural dimensions, Denmark is characterised by high individualism, low power distance between subordinates and managers, low risk aversion in terms of rule reliance, job stability and stress level, and low masculinity (feminine leadership style and life quality care).

High individualism. General characteristics of high individualism are that every person has his (or her) own identity and personal characteristics, and that people handle things on their own and take care of themselves and their closest relatives. However, the Danish social system secures a relatively high average standard of living. Employment is thus a business matter rather than a life insurance. Danes generally feel that the job should offer time for personal life, and the ability to influence the job as well as challenges and recognition.

Denmark has the second shortest working week among the IMSS countries (Economic and Social Research Foundation, 1994). Incentives used by the companies were largely based on group incentives, which are inconsistent with individualism, unless they are intended as a countermove to promote teamwork. Payment systems with

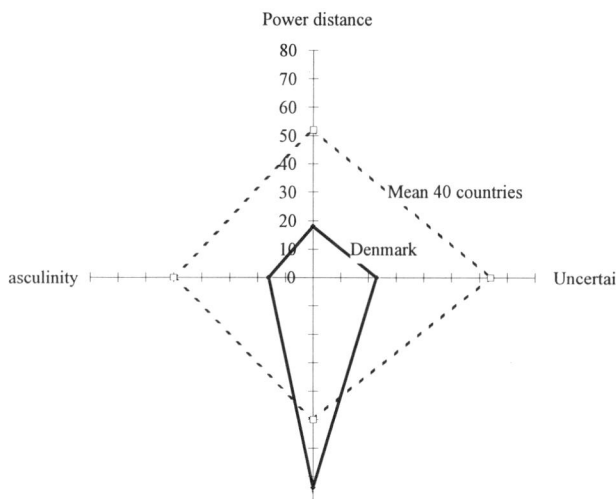

Figure 5-5. Cultural Dimensions for Denmark Compared with the Total Average (Based on Hofstede, 1991)

more emphasis on individual incentives are currently being discussed and introduced in many companies.

Low power distance. Low power distance is often related to less hierarchy and higher individualism. Low power distance between subordinates and managers results in a democratic leadership style with interdependence between partners. This seems to be reflected in the low number of organisational levels (3.3 from plant manager to operators), which is the lowest among the IMSS countries. Although we expected a high number of suggestions per employee, it is only in the top half of the sample (2.73 suggestion per employee per year).

A rather symmetric power position of the labour market parties may play a role as a stabilising factor. Denmark has a very high percentage of organised labour and a centralised union with a tradition of good relationships with the political parties and a strong negotiation position. In general, the unions and management are co-operative at a local level and there is a trend towards decentralisation of negotiations.

Low uncertainty avoidance. Low uncertainty avoidance and tolerance of ambiguity indicate that people do not rely on rules, do not need high job stability and have a low stress level. It may mean that people work best under pressure, which means that they have to improvise as they go along and thereby experience challenges in unexpected events. It might promote innovative activities but not the character needed to implement the ideas. The Economic and Social Research Foundation (1994) reports that Denmark has the highest rates of hiring and firing and employee turnover among the IMSS countries.

Low masculinity: Denmark is reported as having a low masculinity index score, that is a 'feminine' leadership style with participation and care for the weak part (equality), and care for life quality. The latter sometimes calls for trade-offs between career and family. This can conflict with individualism, for example, in the case of payment systems with individual incentives, which may conflict with such values as equality.

It is likely that low power distance, individualism, low uncertainty avoidance and femininity match or promote such manufacturing characteristics as participatory management, *Kaizen*, team work, multiskilled workers, and job rotation, though we did not find sufficient evidence for this in the survey. However, we do believe that Hofstede's national dimensions can contribute to the understanding of differences on the company level as long as we avoid a stereotyped way of applying the theory. Other have offered paradigms, such as the 'pluralists interest-based democracy' (Enderud, 1987)—a somewhat terrible phrase, but it captures the way modern western organisations make decisions, pursuing democracy whilst influenced by the various interests of individuals and groups that are neither in harmony nor in conflict. In Hofstede's work, Denmark is clustered with Sweden, Norway, the Netherlands and Finland.

Within the European Union, Denmark is known as a 'critical' partner. Perhaps the national character was revealed when the people's scepticism and ambivalence towards the European Union were exposed in 1993, and the Danes turned down the Maastricht Treaty, and afterwards entered a compromise agreement in Edinburgh. Fear of losing control and the right to choose on our own (individualism), of a threat to democracy (power distance) and of having too many rules imposed (low uncertainty avoidance) seemed to be factors that came into play.

Cultural values point to an ideal Danish organisation where decisive leaders are accepted (individualism), if they consider the weak parts (femininity), consult subordinates (low power distance) and allow employees to cope with problems and challenges (low uncertainty avoidance). Such an organisation can foster innovation, differentiation and customisation. Finally, let us not forget that companies can certainly also achieve advantages by creating cultures different from the national ones.

5.3. Findings of the Danish IMSS Study

The Danish data were collected in the spring of 1993 from eighteen major Danish manufacturers of metal products, machinery, and equipment, representing 17% of all Iron and Metal Industries (ISIC 38 industries) firms with more than 200 employees. The sample does not represent all ISIC 38 segments, although it includes 25% of its employees (31,000). This reflects the structural characteristics of Danish trade and industry. The average size of the surveyed companies is about twice as big (1780 employees) as the average of the total IMSS sample.

5.3.1. COMPETITIVENESS, ECONOMICS AND KEY DRIVERS

Competitiveness and drivers

Company goals. The Danish IMSS companies emphasised quality, customer service and dependable deliveries as their most important current strategic goals (Figure 5-6). The reduction of manufacturing costs was rated of low importance compared with most other countries, consistent with the lower priority on reduction of unit costs.

Questions about the future importance of these strategic goals indicated that the Danish companies will give high priority to cost reduction in various areas. This suggests a shift in focus from quality and delivery to cost. In a macroeconomic environment with a strong currency this seems reasonable to remain price competitive. However, quality and delivery performance remain of nearly equal importance, and analyses of the relationship between improvement of cost, quality and delivery indicated that there is no trade off, but they instead improve simultaneously.

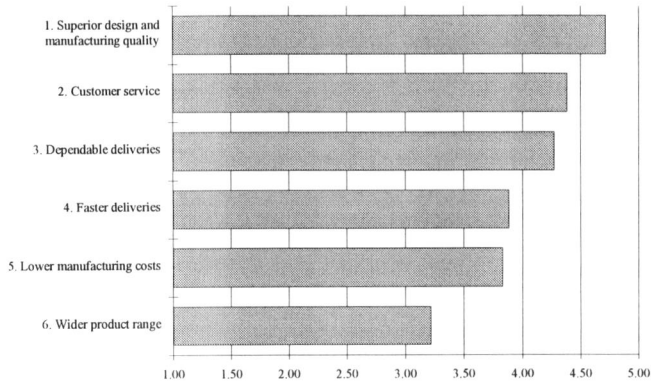

Figure 5-6. The Priority of Current Goals (1-5 scale from not important to very important).

Products and Markets

Market share. Danish companies tend to be small players in larger markets. The market share of the companies' dominant product line averages 20%, a low relative market share compared with study average of 33.9%. We found the following common characteristics in the way the companies dealt with their markets:

- Market coverage varies from company to company but follows a normal distribution. Normally, as the size of a company increases so does the number of markets (Andersen, 1993).
- At least half of the companies have few customers.
- About 75% of the companies claimed that their market is international. This possibly implies that most of the Danish companies have to compete with international companies.
- 56% of the companies reported that stable markets, 33% declining markets, and only 11% increasing markets.

Products. The number of different products was quite low, 280 compared with a study average of 706. This put Denmark at the same level as Sweden, but far below USA, Japan and Germany. It has grown considerably (72%) during the past 5 years and the expected increase for the next 5 years is 93%, almost twice the IMSS average. The product variety (minor variations) during 2 years also exceeds most IMSS countries.

In 1991, the percentage of revenue from new products was high (34%). This grew by 91% over the past 5 years, reflecting the large amount of R&D investments, but is expected to grow only 26% over the next 5 years, well below average. However, it takes some effort just to maintain 34%.

Table 5-1. Revenue from New Products

	% of revenue from new products			
	0-20%	21-50%	51-80%	81-100%
Small companies	13.8	**36.2**	33.3	16.7
Medium companies	**34.0**	24.0	30.0	12.0
Large companies	**42.9**	28.6	20.4	8.1
Total	31.8	28.9	27.4	11.9

Source: Ministry of Trades and Industry, 1990 (Source: Andersen, 1995)

Table 5-1 shows the general pattern of the overall manufacturing industry. Smaller companies relied more on new products than larger ones. The survey indicates a trend of increasing percentage of revenue from new products for the large companies.

This may reflect the high level of R&D in Denmark. Companies in Denmark increased their business R&D expenditures more than those in most IMSS countries, with an average annual rate of increase of 6.90% from 1988-1992 (Economic and Social Research Foundation, 1994). This is considerably more than the rest of the Danish manufacturing industry, even if we consider that the large companies in general spent more than the smaller ones (see Figure 5-7). The figure shows the distribution of R&D investments by company size for all Danish manufacturing industries compared with that of the Danish IMSS companies.

The IMSS companies do not fit in with the general pattern of Danish manufacturing's R&D distribution by size. Beside the different pattern, they also have much higher R&D expenditures than the average of the manufacturing sector (however, it should be noted that some of the difference is due to the four-year time span).

Economics of the firms
The average of 3% Return on Investment (ROI) was very low. In Figure 5-8 the Danish companies are benchmarked against the IMSS. Even though the low ROI is due to a few very bad performers the results are not impressive.

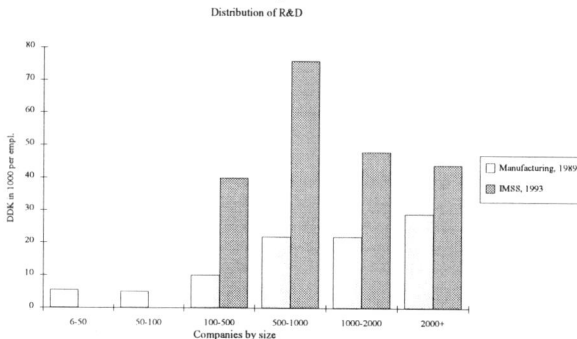

Figure 5-7. Comparing the Distribution of R&D in Danish Manufacturing 1989 (Reproduced from Andersen, 1993) and the Danish IMSS Companies

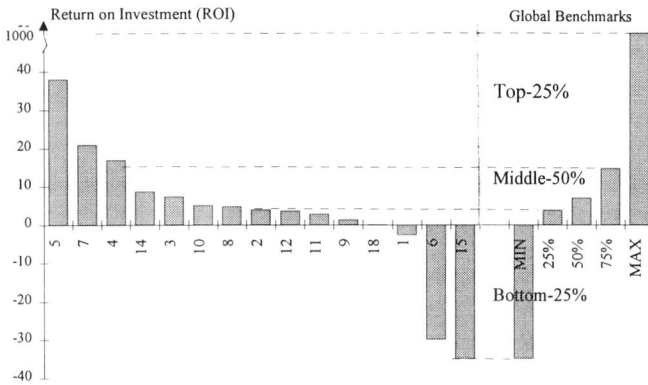

Figure 5-8. Benchmarks for ROI (Profit Before Tax/Total Assets, Percentage, 1992) (The numbers identify the Danish companies)

From case studies of Danish companies, we sometimes get the impression that somehow profit is not the prime concern, for example, the Lego mission states that the company should 'ensure a profit that allows for continuous growth and financial independence', indicating that profit is just a necessary means to an end.

Conclusion

The Danish companies emphasised quality, customer service, and dependable deliveries as the most important strategic goals, but in the future will place a higher priority on cost reduction. Companies believe in and depend on a narrow product range, but do produce variants to meet customer demand. They have a low market share, both relative to the total market and to competitors, few customers, high global market/international competition, and stable/slightly declining markets. R&D expenditures have grown There are few different products, but the number is growing fast and the proportion of revenue from new products is high.

A possible explanation for this pattern is that most of the companies are unable to compete based on scale economies in a relatively small economy. As a result, they focus on the specialised needs of a few companies (niche strategy), producing differentiated customised products characterised by high quality, customer service, dependability, and high prices. To sustain this strategy, they must compete in international markets, but obtain only a small share of individual markets.

Danish trade compared to IMSS average

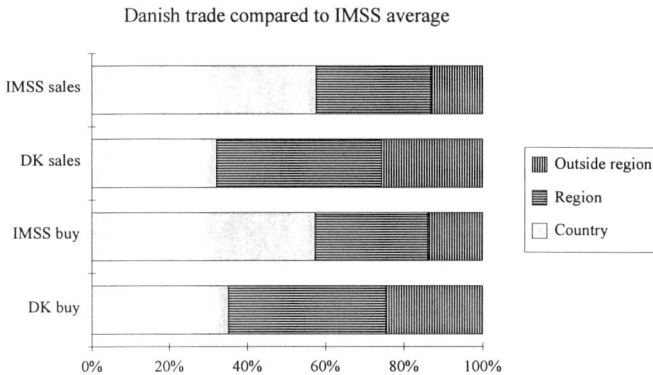

Figure 5-9. Distribution of Danish Sales and Purchasing from Denmark, EU and Outside EU, Compared with the IMSS average

5.3.2. SUPPLY CHAIN MANAGEMENT

Characteristics of External Logistics
Denmark has traditionally been a 'trading country'. Figure 5-9 shows that we buy more abroad and less in Denmark. Fortunately, the distribution of sales is similar, so that the trade balance with external partners is slightly positive, a trend that has continued after 1992.

As mentioned earlier, the Danish trade balance and export have developed very positively recently. The external trade is illustrated in Figure 5-10.

Figure 5-11 shows the trade with various constellations of trade partners. OECD and EU constellations are definitely the largest, whilst eastern Europe and ASEAN countries are not yet significant.

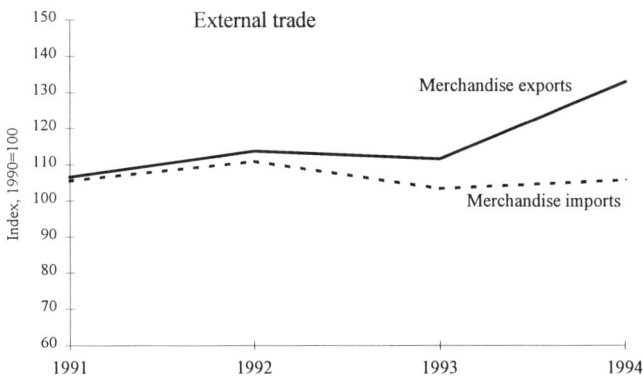

Figure 5-10. External Trade of Merchandise (based on Danish Statistics, *Monthly Review*, 1995/4)

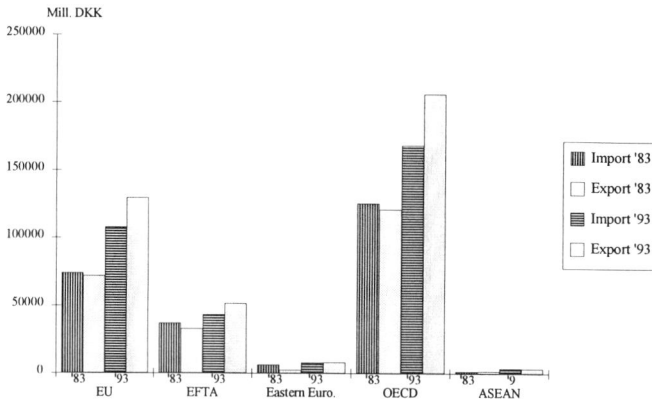

Figure 5-11. External Trade with Various Partners (Source: Danish Industries, 1994).

Suppliers

The average number of suppliers per firm is rather high (1147), which may be due to the presence of large companies in the sample. The percentage change over the past 5 years is high (+23.6), whilst the estimated percentage change over the following 5 years is -9%, which is close to average.

Danish companies had somewhat closer relationships with principal parts/material suppliers than average. However, other studies in Denmark point to some trends at the supplier level (e.g., Andersen, 1995). Within the EU, outsourcing has increased, resulting in closer relationships with suppliers. Development/knowledge-dependency, and focusing on fewer suppliers and longer-term relationships seem to point to a compact supplier network. This indicates a change from standard to partner-based relationships with suppliers. It may be critical for the suppliers to achieve a 'first tier' positions, occupying a strategic position in the customer's business. Suppliers seem to become more like the companies they supply, in terms of more product development, specialisation, dialogue, and ability to rationalise production when the product matures.

Downstream and Internal Logistics

29% of the production orders were forecast orders and 71% were customer orders. For the majority of companies (87%), large and small orders were produced on the same equipment, indicating non-specialised and/or flexible production facilities.

The utilisation of capacity in main processes is quite high, on average 17.4 hours per day, and the utilisation of planned capacity is 81%. Except for work in process, the size of inventories was below average, consistent with a slightly higher percentage of JIT deliveries of raw materials and components (36%). Lead time from customer order to delivery of product varied from 7 to 360 days with an average of 98 days, nearly double the average. These long lead times may be due to companies manufacturing large and/or complex customised products such as ships or dedicated electronics to order.

5.3.3. ORGANISATION, PEOPLE AND TECHNOLOGY

Organisation and people management
Danish companies reported a surprisingly low share of salaried employees compared with most other countries (38%, 17% below the IMSS average). The national average was 31% in 1988 (26% in 1976), which varied depending on industry and location. The highest salaried employment percentage was in the measuring and controlling equipment industry (ISIC 385, 38%), the lowest in transportation (22%) and iron and metal work (26%). Salaried employees seemed to prefer large cities, whereas the direct labour-intensive companies were clearly located in the country side (Maskell, 1992). However, Denmark has a relatively large percentage of skilled workers in direct labour positions who undertake functions as part of their jobs that would be carried out to a larger extent in other countries by salaried employees. This seems to be also reflected in fewer organisational levels.

The education of the regular work force was above average, although the total amount of turnover spent on education is below average. Fewer salaried employees (with expensive educations) and the provision of free workforce education from the publicly-supported AMU system may explain this inconsistency. The duration of training of newcomers was average. The Danish companies are below average regarding absenteeism and direct employee turnover (although it should be high, according to the general picture provided by Economic and Social Research Foundation (1994)).

Table 5-2. Comparison of Organisational Measurements Between Denmark, Sweden, Italy, Great Britain, and the IMSS average

Organisational measure	Denmark	Swe-den	Italy	Great Britain	IMSS average
Salaried employees in percentage of total number of employees	39	53	60	57	56
Number of organisational levels (plant manager through to operators)	3.3	3.8	3.4	3.9	4.1
Percent of workforce that work in teams (average)	32	51	30	42	37
Hours of training per year given to regular workforce (average)	45	26	27	23	36
Personnel turnover for direct employees within the factory (% per year)	5.4	3.7	6.4	1.8	8.1
Short term absenteeism for direct employees within the factory (% per year)	3.8	5.3	7.3	2.9	4.7

The tradition of workmanship and the apprenticeship system enhance an understanding of the way work is organised in large parts of the manufacturing companies, especially in traditional industries, such as shipbuilding and machine fabrication, which lead even further back to the small-holder economy of agriculture and industrialisation (e.g., Kristensen, 1993).

Manufacturing technology
The adoption of manufacturing technology was high compared with the IMSS average, partly because the Danish companies were larger than average, and there is a significant correlation between size and amount of technology. However, in terms of density (machines per 100 employees) the Danish companies were below average and considerably below Japan and USA.

Whilst many of the Danish companies are familiar with the technologies, indicated by a high adoption rate, it may be concluded that they do not utilise them extensively, instead using dedicated technology that does not fall into the above categories. The investment in process equipment (in percentage of turnover) was significantly lower than the IMSS average (4.2% vs. 8.5%), although some of the development of process equipment may be registered as R&D, which was higher than average.

Even utilising a relatively low level of automation, the Danish companies managed to have a surprisingly low share of direct salary/wages (*cf.* Table 5-2). However, the above-average value-added in assembly is above average, which could indicate that more manpower is utilised for 'putting things together'.

Change programs
Previous work on the balance of technology and organisation indicated that the best performance is obtained by those companies able to keep a delicate balance between organisational and technological effort (e.g., Frick et al., 1992). Total Productive Maintenance (TPM), *Kaizen*, JIT/Lean Production and CAD were assessed by Danish companies as some of the most beneficial programs in terms of payoff. On the other hand, ISO 9000 was rated low despite a considerable effort. Figure 5-12 illustrates the

Table 5-3. Adoption and Density of Manufacturing Technologies within Danish Companies and IMSS

Technology	Adoption rate (% of companies adopting)		Machines per 100 employees			
	Denmark	IMSS	Denmark	IMSS	Japan	USA
FMS	39	18	0.03	0.40	0.15	0.41
NC	100	55	3.33	5.39	23.21	4.83
Conventional machines	94	67	14.46	41.76	168.56	40.68
Machine centre	67	34	0.39	1.40	1.62	1.90
Robot	39	29	0.61	2.19	34.30	0.72
Assembly robot	39	15	0.41	1.90	5.61	4.38
Assembly FMS	39	18	0.10	2.19	0.49	4.72

Source: Based on Voss (1994) and Gertsen, Sun, Riis (1994)).

manufacturing history and fashion in terms of programs that have required the attention of managers.

Conclusion
Comparatively fewer salaried employees than in other countries may be due to a relatively large number of empowered, skilled workers and the redeployment of salaried job functions as well as to the Danish industrial structure with large number of non-salaried consuming medium-sized companies with between 10 and 99 employees. Fewer organisation levels in manufacturing shows flatter organisations. The education of regular workforce is rather extensive, although this is not true for new employees. These and other aspects of people and management are illustrated in the Bang and Olufsen case.

The large Danish companies are familiar with the advanced manufacturing technologies, but they do not utilise them extensively, instead using more or less integrated or automated dedicated technology. TPM, Kaizen, JIT/Lean Production and CAD were assessed as some of the most beneficial programs in terms of payoff, ISO 9000 was rated low despite a considerable effort. Improvement programs are often employed sequentially, providing over the years a multi-focused approach with a balance of organisational and technological means adopted.

5.3.4. PERFORMANCE IMPROVEMENT

Market Indicators
Percentage change in market share from 1990 to 1992 was negligible, below the IMSS average (11.3%). This is probably below the overall average of Danish industry, which increased its export market share 7% from 1980 to 1993, especially after 1987 (Danish Industries, 1994). Customer service improved 10% (below IMSS average, 20

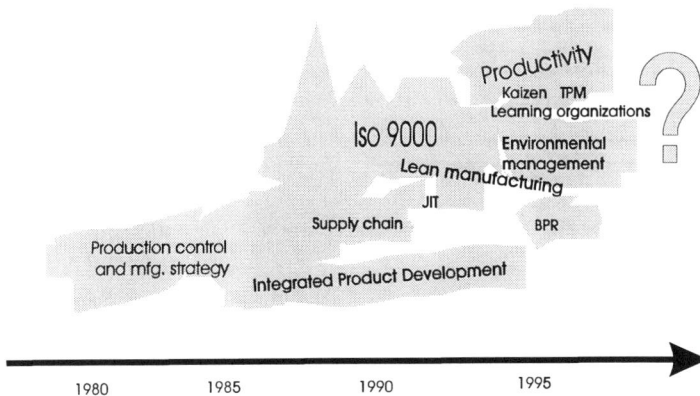

Figure 5-12. Past and Current Programmes in Manufacturing.

percent), but product variety improved 33%, 15% above average. We believe that this is due to 'customerization' efforts. On-time delivery improved 21%, 6% below average, but manufacturing lead-time improved significantly.

Cost Indicators

Unit manufacturing costs were reduced by 10%, 4% below IMSS, and inventory turnover was 6% below IMSS even when it improved 21% (in absolute figures it was also below average). Profitability improved 2.5%, above average, but in absolute terms was the lowest ROI among the countries surveyed.

Manufacturing/Technology Indicators

Speed of product development increased by 13% but did not reach the 20% average increase rate; thus increased spending has not shortened the product development cycle. Manufacturing quality increased by an estimated 20 percent, 9% below average. Equipment changeover (127 vs. 118), manufacturing lead time (42 vs. 28 percent) and delivery lead-time (32 vs. 24 percent) increased more than average, indicating an improved flexibility in manufacturing.

Overall performance was not impressive. However, it should be seen in the light that 1993 to some extent was a turning point towards a better economic climate, indicated by the export rate (*cf*. Figure 5-10) and change of the general business indicator from minus to plus in the middle of 1993 (Danish Statistics, 1995:4). The areas in which the Danish companies improved above the average point to emphasis on flexibility and customerization in manufacturing of products. The relatively heavy investments in R&D may have contributed to the recovery and may also give promise for the future.

5.3.5. CONCLUSIONS

The picture of Denmark has been developed based on both the study results and other socio-economic data. Denmark has an outwardly-oriented economy. Its macroeconomic development has been positive since 1993. The GDP has grown moderately and the GDP per capita is high. Inflation is low, the Danish krone is strong and the trade balance is positive, particularly due to the growth of exports. Lacking natural resources, the Danish economy has always been fairly open, with an international trade that exceeds most countries but still closely tied to the strong German economy. The total central government debt is quite high and growing due to the interest burden, but the major national problem is considered to be 11% unemployment rate.

Industrial manufacturing accounts for only 15% of GDP and the iron and metal industries (ISIC-38) for only 5%. The landscape of companies is dominated by SMEs from 10-500 employees, especially those from 10-99 employees, when comparing to the EU average.

The manufacturing strategies of the Danish companies surveyed were consistent with their marketing strategies. Most of the companies chose to compete not based on scale economies, but to focus on the specialised needs of a few customers (niche strategy), differentiated customised products, with high quality, customer service, dependable deliveries, and high prices. To do this, they have to operate on the global market, with only a small share of individual product markets. This is reflected in the narrow product range, high number of product variations, high R&D spending, and high revenues from new products.

Reflecting these strategies, the Danish companies emphasised quality, customer service and dependable deliveries as the most important strategic goals, but a shift towards cost reduction in various areas was indicated. The large Danish companies are familiar with the advanced manufacturing technologies, which were indicated by at high adoption rate, but on the other hand, the density showed that they do not use them extensively. They instead use dedicated technology more or less integrated or automated. TPM, Kaizen, JIT/Lean Production and CAD were assessed as some of the most beneficial programs in terms of payoff, ISO 9000 was rated low despite a considerable effort.

The low share of salaried employees compared with most other countries may be because the relatively large number of skilled workers is undertaking traditionally salaried job functions and because of the many SMEs' that need less and can only afford few salaried employees. The number of organisation levels in manufacturing showed that the organisations are more flat than in most countries. The education of regular workforce was more extensive. These, and other aspect of people and management, have been illustrated in the Bang and Olufsen case.

5.4. Case Study: Bang & Olufsen A/S

The spectacular design of Bang & Olufsen audio/video products is embellishing an increasing number of homes again, after three years of loss and stagnation in the company. The stock value set a growth record in 1993 and profitability improved significantly during the fiscal year 1993/1994. B&O produces integrated systems including TV, video and telephones, within a niche at the high end of the market. The company employs 2,400 people and has annual sales of $450 million. The recovery has not been coincidental or surprising to the firm, but the result of the continuous development of new products and a comprehensive set of various change programs that started back in 1987–88. These changes have been accompanied by a substantial change in the management and organisation paradigm, new coaching-style management, removal of two organisation levels, employee participation, empowerment and delegation at all levels. The trade mark of B&O's change has been 'people first'.

Though many factors have contributed to the recovery of the firm, this case will focus on the people and the organisational aspects. However, the change of organisation and management was preceded by a 4 to 5 year period of down-sizing and out-sourcing, which was mainly the responsibility of top management. A restructuring of the organisation removed two levels involving a reduction of sixteen middle managers. As one of the directors put it: 'Each organisational level adds six months to the time it takes to make changes'.

The keywords to survival were to improve customer focus, flexibility and productivity. In order to do this, it was necessary to focus on value-added activities through participation, goal-setting, benchmarking and incentive system. The short term goal was a 10 percent productivity improvement per year over a three-year period to reach world class level in specific areas. The process of changing the management's attitude began with internal discussions in the top management group. At first, traditional changes occurred in the sequence: investment, then structural changes and then people involvement. The management, however, had a strong belief that the best way of pursuing the goals was through participation, thus the blueprint was turned upside-down, starting with the employees. The process was managed by a project team including the managing director, six managers and one union representative. They came up with a plan that was voted through by production. The supervisors took off-site courses and were trained to facilitate their new coaching role, helping them to delegate. The result of all this activity, then, was that 125 production groups were formed in order to help the operators to become more decisive, to deal with scheduling, to set goals and to show more initiative. The autonomy also included flexible working hours, vacation scheduling and a minimum of one week of education per year. The process was also supported by an extensive program of training courses.

Furthermore, internal and external benchmarking were used as a driver for stepwise improvements. The benchmarking projects included the operator level and thus fit into the new responsibilities they had been given.

The next major cornerstone was a new incentive system supporting the objectives that B&O wanted to pursue, such as education/flexibility, breaking boundaries across functions and promotion of both groups and individuals. The system included all operators (skilled and unskilled). They got a monthly payment divided into two parts;

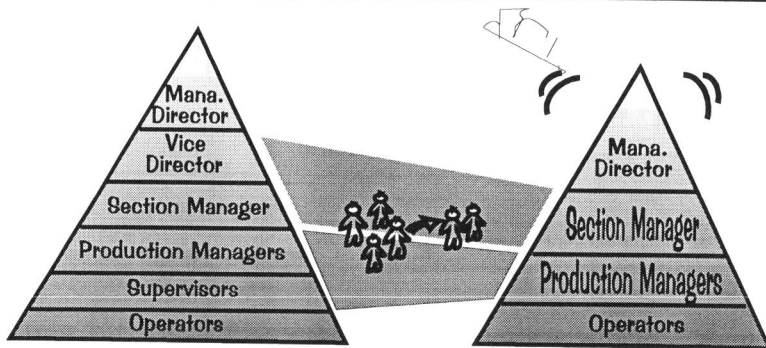

one was a fixed salary according to the type of job (job description) and one was an individual salary according to their working results based on a personal annual interview. The fixed part depended on the ability to solve a problem, competence/freedom of action, contact and communication, education and experience. The individual part included quality of work, job ability, initiative and autonomy, flexibility, and overall working results.

As a result of these changes, little supervision is now needed. Employees are not only motivated to achieve specified goals but also competent in their own job domain as well as having a broad insight into the dynamic operation of the section or department. This vision relies on the willingness of employees to take the initiative and will be strongly supported by IT tools and systems.

5.5. Pieces of a Portrait of a Danish Industrial Company

In their wide range of practice and performance, the sample companies reflect very well the diversity of companies within Danish industry. Nevertheless, on the basis of the IMSS results combined with other industrial studies and statistics about the Danish economy, structure and culture, we will draw a portrait of a typical Danish industrial company, which does not display any specific company in the sample, but is hoped to capture characteristics of the 'Horse Sense' features of Danish industrial enterprises:

- *Profit is not a prime objective, but a necessary means to secure freedom of action.* Instead, the creation and maintenance of jobs are a prime concern, which is understandable in the light of the relatively large unemployment rate and the perception of social responsibility. Also, the positioning on the market place represents a great challenge to industrial management in its effort to secure competitiveness
- *Internationally-oriented industrial companies.* The domestic market is very limited and unable to sustain a large turnover. Combined with the long tradition of international trade, most Danish industrial enterprises are internationally oriented and typically export more than 80% of their production.
- *A solid understanding of the business, obtained by close contact with customers, guides the direction of development.* The domestic market, however, plays an important role for securing close contact to customers during product development. Most products manufactured in Denmark thus originate from a domestic need. The survey portrays the Danish industrial companies as having a rather narrow product program, but the number of variations has increased drastically in recent years pointing to a larger degree of customerization.
- *A balanced market, technological and organisational development are important.* This is likely to result in a 'follower strategy' for the introduction of manufacturing technologies, product technologies and new materials, as well as the employment

of soft, organisational means. Inspired by best practice, methods and tools are adapted to the organisational culture and management attitude of the individual company

- *A long tradition of constructive co-operation between unions and employers' associations.* Supported by the acceptance of a very low power distance, the constructive co-operation in the individual company between workers and management has offered excellent opportunities for introducing new organisational forms based on a flat organisational structure as well as confidence in the loyalty and responsibility of workers. Several industrial companies, small and large, have utilised this opportunity with great success.
- *Sequential attention to conflicting goals - towards multi-focusedness.* Previously many Danish industrial companies would focus on a single goal for a long period of time with the risk of neglecting other goals. However, in recent years companies have developed a capability to shift focus between conflicting goals without losing sight of the other goals. Several companies are aware of the challenge to let every operator be capable of attending to conflicting goals in a concurrent way, depending on the actual situation. This represents the vision of multi-focusedness. The Bang & Olufsen case illustrates this point.

Acknowledgements
    The authors are grateful to the companies that contributed to the IMSS survey and to Bang & Olufsen for providing case materials. Also, thanks to those of our colleagues who provided valuable comments on the drafts of this chapter.

References
Andersen, P.H. (1995) 'Study of suppliers'. Aalborg University.
Andersen, P.H. (1993) 'Globalization of economic activities and small and medium-sized enterprises (SME) development'. Country Report. Aarhus School of Business. Denmark.
Danish Industries (1993) 'Key Figures of Danish Manufacturing'. Danish Industries.
Danish Industries (1994) 'Key Figures of Danish Manufacturing' (Industriens Hovedtal 94). Danish Industries.
Danish Statistics (1989) 'Industrial Statistics' (Danmarks Statistik, Industristatistik, 1989).
Danish Statistics (1995:4) 'Statistical Monthly'.
Economic and Social Research Foundation (1994).
World Competitiveness Report (1994) World Economic Forum.
Enderud, H. (1987) 'Conflict decisions (Konfliktbeslutninger)'. Copenhagen School of Business.
Eurostat & Commission of the European Communities (1992) 'Enterprises in Europe', Second report 1992.
Frick, J., Gertsen, F., Hansen, P.H.K., Riis, J.O., & Sun. H. (1992) 'Evolutionary CIM implementation - An empirical study of technological-organisational development and market dynamics', In: *The Proceedings of CIM-Europe 8th Annual Conference*, Birmingham, May.
Gertsen, F., Sun, H. & Riis, J.O. (1993) 'Compare Your company with the best - A preliminary report of the International Manufacturing Strategy Survey (IMSS)'. ISBN 87-89867-34-3. University of Aalborg.
Gertsen, F, Sun, H. & Riis, J.O. (1994) 'Compare your company with the world's best - Danish report of the International Manufacturing Strategy Survey (IMSS)'. ISBN 87-89867-15-7. University of Aalborg.
Hansen, P.E. & Lunding, C. (1992) 'Danish trades and industries (Dansk erhvervsliv)'. Handelshøjskolens forlag.

Hill, T. (1993) *Manufacturing Strategy: The Strategic Management of the Manufacturing Function.* Macmillan.

Hofstede, G. (1980) *Cultures' Consequences: International Differences in Work-Related Values.* Sage.

Hofstede, G. (1991). *Cultures and Organisations* (Kulturer og organisationer). Schultz.

Hörte, S.-Å. et al. (ed.) (1991) 'Manufacturing strategies in Sweden - 1982-1989'. (Partly Swedish), IMIT, Chalmers Tekniska Högskola.

Kristensen, P.H. (1992) 'The Small-holder economy in Denmark: The exception as variation'. In C.F. Sabel & J. Zeitlin (eds.), *Worlds of Possibilities: Flexibility and Mass Production in Western Industrialization,* Paris.

Maskell, P. (1992) 'New establishment of industrial companies (Nyetableringer i industrien - og industristrukturens udvikling)'. Handelshøjskolens Forlag/Nyt Nordisk Forlag Arnold Busck.

Møller, K. & H. Pade (ed.) (1988) 'Industrial success - competitive factors in 9 Danish Industries (Industriel Succes - konkurrencefaktorer i 9 danske brancher)'. Samfundslitteratur.

Voss, C.A. (ed.) (1992) *Manufacturing Strategy: Process and Content.* Chapman & Hall.

CHAPTER 6

FINLAND: CHANGING FROM TECHNOLOGY-BASED TOWARDS PROCESS-BASED MANUFACTURING STRATEGY

Magnus Simons, Kari Pietiläinen, and Raimo Hyötyläinen, VTT Automation, Industrial Automation, Espoo, Finland

Manufacturing strategy is changing in the Finnish manufacturing industry. For decades, it has been based on hard investment in machines and technology, but in the 1990s more emphasis is beginning to be put on the organisational and human aspects of manufacturing. Now companies are trying to leverage both the existing human and financial resources. One of the driving forces behind the change seems to be the economic development of the country.

Manufacturing strategy and the use of manufacturing practices in the Finnish manufacturing industry also differ slightly from what can be considered the mainstream in manufacturing. One explanation can be found in the structure of Finnish industry and in the role of the manufacturing industry. Business-to-business type production is common, and it stresses different means of competition from those used in mass production or mass customisation.

In this article we show how the rapid economic growth resulting from the need for capacity expansion and the need for technical development and innovation has led to an industrial strategy based on hard investment, and how industrial development has favoured business-to-business production, mainly in small-volume niche markets. At the end of this article the results from the IMSS survey on the use of lean manufacturing technologies in Finnish industry are interpreted through this historical development of Finnish industry.

6.1. The Finnish Economy and Industry

Finland started its development towards becoming an industrial nation rather late, well behind the countries leading the industrial revolution. Finland could still, in modern terms, be considered a developing country in the middle of the nineteenth century. During the latter part of the century, however, the economy started to grow, and for most of the time since the 1860s the Finnish economy has grown faster than the economies in other OECD countries. In 1913, four years before Finland was declared an independent nation, the GDP per capita was about half of the mean level in the OECD countries, but by late 1980s the economy had grown to a level well above the OECD

P. Lindberg et al. (eds.), International Manufacturing Strategies, 103-119.
© *1998 Kluwer Academic Publishers. Printed in the Netherlands.*

mean. During the 75 years before the 1990s the GDP growth rate was about 3% per year. Since World War II growth has been constant and industrial production has declined only once during the post-war period—in 1975. Between then and the end of the 1980s the Finnish economy experienced exceptional growth of about 5% per year (Vartia and Ylä-Anttila, 1992). The role of industry has been central in this growth. During the last 100 years industrial production has grown 1.5 times faster than total production. This means that the pace has been fast even by international comparison.

Throughout this time of growth and structural change, the investment rate was high, an average rate of about 26% of GDP from 1960 until 1990. This provides a central explanation for the growth in production and productivity during this period. Unfortunately for Finland, investments were not always profitable. Although the growth in total production in Finland was faster than in most other countries, it was not as fast as would be predicted from the investment rate. This is due to two factors. First, in Finland a large sector of the economy is protected from competition, mainly agriculture and services. Companies operating in this sector have been able to transfer their costs directly into the prices of the products. Secondly, for a long time regulation of the financial markets kept interest rates and expectations for investment profitability at a very low level.

In the 1980s financial markets in Finland were deregulated. This created a market for short-term capital both between companies and between companies and the banks. During this time credit regulation was also changed; for example, interest rates were no longer controlled by the Bank of Finland (ETLA, 1986).

The beginning of the 1990s brought a steep downturn in the Finnish economy. In 1991 alone GDP dropped by 6.5 % (Vartia and Ylä-Anttila, 1992). A central reason for the recession was the overheating of the economy in the 1980s. Despite the financial market deregulation, there was no change in the thinking that had characterised behaviour in the era of regulation, that is, the assumption that inflation always takes care of one's debts. Many companies and households continued to run up debts, with disastrous consequences (Hämäläinen, 1994). In addition to the domestic economic problems, the global economic recession also affected the Finnish economy. The changes in the former Soviet Union especially, Finland's biggest trading partner in the 1980s, had drastic effects on trade. In 1987, just a few years before the fall of the Soviet Union, its share of Finnish exports was about 15%.

In 1993 the economy bottomed out and in 1994 it started to recover. During the first half of the year GDP grew by about 3.3 % compared with the same period in 1993. A central role in the recovery was played by exports, which, after a devaluation in 1991 and the decision to float the Finnish mark in 1992, increased Finnish competitiveness internationally: exports increased in 1992 by almost 10% compared with the previous year and by 16.6 % in 1993. The economy was also improved through the strict financial policy of the government and of the Bank of Finland, which kept inflation and interest rates low. The main problems remaining are the unemployment rate, the budget deficit and the task of keeping inflation at a low level in an environment of economic growth. Although dropping somewhat, the unemployment rate was still at a very high level—18.1% in September 1994 (Bank of Finland, 1994). In the future, Finland also has to reduce spending in the public sector, which for a long time has been growing faster than the rest of the economy; it also has to cope with new increasing pressures from the environment and with an ageing workforce (Vartia and Ylä-Anttila, 1992).

6.2. The Role of Manufacturing in the Finnish Industry

To understand the production strategies employed in the manufacturing industry, we take a look at the whole of Finnish industry. To describe this we use a concept framed by Michael Porter—the industrial cluster (Porter, 1990).

The forest industry has always played a central role in Finnish industry and, whilst the production of sawn timber has decreased, the production of paper and pulp is still at the core of the Finnish economy. This industry has also offered an environment suitable for the production of necessary production equipment and machinery, of chemicals needed in the processes, and of services. Thus, the forest industry has grown to become a complete industrial cluster. In parallel with this cluster, eight other industrial clusters have developed: construction, basic metals, transportation, telecommunications, energy, foodstuffs, health, and environmental clusters (Hernesniemi et al., 1995).

Of the nine Finnish industrial clusters only the forest cluster can be considered a strong cluster; that is, one having a strong domestic market and competition in all parts of the cluster, as well as support from the government, from research institutes and from other organisations. The basic metals and energy clusters are considered semi-strong clusters, the former relying heavily on foreign producers of vital production equipment whilst the energy cluster has to import most of its raw material. The fastest growing cluster is telecommunications, but the health and the environment clusters also have potential for future growth. The construction and foodstuff clusters are defensive or latent, meaning the structures for a cluster exist, but the capacity runs the risk of being partly unused in the future (Hernesniemi et al., 1995).

As can be seen from Table 6-1, the manufacturing industry is present in all the Finnish industrial clusters. In eight of the nine clusters the role is to *produce equipment for other industries within the cluster.* The only exception is the manufacturing segment in the fast-growing telecommunications cluster. Here, the manufacturing segment not only produces inputs for other industries, but also produces mass-produced consumer goods.

Many of the industries for which the manufacturing industry provides equipment operate in mature markets competing in the traditional mass-production way. The strategy in the forest industry has been described as 'more, faster, bigger and more

Table 6-1. The Industrial Clusters in Finland (Source: Hernesniemi et al., 1995)

Industrial Cluster	Major Products in Cluster	Role of Manufacturing Industry
1. Forest	Paper, pulp, cardboard, wood articles	Machinery and equipment for pulp & paper industry
2. Construction	Buildings, construction	Material and equipment for construction industry
3. Basic Metal	Refining foreign ore resources	Machinery and equipment refining of ore resources
4. Transportation	Transportation equipment, car ferry traffic, transit transportation to Russia	Transportation equipment
5. Telecommunication	Telecommunications, telecommunication equipment	Telecommunication equipment, consumer electronics
6. Energy	Energy distribution	Equipment for production, distribution and use of energy
7. Well-being	Public health services, drugs	Health service equipment
8. Food	Foodstuff	Equipment for production of milk products and refrigerated transport
9. Environment		

inflexible' (Eloranta et al., 1994). This strategy stresses the need for ever more efficient production equipment, and this is where the domestic manufacturing industry has been helpful in many clusters.

Through the co-operation between producers and users of production equipment, highly competitive manufacturing industries have emerged within the different industrial clusters. For instance, the producers of equipment for the paper and pulp industry, energy production and distribution and for construction material are today successfully operating globally within their special market niches. Through the internationalisation and globalisation of these operations, dependence on the domestic user industries is decreasing.

In clusters such as basic metals and foodstuffs, domestic equipment production plays a smaller role, but in these clusters the Finnish manufacturing industry has also found its own niches of excellence.

The telecommunications cluster is the fastest growing cluster in Finnish industry. This is also the first industrial cluster in Finland where raw material supply plays a minor role and the main role is played by information and know-how. Telecommunications is considered a central part of the infrastructure of Finnish society. There has always existed a large number of competing telephone companies in the country, and in the 1980s free competition among these was allowed. As a consequence, Finland is one of three countries in Europe where the country's telecommunication services are produced by private companies. The other two countries are Sweden and the UK (Hernesniemi et al., 1995).

Highly developed telecommunication services in Finland have also provided a basis for the production of telecommunication equipment. By 1992 this business had grown to a volume of FIM 4.1 billion and the growth rate is very high. For instance Nokia, the leading producer of telecommunication equipment in Finland, increased its production volume in this field by some 50 per cent in 1993 (Hernesniemi et al., 1995). The manufacturing industry in this cluster differs from most manufacturing companies in Finland, in that a large amount of the products are consumer products and they are produced in large volumes.

6.3. Manufacturing Strategy in Finland

The IMSS database gives us a picture of manufacturing strategy in the Finnish manufacturing industry in comparison with that in other countries. Three characteristics of the Finnish manufacturing strategy stand out; *(1) Finnish companies are small and flexible, (2) they have a high level of automation,* and *(3) they have a low level of use of so-called 'softer' manufacturing practices.* Here we try to plot these facts against the background of the manufacturing industry given in the previous section. The survey results originate from companies answering the IMSS questionnaire initially sent to roughly 100 Finnish metal companies. The 17 companies in the Finnish sample represent traditional Finnish manufacturing industries from the forest, transportation, energy and basic metals clusters. The fast-growing telecommunication cluster is not represented in the sample.

The Finnish companies in the survey are rather small by global comparison in terms of employees, production volume and assets, and they operate in niche markets, where the market share of the dominant product line is considerable, but smaller than the average of the total IMSS sample (Table 6-2). Within these niche markets, companies strive to become global market leaders, and also to lead the technical development of the products and production. This is somewhat in contradiction to the results from the IMSS survey, where the Finnish companies report very low figures for R&D spending, process equipment investment and training and education. The explanation for this can be found in the markets served. The business-to-business niche markets served by Finnish firms are often mature and stable.

Table 6-2. Company and Production Characteristics

	Finland	World
No. of Employees	470	865
Production Volume (1 000 units)	55	6,655
Assets (Million $)	638	7,031
Market Share (%)	26.4	33.9
No. Different Prods.	909	706
Customer Demand Variations (scale 1-5)	3.18	2.99

Production orders

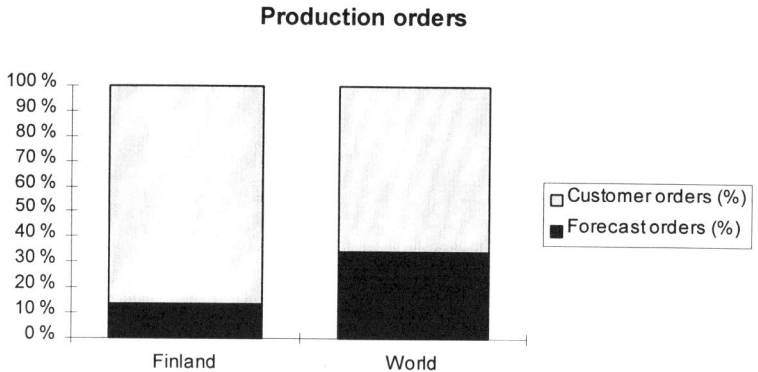

Figure 6-1. Production Orders

Production in Finnish companies is highly flexible. Whilst the average number of different products in Finland is only slightly larger than in the whole sample, if the low production volume of Finnish companies is taken into account, product variety is very high. Products are often tailored to customer specifications, and are typically produced to order with a very high variation in customer demand (Figure 6-1).

The external and internal logistics of the supply chain must match the markets' demand for flexibility and delivery reliability. The external logistics of the Finnish companies in the sample are, in terms of volume and number of suppliers and markets served, smaller than the average, but in terms of geographical distribution of purchase and sales, the supply chains of the Finnish firms are widely spread. This puts high demands on external logistics.

The demand for flexibility can also be seen in the structure of the internal logistics of the manufacturing process. Inventory holdings of the Finnish companies consist mainly of raw materials. Here, the level of inventory is on a par with the total IMSS sample (Figure 6-2). Finnish companies show a slightly lower level of WIP holdings, but in final goods the level is less than half of the level of the total sample. Also, the average lead time of the Finnish companies is shorter than the average lead time. Short lead times and small inventories are achieved through one-of-a-kind and batch production. Line production is rare both in part manufacturing and in assembly. In the Finnish companies, part manufacturing is mainly performed in batches, whilst the bulk of assembly work is done in one-off processes (Figure 6-3). Usually a pull-type production control system (e.g., a two-box system) is used to link assembly and manufacturing to each other. Cellular layouts are common both in part manufacturing and in assembly.

Inventory Holdings

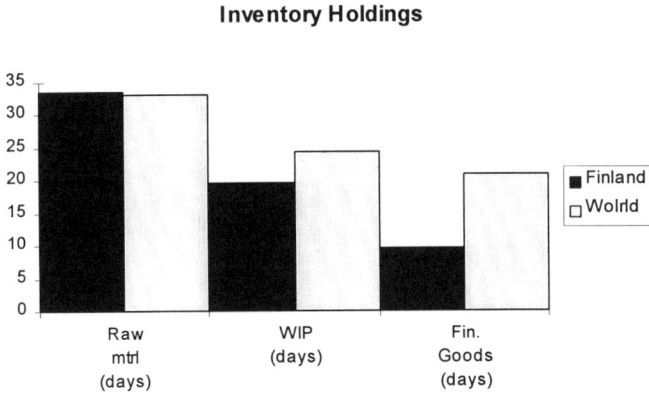

Figure 6-2. Inventory Holdings

The cost of logistic functions in Finland was on average 11.6 per cent of the turnover whilst the corresponding figures for all Europe in 1991 were 4.6-7.1 per cent (Liikenneministeriö, 1993).

The data in Table 6-3 show the use of automation in Finland and in the whole IMSS sample. It shows that *part manufacturing in the Finnish companies is highly automated.* In absolute terms, the numbers of automatic machines and systems do not differ very much from the average of the whole sample but, compared with the size of the companies, the numbers from Finland show that flexible manufacturing systems, machining centres and numerically controlled machine tools are more common in the

Process Type

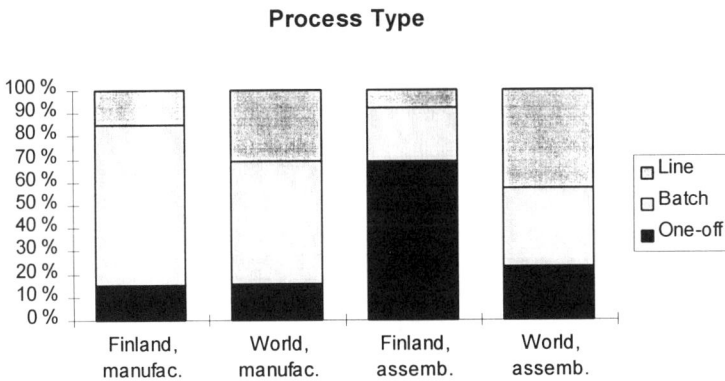

Figure 6-3. Process Types in Finland and in the IMSS Sample

Table 6-3. Use of Automation

	Number of machines			
	Average		Per 1000 employees	
Use of Automation	Finland	World	Finland	World
FMS/FMC	1.9	1.6	4.0	1.8
Machining centres	4.2	3.9	8.9	4.5
NC	9.8	14.4	21	17
Conventional machines	39.4	106.3	83	120
Manufacturing robots	3.7	4.3	7.8	5.0
Assembly robots	0.1	4.8	0.0	5.5
FAS (Flexible Assembly Systems)	0.0	2.4	0.0	2.8

Finnish sample than in the total sample. The table also shows that the investments are directed towards the production of parts and components, that is, the machining of parts is automated. In assembly, the Finnish investments are smaller than the average for the whole sample. This is due to the need for extremely high flexibility in the assembly process, where products are made to specific customer requirements.

The high level of use of automation technology in the Finnish manufacturing industry is related to the development trajectory of the Finnish economy. In a survey among FMS users in Finland in 1989, the most important reasons for investing in flexible systems were: the need for increased capacity, productivity, and higher flexibility (Mieskonen, 1989). Highly automated machinery was used to replace older less efficient equipment or simply to add new capacity. Another argument often used for investing in FMS in the 1980s was the lack of skilled workforce. In a way this was also a consequence of the fast-developing economy of the country. Young people in Finland were more interested in working in the growing service sector than in the manufacturing industry. Many of the older people working in the industry lacked the required skills to work with highly automated equipment.

The level of use of different manufacturing practices in Finland is shown in Figure 6-4. The level of use of the Japanese, Western and general manufacturing practices listed in the IMSS survey (see Voss et al., Chapter 10) is lower in Finland than in the total IMSS sample and especially compared with Japan. The difference is larger in the groups of Japanese and general management techniques, but somewhat smaller in the group of so-called Western techniques. In the perceived payoffs for the different management techniques, the Finnish companies also show a lower level than in the whole sample and in Japan in particular. In payoff from the Japanese techniques the Finnish results are at the same level as the total sample.

Why is the diffusion of manufacturing practices slower in Finland than in other countries? A central reason for this is the structure of the industry. As described in the previous chapter, the Finnish manufacturing industry is to a great extent business-to-business based manufacturing companies. The manufacturing practices surveyed here come from the Japanese mass-customisation paradigm or the American mass-production paradigm.

The Level of Use

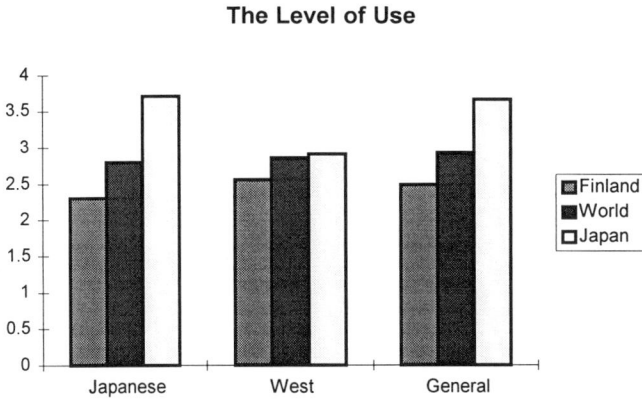

Figure 6-4. The Average Level of Use of Manufacturing Practices

Statistical analysis of the IMSS database shows that there is a structural bias in the level of use of manufacturing practices. Figure 6-5 shows that there is a difference in the level of use of these practices between different types of processes. One-of-a-kind producers use the manufacturing practices less than batch or line producers (p = 0.000). In Table 6-4, the level of use of the manufacturing practices is compared with a rough flexibility measure (no. of units produced/no. of different products) for the company. The table indicates that the level of use increases as the number of units per product type grows; that is, companies with high flexibility (producing a few units of a product type) use most of the manufacturing practices less than companies producing larger amounts of their different product types. The phenomenon is rather evident in the group of companies producing very small numbers of units (<200 units/product/year) and this

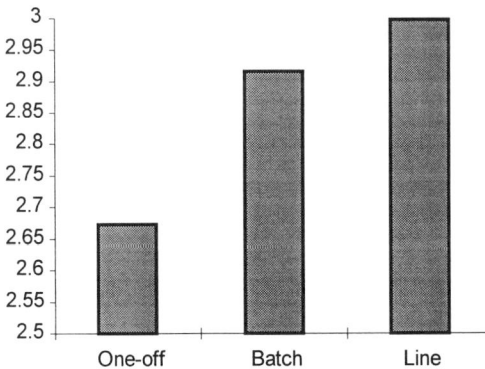

Figure 6-5. Average Level of Use of Manufacturing Practices in Companies with Different Assembly Processes.

Table 6-4. Regression Analysis on Correlation between the Number of Units Produced Yearly of a Certain Product Type and the Level of Use of the Listed Manufacturing Practices.

Manufacturing Activities	Type[1]	No. Units/Prod.[2]		
		<200 (R Square)	<2000 (R Square)	<20000 (R Square)
TQMP	J	0.091	0.040	(-0.000)
SPC	J	0.039	0.034	0.019
JIT / Lean production	J	0.116	0.084	(0.004)
JIT/customer deliveries	J	0.095	(0.014)	0.019
SMED	J	0.093	0.062	(0.000)
Kanban	J	0.205	0.061	(0.010)
Zero defects	J	0.143	0.057	(0.001)
QFD	J	(0.017)		
QPD	J	(0.003)	0.020	(0.001)
Simultaneous Engr.	J	0.049	(0.001)	
Teams	J	0.043	(0.000)	(0.005)
Kaizen	J	0.161	(0.006)	
TPM	J	0.098	0.037	(-0.000)
ISO 9000 cert.	W	(0.006)	(0.001)	(-0.003)
MRP	W	0.033		(-0.001)
MRP II	W	0.049	(0.004)	(-0.005)
Plant-in-Plant	W	0.099	0.055	(0.000)
ABC	W	(0.015)	(0.005)	
CAM	G	(0.019)	(0.002)	-0.018
CAD	G	(0.000)	(-0.013)	(-0.009)
DFA/DFM	G	(0.012)	(0.001)	(0.004)
Benchmarking	G	0.087	(0.005)	
Value analysis	G	(0.012)	(0.006)	
Manufacturing Strategy	G	0.047	(0.003)	
Energy Conservation	G	0.052.	0.024	(0.002)
Environmental safety	G	(0.019)	0.029	(0.001)
Health & safety	G	(0.001)	(0.006)	

Notes:

1. Type: J= Japan; W= West; G= General

2. The bold figures indicate a significant (< .005) correlation. A negative figure indicates a declining trend.

is especially true in the case of Japanese and Western-type practices. Of 13 Japanese practices only two are not affected by the flexibility of the manufacturing process. This is, perhaps, a somewhat confusing result, since the Japanese production is famous for its high flexibility. The results, however, indicate that the Japanese manufacturing practices are used less in one-of-a-kind production than in mass customisation.

Roughly half of the Finnish companies in the sample operate in the range of less than 200 units per products yearly, which at least partly explains the low average level of use of the manufacturing practices in the Finnish sample.

Although the data from the IMSS database indicate that the average level of use and payoff of different manufacturing practices are low in Finland, there are some

Actions within last two years

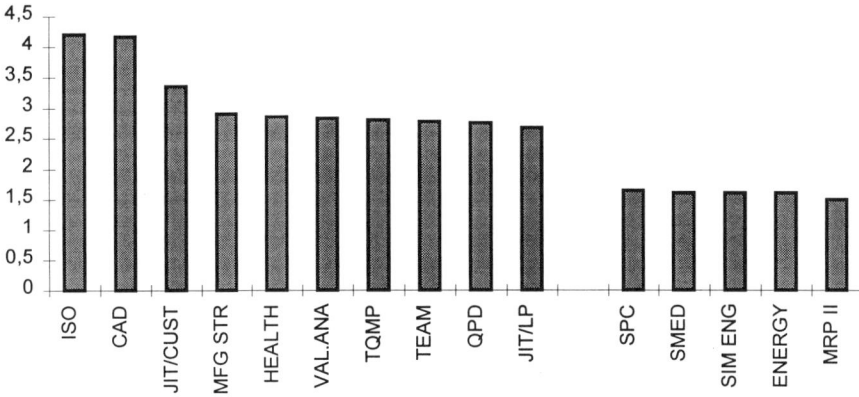

Figure 6-6. The Highest and Lowest Ranked Actions Undertaken Within the Last Two Years in Finnish Companies.

practices that have an extremely high level of use in the Finnish sample. These are ISO 9000, CAD, and JIT deliveries to customers (Figure 6-6). In these three practices the Finns are among the three leading countries in the total sample. This also indicates that a difference in manufacturing strategy exists between small, niche market players and mass producers. Of the three practices given the most attention by the Finnish companies, two belong to the group of manufacturing practices that report a higher than average level of use in highly flexible manufacturing processes (Table 6-4). Typical mass production methods such as SPC, SMED and MRP II were given the lowest rankings by the Finnish companies.

6.3.1. THE CHANGE IN MANUFACTURING STRATEGY IN FINLAND

The drastic changes that occurred in the early 1990s in the Finnish economy, as well as in the markets served by the industry, did not leave the manufacturing strategies of the manufacturing companies untouched. As the markets demanded increasing flexibility, higher quality and faster deliveries, the companies responded by changing their organisations, scheduling, logistic systems, management or other production facilities. Whilst the structure of the industry sets the frame in which companies operate, the economic situation in the country does affect the means and resources available in manufacturing.

We can see two trends in the Finnish manufacturing strategy. First, the focus of manufacturing strategy is seemingly changing from production, machine tools and management towards management of a larger whole: *the business processes of the company or the logistic chain.* Secondly, *the role of technology is being challenged by softer methods as a means of production and productivity improvement.*

During the period of fast market changes and of rapid economic growth in Finland in the 1980s, Finnish companies frequently used hard *investment in technology* in response to increasing demands from customers. During this decade the number of NC machine tools, machining centres, flexible manufacturing systems and robots grew quickly in part and component manufacturing. Not only did the companies renew their machinery, but they also increased their manufacturing capacity. Through these investments, the competitiveness of the companies was secured. Spending on product development and R&D, on the other hand, can be considered moderate (Vuori and Vuorinen 1994).

The diffusion of Flexible Manufacturing Systems (FMS) is an example of the use of technology in Finnish industry. Flexible manufacturing systems were introduced in Finnish workshops in the early 1980s. Although this new technology was more expensive than conventional machine tools and the technology was new to most managers, the diffusion of these systems was rapid in Finland. In 1982 there were two FMS systems in use in Finnish industry, but by the end of the decade the number was 25 (Figure 6-7). In comparison to the situation in most other countries, this is a very high number of systems. The growth of the number of FMS in use culminated in six new systems in 1990. After this, only four new systems were implemented by the end of 1993.

The diffusion of FMS is only an example, but it reflects the situation in the economy. The total volume of investments in Finland peaked in 1989, dropped by roughly 5 per cent in 1990 and 15-20 percent in the following three years (MET 1994).

As the economic situation shifted in the beginning of the 1990s, the behaviour of companies also changed. During the first half of the decade, the number of investments in machines or new production facilities dropped drastically and the companies, who in the 1980s had felt free to run up debt, had to pay back their investments. This had to be

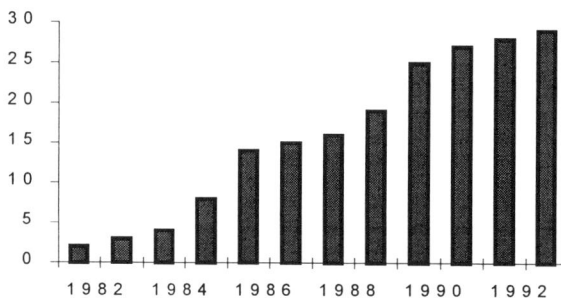

Figure 6-7. The Number of FM Systems in Use in Finland (Source: VTT FMS database)

done even though production volumes had dropped drastically and the new capacity was run at a very low utilisation rate. As sales started to increase again after 1991, companies concentrated on improving their capital self-sufficiency, a part of the business that was very often neglected before the deregulation of the economy.

Investments in hard technology and machines were not the only means of manufacturing development in the 1980s. Also used were *softer means of production rationalisation*. At the beginning of the decade JIT production was introduced in Finnish industry. One of the first companies to adopt these ideas was Saab-Valmet, the only producer of passenger cars in Finland. JIT was often used as a means for capital rationalisation, that is, for cutting down work-in-progress and final product storage. Between 1980 and 1985 the value of fixed capital in stock in Finnish industry represented on average 53 per cent of the value added, but in the period between 1986 and 1990 it was only 45 per cent (Eloranta et al., 1994).

In the early years of the 1990s, many companies once again looked for means of capital rationalisation and lean manufacturing was often perceived to be a means of cutting down cost through reduction of personnel and fixed capital. This was perhaps a natural, but in the long run not always healthy, reaction to the situation in which most companies were during the early years of this decade. But, it also reflects another characteristic of Finnish industry: the management itself. Although there was a lot of talk and even some concrete examples of changes in management in the 1980s, the traditional authoritarian style still prevailed.

There is, though, some evidence that there is a change in attitude towards management in the companies. An example of the trend towards *leveraging human resources* is the growing interest in cell production and teamwork. Most of the cells implemented in the 1990s included characteristics such as job enlargement and support activities (Table 6-5), which indicates that the cell has an independent role within the production department. Group work is performed in one out of four cells and roughly half the companies using cell production report that their cell teams are performing or preparing to perform continuous improvement activities. This should be seen in the light of quality activities performed in Finnish industry in the 1980s.

In early 1980s Quality Circles were introduced in the Finnish metal industry. The prime motor behind the introduction was FIMET, the central organisation for the metal industry in Finland. Quality circles were started in a set of companies, but their

Table 6-5. The Characteristics of Finnish Production Cells (Source: Hyötyläinen and Simons, 1997).

Characteristic of the cell	Number of cases (included/planned) (n=26)
1 Job enlargement	22/3
2. Support activities	24/1
3. Group work	6/4
4. Network activity	12/2
5. Development activity	10/5

activities soon died out. For one reason or other the companies were not ready for these kinds of activities, where the workers play a central and largely independent role. Statistical Process Control and other statistical quality management tools were tested and used, but the diffusion of these tools was also slow.

The use of FMS, JIT and flexible cell production led to a situation where the production of parts and components was more or less independent of the size of the production batch, but was still dependent on the size of the production series of the product for overhead costs such as product and production planning (Eloranta et al., 1994). This can be considered the starting point for the second trend: changing the scope from local to global.

In the 1980s the rationalisation of production processes, flexible automation, and the need for faster throughput led to the use of cellular layout in many workshops, and, through group technology–based techniques, formerly functionally-organised production processes were turned into product workshops. The next step was the so-called focused factory or product factory, where functions such as product planning, production planning and purchase were decentralised and integrated into the workshop organisation. Through this organisational structure, co-operation between the functions was made easier in order to improve productivity through lower manufacturing costs, better quality and a shorter lead time.

From the IMSS data we can see that the highest-ranked actions undertaken in 1991-1992 were the introduction of ISO 9000, CAD and JIT deliveries to customers (Figure 6-6). Whilst the use of the ISO 9000 quality standard and of JIT delivery to customers are central means of increased customer service, they also indicate that not only production but the manufacturing process as a whole is important to business performance. Through the use of CAD, product planning and design are automated, which is a precondition for technical integration of these functions and the production.

In the ISO 9000 standard the quality of the product is seen as a result of all the operations of the company. To describe the total operation, the activities are grouped into business processes. A condition for the ISO 9000 certification is that these processes are documented. In the summer of 1995 there were some 600 certified ISO 9000 users in Finland (Moisio, 1995). Most of these companies see the certificate only as a quality guarantee for the customer, but in some companies the business process concept is used for continuous improvement of operations. Large Finnish companies such as Kone Oy, Valmet Oy and Neles-Jamesbury are pioneers in business process development and re-manufacturing in Finland.

Business Process Re-Engineering has been in the 1990s introduced at different levels within the companies. In large corporations business process re-engineering concepts are used to restructure the whole organisation including all its product lines and organisational units, but they are also used to re-engineer the operations in focused factories or SMEs.

The concept of business processes in the 1990s is often also broadened to include the logistics chain and component suppliers. Trends in logistics are the outsourcing of transportation functions and co-operation between companies in warehouse and terminal functions (Liikenneministeriö, 1995).

6.4. Case Study: Roll Manufacturing at Valmet Papermachine

In 1991 the production of cardboard machinery at the Tampere works was part of Tampella Papertech, a part of the Tampella corporation. Due to financial difficulties of the mother corporation, the new company was taken over only one year later by another Finnish producer of paper machinery—Valmet Oy. Through this merger, Valmet became the global market leader in cardboard machines, and a major player in the global paper machine market. The merger, however, meant the restructuring of the manufacturing operations of the 'new' company.

The manufacturing process at the Tampere works had been divided into five focused factories before the merger. Each factory produced a different part of the cardboard machine. To streamline the organisation the formerly functionally divided organisation was decentralised to include functions such as product planning, production planning and purchase into the organisations of the focused factories. The factories were also given the freedom to operate as free agents within the whole Valmet network. For the roll factory—the first focused factory to be organised at the Tampere works—this meant that instead of one customer it got three, producing rolls not only for cardboard machines but also for paper machines. Instead of producing parts only for machines manufactured in Tampere, it now delivers parts for machines designed and produced in Sweden and USA. This had a drastic effect on the product family. Whilst some 80 per cent of the products produced before the merger were produced according to Tampella standards, the corresponding figure after the merger was about 10-20 per cent. The rest of the products were produced according to demands set by the new customers.

In addition to the need for higher flexibility, the fusion brought rationalisation of the production facilities within Valmet, that is, the production of rolls was concentrated in only a few factories. Due to this change the volume of roll production at the Tampere works increased from about 600 rolls per year in 1993 to some 1000 rolls in 1995 and there is still a need to double the volume.

One of the driving forces behind the division of the manufacturing process into focused factories was the need to cut down delivery time for paper and cardboard machines. Over the last ten years the delivery time had already been cut from 2 years to 11 months. Although the rolls are not critical to the total lead time of these projects, lead time reduction is also a crucial target within the roll factory, because it is considered the most efficient path to productivity increases.

The production process in the roll factory is focused around the larger rolls, those with a length of up to 10 meters. The machining of the roll cylinders is performed in ten stages, mainly on manual machine tools, but NC machines are also used. In 1991, an

FMS was implemented for the machining of the smaller parts for the roll bearings. This new system had a positive effect on lead time. At the same time, the number of NC machines was increased in roll cylinder production, and the company also made investments in CAD/CAM to shorten the time spent on NC programming and to improve the quality of the programmes. Assembly is performed manually.

To cope with the increased complexity of the production process and to meet the need for an ever-shorter lead time, a project was started in 1992 to turn the production process into cells. Following the promising results from the pilot cell, it was decided to turn the total process into seven cells. In 1995, three cells had already been implemented and four more were being implemented. The central characteristics of the cells at the roll factory were job rotation, supporting functions, and continuous improvement (see Hyötyläinen et al., Chapter 22).

Having made changes in the workshops, the roll factory management also turned to look for changes and improvement elsewhere in the organisation. Manufacturing practices such as target management, activity-based management and ISO 9000 had been tested and implemented earlier, but in 1994 it was decided to start a project to develop business processes stretching across all functions in the roll factory. Business processes had earlier been implemented on a macro level within the Valmet corporation. The target of this project, which started in April 1995, was to improve delivery reliability through cutting down lead times both on the shop floor and in product design.

6.5. Conclusions

Manufacturing strategy in Finland stresses high flexibility, and the use of automation and hard technology is often a central means. On the other hand, the use of so-called softer manufacturing practices is scarce. It seems as if the factors influencing manufacturing strategy in the country can be found both in the structure of the industry and in the structure and the state of the economy. As these conditions are changing, the focus of the manufacturing strategies also seems to be changing.

The Finnish manufacturing industry mainly produces machinery and equipment for other industries, within or outside the country. This, and the fact that most of the products are produced on a one-of-a-kind basis, puts manufacturing in Finland in a category quite far away from mass production or mass customisation. Here, the means of doing business as well as of manufacturing differ from the mainstream.

During the era of rapid and continuing economic growth, practices in the Finnish manufacturing industry, as well as in other parts of the industry, were focused on hard, technical investments. Up to and even after deregulation of the Finnish economy in the early 1980s, this trend was reinforced by an economic system allowing low returns on investment and liberal use of liabilities. The downturn in the economy and the changes made in the economic system now force the companies to look over their capital structure. This seems to stress the need for implementing softer manufacturing practices.

This article gives an indication of the effects of a small set of macroeconomic and cultural characteristics of a country on the manufacturing strategy used in the industry. These results are preliminary and leave many questions and aspects unanswered. To get further answers we need more research and continued international collaboration.

References

Bank of Finland (1994) *Bulletin*, 68, 11, November.

Eloranta, E., Ranta, J., Ollus, M., and Suvanto P. (1994) *Uusi teollinen Suomi* (The New Industrial Finland), WSOY, Helsinki.

ETLA (1986) *Kansantalouden kehitysnäkymät 1986-1990* (Development trends in the Economy 1986 - 1990), ETLA, Helsinki.

Hernesniemi, H., Lammi, M., and Ylä-Anttial, P. (1995) *Kansallinen kilpailukyky ja teollinen tulevaisuus* (The National Competitiveness and the Industrial Future), ETLA & SITRA, Helsinki.

Hämäläinen, S. (1994) 'National Economic Policy and Its Credibility', Bank of Finland Bulletin, Helsinki, 68, 11, November, 3-7.

Liikenneministeriö (1993) *'Logistiikka selvitys* (Logistics survey)', *Liikenneministeriön julkaisuja 6/93*, Liikenneministeriö, Helsinki, 85.

Liikenneministeriö (1995) 'Vaihtoehtoiset logistiikkajärjestelmät Suomessa (Alternative logistics systems in Finland)', *Liikenneministeriön julkaisuja L 10/95*, Liikenneministeriö, Helsinki.

Metalliteollisuuden Keskusliitto (MET) (1994) *Metalliteollisuus - Metallindustriin* (Metalindusty), Metalliteollisuuden Keskusliito raport.

Mieskonen, J. (1989) *Suomalaiset FM-järjestelmät - Havaintoja kentältä ja vertailuja kansainväliseen aineistoon* (Finnish FM-systems - Empirical observations and international comparison), IIASA TES Program in Finland, keskustelualoite 11. SITRA, Helsinki.

Moisio, J. (1995) 'Laatujärjestelmät - 250 miljoonan hukkainvestointiko?' (Quality management - 250 miljon investment wasted?). *Talouselämä*, 21, 95.

Porter, M.E. (1990) *The Competitive Advantage of Nations*, The Macmillan Press Ltd., New York.

Vartia, P. and Ylä-Anttila, P. (1992) *Kansantalous 2017* (The Finnish economy in 2017), ETLA, Helsinki.

Vuori, S., and Vuorinen, P. (1994) *Explaining Technical Change in a Small Country - The Finnish National Innovation System*. ETLA, Helsinki.

CHAPTER 7

MANUFACTURING STRATEGY IN THE NETHERLANDS: CONTEXT AND PERSPECTIVE

Harry Boer and Domien Draaijer, University of Twente School of Management Studies, Enschede, Netherlands.

7.1. Introduction

The purpose of this chapter is to present some of the context of the Dutch manufacturing industry and to use this background to put into perspective what we think are the most salient results of the Dutch part of the IMSS. Firstly, the chapter will address the macro-economic context and the structure of the Dutch industry, and put both into an international perspective. Next, some key sociocultural dimensions will be described and illustrated with reference to the organisation and functioning of politics, religion, industrial relations and education in the Netherlands. Then, the most interesting IMSS results are presented in the light of other observations on the Netherlands and its industry, and discussed in the perspective of the macro-economic, industrial and sociocultural context introduced before. The chapter concludes with some lessons for Dutch industry.

7.2. The Netherlands

7.2.1. MACRO-ECONOMIC BACKGROUND OF DUTCH INDUSTRY[1]

Population and Gross Domestic Product
With a population of some 15.3 million (CBS, 1995, 1995, p. 383), the Netherlands is among the smaller countries involved in the IMSS survey. Between 1990 and 1993, GDP grew from US $277 billion (US $18,500 per capita) to US $308 billion (US

---

[1] Most figures mentioned in this section are based on the Statistisch Jaarboek 1995 (Statistical Yearbook 1995), a publication of the Dutch Centraal Bureau voor de Statistiek (CBS; Central Statistical Office) and Cijfers & Trends '94-'95 (Figures & Trends '94-'95) of the Rabobank. Unfortunately, many of the figures we needed were either not readily available or less unambiguous than we had hoped, so we had to calculate several figures ourselves and also to make assumptions to be able to make some of the calculations at all. However, we trust that the figures presented here are good estimates and representative of the Dutch industry and its context.

P. Lindberg et al. (eds.), International Manufacturing Strategies, 121-143.
© 1998 *Kluwer Academic Publishers. Printed in the Netherlands.*

$20,100 per capita), a growth of 3.6% per year.[2] In terms of both GDP per capita and annual growth, the Netherlands is average in the sample.

Inflation
Between 1990 and 1993, price inflation of consumer goods was 3.5% per year, whereas the prices of industrial goods fell by some 0.6% per year during the same period.

International trade: import and export figures
The perception of the Netherlands as a country of traders is correct. With some 45% of its GDP exported in 1993, the Netherlands is topped only by its neighbour Belgium. In absolute figures, imports were US $126 billion, and exports US $139 billion, resulting in a positive trade balance of US $13 billion (CBS, 1995, p. 327). The Netherlands had a positive trade balance of US $11 billion with Germany, US $14 billion with the rest of the EU, and just US $1 billion with Europe minus the EU. Trade with the US and Japan resulted in negative balances of US $5 billion and US $4 billion, respectively (CBS, 1995, p. 329).

As the above figures indicate, the Dutch economy is closely tied to the German economy. In 1993, Germany accounted for 29% of Dutch exports, and in the same year 23% of imports came from Germany. Other important trading partners are Belgium and Luxembourg, with 13% of exports and 12% of imports, followed by the US, with 4% of exports and 8% of imports. A remarkable point is that the Netherlands, in spite of its reputation of internationalism, really trades regionally rather than globally, with 83% of exports and 71% of imports going to and coming from European countries. The rest of exports and imports are divided between North America (4% and 9%) and Asia (7% and 15%). Trade with South America, Africa and Australia is negligible (CBS, 1995, p. 329).

Government debt, financing deficit and public spending
Total central government debt increased from 61.5% in 1985 to 63.6% in 1992. Over the same period, financing deficit fell from 4.3% to 3.7% (CBS, 1995, p. 262). In 1993, public spending reached a level of US $99 billion, that is, 32% of GDP (1992: US $98 billion = 33% of GDP) (CBS, 1995, p. 238). About 18% of this goes to social security, 17% education, 9% defence, 7% health, and 5% trade, industry and infrastructure (CBS, 1995, p. 241). Compared with other IMSS countries, expenditure on both health and education are high. Investment in infrastructure is more average.

(Un)employment
Between 1985 and 1992, unemployment fell from 9.7% to a low of 5.3%, and then increased again to 6.5% in 1993. With 7.4% of the working population unemployed, the figure for 1994 is even bleaker. Over the same period, the working population, those who have a job of 12 or more hours per week, increased from 5.3 million to 6.5 million

---

[2]   Calculations based on exchange rate of 1US$ = Dfl 1.86

employees (39% women, 61% men) or 4.4 million and 5.4 million full time equivalents (FTE) (31% women, 69% men), respectively.

Participation of women in the workforce is rather low, as these figures indicate, and indeed much lower than in Germany, the Scandinavian countries, France, Portugal or the UK. Recent figures, however, suggest that the number of female employees is increasing rapidly. In 1992, 61.1% of the women 15-24 had a job or were registered as unemployed, whereas a slightly lower 60.7% of young males belonged to the Dutch working population (CBS, 1995, p. 102-103). Furthermore, a calculation based on Table 7-1 and Table 7-2 shows that the average job is 0.82 FTE. However, most men work full-time (average job 0.93 FTE), but many women work part-time (average job 0.65 FTE).

7.2.2. INDUSTRIAL STRUCTURE

Table 7-1 and Table 7-2 show how employment is divided over the different economic sectors. Agriculture and industry account for 30% of employment, the service industry for 70% of the jobs. As in most Western countries, the private service sector is still growing, whilst employment in the public sector is decreasing.

A further breakdown of Table 7-2 gives the picture in Table 7-3. A comparison of employment and turn-over in the different manufacturing sectors (Table 7-4) confirms

Table 7-1. Working Population in Different Economic Sectors (Employees x 1,000) (1993).

	Agriculture	Industry	Comm.svcs.	Other svcs.	Total	Women
Employees						
No	104	1483	2170	1996	5753	2267
%	2%	23%	33%	30%	88%	34%
Self-employed						
No	187	86	338	187	798	311
%	3%	1%	5%	3%	12%	5%
Total						
No	291	1569	2508	2183	6551	2578
%	5%	24%	38%	33%	100%	39%

Source: CBS, p. 103.

Table 7-2. Full Time Equivalents in Different Economic Sectors (FTE x 1,000) (1993).

	Agriculture	Industry	Comm. svcs.	Other svcs.	Total	Women
Employees						
No	86	1361	1840	1500	4787	1538
%	2%	25%	34%	28%	89%	29%
Self-employed						
No	150	67	263	104	584	142
%	3%	1%	5%	2%	11%	2%
Total						
No	236	1428	2103	1604	5371	1680
%	5%	26%	39%	30%	100%	31%

Source: CBS, p. 103.

Table 7-3. Employment Figures by Sector (FTE x 1000) (1993).

Sector	Employment	
Agriculture	4%	
Construction industry	7%	
Manufacturing industry	19%	
Chemical/oil		-3%
Food		-4%
Metal/steel + electronics		-8%
Other		-4%
Trade	29%	
Service industry	41%	
Transport/distribution/logistics		-12%
Other private services		-15%
Public services (central, regional and local government)		-14%

conventional wisdom that productivity depends a great deal on technology: the higher the process character, the higher the turnover per employee.

Stock market
In 1993, only 110 Dutch companies were listed on the stock market: 66 industrial, 39 service and 5 international companies. Together these companies employed 1.2 million people (CBS, 1995, p. 270).

Fortune 500
Considering its small size, the Netherlands has a surprisingly high number of large multinational companies. Well-known examples are the seven companies representing the Netherlands in the Fortune 500 (ranked by sales), including Shell, the leading oil company and Unilever, a food and chemical industry (both Anglo-Dutch but with headquarters in the Netherlands); Heineken (Dutch), one of the top three breweries in the world; Philips (Dutch), one of the few Western electronics companies to have survived the Japanese conquest; and AKZO-Nobel (Dutch-Swedish), a chemical

Table 7-4. Employment and Turnover in Different Manufacturing Sectors

Manufacturing industry	Employment		Turnover[1] (% of total turnover of manufacturing companies of > 20 employees)
	(% of total employment)	(% of employment in manufacturing)	
Total	19%	100%	100%
Chemical/oil	- 3%	- 15%	- 25%
Food	- 4%	- 20%	- 30%
Metal/steel + electronics	- 8%	- 44%	- 29%[3]
Other	- 4%	- 21%	- 17%

1. CBS, 1995, p. 167

---

[3]   Metal/steel: 22%; electronics: 7%

Table 7-5. Manufacturing Companies by Number of Employees

Number of employees	<10	10-19	20-99	100-499	>500
Number of companies	37,200	4,400	5,300	1,300	200
%	77%	9%	11%	3%	< 1%

Sources: CBS, 1995, p.167; Rabobank.

company. Only one of those, Philips, is an ISIC 38 company. Other well-known ISIC 38 but non-Fortune 500 companies are DAF Trucks and Stork.

Small and medium-sized enterprises (SMEs)

SMEs, companies employing 1-99 people, play an important role in Dutch industry in several respects. Firstly, SMEs account for some 58% of total industrial employment, as opposed to 42% provided by companies employing 100 or more people (Rabobank, 1994). SMEs are also important in terms of turnover: 55% versus 45% by large companies. Expressed in number of companies, however, the contribution of SMEs is much more impressive: 98% of the Dutch companies employ less than 100 people, and 90% even less than 10 employees. For the manufacturing industry, which includes about 48,400 companies, these figures are only slightly less dramatic (Table 7-5).

A current debate in the Netherlands, as in many other countries, concerns the role of SMEs in industry. SMEs are increasingly recognised as an important engine of industrial development and product innovation. A subsequent section will discuss the subject in much more detail.

7.2.3. STRONG AND WEAK(ER) SECTORS[4]

Industry

The Netherlands is strong in energy exports (gas; oil refineries in the Europoort near Rotterdam, the home of the oil spot trade). Shell, for example, is the Netherlands' third largest company in terms of number of employees, but by far the biggest in terms of turnover and assets. Also relatively strong are Dutch agriculture and related industries such as the food processing (e.g., Unilever) and brewery industries (e.g., Heineken and Grolsch). Much less strong are the heavy, mechanical engineering, and other traditional industries except chemicals. This contrasts sharply with neighbouring Germany. Even Philips is not involved in the heavy end of electrical engineering, for example, power generation and transmission equipment. Most of the (mainly smaller) companies producing this type of equipment are subsidiaries of American or German companies.

Transportation and logistics

A particularly strong point is the Dutch infrastructure and also certain parts, but not all, of the transport industry. Rotterdam is the world's largest harbour and, with 22,000

---

[4]   Much of this section is based on Lawrence (1991)

employees, the shipping company Nedloyd is among the country's top twenty employers. Between 25% and 30% of inland goods transport, as measured by ton-kilometres, is by canals and rivers, some of which (the Meuse and the Rhine) cross international borders and give the country an important stake in European commercial transport. The road system is excellent, well-endowed with motorways and well-integrated with the European road system.

A well-known name is KLM, one of the world's pioneering airlines and in a sense the oldest as it has operated under the same name, from its inception in 1919. The volume of domestic air transport is very modest, but KLM is a strong international and intercontinental carrier, and was voted top airline in the late 1980s.

Furthermore, there is a dense railway network, with the entire railway timetable repeating itself every hour (some portions every half hour). In 1987 rail travel was 69,000 passenger-kilometres per person, compared with 48,000 in France, 31,000 in Germany and 21,500 in Belgium. The Nederlandse Spoorwegen (NS; Dutch Railways) run about 4,400 passenger trains daily, and nearly 100% arrive and depart on time. The NS is in fact the second largest industrial employer in the Netherlands, behind Philips but ahead of Shell.

Finally, road haulage is a substantial industry, with seven or eight major hauliers and some 2,000 smaller ones. There is a strong European orientation, with some names (e.g., Frans Maas) well-known abroad. However, this sector is in deep trouble. With a negative ROI of about 2% on average, caused by fierce price competition and growing competition from the East-European countries, there is a severe need to restructure the whole industry. Perhaps the best opportunity is collaboration between the mostly small companies, through the development of networks offering total logistical services, based amongst others on the application of telematics, and pursuing a differentiation strategy rather than engaging themselves in price competition. However, this sector is very traditional, with many family-owned businesses, and much will depend on the willingness of these companies to collaborate with their competitors. Some of the few attempts made so far to establish such networks have been very successful, others have failed, however.

7.2.4. SOCIO-CULTURAL DIMENSIONS

Hofstede's cultural dimensions
In terms of Hofstede's cultural dimensions (Hofstede 1980), the Netherlands score:

- 80 on *individualism*; this is relatively high: lower than the US but higher than Japan and all EU countries except the UK, and also higher than the world mean;
- 38 on *power distance* between managers and subordinates; this is much more average, albeit (slightly) lower than Japan, the US, the other EU countries and the world mean;

- 53 on *uncertainty avoidance*; also rather average, but now higher than the US, much lower than Japan, and slightly lower than the rest of the EU and the world mean;

- 14 on *masculinity/femininity*; this is very low, meaning that the emphasis is on feminine aspects of leadership, much more so than in Japan, the US and indeed most of the world; of the EU countries only Sweden scores lower, as does Norway.

Overall, the Dutch culture seems to combine more elements of the Scandinavian and the Anglo-Saxon (UK and Ireland, US and Canada, South Africa, New Zealand and Australia) cultures than of the other Western-European cultures, including those of Germany and Belgium.

The remainder of this section will not only illustrate the Hofstede dimensions but also suggest that they tell only part of the story. Much of the following is based on Lawrence (1991).

Regional differences

In spite of the Netherlands being a small country, there are considerable regional differences between the happy and highly industrialised Catholic south; the reliable, but less wealthy and largely agricultural Calvinist north; the highly urbanised and heavily populated West with its process (Rotterdam—the world's largest harbour), public service (The Hague), and private service (Amsterdam, Utrecht) industries; and the East, attractive not only for its countryside but also for its potential as a nodal point in the transport and distribution between the Netherlands and the Ruhrgebiet as well as the northern parts of Germany, and Scandinavia.

These regional differences probably lead to differences in the workforce climate in companies, in terms of deference, radicalism, company loyalty, work-centredness and propensity to stay or leave. And, coupled with religious differences (see below), they may also affect labour mobility and managerial appointments as well, especially in smaller family-owned businesses, which tend to appoint co-regionalists and/or co-religionists.

Politics and society

The high level of individualism is reflected in the number of different parties participating in the political system. A feature distinguishing the Netherlands from many other countries is the number of political parties represented in central and local authorities. For example, in the Dutch parliament thirteen parties are represented, including not only the mainstream Christian-Democratic, Liberal-Democratic, Conservative and Socialist parties, but also sectoral ones, for example, those representing elderly people or farmers. At local level, there are many parties that are not active on national level but aim to represent the interests of the local population. In the local elections of 1994, those parties attracted about 21% of the vote.

As a result, never in history has any party been able to win an absolute majority in parliament. Consequently, the typical Dutch government is a coalition of two or three parties. Although the Dutch parliamentary system is a system of differentiation, it has always forced political parties to negotiate and search for compromise and coalition. The result is a deeply ingrained, typical Dutch way of doing things. Dutch society is not a 'winner takes all' society, but rather a bargaining-wrangling forum which proceeds by adjustment. This may lead to conservatism and risk-avoidance but also to more progressive behaviour, depending on how much momentum new ideas can win. Examples of the latter are the role of the Netherlands in the European debate on safety (e.g., the accusation of 'Hollanditis' in the 1980s) and environmental issues (e.g., the pressure exerted together with Denmark to get the pollution of the North Sea higher on the agenda). Another example is the tolerant government policy and societal attitude towards (soft) drugs, abortion and euthanasia. This has given the Netherlands a bit of a bad press in Europe. However, this is partly due to a lack of appreciation of what is really happening. Dutch society and politics are rather more conservative than many people think, and the political and societal debate aimed at reaching consensus legislation on these forms of individual freedom has taken several decades and is still going on.

Religion

The number of religious denominations is extremely high in the Netherlands. The reformation in the 16th century, resulting in a split between the Roman Catholic and the Protestant (Calvinist and Lutheran) churches, initiated a range of further secessions, mainly within the Calvinist churches, resulting in over 20 different denominations. Only recently, over the past 20 years or so, have relationships between the major Protestant churches, that is, the Reformed, Presbyterian and Lutheran churches, improved. At present, these churches are even in a process of reunification that is expected to be finished by the end of this century. Again, individualism (many denominations) *and* collectivism (good relationships between, and even re-unification of, the different churches) in action at one and the same time.

Labour relations

Relations between management and the workers are characterised by the same mixture of a high degree of individualism and a tendency to seek consensus and compromise rather than conflict. There are numerous federations of employers and unions, mostly organised around branches and/or religious denominations, rather than trades. The unionisation rate of employees is about 25%. Most of the employers' federations and the trade unions as well are organised in confederations like the non-religious *Verbond van Nederlandse Ondernemingen* (*VNO*: Federation of Dutch Enterprises) and *Federatie Nederlandse Vakverenigingen* (*FNV*: Federation of Dutch Trade Unions), and the Christian *Nederlands Christelijk Werkgeversverbond* (*NCW*: Dutch Christian

Employers' Association) and *Christelijk Nationaal Vakverbond* (*CNV*: Christian National Trade Union Federation).

Annually, some 700 collective work agreements are negotiated, some for one year, many for two or three years, usually between one union or confederation of unions and one (very large) firm or a group of (often smaller) firms. In the latter case, the firms are often represented by their employers' federation. The negotiations involve subjects such as: pay and pay rises, the length of the working week, (early) retirement conditions, sick pay, level of employment (including youth employment), training of personnel, holiday bonuses, and equal treatment of and opportunities for men and women. The relationships between employers and employees are usually very good, resulting for example in a very low level of days lost through strikes since World War II.

Within the firm there is another typical mechanism affecting employer-employee relationships, the so-called 'works council'. These councils, which are elected by the employees, have:

- *the right to consent* on matters such as policies on pay, working hours and holiday; health, welfare and safety; and appointments, promotion and dismissal;
- *the right to advise*, for example, on substantial changes in work and working conditions; and the termination, substantial increase as well as the location of business activities;
- *the right to receive information* on virtually anything else, including financial statements; personnel plans; social policy; and past and anticipated business activities and the results thereof.

The works councils are fairly well accepted by Dutch managers and indeed they seem to make a difference. Many managers find that the existence of the works councils and their attendant rights condition management's thinking, put a brake on their proposals, and lead to a more reflective and conscientious approach to possible change. Management is much more likely to ask itself: how will this look to the works council, will they see objections we don't see, is it possible to sell it to them, and if so how? Works councils may well have induced an element of humane realism, whereas previously intended initiatives were only evaluated in relation to the question of how the shareholders and the stock market would react.

Works councils, and especially the way they exercise their rights to consent, advise and information ensure that the power distance between managers and workers will never become high. The relatively high level of education of both managers and the workforce described in the next subsection will have a similar effect.

Education
In the Dutch educational system, three levels can be distinguished. The primary school is attended between the age of 4 and 12. The secondary system encompasses various school types: lower vocational education (LBO: 13-16), middle general continued

education (MAVO: 13-16), higher general continued education (HAVO: 13-17), and preparatory academic-scientific education (VWO: 13-18). The third level starts at the age of 17 or 18. Students can attend middle (MBO) and higher (HBO) vocational education which lasts 3-4 years, or go to university (4-5 years), depending on the secondary type of school they went to. There are three technical universities, nine general universities, one open university, thirty-four technical colleges on HBO level, and numerous other institutes of middle or higher vocational education.

Most blue collar workers are educated at (technical) LBO level. Due to the increasing application of advanced manufacturing technology on the shop floor, the number of operators educated at a (technical) MBO school is increasing. Office workers typically have had an education at MAVO, HAVO or (administrative) MBO level. Most Dutch managers are either university or HBO graduates. By and large, those who have been to university have studied engineering, economics or law. In ever more cases an engineering course has been 'topped off' with some management education. Undergraduate courses in management are relatively recent (early 1980s) and are taught at two general and two technical universities. Furthermore, there are numerous management courses at HBO level.

Conclusion
Broadly speaking, Dutch society is characterised by high levels of individualism and femininity and average levels of power distance and uncertainty avoidance. However, within Dutch society the regional differences are substantial. Furthermore, the level of individualism (high diversity of goals, opinions, values) may be high, but at the same time coalition formation (joint discussion, looking for consensus) plays a dominant role. This is not only the case in politics, but also in society as a whole and within companies between managers and employees. The consequence is a peculiar mixture of progressiveness (based on individualism) and conservatism (group-think by coalitions), which does not make it easy to explain the functioning of Dutch industry.

7.3. The IMSS in the Netherlands

The Dutch contribution to the IMSS took place in Spring 1993: 115 questionnaires were posted; to 57% of the 203 ISIC 38 companies employing 200 people or more. 26 companies returned processable questionnaires (23%). All ISIC 38 industries were represented (Table 7-6).

Table 7-6. Division of the Dutch IMSS Sample by ISIC Codes 381-385.

ISIC 38	ISIC 381	ISIC 382	ISIC 383	ISIC 384	ISIC 385
26	9	4	3	6	4
100%	35%	15%	12%	23%	15%

The ISIC 38 industries account for 20.5% of total manufacturing turnover. The majority of the companies are SMEs; a very small number employ more than 500. The sample represents the bigger of the ISIC 38 companies. The average workforce of 545 is less than the 867 employees of the average IMSS company, but much larger than the average size of the Dutch ISIC 38 companies. About 37% of the respondents were employed in a division (15%) or a plant (of a larger company 22%).

The purpose of the subsequent sections is to present and discuss some of the more salient results of the survey, with reference, if appropriate, to the macro-economic, industrial and sociocultural aspects presented in the previous sections, as well as to some secondary data.

7.3.1. MARKET PERFORMANCE AND STRATEGY

Market performance
With an average market share of the dominant product or product line of 37%, the Dutch IMSS companies score higher than the 23% share of their leading competitors. This does not differ much from the IMSS average. Most companies expect their market to grow considerably. Whereas the last 5 years saw an average growth of 39%, production volume is expected to grow by as much as 91% over the next 5 years.

Product variety varies widely between the Dutch IMSS companies. Some companies produce as many as 5,000 different products, others a mere 30. On average, the product envelope encompasses some 930 active products. This is comparable to countries like Australia and Finland, higher than Denmark and the UK, but much lower than Japan, Norway and the US. As in many other countries, average product variety increased between 1989 and 1993, by some 15%, and for the next 5 years a further increase by 25% is expected.

Market strategy
An analysis of how Dutch ISIC 38 companies win orders[5] in the market place reveals an emphasis on customer service, product quality and delivery speed. The companies indicate that for the next two years they will give high priority to improving delivery reliability,[6] product quality, delivery speed, unit cost reduction and, finally, time-to-market (Table 7-7).[7]

It is not so much the order winners that are in the present or future top three that are interesting, but rather those that are *not* in the top three, namely:

---

5   The IMSS questionnaire used the term goals but expressed them in terms such as 'offering faster ...' or '... more dependable deliveries ...'or '... a wider product range ...' '... than our competitors'; this means that the questionnaire in fact asked for present and future order winning criteria.
6   This could also be regarded as a component of product quality.
7   Due to the structure of the IMSS questionnaire, no one-to-one match could be made between the present and the future goals; therefore, the data in the table should be interpreted with some caution.

Table 7-7. Present and Future Order Winners.

| Order winners | Degree of importance | | | | Quantified goal[4] |
	Present[1]	Rank	Future[2]	Rank[3]	
Customer service	4.44	1	—	—	—
Product quality	4.33	2	—	2	—
Improved conformance quality	—		4.25	—	83.3%
Improved supplier quality	—		3.71	—	72.7%
Delivery speed	4.00	3	—	3	—
Reduced manufacturing lead time	—		3.87	—	65.2%
Increased delivery speed	—		3.80	—	87.5%
Unit cost	3.67	4	4.00	4	88.5%
Reduced overhead cost	—		3.74	—	70.8%
Improved direct labour productivity	—		3.65	—	80.8%
Reduced materials cost	—		3.50	—	75.0%
Delivery reliability	3.52	5	4.14	1	75.0%
Product range	3.44	6	—	—	—
Time-to-market	—		—	5	—
Reduced new product development cycle	—		3.58	—	57.1%
Rapid design changes	—		3.29	—	31.8%

Notes:
1. On a scale from 1 = not important to 5 = very important
2. On a scale from 1 = low to 5 = high
3. —: not asked/not available
4. Percentage of companies saying they have quantified this goal

- 'having lower manufacturing costs than our competitors' and related issues, such as overhead cost, materials cost and labour productivity affecting product cost, score only fourth, both now and in the plans the companies have for the near future;
- 'offering a wider product range than our competitors' and related criteria such as having a short time-to-market through the rapid development of new or changed product designs, are ranked even lower, again both now and for the future.

The fact that product cost and product design are given little attention on the companies' goals seems to be at odds with other results of the survey as well as other studies conducted in the Netherlands. Subsequent sections will go more deeply into these issues.

7.3.2. PRODUCT CREATION PERFORMANCE AND STRATEGY

Performance
The Dutch IMSS companies appear to spend about 8% of turnover on R&D. This is not poor, compared with the IMSS average. However, other research (AWT, 1994) suggests that the pattern of R&D expenditure is changing. Large multinationals are moving their R&D abroad, in particular to Asia but also to North America. In effect, although SMEs are maintaining their R&D expenditure, the overall level of R&D spending is falling, though slowly. Some 29% of R&D expenditure is directed towards production, and

71% towards product innovation. In Japan these figures are just the opposite. In the US they are 16% and 84%, respectively (AWT, 1994). Anyway, given the situation that R&D expenditure is at best stagnating, it is rather surprising to find that companies expect not only the number of, but also the contribution to revenue by, new products to increase so much. There are two possible but not very satisfactory explanations.

Firstly, one could put forward the hypothesis that the effectiveness of the product creation process is high in the Netherlands and perhaps even increasing. However, the results of further analysis are ambiguous. For example, the number of product designs subject to engineering change orders is very low, 9.5% on average, the lowest percentage in the IMSS sample. On the other hand, for 17% of the orders delivered late, the reason was design changes, which is much higher than the IMSS mean (only Italy and Finland score worse). The question remains whether this is due to poor management of the product creation process, or the result of creating higher customer service by rewarding specific customer demands.

Secondly, an alternative explanation could be that companies will put a lot of effort into improving the product creation process. This would require that:

- time-to-market of new and changed products become important goals for the future; and
- R&D and also Engineering play a prominent role in the companies' operations strategy.

Table 7-7 clearly shows that the former is not the case.; unfortunately, this also applies to the latter. As Table 7-8 shows, the use of formal design/engineering technologies and methods is relatively poor. CAD systems, for example, are not wide-spread (only Argentina, Chile, Mexico and Portugal make less use of CAD) and then their application is mostly limited to drawing and some calculative functions. Also as regards the use of methods like DfA/DfM, QFD and VA, the Netherlands are below the IMSS average. The situation with respect to the use of simultaneous engineering is even

Table 7-8. The Present and Planned Use of Design/Engineering Technologies.

Technology in design and engineering	Currently used[1]		Adopted within two years[2]	
	Netherlands	IMSS	Netherlands	IMSS
Computer-Aided Design	3.35	3.60	11.5	15.3
Design for Assembly/Design for Manufacturing	2.42	2.49	11.5	16.1
Quality Function Deployment	2.33	2.56	11.5	15.5
Value Analysis	2.67	2.64	11.5	13.9
Simultaneous Engineering	2.09	2.56	11.5	15.5

Notes:
1. On a scale from 1 = no use to 5 = high use
2. Percentage of companies indicating that they plan to implement the technology within the next two years

Table 7-9. Co-ordination between Design and Manufacturing.

Co-ordination between design and manufacturing	Currently used[1]	
	Netherlands	IMSS
Rules and standards	22.5	25.1
Formal meetings	25.0	20.7
Informal meetings	15.0	14.8
Cross-functional task forces	22.5	27.5
Personal contacts	15.0	10.8

Notes:
1. On a scale from 1 = no use to 5 = high use

worse, with only Argentina, Chile and Finland scoring lower.[8] The greatest point of concern, however, is that in terms of plans for adopting design and engineering technology, the Netherlands scores lower than most other countries on each and every individual technology mentioned in Table 7-8.

The mixture of different mechanisms to achieve co-ordination between design and manufacturing is a bit puzzling, too. A high content of formal (46% by rules and formal meetings) is combined with considerable use of informal (26% by informal meetings, personal contact) arrangements and task forces (28%) (Table 7-9).

Conclusion

Given the ambition of the companies, a lot needs to be done in the area of product design and engineering. Unfortunately, Dutch industry hardly seems to realise this. Not only are time-to-market and its constituents (new product development cycle and rapid design changes) ranked too low in the list of future goals, industry also fails to put sufficient effort into up-dating the design and engineering functions. It appears that Dutch companies do not sufficiently understand that expressing expectations is one thing, but setting (strategic) objectives and creating the conditions to achieve those objectives is another.

7.3.3. MANUFACTURING PERFORMANCE AND STRATEGY

Performance

On average, ROI of the Dutch IMSS companies reached a level of 23.3% in 1992, which is high when compared with both manufacturing in general and the whole IMSS sample as well. A breakdown of manufacturing costs in comparison with Japan, the US and the other countries suggests that direct wages in the Netherlands are relatively high compared with the IMSS sample as a whole, whereas overhead costs are relatively low (Table 7-10).

Labour costs are not only relatively high but also high in an absolute sense. According to A.T. Kearney and Knight Wendling (1994):

---

[8]   Recent evidence, however, suggests that Concurrent Engineering is rapidly becoming a 'hot issue'

Table 7-10. Relative Cost of Direct Labour, Direct Material and Overhead in the Netherlands and Elsewhere.

	Direct wages	Direct material	Overhead
Netherlands	29.2	47.6	23.2
Japan	14.6	64.4	21.0
US	15.9	53.6	30.5
Total sample	20.3	54.6	25.1

- the amount of hours worked in the Netherlands is 97% of the American and 79% of the German amount, respectively, and only 70% of the hours worked in Japan;
- labour productivity of Dutch industry in general is quite high, but productivity improvement is at the same level as in the US, and lower than in Japan, Germany, the UK and France;
- in the ISIC 38 industry, only Germany is more expensive than the Netherlands in terms of labour costs per hour (9% in the metal and steel products sector and 18% in the machine tool industry); labour in Japan and the US is considerably cheaper (between 7% in the American machine tool industry and 36% in the Japanese electronics industry);
- labour productivity in the ISIC 38 industry is already lower than in the US, Japan, Germany, and France.

In addition, the same report also concludes that the top five competitive weapons of the Dutch manufacturing industry are product quality, delivery reliability, customer relations, company image, and flexibility, with product cost ranked only 11th.

So, one would expect manufacturing cost reduction, through reductions in overhead and materials cost and an increase in labour productivity, to play a prominent role in the companies' manufacturing strategy. However, this is not the case, that is, not as far as the IMSS companies are concerned, as Table 7-7 clearly shows. Interestingly, this seems to contradict the results of a survey reported in A.T. Kearney and Knight Wendling (1994). This survey of 39 mostly ISIC 38 companies not surprisingly revealed that unit cost reduction was mentioned by nearly 60% of the companies as the most important goal in order to improve performance in the market place, followed by productivity improvement through improved quality of business processes (50%), improved delivery reliability and speed (35%), and time-to-market (25%).

Strategy
In an attempt to explain why the Netherlands is lagging behind better-performing and faster-improving countries, and, at the same time, to indicate directions for achieving the above goals, A.T. Kearney and Knight Wendling (1994) conclude that Dutch companies do not invest enough in R&D, especially in R&D aimed at developing new production processes and technologies. This is confirmed by the IMSS survey (see

previous section). Another finding is Dutch companies fail to invest sufficiently in training and education. The IMSS data support this conclusion as well. Although the companies spend 2% of turnover on training and education, which is average in the IMSS sample, the training given to new production workers (72 hrs.) and the regular workforce (27 hrs.) is rather lower than the IMSS averages (110 hrs. and 36 hrs., respectively). The result: 37% of the workforce are multiskilled, versus 46% for the whole sample.

Furthermore, the report concludes that Dutch industry:

- tends to adopt modern production methods and forms of organisation (e.g., lean production and just-in-time manufacturing) too slowly and then only partially;
- is relatively slow in adopting advanced manufacturing technology (e.g., robotics and CNC);
- should, more generally formulated, put much more effort into continuous improvement, using both 'hard' technology and 'soft' production methods and organisational concepts.

The IMSS survey confirms only part of this. Approximately 11% of turnover is spent on new process equipment. This is similar to the 10% for the whole IMSS sample. And, as Table 7-11 shows, the Dutch IMSS companies are doing rather well in terms of the current and planned use of incremental improvement programmes like *kaizen* or Total Preventative Maintenance. Also, strategic thinking on manufacturing will be strengthened in the immediate future.

Table 7-11. The Present and Future Use of Software Technologies and Forms of Organisation in Manufacturing and Assembly

Software technology and organisation in manufacturing	Currently used[1]		Adopted within two years[2]	
	Netherlands	IMSS	Netherlands	IMSS
Total Quality Management	3.10	2.95	41.0	34.2
Statistical Process Control	2.61	2.79	22.2	22.4
ISO 9000	3.82	3.25	52.0	40.5
MRP	3.88	3.12	14.9	10.8
MRPII	2.65	2.58	11.1	12.4
Just-in-time manufacturing	3.44	2.92	33.3	25.8
Pull scheduling	2.53	2.22	14.8	17.4
SMED	1.83	1.99	11.1	14.2
Just-in-time delivery	3.50	2.75	29.6	18.9
CAM	2.66	2.70	7.4	12.6
Plant-within-a-plant	2.25	2.79	14.8	17.4
Team approach	3.41	3.27	3.7	24.7
Kaizen/Continuous Improvement	3.00	2.74	22.2	23.7
Total Preventative Maintenance	2.29	2.18	22.2	17.1
Defining a manufacturing strategy	2.86	3.15	29.6	17.1

Notes:
1. On a scale from 1 = no use to 5 = high use.
2. Percentage of companies indicating that they plan to implement the technology/form of organisation within the next two years.

The same holds for the present or planned adoption of (software) technologies and organisational solutions supporting the management of quality and logistics. TQM is spread widely and its popularity is even growing. This is mainly due to what some people like to call 'ISO-mania', the adoption of ISO 9000 imposed upon many SMEs by their large industrial customers. As for the use of logistical concepts, it should be noted that there are only a few companies that can afford a full-blown MRP or MRPII system, and the number of OPT implementations is even smaller. The adoption of such systems is concentrated in some of the larger companies. Small companies do not use MRP or even one or two modules. JIT manufacturing is mostly limited to a few large batch and mass manufacturers and assemblers. However, the ideas behind MRP, OPT and JIT are spread much wider.

The Netherlands does not have a history of mass production. Among the few exceptions are companies like Philips and, to a certain extent DAF and also Scania (trucks) and, more recently Nedcar (car manufacturing; a joint venture of Volvo, Mitsubishi and the Dutch state). The vast majority of Dutch manufacturing companies are SMEs, and the majority of these companies use batch production as a dominant mode of fabrication and assembly (Table 7-12). This seems to reflect the variety of market segments and products, but has certain disadvantages, in particular with respect to inventory levels and possible also delivery speed. Broadly speaking, the solution is focused operations, teamwork and other forms of organisational and also technological integration (see the next section).

Even more surprising is the limited role of self-managed teams in the companies' manufacturing strategy. This is difficult to explain, considering the current use of teams, future plans as regards for example Total Quality Management, just-in-time manufacturing and the incremental improvement programmes, in which teams play an important role, and more generally the widely held belief that self-managed teams are

Table 7-12. Process Choice in Manufacturing and Assembly (% of Companies).

	Netherlands	Japan	US	Total
Manufacturing				
One-off process	6.7	6.5	23.3	15.6
Batch	73.3	29.0	50.0	53.6
Line	20.0	64.5	26.7	30.8
	100.0	100.0	100.0	100.0
Cellular lay out	62.2	40.8	56.4	55.4
Assembly				
One-off process	28.0	5.6	17.9	23.2
Batch	32.0	5.6	28.6	33.7
Line	40.0	88.9	53.6	43.0
	100.0	100.0	100.0	100.0
Cellular lay out	54.2	50.0	61.2	61.0

Table 7-13. The use of Hardware Technologies in Manufacturing and Assembly .

Hardware technology in manufacturing	Average number of machines/systems used in Dutch companies, compared with companies in[1]			
	Germany	US	Japan	IMSS
(Computer) Numerical Control	1.2	0.9	0.3	0.7
Machining Centres	0.7	0.4	0.6	0.6
Manufacturing Robots	0.9	0.9	0.1	0.4
Assembly Robots	0.4	0.1	0.1	0.2
Flexible Manufacturing Systems	1.0	1.9	0.4	1.2
Flexible Assembly Systems	1.6	4.6	1.9	1.7

Notes:
1. The relative number of machines/systems, corrected for company size (number of employees); read: on average, Dutch companies have 20% more CNC tools and 60% fewer assembly robots than German companies.

one of the key building blocks of state-of-the-art, that is, much more horizontal than hierarchical organisations.

As for the use of advanced manufacturing technologies, the IMSS survey confirms the conclusion drawn in the A.T. Kearney and Knight Wendling report. In absolute figures, the more complex and integrated the technology, the fewer companies that have adopted them. The use of CNC equipment, for example, is much wider-spread than that of FMS. This is probably related to the relative small scale of Dutch companies and the expense and flexibility of the technology itself. CNC machining centres are more flexible than FMSes. Consequently, the few FMSes that are found in the Netherlands are found in larger companies, such as DAF, Fokker, and some of the larger SMEs (Stork, Grasso), although there are some exceptions. Yet, as Table 7-13 shows, the Netherlands lag behind in the adoption of stand-alone applications of micro-electronics (CNC, machining centres, and manufacturing and assembly robots). The rate of adoption of systems of equipment is much more average (flexible manufacturing systems) or even high (flexible assembly systems).

Conclusion

The Dutch IMSS companies emphasise customer service, product quality, delivery speed and delivery reliability as the most important goals for today and the near future. Most companies seem, or would like, to compete on the basis of a differentiation strategy rather than cost effectiveness. However, their product creation strategy is not consistent with this ambition.

Furthermore, product costs are high, mostly due to high relative and absolute labour costs, especially in manufacturing. And, in terms of productivity improvement, Dutch industry is lagging behind several other industrialised countries. According to the A.T. Kearney and Knight Wendling (1994) survey, cost reduction and productivity improvement are high on the industrial agenda. However, this is not confirmed by the IMSS data.

In order to keep pace with international competition, Dutch industry appears to emphasise software technology, mainly to support quality and logistics management. This is consistent with the intentions expressed by the Dutch IMSS companies. It is less clear though, how this contributes to the required cost reductions and productivity improvement. In this respect, the intended use of self-managed teams is surprisingly low, as is the use of state-of-the-art design and engineering software and manufacturing and assembly hardware.

7.3.4. LOGISTICS PERFORMANCE AND STRATEGY

External logistics

Performance. Due to its geographical position, the Netherlands has traditionally been a country of traders. As noted before, Germany is the most important trade partner, accounting for 29% of exports and 23% of imports. Other important trading partners are Belgium and Luxembourg, with 13% of exports and 12% of imports. These figures explain a great deal of the effort the Netherlands puts into improving its infrastructure and positioning itself as 'Nederland Distributieland', the Netherlands as a nodal point in European logistics.

Strategy. For example, over the decades Schiphol International Airport in particular has become an important thoroughfare to mainland Europe airports. Elected the world's best airport in 1995, its present strategy is to become one of Europe's top three hub-and-spoke airports. Another recent example is the decision to build a new rail link between Rotterdam and Germany. This will add to Rotterdam's inbound and outbound capacity and also its portfolio of transportation modes, which has been dominated by inland shipping (the river Rhine). A third effort worth mentioning is the high speed railway that will be built to link Amsterdam and Rotterdam to the European high speed network. In addition, many other infrastructural initiatives are being carried out, all of which are aimed at further improving the ship, road, railway and air connections with Germany and the rest of Europe. And indeed a whole range of new logistical service centres are being established along the motor ways to Germany and Belgium. All these initiatives enhance the increasing emphasis on JIT manufacturing and delivery (see Table 7-11).

Finally, there is a tendency towards more concentration of, and collaboration between, road haulage companies, not so much geographically, but rather in terms of company size. Some companies are trying to 'buy' market share through a process of take-overs and mergers. Others prefer to join forces, even with competitors, in order to be able to escape the fierce price competition by bringing their individual, generally small, family-owned companies into a network offering door-to-door, one-stop logistics. Obviously, new computer-based communication and information technology (e.g., Electronic Data Interchange) plays an enabling role here.

Internal logistics

Performance. The so-called customer-order-decoupling-point (CODP) is increasingly regarded as an important issue when considering the internal logistics of a company. The CODP decouples the upstream, forecast-driven part of the primary process from the downstream part which is driven by customer orders. A distinction can be made between engineer (or develop or even design) to order, manufacture to order, assemble to order, and produce to forecast. It is often assumed that the majority of companies produce to forecast, that is, deliver from stock. However, the IMSS companies show a somewhat different picture (Table 7-14).

The position of the CODP is a direct reflection of a company's flexibility. That is, if the CODP is in engineering, the product and modification flexibility must be high and the factory must obviously be flexible, too, in terms of both process flexibility and possibly volume flexibility as well. If the CODP is in manufacturing or assembly, (manufacturing or assembly) process and possibly also volume flexibility must be high.

Strategy. There are several ways in which these flexibilities can be achieved. One approach is to create delivery and volume flexibility by building slack into the company's operations through holding stock and/or long customer order lead times. Table 7-15 suggests that Dutch companies prefer this approach. Compared with companies in other countries, stocks expressed in days of inventory are high, and lead times long. One reason is that suppliers have sufficient bargaining power over their mostly small customers to force them to accept larger order quantities (of raw materials stock) than they would strictly need. Another is that the majority of companies use batch production as the dominant mode of production. This has certain advantages but disadvantages include high work-in-progress, high finished goods inventories and long manufacturing lead times. Customer order lead times need not be long if products can be delivered from stock.

Table 7-14. Location of the Customer Order Decoupling Point (CODP)

(average % of forecast and customer-driven production orders).

	Netherlands	Japan	US	Total
Forecast orders	29.4	30.7	41.7	34.6
Customer orders	68.3	69.3	61.2	66.5

Table 7-15. Days of Inventory

	Netherlands	Japan	US	Total
Raw material	39.4	10.9	24.8	33.1
In-process	42.7	12.1	26.3	24.3
Finished goods	54.5	20.5	21.5	20.9
Lead time	63.1	32.5	41.8	54.7

An alternative approach is to increase the flexibility and speed of the whole primary process, from design and engineering to manufacturing and assembly, by integrating these functions as much as possible. This can be achieved by adopting a suitable mix of hardware and software technologies, including those mentioned in Table 7-8, Table 7-11 and Table 7-13. These technologies contribute directly to the flexibility and speed of operations and also provide the building blocks for technological integration between functional departments. In design and engineering, simultaneous engineering and underlying methods such as design for manufacturing and assembly, and quality function deployment have the same integrating effect. Organisational solutions include self-managed teams and focused factories or plant-within-plants. The discussion in previous sections on the present performance of, and the future strategy for, product creation and manufacturing has indicated that Dutch companies still have some work to do as regards the flexibility in these areas. This is becoming ever more urgent, as many Dutch companies are increasing their active product base and, hence, their product variety. This creates a tendency to move the CODP upstream. In effect, however, stocks will increase even further, unless the flexibility of the organisation is increased as well.

A final interesting and partly logistics-related development concerns the changing relationship between customers and suppliers. Compared with other countries, companies in the Netherlands have a relatively small supplier base and the number of suppliers of raw materials and components will decrease even further. Three trends are visible (Table 7-16).

Firstly, different studies in the Netherlands have concluded that there is a necessity for, and indeed a trend towards, 'back-to-core-business'. This means that ever more companies and in particular the larger ones, are in a process of outsourcing activities, in particular support activities and also the less strategic primary activities. Furthermore, a growing number of companies are collaborating in strategic alliances. As noted before, it is not only the large companies that are looking for and finding partners (sometimes abroad, sometimes even competitors) for example for R&D joint ventures; very small, family-owned transport companies are also collaborating in networks as the only way to escape the 'price-battle' between these companies and to increase and complete their product portfolio. Finally, the number of suppliers providing the manufacturing and assembly processes with materials and components is decreasing (Draaijer 1993). This has mainly to do with establishing co-design and co-logistics relationships with a limited number of very reliable suppliers in order to make the operational processes more effective and efficient.

Table 7-16. Number of Suppliers and Closeness to Suppliers

	Netherlands	Japan	US	Total
Average	201.2	412	1718	469.5
Past five year change (%)	24.4	54.9	-1.9	4.2
Next five year change (%)	24.6	43.5	-4.0	-5.1
Closeness	66.2	62.2	72.2	66.4

Conclusion

The Netherlands is fairly active in facilitating physical distribution activities within the country itself and with other countries. This is necessary in order to let its open economy function as efficiently as possible. Due to the high share of batch production in industrial processes and the bargaining power of suppliers, the stock-positions held in Dutch companies are rather high compared with other countries. It is therefore also noteworthy that Dutch companies are very active in restructuring their supplier base in order to go back to core business and to establish a reliable supplier base that can used for co-design and co-logistics, thus enabling lower stock-positions and higher quality products.

7.4. Conclusion and perspective

Over the past decades, the emphasis in the operations strategy of Dutch companies has changed. During the late 1960s and the 1970s, work experiments and semi-autonomous groups were very popular. TQM was discovered in the late 1970s, really took off in the 1980s, and is still seen as one of the cornerstones in the competitive battle. In logistics, software packages based on MRP I and II have sold well since the 'MRP crusade' reached the Netherlands in the mid 1970s. However, full-blown implementations are very rare and most companies are using some modules combined with proprietary software for other functionalities.

Factory automation started with the adoption of NC equipment, followed later by CNC machining tools and centres, robots and other types of handling equipment, and combinations of all such pieces of equipment in FMS and FAS. However, the Netherlands are still lagging behind in terms of the adoption rate of AMT. This may partly be due to the lessons learnt by early adopters that technological innovation through automation needs to be integrated with social and organisational innovation in order to be successful (e.g., Boer 1991). Possibly stimulated by previous experience with quality circles in the 1980s, this led to the re-discovery and renewed popularity of semi-autonomous groups and also cellular manufacturing.

An emerging, but not yet widely spread interest, hence not visible in the IMSS data, is in concurrent engineering aimed at the design of products that meet customer expectations and can be manufactured, assembled and distributed easily, quickly and cost effectively. Based on experience with TQM, but much more recent as far as the Netherlands is concerned, is the growing number of attempts to try continuous improvement. Interestingly, Business Process Redesign is enjoying some interest as well, however mostly at seminars rather than in practical application. This is probably a reflection of the cautious nature of the Dutch and their preference for incremental, stepwise change rather than radical innovation.

Technologies, organisational concepts and approaches towards change are subject to fashion. However, whatever the current fashion is, more Dutch companies should realise that production is the key process and market orientation the key performance criterion. Key words in the required migration to market-oriented production are integration and consistency. Internal organisational and technological integration of the whole internal value-adding chain, from marketing, through product and process design, manufacturing, assembly, (again) marketing, sales, to distribution. External integration, also technologically, through closer relationships with suppliers and customers, based on trust and a pursuit of win-win situations. Internal consistency, by aligning all technological and organisational design decisions properly. And external consistency by ensuring that whatever the decisions made, they contribute to the company's competitiveness in the market place. The IMSS results suggest that Dutch industry is well underway, but there is still a long way to go.

References

A.T. Kearney en Knight Wendling (1994) Produceren in Nederland. Aanpak om de produktieperformance van de Nederlandse maakindustrie te verbeteren, July.

AWT (Adviesraad voor het Wetenschaps- en Technologiebeleid - Council for Scientific and Technological Policy) (1994) *Technologie en Economische Structuur*, 's Gravenhage, April.

Boer, H. (1991) *Organising Innovative Manufacturing Systems*, Gower/Avebury, Aldershot.

CBS (Centraal Bureau voor de Statistiek - Central Statistical Office) (1995) *Statistisch Jaarboek*.

Draaijer, D.J. (1993) *Market-oriented manufacturing systems. Theory and practice*, PhD thesis, University of Twente, Enschede.

EZ (Ministerie van Economische Zaken - Ministry of Economic Affairs) (1995) *Kennis in beweging. Over kennis en kunde in de Nederlandse economie*, 's Gravenhage, June.

Hofstede, G. (1980) *Culture's Consequences - International Differences in Work-Related Values*, Sage Publications, Beverly Hills.

Lawrence, P. (1991) *Management in the Netherlands*, Clarendon Press, Oxford.

Rabobank (1994) *Cijfers & Trends. 75 visies op het Nederlandse bedrijfsleven*.1994/1995, 18.

CHAPTER 8

ITALIAN ASSEMBLY INDUSTRY: CHALLENGES AND RESPONSES TO
GLOBALISATION AND INNOVATION

*Emilio Bartezzaghi and Gianluca Spina, Politecnico di Milano, Dipartimento di
Economia e Produzione, Milano, Italy*

8.1. Introduction

Since the beginning of the 1990s, both global and country-specific trends have put
pressure on the Italian economy, the fifth or sixth largest in the world. The process of
the internationalisation of the economy and integration into Europe, though jerkily in
progress, presents challenges to the industrial sector, the first to feel the impact of these
trends. In combination with macroeconomic, political, and sociocultural changes, the
above mega-trends present opportunities for Italy, but also loom threateningly over the
prospects for the long-term prosperity of the country and its ability to keep up with the
more advanced industrialised economies.

This chapter outlines how the Italian manufacturing industry is facing internal and
external challenges in the light of recent history. Strategic drivers and patterns of
innovation are also explored to show how Italian manufacturers are approaching and
implementing some paradigms of change that are emerging world-wide in the way
production activities are organised and managed.[1] In this chapter, challenges and
responses, threats and opportunities, and strengths and weaknesses of Italian
manufacturers are described. This should be of interest to foreign competitors, foreign
direct investors in Italy, companies increasingly involved in global competition
generally, and anyone interested in the prospects for Italian manufacturing systems,
beyond the myths and the commonplace, in the light of country-specific factors and
global trends.

The IMSS Italian sample is the primary source for the analysis, but further evidence
is used from other surveys of manufacturing strategy and 'best practices' in Italy, in
order to detect evolutionary trends and to provide further insights on the small company
environment. Two case studies of large multinational corporations also provide an

---

[1] See also Chapter 26, which describes the rise of a new manufacturing paradigm world-wide, based on
(1) multifocusedness of manufacturing systems and strategic flexibility; (2) process focus and
integration across functions; and (3) process ownership, that is, delegation and involvement of the
work-force and empowerment through training and increased knowledge of the processes. In this
chapter, we describe and comment on the positioning of Italy with respect to these concepts, which are
described in more detail in the following chapter.

P. Lindberg et al. (eds.), International Manufacturing Strategies, 145-161.
© *1998 Kluwer Academic Publishers. Printed in the Netherlands.*

additional perspective on the strengths and weaknesses of the manufacturing sector in Italy.

8.2. General Background

8.2.1. MACROECONOMICS[2]

Over the past decade, great changes have occurred in the Italian economy. The steady growth of the GNP at an average yearly rate of 2.7% from 1983 to 1988 brought the country out of the recession of the 1970s and into the club of the most industrialised nations. Industrial production in particular grew during that period at a yearly rate of 4.5%. The expansion of domestic demand supported this growth until 1989, when Italian economic performance plummeted due to the international recession. The crisis peaked in 1993, with negative growth rates for both GNP and industrial production, but in 1994 the recovery started, with further consolidation in 1995 strongly supported by booming exports, whilst internal demand remained stagnant.

The boom in exports can be traced back to the huge currency depreciation—40% against the Deutschemark—that began in 1992, largely because the financial markets perceived political uncertainty and were frightened by the dramatic increase in national debt. The national debt had increased since the 1980s to the point where it exceeded GNP at the beginning of the 1990s, thus jeopardising the process of monetary integration into the EU. Recent intervention by the government to balance public accounts has resulted in a period of low interest rates, but the political crisis has hampered the launch of a long-term effort to bring national debt within the limits fixed in Maastricht. On the other hand, the battle against inflation has had positive results, and over the past three years inflation has ranged between 4% and 5%, a dramatic decrease from the double-digit inflation of the early 1980s. Important wage settlements between the unions, industrial associations, and the government have also contributed to controlling inflation.

Unfortunately, regional differences within the country have been on the increase. Northwest Italy, which has the longest history of industrial development, has suffered through the recent global recession, but has bright prospects for recovery; Northeast Italy and the North Adriatic coastal regions developed more recently, during the 1950s and 1960s, but are now experiencing fast growth and booming exports; on the other hand, Southern Italy suffers high unemployment and underdevelopment in industrial activity and infrastructure, notwithstanding the huge amounts of money that have been invested in the region by the government during the post-war era. Today, some districts in Lombardy and the Northeast rank among the most developed and richest areas in

---

[2]   Most of the data are taken from the annual reports of the Bank of Italy over the past decade.

Europe, whilst some in Southern Italy are among the poorest, with a per capital GNP half that of the North.

8.2.2. INSTITUTIONAL CRISIS

At the beginning of the 1990s, the political coalition that had ruled over Italy for forty-five years began to fall apart in the wake of the revelation of scandals and corruption. The political system has been transformed from a proportional electoral system to a majority one, although the transition is not yet complete and the country still suffers from political uncertainty. Strong pressures for federalism, decentralisation, and regional autonomy have begun in the North, but at the moment no reform towards federalism has yet taken place. The crisis of the welfare system, the burden of the national debt, and the incomplete institutional reforms have hampered the emergence of a stable government, one that is able to formulate a long-term strategic plan.

With many notable exceptions, productivity and customer orientation in the public sector is markedly lower than in private industry, and is not up to the standards of the most advanced industrialised countries. Apart from its negative effects on the domestic scene, public services and the underdeveloped infrastructure burden the industrial system, since they create additional costs, time and inflexibility for companies.

8.2.3. SOCIO-CULTURAL TRENDS

Italy clearly differs from other countries in a number of key sociocultural aspects. A pervasive entrepreneurial attitude and strong individualism have survived and dominated economic change during the post-war period. Conversely, organising is difficult in Italy. Italy is a country of thousands of parishes, where it is hard to design big projects for the public good and even harder to implement them successfully, despite the centralisation of political power. The dominant individualistic approach has strong implications for the business environment, since it favours fragmentation and small companies. It hinders the development of a managerial culture and broad managerial class separate from the owners, which is happening even so, and it does not favour the delegation and involvement of workers in decision-making at any level inside companies.

Industrial relations have been characterised by arm's length approaches, with even sharp conflicts in the 1970s. Unions have retreated over the last decade, after losing consensus within the factories, and a heated debate is now taking place within the national federated unions as to whether to renew the adversarial approach or to take a co-operative stance in the private sector.

8.2.4. INDUSTRIAL STRUCTURE AND TRENDS

Even though Italy has the fifth largest industrial system in the world, it is extremely fragmented. Thousands of small companies form the framework of the industrial system; there are few large groups and few medium-sized enterprises compared with the most advanced economies. This also reflects an industry bias in the industrial system. In many 'supplier-specialised' and 'supplier-dominated' industries such as OEM and textiles, Italian manufacturers are export-oriented and often market or at least niche leaders. In 'scale-intensive' and 'science-based' industries, Italy is relatively weaker than other European countries, with some exceptions such as the automotive industry. The mechanical, textile, and some other industries have typically concentrated their production regionally. These industrial districts have their pluses and minuses, but these different sorts of networks have helped small companies to face increasingly complex and internationalised markets.

The ownership structure of Italian industry has two main characteristics: (1) most companies are owned by individuals or families; and (2) a large part of Italian industry is state-owned, especially in the South and in scale-intensive industries such as steel and oil. Family ownership is linked to the average small size of companies, the underdevelopment of the stock exchange, and the low reliance of company growth on expansion of the ownership base. Italian industry tends to be family or state-owned, with little space for public and managerial capitalism. However, this state of affairs is changing fast. For example, the process of wide-spread privatisation has recently been initiated, although it has created hundreds of financial and political problems, and many sectors are rapidly becoming more concentrated.

Traditionally, Italian industry has been less internationalised than the most advanced economies. Italian industry has tended to be oriented towards exports rather than foreign production, which is also influenced by the large number of small companies. The heavy currency depreciation has supported the export trade, and in 1993 Italy reached a positive balance of foreign payments (+1.2% of GNP), with a further increase to +3.5% in 1994. In addition, foreign direct investment by Italian companies has been lower than direct investment in Italy from abroad. However, Italy has begun to bridge the gap in international production. A number of medium-sized Italian companies have recently begun to internationalise their manufacturing base: the number of employees in Italian-owned manufacturers abroad has increased by 148% since 1986 (see Mariotti, 1994).

Finally, manufacturing has had a limited role in corporate and business strategy in Italy. Production and operations management has had little influence compared with marketing and financial management. Most of the top managers in Italian companies, even in high-tech industries, come from sales or finance. In addition, the number of graduate engineers is lower than abroad, though their quality is generally high. Even so, the management of production, operations, and technology is gaining momentum, as many companies have begun to rediscover manufacturing as a source of competitive advantage and have started a number of programmes to implement innovative manufacturing activities.

As we have shown above, the macroeconomic and sociocultural aspects of Italy and the industrial structure have been influenced by historical and country-specific factors, but the situation is evolving in the 1990s. These evolutionary trends present many challenges to the Italian industrial system, since new opportunities emerge and old and new threats now impend. In summary, the growth of industrial production presents opportunities for innovation and improved global competitiveness, but problems with the institutional level, political system, and public finance make the renewal of the industrial system more difficult and uncertain. In turn, the growth and health of the economy and the industrial sector seem to be necessary for solving the financial and institutional crisis.

8.3. The challenges in more detail

Italian manufacturers now face various challenges to improving their global competitiveness. The Italian assembly sector has traditionally been known for niche players, mainly small- and medium-sized enterprises (SMEs), who emphasise customisation, flexibility, and low cost, and offer high design quality, but sometime suffer from low levels of manufacturing quality. Up to now, their ability to seize volatile market opportunities in the international marketplace has not been fully supported by management and operations on an international basis. On the other hand, the few large companies have not performed with much flexibility or customisation, instead emphasising volume and cost-based competition, whilst they generally have only been able to reach levels of quality conformance well below those of the best foreign competitors. The challenge for SMEs is to maintain flexibility and customisation, and to develop their management and internationalisation. Large companies should move from bureaucratic organisation to leaner and more agile structures; they should maintain efficiency but improve quality and customer service substantially.

The strategic challenges for all companies have strong implications for manufacturing. To some extent, they have to do with the emergence of a new manufacturing paradigm (see Chapter 26 for details and operationalisation), which has arisen world-wide and driven companies away from the principles of Fordism. First, a multi-focused approach to manufacturing strategy is required, because focusing on single goals (generally regarded as antithetical, for example quality and cost) is no longer effective. As a consequence, the challenges in particular for large companies are to design multi-focused and strategically flexible manufacturing systems and to move from the management of trade-offs to shifting trade-offs.

Second, a resolute process focus is required, especially concerning those processes involved in the value chain. More process integration should be pursued at various levels: across internal functions, with both customers and suppliers, and between manufacturing and the business strategy level. The past emphasis on functional optimisation should be abandoned in favour of corporate redesign, leveraged by the concepts of operating continuity and process integrity across organisational barriers.

Third, process ownership is needed. This aims at involving employees at every hierarchical level in decision-making and problem solving. Delegation of responsibility, employee involvement, and knowledge of the process are embodied in this concept. The ultimate goal is to develop at least some degree of local problem-solving capability, in order to detect and resolve process anomalies as soon as possible, and to avoid time-consuming hierarchical referral. The third element appears to be crucial in Italy, and it is essential for both the large companies, who need leaner and more agile organisations, and the SMEs, for whom international and managerial growth is difficult without delegation, involvement, and empowerment of the workforce.

8.4. The responses

8.4.1. STRATEGIC DRIVERS AND PERFORMANCE

In this section, data from the IMSS, other surveys, and empirical evidence from case studies are used to figure out (1) how Italian companies compete in response to the macro-trends, both domestic and international, (2) how they perform, and (3) what the implications and drivers are at the manufacturing level.

'Quality first, cost second, and service third' is the ranking of business unit goals and competitive priorities for Italian manufacturers at the beginning of the 1990s, as revealed by the IMSS. Quality is a must, regardless of industry or company size, and it is such an important goal that the scope for differentiation on the basis of superior quality seems to be limited. 'Having lower manufacturing costs than competitors' is ranked next, though there are differences across industry segments. Goals related to service have been given a significantly lower priority than quality and cost, with two exceptions. The transport equipment segment, mainly automotive component suppliers, emphasises time-based competition, which has not received much attention in other segments. Although the IMSS does not provide longitudinal data, a previous survey (Bartezzaghi et al., 1992) at the end of the 1980s detected a quality-service-cost ranking in a sample of 173 Italian manufacturers, mainly in the assembly industry. On the whole, time-based competition seems to have been overtaken by the new emphasis on cost reduction.

How does the importance placed on various strategic and manufacturing drivers in Italy compare with those of the most advanced economies? The matrix in Figure 8-1 allows us to simultaneously compare the strategic drivers and the manufacturing goals of the Italian companies on an international basis. Quality and cost are the foundations of global competition, and Italy is not an exception. Since the Italian response lies in the top right hand quadrant of the matrix (high importance at both business and manufacturing level) for all of the areas, manufacturing goals and competitive priorities seem to be aligned. It is noteworthy that quality dominates cost at the level of business unit goals, whilst they have approximately equal rankings at the competitive priority level in nearly all countries. Italian companies have noticeable difference in the ranking of the service-related drivers. Delivery speed has been ranked much higher by foreign companies, where it is an order winner, than in Italy. On the other hand, product variety is ranked in Italy as being more important as a business unit goal than elsewhere, but receives an average ranking as a manufacturing goal.

Italian small and medium-sized companies
However, relevant differences in goals and competitive priorities emerge when comparing the SMEs (fewer than 500 employees) and large companies. The smaller Italian companies in the IMSS sample ranked improving delivery speed second only to quality. Manufacturing priorities related to cost were ranked lower than volume flexibility and design changes. In addition, the SMEs have a broader product/market focus than foreign SMEs. In fact, they cover more market segments, focus on more customers and operate more internationally. They also show higher performance in customer service, since they are more reliable in product delivery.[3]

Figure 8-1. Competitive Priorities and Manufacturing Goals in the Global Assembly Industry

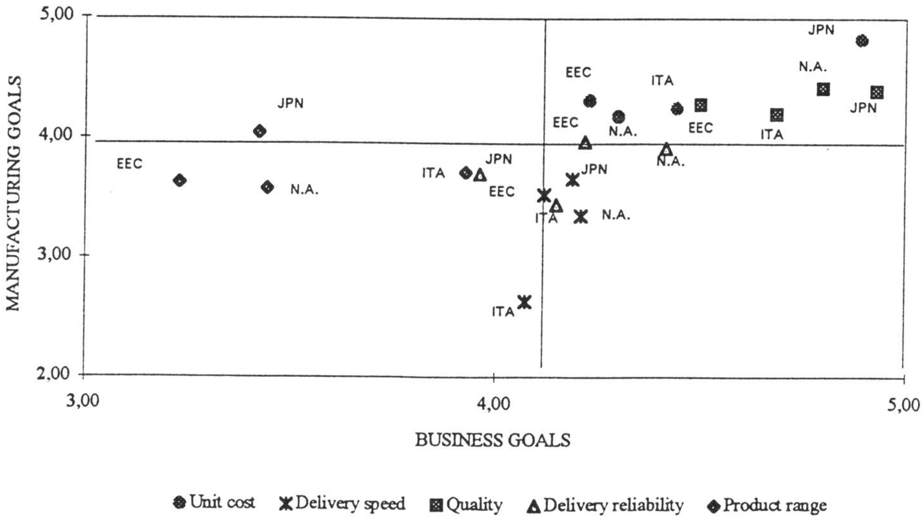

---

[3]   Italian SMEs report 3.9% delivery delays out of total deliveries, compared with 11.3% of foreign SMEs and 15.4% of Italian large companies.

To investigate the flexibility of Italian manufacturers, we merged data from the IMSS database, which includes a number of medium-sized companies, with additional data from a sample within the ISIC 38 segment of very small manufacturers (fewer than 100 employees) from the East Lombardy industrial district, an area that has been widely studied by researchers. The evidence from the East Lombardy district supports the findings from the IMSS. In fact, the current drivers are quality and delivery dependability, followed by wider product range and delivery speed, whilst cost is much less crucial. At the manufacturing level, this order—quality plus time and then cost—is substantially the same.

On the whole, both the IMSS data and the complementary analysis of some typical Italian small manufacturers confirm that Italian SMEs seek to be flexible and seize volatile market opportunities. They still focus on flexibility and generally provide delivery dependability higher than foreign small competitors and national large ones.

Italian large companies
Looking at the Italian larger companies, it appears that they focus on cost reduction and also quality, whilst flexibility is relatively less important. These strategic drivers actually influence manufacturing strategy, since the most important goals at the manufacturing level are reducing factory overhead and materials costs and improving white collar productivity. As a consequence, the quality issues are now on the company strategic agenda at the manufacturing level.[4] The Fiat case (see Section 8.5) provides an exemplar illustrating the huge effort required for large Italian manufacturers to close the quality gap. On the whole, the large Italian manufacturers seem to be striving hard to achieve better quality performance. Their rate of improvement is adequate now, though they still pay for the lower starting level.

Lack of multi-focusedness and strategic flexibility
Although Italian companies compete on a number of dimensions—quality, cost, product range—their manufacturing base seems not to be fully consistent with a multi-focused strategic intent. In fact, both large companies and SMEs seem more single-focused than foreign competitors at the manufacturing level, and accord relatively lower emphasis to time compression, which can potentially improve different performance areas—for example, service and working capital productivity—simultaneously (see Chapter 26 for further details). This also reveals a missing link between business and manufacturing strategy. Italian manufacturers seem not to recognise properly the role of a multi-focused production system in supporting global competition along several dimensions.

---

[4]   In terms of performance, larger companies show a higher proportion of external quality costs when compared with the average for the total sample (22% vs. 19%), and are less concerned with prevention (16% v. 22% of the total quality cost). Yet, their quality improvements are remarkable (26.5% over the past three years) basically alighted with the 28% of the large companies in the total sample.

Performance

Major improvements in the performance of the Italian companies have been made in product flexibility and customisation. The focus on these issues was already revealed through the international comparisons and appears to be successful, since the improvements in product variety have been 20.4% vs. 17.4% for the total sample. Also, working capital productivity has increased noticeably, but not enough. In fact, raw materials and finished goods turnover are still higher than in the total sample, whilst WIP turnover is now aligned with the global average.

Italian companies suffered low profitability in 1992–1993—within the Italian sample ROI was less than half the average world-wide, 5.8% vs. 10.2%. However, during that period they maintained and also improved their global market share (average improvement was 12.7% vs. 11.3%).

Subsequently (1994–1995), the heavy currency depreciation and the international recovery gave new vigour to the industrial system. Initially, depreciation was used to regain profitability, first to immediately increase prices, and later to expand global market share. In particular, SMEs from the Northeast benefited greatly from the depreciation, combined with their flexibility, in penetrating the international marketplace.

At the moment, it is hard to say if the opportunity to increase market share abroad is driving companies, particularly the small ones, to improve their operational capabilities abroad, or if it is going to be wasted due to the lack of managerial capabilities. Indeed, some aggregate data show that a number of medium-sized companies have recently started to internationalise their manufacturing base and to develop their managerial capabilities on an international basis (Mariotti, 1994).

8.4.2. INNOVATION

Expenditures

Italian managers are generally believed to underinvest in product and process innovation. On the basis of the IMSS data, this perception should be revised. In fact, the data on R&D expenditures as a proportion of sales are aligned with the other industrialised countries. As a consequence, revenues from new products as a proportion of total sales slightly exceed the average for the total sample (21.4% vs. 18.9%), and expectations growth for the next five years far exceeds the average (75% vs. 31%). On the other hand, Italian manufacturers take a poor attitude to the development of human resources. The expenditure for training is only 0.81% of sales, vs. 2.37% for the total sample. Also, the degree of employee involvement and task delegation of the workforce (see Chapter 26 for further details) appears to be lower than in the other industrialised countries. Finally, capital expenditures to renew plant and process equipment as a proportion of sales are far below the average (11.6% vs. 17.3%). So, on the whole, Italian companies appear to allocate adequate resources to product innovation, whilst process innovation, both *hard*—process equipment and plants—and *soft*—human resources development—seems to be underestimated.

Improvement Programmes

Italian companies have been adopting a number of programmes to improve manufacturing activities and implement 'best practices'. In particular, they have worked hard to streamline operations through cellular manufacturing, 'plant within a plant' approaches, statistical process control, single minute exchange of dies, and so on. Also, computer aided technologies have been massively implemented to support design and planning activities.

Still, Italian firms have reached a lower level of integration of manufacturing process because they have yet to achieve horizontal labour despecialisation via teamwork or multi-skilled workers. They have concentrated more on 'hardware' integration—layout, set-up, quality control—than 'software' integration—organisational design and human resources management. This is especially relevant for the large companies, where the Tayloristic approach is hard to eradicate.

Supply chain management is quite good, comparatively. Upstream integration with suppliers is in line with foreign competitors from the most advanced countries. On the other hand, production engineering integration is lower than in the US, Scandinavia, and Japan, since there are more information loops and cycling of engineering activities and less job rotation between production and design roles. In addition, Italian companies show a poor uptake of the approaches devoted to improving the new product development process, especially quality function deployment and design for manufacturing/assembly. Finally, a poor link between business and manufacturing strategy clearly emerges, particularly in the SMEs, who have not fully recognised operations and production management as a competitive weapon.

On the whole, Italian companies seem to be less aware than the most advanced foreign competitors of the key role of human resources at every organisational level in facing the challenges of globalisation and innovation. Delegation of responsibility, knowledge of the process, and employee involvement are lower than average. In particular, within the SMEs, workforce involvement is very low, whilst knowledge of the process is about average. Within the larger companies, the most critical aspect is delegation, since the organisation is more hierarchical than abroad and production planning and control is more centralised.

8.4.3. LINKING INNOVATION TO PERFORMANCE

It is interesting to link recent innovations in manufacturing to the improvement of manufacturing and business performances over the past few years, in order to address future practices according to the IMSS results in competitive priorities and manufacturing objectives. Table 8-1 summarises the average improvement of various

Table 8-1. Manufacturing Performance and Related Manufacturing Activities

Performance Area	Average improvement for the total setup (%)	Manufacturing activities	Average improvement for the adopters	
			(%)	sig.
Average unit cost	10.3	Team approach	18.9	*
		Single-minute exchange of dies (SMED)	20.9	*
Conformance quality	22.7	Single-minute exchange of dies (SMED)	35.0	***
		Zero defects	38.1	*
		Value analysis	37.0	*
		Team approach	32.3	:
		Quality policy deployment	41.5	*
Inventory turnover	19.7	JIT/Lean production	38.6	***
		Team approach	30.6	*
		MRP	32.6	*
		CAD	26.6	*
		SMED	32.3	*
Delivery speed	16.9	SMED	27.0	**
		Team approach	25.2	*
Product variety	17.2	CAM	26.2	**
Manufacturing lead time	21.5	Team approach	25.2	*
Procurement lead time	16.2	Plant within a plant	26.4	****
		Value analysis	27.5	***
		Quality policy deployment	38.3	****
		TPM	29.4	***
		JIT/Lean production	27.0	***
		ISO 9000	27.0	***
On-time deliveries	17.7	None		
Product development speed	15.2	None		

Significance level of the difference between adopters and non-adopters:
* $p < 0.10$. ** $p < 0.05$. *** $p < 0.025$. **** $p < 0.01$.

areas of manufacturing performance and those techniques primarily related to individual performance areas. On the whole, managerial and organisational techniques play a dominant role. In particular, the team approach positively affects a number of different performances and is the only item significantly related to the reduction of manufacturing lead time. It is a key element, since there is a strong inter-item correlation (Cronbach's alpha = 0.81) between manufacturing lead time reduction, unit cost reduction, service improvement, and profitability increase.

Retrospectively, lead time reduction via a team-based approach seems to be the main route to the implementation of multi-focused strategies to simultaneously improve cost efficiency and customer service. Its effectiveness is confirmed by superior improvements in profitability, as manufacturing strategies are ultimately evaluated by their impact on the bottom line of business performance. It is noteworthy that no significant correlation was found between the use of techniques from concurrent engineering and superior performance in terms of time to market, although the adopters generally declare that they are satisfied with the results of these innovations. This is quite alarming, since the increase of market share appears to be positively affected by superior performance in the speed of product development, whilst the implementation of specific manufacturing techniques in this areas seems to have had no effect.

8.5. Case Study - Fiat-Melfi - The 'Integrated Factory' in a Green-Field Site[5]

Fiat, headquartered in Turin, is the largest private Italian group, with 1994 global sales of over $40 billion, and 250,000 employees. It is one of the big players in the automotive industry at the European level. During the 1980s, Fiat benefited from holding a nearly 60% domestic market share, which was supported by import quotas for cars produced outside the EEC. This comfortable situation prevented the company from becoming aware of the growing competition in the international arena. At the beginning of the 1990s, Fiat faced a deep crisis, since it had lost its shared leadership of the European market, dropping from a 15% share in 1989 to 12.8% in 1991, mainly to the advantage of Volkswagen, who reached a 16.5% market share.

Just at the end of the 1980s, when market share and profits were still high, top managers recognised that poor quality and an outdated product range were the main causes of the coming crisis. In 1989, a Total Quality campaign was initiated, but the company's breakthrough was achieved with the building of a green-field factory in Melfi, in Southern Italy. This has been a radical change in many respects, considering that a total different factory organisation and a new product (the Punto) were developed together. It took 28 months to build the new factory, which has the capacity to produce 450,000 compact cars per year; over 2.7 million square metres of work space; 21 core suppliers located next door; and 7,000 employees; and represents a combined investment of $6 billion in product and factory.

Strong upstream integration with suppliers was realised through long-term contracts, free access to the assembly lines, and parts and subassembly delivery based on electronic *Kanban*. A massive 'plant within a plant' approach has been implemented, with autonomous work groups ranging from 20 to 60 people controlling discrete parts of the production process. Each group, or elementary technology unit (UTE), has been delegated responsibility for budgets, inventory, quality, on-line production and delivery scheduling, and local maintenance. The engineering staff has been partly decentralised to the factory and group level. The organisation structure has been completely revamped—very flat hierarchies, few white-collar workers (less than 10% of the workforce), interfunctional co-ordinating mechanisms, continuous improvement, and delegation at the shop-floor level. The workforce received extensive training, though this was selective and involved only one-quarter of the workforce extensively. Special jobs such as integrated process operator, line technology operator, line maintenance

---

5   The assistance of Luciano Massone, Personnel Director of Fiat-Melfi, in the development of this case
       study is gratefully acknowledged.

operator have benefited from up to 2000 hours of training. All this has resulted in a very productive plant and a huge increase in conformance quality that has allowed Fiat to eliminate having a final rework area, common in traditional car assembly plants.

The integrated factory represents a rethinking of Fiat's manufacturing strategy. In the 1980s, the company pursued a fully computer-integrated-manufacturing (CIM) approach through huge investments in robotic manufacturing and assembly systems in a quest for the 'unmanned factory', but it experienced many quality and reliability problems. Melfi, though benefiting from the technological know-how that was developed through this project, is a step backwards in terms of automation, and relies on highly-skilled employees to achieve quality, flexibility, and productivity. 57% of the work-force has graduated from high school, and 5% hold a university degree. Unions took some credit for the success, since they chose a strategy of co-operating with management, negotiating for training and upgrading of the workforce, and accepting additional constraints and responsibility for the workers because of the streamlining and the vulnerability of the integrated factory. Now Fiat is disseminating the lessons of the integrated factory in its established 'brown-field' sites, expecting it to become a shared corporate vision; indeed, Melfi can be seen as pointing the way to an Italian paradigm of Lean Production.

In summary, Fiat's Melfi factory shows us how a large Italian company can move successfully to world-class manufacturing, whilst also bridging the quality gap. However, a quantum leap is needed to create a lean and agile organisation, and the key element here has been a radically new approach to human resources management. Finally, it should be noted that all of this was achieved in the South of Italy, overcoming significant sociocultural and infrastructural hurdles.

8.6. Case Study - Hewlett Packard - Bergamo Hard Copy - from Plant to 'Knowledge-Based Production Site'[6]

Hewlett Packard opened a plant in 1990 to assemble and test Printed Circuit Boards (PCBs) for printers in Bergamo, Central Lombardy. The plant supplies a final assembly centre in central Europe where a plastic chassis, minor components, PCBs from Bergamo, and the laser device are assembled into a printer, packaged, and distributed all over Europe. This production system is aligned with Hewlett Packard's global location strategy, which follows the principle of 'thinking globally, but acting locally', to balance scale and logistics costs and service. Bergamo is one of three similar plants, one in each of the three major geographic regions: the Americas, Europe, and the Far East. Internal competition between these plants is also used to stimulate new product or technology development and implementation.

---

[6]   The assistance of Gianni Contini, Real Estate Manager of HP Italy, in the development of this case study is gratefully acknowledged.

Hewlett Packard has marketed its products in Italy for many years. At the end of the 1980s, the company decided to locate a production plant in Italy because of the country's sales growth and contribution to corporate profits, and because having a domestic presence would be useful to further development. This choice emulated IBM, who have had a production facility in Italy since 1966, and is thus perceived to be a domestic company because of local employment and technology development. The choice of Bergamo as a plant location was influenced by infrastructural factors, especially (1) proximity to international airports, for managing a global supply chain and 'knowledge chain' to and from the US head office and research centres; and (2) proximity to a large engineering university—the Politecnico di Milano—as a source of technical know-how and top technicians to recruit.

From the beginning, Bergamo Hard Copy (BHC) performed well. Once the plant was running, successful programmes to reduce inventories and cut lead times were initiated. Excellent quality and productivity standards were achieved in manufacturing activities, and BHC was able to attract more engineering product development activities beyond simple 'screwdriver' assembly of PCBs. External pressure on time-to-market and shortening product life cycles raised plant awareness of the importance of bridging the gap between design and manufacturing. The transition from 'plant' to 'knowledge-based production site' was to some extent foreseen from the beginning, but it was made possible by operating performance, and by winning the internal competition with the other two plants. Now HP develops and manufactures new products or subassemblies in BHC. The company built up efficient operations management first, and then integrated engineering and production effectively.

But the story does not end there. In 1994, HP decided to produce ink-jet printer cartridges in Europe. BHC was a candidate for absorbing the new production, to benefit from scale and synergy with existing products. HP finally chose a green-field site in Dublin, Ireland. Although fiscal incentives from the Irish government played a part in the decision, the inefficiency of Italian public administration created delays in granting permission and bureaucratic procedures, partly due to the fragmented number of different public offices and departments, and prevented the plant from receiving permission in time to meet market demand for ink-jet printer cartridges.

BHC is a story in three acts: plant installation, the transition from 'plant' to 'knowledge-based production site', and the missed opportunity for further growth and differentiation. It is an example of world-class manufacturing in Italy, focused on streamlining manufacturing and production engineering integration. However, it also demonstrates the growing role of specific determinants for locations beyond traditional labour and logistics costs. In fact, the emerging paradigm turns manufacturing into knowledge work, and knowledge work locates in a knowledge-intensive environment. However, time-consuming bureaucratic procedures and in general all the burdens from public administration actually hamper the development of integrated manufacturing activities to support time-based strategies.

8.7. Conclusions

Companies in the Italian assembly industry are niche players to a large extent with good international competitiveness in many 'supplier-dominated' and 'specialised supplier' sectors. In the 1990s, the industry is challenged sharply by megatrends, both global, such as the internationalisation of competition, European integration, and shortening product life cycles, and national, such as the political crisis and the size of the public debt, which hampers a long-term view and nation-wide projects to develop the global competitiveness of the country. There are basically two challenges:

- *internationalisation*, not simply via the traditional route of increasing exports, but also through the development of effective production and operations management on a global basis;
- *innovation in the manufacturing system*, not only product innovation, but also improvements in the manufacturing system to improve the effectiveness and efficiency of both production and product development. This requires the implementation and adaptation of the emerging manufacturing paradigm, which entails:

  1. multi-focused and strategically flexible factories;
  2. process integration and streamlining across functional boundaries and along the supply chain, and throughout the new product development process;
  3. a new attitude to human resource management based on employee involvement, delegation of responsibility, and knowledge of the process, in other words, process ownership (Spina et al., 1997).

With respect to these two challenges, some country-specific factors should be kept in mind, which concern the industrial structure, some sociocultural aspects and strengths and weaknesses. In light of Italy's historical background and current context, the actual responses of Italian industry can be summarised as:

- *Globalisation* . The international economic recovery of 1994–1995 and the heavy currency depreciation gave new vitality to the tough job of globalisation. Although Italian foreign direct investment is lower than that of the most advanced foreign counterparts, Italian industry is filling the gap. Good news concerns the recent internationalisation of production through investment in the foreign manufacturing activities of rising medium-sized companies, groups and networks of small firms. IMSS data in part support good news, since SMEs show a broad geographic focus. Further research should be directed to investigating if and how these companies are building up capabilities for managing global operations effectively.

- *How to Compete.* In terms of strategic drivers, including business unit goals, manufacturing competitive priorities, and manufacturing activities, and manufacturing performance, four issues emerge:

  1. Italian companies—and Italian SMEs in particular—rank flexibility and customisation as important.
  2. Both large and small companies have joined the quality revolution and are striving hard to close the quality gap, thus challenging the traditional perception of lower quality levels. On average they are still one step behind the best foreign competitors, but the best companies reach world-class levels (see the Fiat-Melfi and HP-Bergamo Hard Copy case studies in this chapter).
  3. Strategic flexibility and multi-focusedness, although emerging at the business level, are not fully supported at the factory level. There is a missing link between business and manufacturing strategy.
  4. Time-based competition is accorded a lower priority than abroad, with the primary internal concerns cost efficiency rather than customer service. The lack of time focus in the new product development process is quite alarming in the light of the differentiation strategy based on wider product range and customisation that characterises Italian niche players.

- *How to support competition.* Italian manufacturers proved to be involved in a number of programmes to improve manufacturing activities and to be able to implement 'best practices'. In particular, they have worked hard to improve the management of the whole supply chain and to streamline manufacturing operations. At the factory level, they have reached high levels of process integration with regards to 'hardware'—layout, set-up reduction, computer-based approaches— whilst the organisation and management of human resources—'software'—is not fully supportive of process integration. More generally, employee involvement, delegation of responsibility, and empowerment of human resources still present critical issues, that are late and hard to implement, also due to sociocultural aspects. This also affects time-based performance. In fact, poor process ownership leads to poor time performance, whilst both the Fiat case and the IMSS Italian sample suggest that lead-time reduction comes from organisational approaches—namely team work—rather than computer-based innovations.

The relatively poor integration of manufacturing and new product development is inconsistent with company expectations of future income from new products and global competitiveness through wider product range and customisation. Again, time-related issues such as time-to-market are underscored , whilst the HP case shows the benefits and the feasibility of moving from 'plants' to 'knowledge-based and integrated production sites'. It also shows two other things: (1) that time focus for industry requires time focus and agility also for the public services; and (2) that the emerging paradigm turns manufacturing into 'knowledge work', and 'knowledge work' needs a knowledge- intensive environment.

In conclusion, there are bright and dark spots for Italian industry in facing the challenges of innovation and globalisation. There are high expectations from world-class exemplars of overcoming weakness, but also urgent messages to improve the public sector, infrastructure, and the national educational system.

Acknowledgements
The financial support from the 'Trasferimento delle tecnologie dei progetti finalizzati' by the CNR (National Research Council of Italy) is gratefully acknowledged. This chapter was jointly written by the two authors. However, E. Bartezzaghi has written sections 8.1, 8.2.1, 8.2.3, 8.3, and 8.4.2; G. Spina has written sections 8.2.3, 8.2.4, 8.4.1, 8.4.3, 8.5, 8.6, and 8.7. R. Cagliano provided support for the data analysis.

References
Bank of Italy, Governor's Annual Report and Appendixes (1980–1994).
Bartezzaghi, E., Turco, F., and Spina, G. (1992) 'The impact of the Just-in-Time approach on production system performance: A survey of Italian industry', *International Journal of Operations and Production Management,* 12, 1, 1992.
Mariotti, S., 'Internationalisation of production within Italian industry: Problems, prospects, and implications for economic and managerial research', in *Proceedings of the Fifth AiIG Conference,* Naples, Italy, November 1994 (in Italian).
Spina, G., Bartezzaghi, E., Cagliano, R., Bert, A., Draaijer, D., and Boer, H., 'The multi-focused manufacturing paradigm: adoption and performance improvements within the assembly industry', in P. Lindberg, C.A. Voss, and K. Blackmon, *International Manufacturing Strategies: Context, Content and Change,* Kluwer Academic, 1997.

CHAPTER 9

JAPANESE MANUFACTURING STRATEGY: TO COMPETE WITH THE
TIGERS

H. Yamashina, Kyoto University, Japan

9.1. Introduction

Japanese policy is to create prosperity through industrialisation, and there is little doubt
that the development of the Japanese economy has been well supported by the growth
of manufacturing industry. In Japan, it is firmly believed that the prosperity of a nation
depends on the excellence of its production capability, and that those who conquer
manufacturing will eventually conquer technical innovation. The Japanese commitment
to continuous technical innovation in the manufacturing industry has allowed it to
become a leading economic power; however, Japanese manufacturing companies are
currently facing very tough competition primarily due to the appreciation of the Yen
and the dramatic improvements in competitiveness from both the advanced countries
and the rapidly growing Asian Tigers.

The objective of this chapter is to provide an insight into the strategies currently
being adopted by Japanese manufacturers to address these very serious issues. Firstly, I
shall discuss the strategies used by Japan in the past; then the problems of the 1990s
including the impact of the appreciation of the Yen, cost structures and offshore
production; and, finally, the strategies being implemented by Japanese manufacturing
firms to help maintain and strengthen Japan's competitive edge.

9.2. Japanese manufacturing strategy prior to the appreciation of the yen

Many Japanese manufacturing companies believe that the two key factors needed to
establish a competitive advantage are attractive products and strong manufacturing
capability. To have attractive products, one must ensure originality and creativity. The
Japanese have long argued that there are two kinds of originality: one is to discover and
the other is to develop—Japan's strengths lie in the latter. For example, the transistor
radio, VCR, TV, and compact disk all originated in the West, but were developed into
commercial products by the Japanese (See Table 9-1). Creative development goes with,
and is as important as, creativity in invention.

P. Lindberg et al. (eds.), International Manufacturing Strategies, 163-178.
© *1998 Kluwer Academic Publishers. Printed in the Netherlands.*

Table 9-1. Invention and Development

Item	Originator	Developer
Transistor Radio	Regency	Sony
VCR	Ampex	Sony, Victor
Television	RCA	Matsushita
Rotary Engine	Vanchel	Mazda
Compact Disc Player	Philips	Sony

Japan has made a concerted effort to strengthen its manufacturing capability over the last three decades, as shown in the general trend of organisational structure (see Figure 9-1). Over the last thirty years different departments such as manufacturing techniques sections, departments, centres, headquarters, and R&D have been added to organisational structures to support manufacturing capability through improvements in manufacturing techniques. At Toshiba or Matsushita there are a substantial number of people involved in such activities and there will continue to be so, as long as they maintain their focus on manufacturing capability.

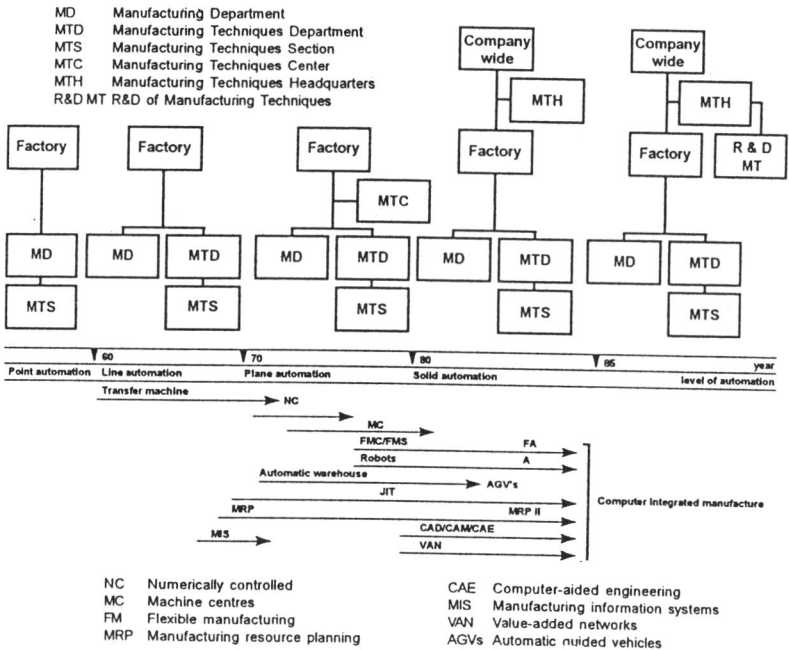

Figure 9-1. General Trend of Organisational Change

Japanese companies employ engineers extensively. Human resource statistics from the 1980s show that Japanese staff with science degrees totalled 12,698 compared with 78,246 in the United States.[1] Even controlling for the difference in population (the US is about double that of Japan) science graduates are far more numerous in the United States. Similarly, in the UK the number of science graduates was 20,151 with half the population of Japan. This, however, is in stark contrast to the situation in engineering— 71,396 in Japan versus 14,616 in the UK and only 91,121 in the United States.

The Japanese tend to recruit scientists primarily for basic research. Compared with Japan, the UK has many more people in basic research, which reflects the concept of 'creativity in invention', inherent in the British economy. Japan, on the other hand, has found its focus and thus, its advantage, in applied research. Japanese manufacturing companies recruit far more engineers and integrate them across the whole company. Japan tends to actively invest and nurture more engineers in product development and design, and hence create better capability. For example, continuous improvement engineers work on the shop floor and often start their careers there. In Japan the ratio of continuous improvement engineers to production operators is about 1:20, compared

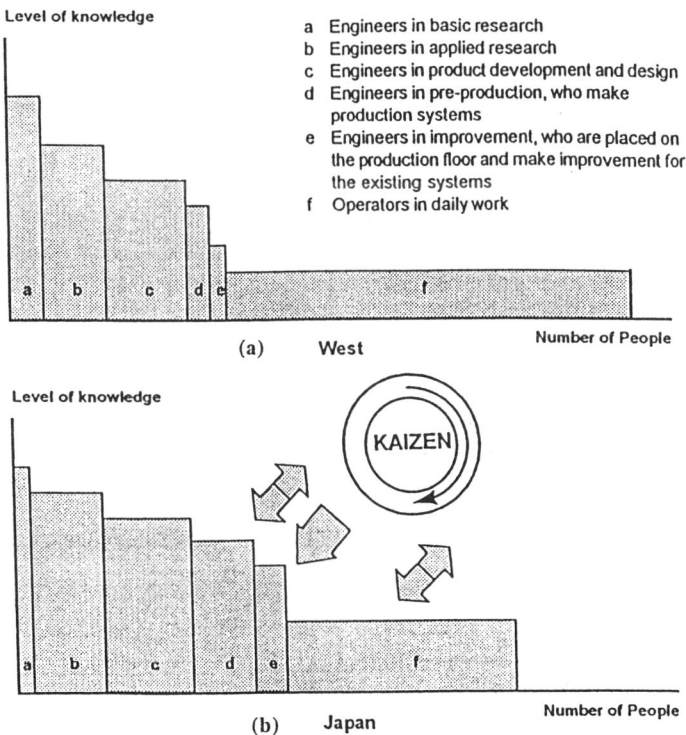

a   Engineers in basic research
b   Engineers in applied research
c   Engineers in product development and design
d   Engineers in pre-production, who make production systems
e   Engineers in improvement, who are placed on the production floor and make improvement for the existing systems
f   Operators in daily work

Figure 9-2. Six Categories of Production Staff

---

[1] Japanese figures are for 1984, US 1980, UK 1981, source: the Japanese Ministry of Education

with an average of about 1:100 in the UK (see Figure 9-2). Interestingly, production operator numbers would, on average, be about equivalent between Britain and Japan, but across the whole manufacturing process (or more aptly manufacturing capability) then Japan, in general, is stronger because of its greater engineering workforce. Today Japanese manufacturers are being forced to rationalise in this area.

From 1945 to 1994, four periods of Japanese manufacturing strategy can be distinguished (refer to Figure 9-3). The first period from 1945 to 1974 was the 'product out' phase, when demand exceeded supply and Japan focused on increased production volume. The measurement of various factors such as output per hour, lost time due to machine breakdowns and defect rates was undertaken to ensure competitiveness in manufacturing and improvements to production capability were made. In 1973–1974 the first crisis struck, as markets for consumer goods such as washing machines, refrigerators and vacuum cleaners started to show saturation. The ensuing second period was known as the 'market in' phase. Japanese manufacturing companies needed new strategies to cope with the fall in demand and diversification theory emerged.

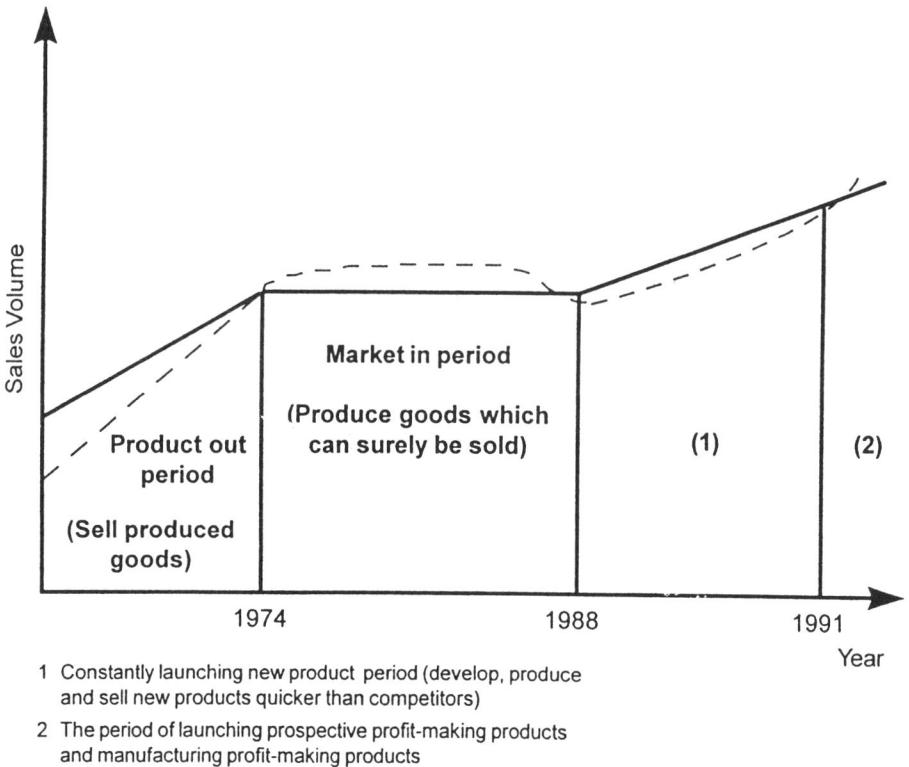

1 Constantly launching new product period (develop, produce
   and sell new products quicker than competitors)

2 The period of launching prospective profit-making products
   and manufacturing profit-making products

Figure 9-3. Four Periods of Japanese Manufacturing Strategy

Diversification theory encompassed the idea that if a certain product can sell a given volume, then if the good is differentiated—produced with many variations to fit every different kind of market need—then there is an opportunity to increase demand. Based on this principle, many companies started to produce goods with many variations, and firms began to develop an appreciation of customer needs and satisfaction levels. As a result, additional performance measurement techniques were introduced such as the number of claims from the customers—if the customer was not happy with what he bought then he would not buy it again. Other measures included the direct going rate, manufacturing lead times, delivery lead times, set-up times, and stock turns. With continuous improvement as the goal and with the increased number of product variations, the complexity factor rose rapidly. Time to market and set-up time because critical—in other words, just-in-time manufacturing had become a necessity.

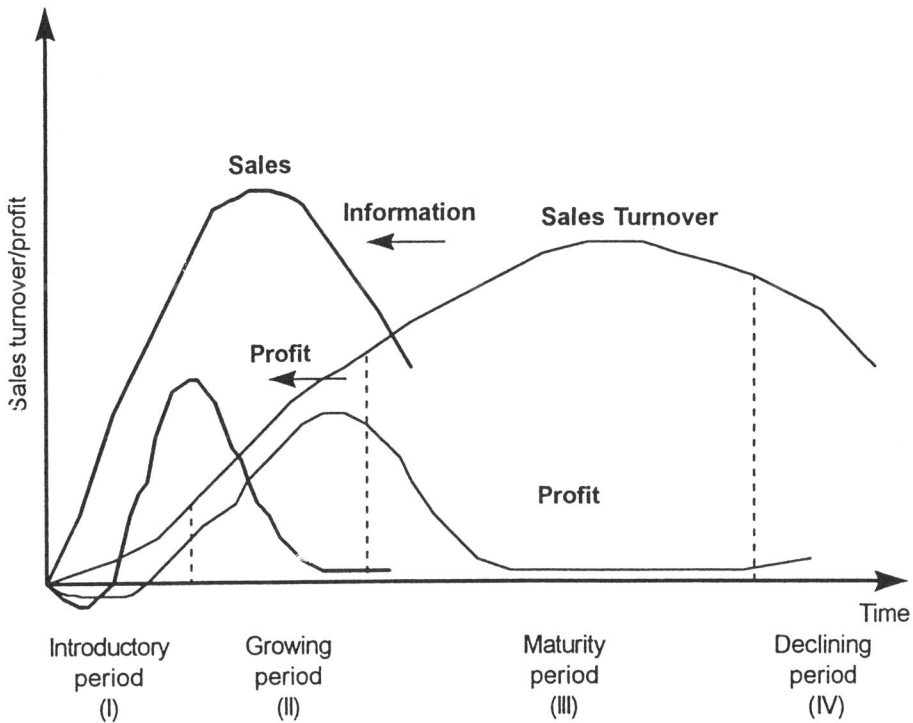

Source: Nomura Research Institute 'Strategy for Creativity'

Figure 9-4. Reduction of Product Life Cycle by Information

During the 'market-in' period, high level industrialisation, strong capitalism, and the maturation of particular markets 'squeezed' the product life cycle. It became imperative for Japanese firms to secure profits earlier than their competitors, and increasing emphasis was placed on shortening the time between manufacture and distribution in order to create productivity improvements. This was exacerbated by the improvements in information technology—companies could launch new products using the media for the rapid and comprehensive transmission of their product information to the customer base, significantly assisting in the maintenance of old and the creation of new markets. In time, however, markets decline, so it becomes critical that the whole process from design, to manufacture, to distribution is shortened. The growth of information technology made productivity improvements a domestic necessity (refer to Figure 9-4).

To align production methods to the changing face of industry, changes in resource allocation were needed. In the 1960s, most people were engaged in either direct

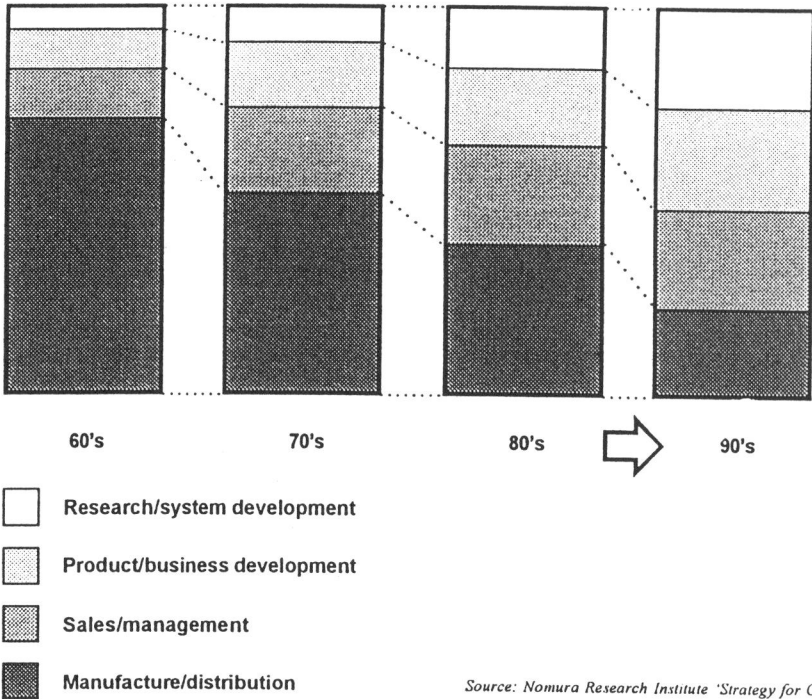

60's   70's   80's   90's

☐ Research/system development

▦ Product/business development

▦ Sales/management

▦ Manufacture/distribution

Source: Nomura Research Institute 'Strategy for Creativity'

Figure 9-5. Change of Human Resource Allocation Toward Period of Creativity

manufacturing or production areas, with only a handful engaged in product development. In the 1970s, fewer people were in manufacturing and distribution areas (see Figure 9-5) and Japanese firms dedicated more and more people to research or product/business development areas of the company (see Figure 9-6).

Japan had been quite successful in supplying and creating market needs up to 1988 based on the theory of diversification, but there was a limit to this idea and in 1988 a new period emerged known as the 'constantly launching new products' period. To improve the process of launching and design new products, measures of product development lead times were introduced. For example, data taken from April 1, 1991 up to March 31, 1992 show that 89 of the 211 television models available during the year underwent model changes—this translates into a life cycle shorter than one year. The product life cycles became shorter and shorter. At Toshiba, for example, the number of employees has remained fairly static since the 1970s, but engineers involved in indirect work rose from 29% to 69%.

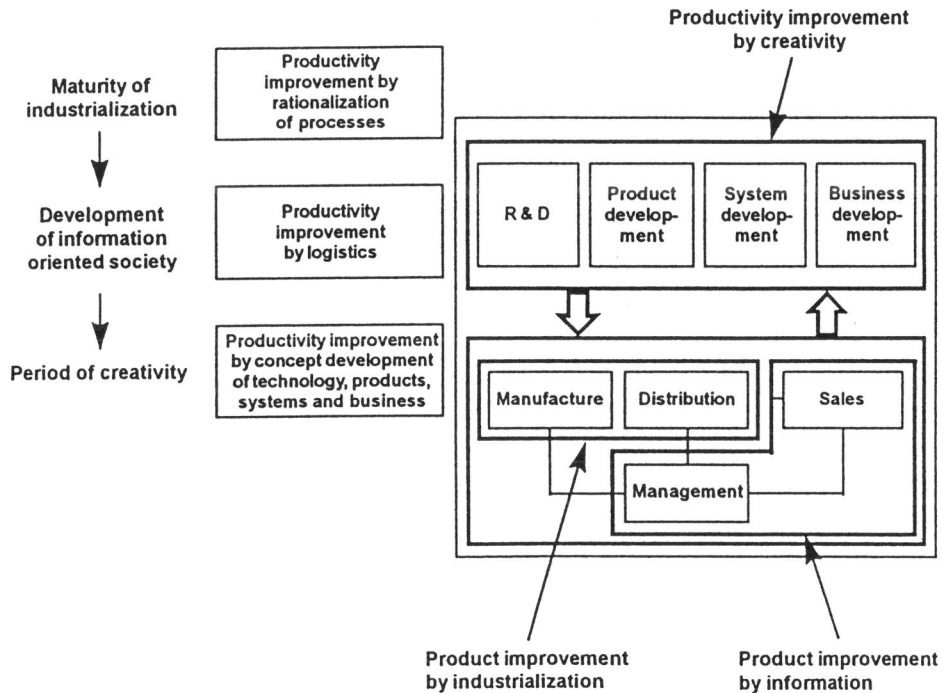

Source: Nomura Research Institute 'Strategy for Creativity'

Figure 9-6. Production Improvement by Industrialisation: Information and Creativity

The company business has changed from one of stable technology to one where the speed of technological innovation is fast—that is, from a labour-intensive assembly industry to a technology-intensive industry of development and application of electronic engineering. This shift is in line with the changing pattern of consumer goods: the first generation was mass production; the second generation the production of many variations to meet diversified needs, with higher quality than is possible through mass production; and the third generation a separate model for every customer; in other words, mass customisation. Industry needs to be ready to meet these market changes with competitive manufacturing capability.

After 1991, the fourth period emerged, the 'period of launching prospective profit-making products and manufacturing profit-making products'. This phase developed in response to various new market trends that emerged during the 1990s. One such change was the need for Japanese manufacturers to face the increasing problems related to the protection of the ecological environment and the disposal of industrial waste by developing resource-saving factories producing less waste and little or no pollutants. Moreover, Japanese firms recognised the need to match market demand by manufacturing products which were energy saving, produced no pollutants and were recyclable at the time of disposal.

During this period, manufacturing companies also faced major fluctuations in demand, both in content and quantity, and the market could no longer absorb all the new incoming products. It was time to reassess the Japanese manufacturing philosophy based on the concept of mass production of better products at lower prices in large volume.

By 1988 intense competition from both advanced countries and the newly industrialised economies started to seriously challenge Japanese manufacturing. This was compounded by intensified economic friction with America, Europe, and Asia, and required an increased focus on international co-operation. To make matters worse, Japan was also beginning to be caught in the grips of a very serious recession.

9.3. The Problems of the 1990s: The Appreciation of the Yen, Cost Structures, and Offshore Production

Japanese manufacturing is currently undergoing a period of severe economic change, exacerbated by the recession and uncertainty surrounding the Yen. Serious problems have been encountered by the strength of the Yen, especially when it rose to higher than US\$1 = ¥100. These problems are different from the ones raised by the Plaza Accord of US\$1 = ¥120 in the latter half of 1980.

The step-by-step increase in exports has been overtaken by the increase in exports, and the appreciation of the Yen has done little to help other areas of the economy, especially as Japan wishes to become a high social welfare country, which costs money. It is easy to identify the problems of a strong Yen—with ¥10,000 you can buy: 25.5 Big

Mac™ hamburgers in Tokyo compared with 32 in New York; 14 Kilos of rice in Tokyo compared with 83 Kilos in America; and 90 litres of milk in New York compared with only 49 litres in Tokyo – gasoline, taxis, TVs, whatever, prices are almost double in Japan, as the stronger Yen translates into higher costs. In 1982 a Japanese manufacturer could make a small car for $4,211 but today, due to the rise of the Yen, it costs $7,313.

The US and Asia have caught up with Japan in the areas where she had a competitive advantage—Japanese manufacturing's competitive edge has been fading away in the 1990s. Japan has been unable to overcome problems in areas where she has always been weak, and is therefore struggling to draw a clear vision of the way ahead in the future. During the years 1980 to 1992, for example, American car makers like GM, Ford and Chrysler made significant efficiency improvements, especially in the area of labour productivity, whilst Toyota, Nissan, and Honda actually suffered reductions in labour productivity because of the introduction of various car models.

In light of the appreciation of the Yen, MITI has undertaken some research to help determine the measures taken by Japanese firms to counter the appreciation.[2] Increased demand for import materials was rated by 42% of respondents as their primary countermeasure—the stronger the Yen, the lower the cost of the imported raw material; increased overseas procurement was cited by 37%; increased end prices of product to compensate for losses attributed to the rise in Yen was the response of 32%; and increased local production by 19%.

Many companies believe that although Japan provides a large consumer market, it is no longer a suitable base for production because of the high level of land prices, taxes and prices of consumer goods. The overall cost structure in Japan may well inhibit efficient production. For example, it costs Canon 100 Yen to produce a given item in Japan, compared with 120 Yen in Europe, 70 Yen in the United States, and less than 70 Yen in Asia to produce the same item. How can Canon USA produce at such a low end cost? The reasons include the appreciation of the Yen coupled with the depreciation of the dollar; high energy and raw materials costs in Japan (US manufacturers can buy steel at half its cost to Japanese manufacturers); and an inflexible labour market—in the US restructuring a business is far easier than it is in Japan because of their flexible labour market. So if the production base is moved offshore as much as possible, the current crisis caused by the stronger Yen and fixed cost structures can be overcome. If Japan relies on producing goods which rely on cheap labour costs offshore, she will eventually lose competitive power.

The primary reasons behind Japanese manufacturing firms increasing their production overseas have been canvassed by MITI[3] and it was found that 56% of respondents felt it was necessary due to the need to properly match local demand; 49% to cope with the appreciation of Yen; 36% to reduce costs of domestic products by expanding re-imports; 27% that overseas demand was a potential high-growth area; 27% due to lower labour costs offshore; and 11% to avoid trade disputes.

---

[2]   The research undertaken was multiple response, hence the percentages do not add to 100.
[3]   The research undertaken was multiple response, hence the percentages do not add to 100.

What then are the disadvantages of moving production offshore? One major difficulty arises from the availability of engineering talent abroad. This is a key point for those of us familiar with manufacturing line techniques—a lack of technical experience will lead to the decrease in product development capability. For example, American industries have not been able to be competitive in high-tech product fields such as VCRs or compact discs because they have shifted their production overseas and have not been able to maintain their engineering capability. The second problem with moving production bases abroad is that, without exception, companies that rely on the sales of OEM products are unstable and dependent on their suppliers. Even if they can obtain a stable product supply they are constrained by the fact that they must produce their stable main products and if they are unable to because of problems derived from offshore production then their profitability is significantly weakened. The third problem is that manufacturing industries can only maintain international competitiveness through continuous improvements to technical innovations, and technical innovation is closely connected to production shop floor activities. The Japanese develop new innovative products primarily through applied research, but support services such as production technology are just as important. Unless excellent production technology and continuous improvement capability exists, then new and attractive products that are produced quickly and efficiently will not result. The shop floor people are integral to this process. The essence of production know-how can only be accumulated through actual production experience and product development will not really be cultivated unless production is done in-house.

Therefore, those manufacturing industries who retain production plants in Japan will concentrate on improving productivity and developing new products. Within Japan, priority will increasingly be given to the fields of R&D, product engineering, production engineering, and marketing.

9.4. Japanese Strategies for Manufacturing in the 1990s and Beyond

Japanese manufacturing strategy recognises that eventually Japan will become an assembly base for highly functional products. Many companies have begun to question whether their business activities are concentrated in the best areas (especially if in Japan); the efficiency and effectiveness of their costs, methods, and processes of production; high fixed costs and the correlation to some inefficient operation; and whether they are really competitive in the field of software. To grow and to prosper, what conditions are needed?

Even though the recession has been severe, some companies have still been experiencing growth and prosperity. Common features of these companies were lofty, but understandable policy; good products with good profit margins (high value-added products); effective education of employees from the top management to the floor level people; and extensive use of cost management.

Most Japanese companies agree that the necessary conditions for prosperity are the need to have growing products in line with market needs and in accordance with mega-trends; very active R&D; needs-oriented product development; very active organisations; progressive development of new businesses; utilisation of ideas and knowledge from outside sources; extensive utilisation of information technology; and clear and easy management policy implemented from the shop floor to the senior management. These are lessons Japan has learnt after the 'continuous launching new product period', post-1988.

The Japanese, because of their language, cannot be leaders in the service sector. Their challenge for economic survival is in manufacturing and the keys to success in this area are high-technology and continuous technical innovation. Revitalising the power of industrial competitiveness lies in the engineering and manufacturing capabilities within each firm in each industry—labour-intensive industry will not survive in Japan, only technology-intensive.

In the scientific fields of quality buildings and market research, Japanese manufacturing industries still lead. Products with static demand *must* be made abroad from a cost reduction perspective, but those that require technological innovation *need* to be made in Japan. For example, exports of Japanese TV sets have outperformed imports for the last twenty years, although this gap has shrunk substantially over the past decade. This has occurred even though the average price of export TVs is higher than the import prices, and is primarily due to product differentiation—the Japanese export big, wide-screen models, whereas imports tend to be small. Interestingly, most of the imports (88%) come from newly developing or ASEA countries, reflecting the significant transition in trade structure with Asian nations (see Figure 9-7).

What then are the causes and effects of de-industrialisation and is this really Japanese policy? Three factors that have been significant in the US's de-industrialisation are low savings and profit rates, which have resulted in stagnant plant investment; investment has been relatively ineffective because it has not been targeted to export- or technology-intensive industries due to restricted employment rules and unsteady labour relations; and increasing intensity of international competition in low-value-added industrial products resulting from the rapid industrialisation of the developing countries. However, the main reason for de-industrialisation stems from the shortage of investment and that product and labour markets lack flexibility. This produces idle resources that are not always channelled into the high-value-added product-making industries.

The ratio of plant investment to GDP in manufacturing industries in the UK in 1985 was 2.8% and in 1990 it remained at 2.8%. In Japan, in 1985 the ratio was 6.0%, rising to 7.7% in 1990. Even more interestingly, in Korea in 1985 the ratio was 8.2% and in 1990 it was 10.4%. The Japanese strategy is to improve competitiveness and maintain the edge by shifting people out of low value added areas to high value-added areas. The growth will be achieved through developing new technology and creating new markets.

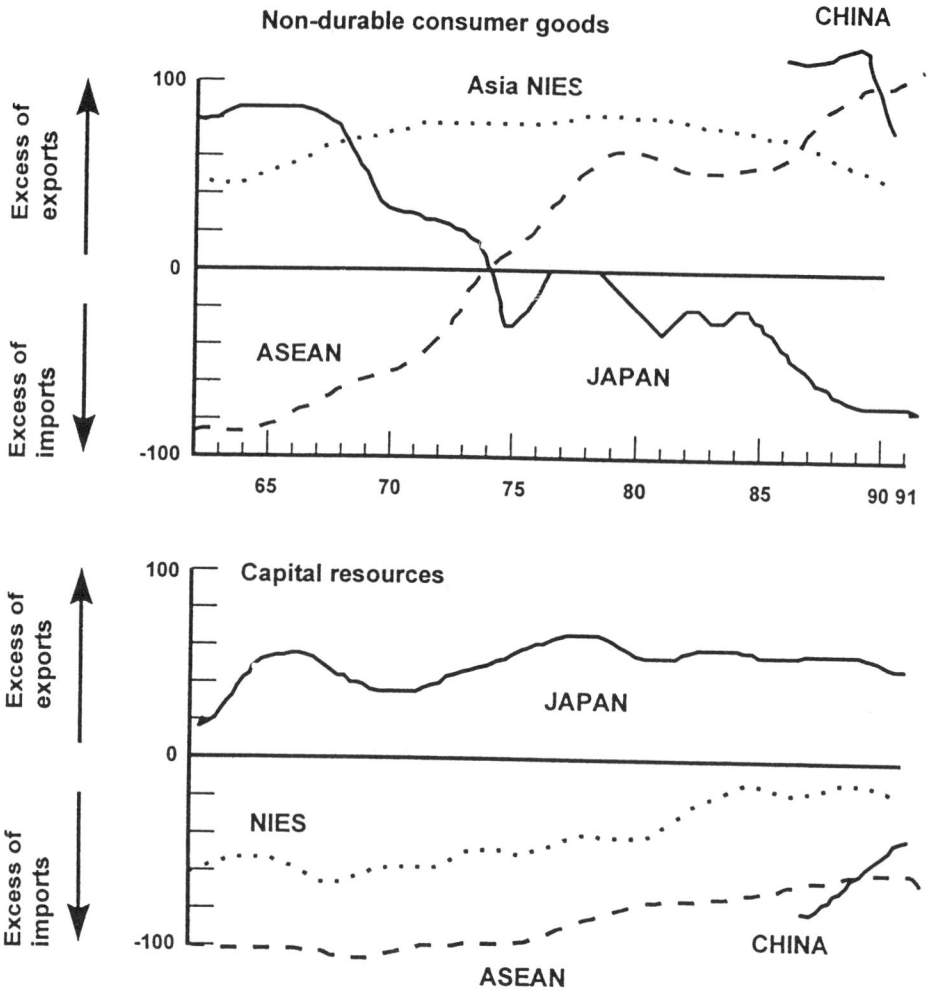

Non-durable consumer goods   CHINA

Asia NIES

Excess of exports

100

0

Excess of imports

ASEAN

JAPAN

-100

65   70   75   80   85   90 91

100

Capital resources

Excess of exports

JAPAN

0

Excess of imports

NIES

-100

CHINA

ASEAN

Note: Trade Index = (export - import)/
        (export + import) x 100
        ASEAN excludes Singapore and Brunei

Figure 9-7. Transition of Trade Structure in Asia

Table 9-2. Ratio of Plant Investment against GDP in Manufacturing Industries

Country	1960	1968	1973	1979	1985	1990	Based on prices in
UK	4.0	4.0	3.3	-	-	-	1975
	-	-	3.2	3.3	2.8	2.8	1985
USA	3.9	4.6	4.1	-	-	-	1970
	-	-	2.3	2.8	2.4	2.5	1985
Germany	5.7	4.6	5.0	-	-	-	1970
	-	-	4.3	3.8	3.7	4.5	1985
France	-	-	-	3.4	3.1	3.9	1980
Italy	-	-	4.4	3.0	2.5	-	1970
Japan	5.2	6.8	6.4	4.4	6.0 (1984)	7.7	1985
Korea	1.9	6.2	7.9	10.2	8.2	10.4 (1991)	1985 for 1990

Source: Ministry of Finance1985

There is little doubt that over time there will be an increasing move in Japan to transfer production offshore, but this does not mean that the Japanese will have to worry about de-industrialisation for some time. The ratio of overseas production remains at 7% and is considered to be low, especially compared with that of the US, being at its highest 26-27%. It is still vitally important that to reinforce a position of economic power, Japan must maintain its large manufacturing presence *within* Japan. Even if overseas expansion does take place, more than 50% of a company's functions *must* remain within Japan, otherwise innovation and business development can no longer take place.

Research by MITI on the domestic production programmes needed to cope with increased overseas production shows that 62% of respondents feel that the correct route to take is shifting domestic production to high value-added products; another 56% maintaining the production levels by further cultivation of the domestic market; whilst 24% the creation of new business areas.

Another feature of the Japanese manufacturing sector is that they import the necessary raw materials and then produce the end product. This is known as 'full-set' policy. In recent years, however, there has been a shift from full-set policy to the 'international division of work' in order to overcome the problems exacerbated by the appreciation of the Yen. An example of this shift is the silverware producer in Tsubame who, in the face of very high labour costs and increasing tough international competition, have changed their policy. The company now import steel from Korea rather than from domestic sources and are making sweeping rationalisations— centralisation and manpower reduction—to compete with other Asian countries. In addition, they have moved those processes in which they must compete on a cost-only basis offshore to China. In return, they have tried to expand their markets to the whole of Asia. Therefore, although a part of the production process has been de-industrialised, the company have compensated by expanding the market for their products within the rapidly developing Asian nations.

MITI has compiled some recent research on the underlying objectives for investment into East Asia.[4] The majority of respondents (69%) stated that their prime objective was to secure and hopefully create new markets in the countries where the investments were made; 40% the establishment of an export base to the third country; 29% to help their local business; 25% to secure overseas labour; whilst 20% as a counter-measure of the appreciation of the Yen.

The Japanese need to establish relationships of coexistence and co-prosperity with the Asian Tigers and to positively support the growth of those countries. Rather than competing directly with them, she should further develop her own engineering capability, resulting in a continuing demand for engineers. The manufacturing bases in Asia, coupled with Japanese manufacturing techniques transferred to the local economy, will assist in the development of Asian home markets and, in the end, will favourable effects to the Japanese economy (see Figure 9-8). Based on the growth of the Asian tigers, Japan should also grow.

When Japan went into this recession, it began to focus on two things: the development of newly attractive products which were durable and cost reduction. This strategy is the current consensus of Japanese management and an integral policy to combat the appreciation of the Yen. Current management thinking comprises eight objectives: the development of original technology; innovation not only in technology but also in new product development; further improvement of manufacturing techniques; the independence of component industries; the creation of a strong domestic market for advanced technology; the attraction of engineering and science graduates into manufacturing by remunerating and treating them well; establishing a reputation that shows concern for the environment and safety; and expansion of export and overseas production without trade disputes.

With these management objectives, some companies can still make profits in the middle of a very deep recession and a high Yen even under the harsh economic climate of $1 = ¥95$. They have developed unique products and/or reduced costs effectively and/or shifted their production bases overseas successfully. Companies with unique products such as Canon, Sharp and Nikon can *raise* prices to compensate for the foreign exchange losses attributable to the stronger Yen. Firms such as KOA, Kyocera and Bridgestone have the structure to *reduce* costs effectively to cover exchange rate losses, whilst companies such as Uniden, Rohm and Mabuchi Motors can use measures such as shifting production offshore or purchasing inputs from abroad, hence using the stronger Yen to their advantage. A selection of case studies is given in Table 9-3 below.

Today, many Japanese companies are focusing their strategy on cost reduction an/or launching more attractive new products that are able to stay in the market longer. In Japan, it is said that in order to be strong enough in manufacturing one has to have good brains which requires total quality management (TQM), but one also needs to have strong muscles, or in other words, strong manufacturing capability, which requires

---

[4]   The research undertaken was multiple response, hence the percentages do not add to 100.

Absolutely advantageous

Co-existence

Preference in Japan
Japan

b **a**

Japan

C •

Asia

Comparatively advantageous

Preference in Asia

Preference in Japan

Asia

Preference in Asia

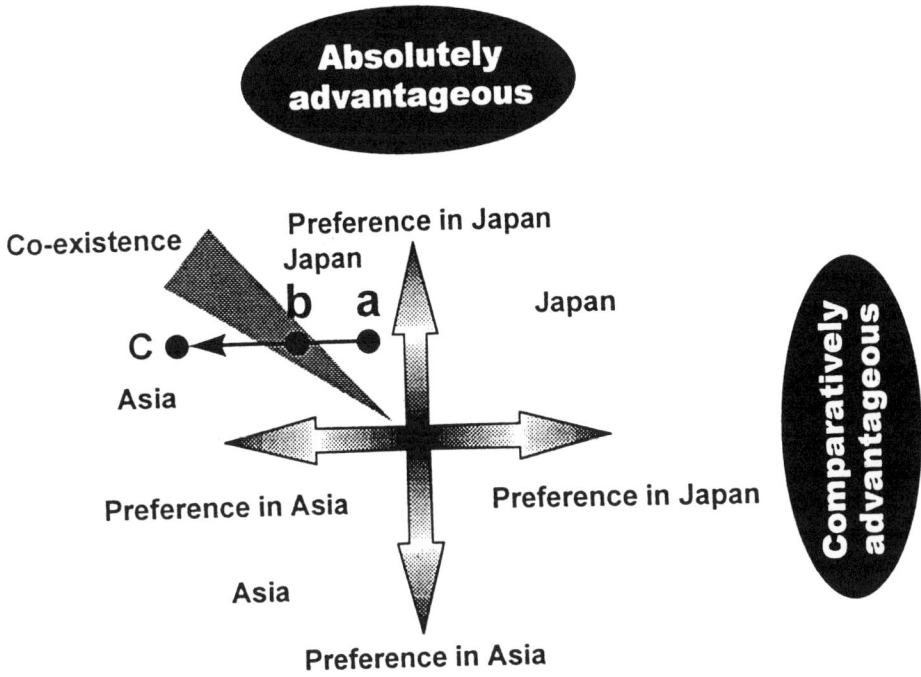

Relation between the location of production site and advanced technology

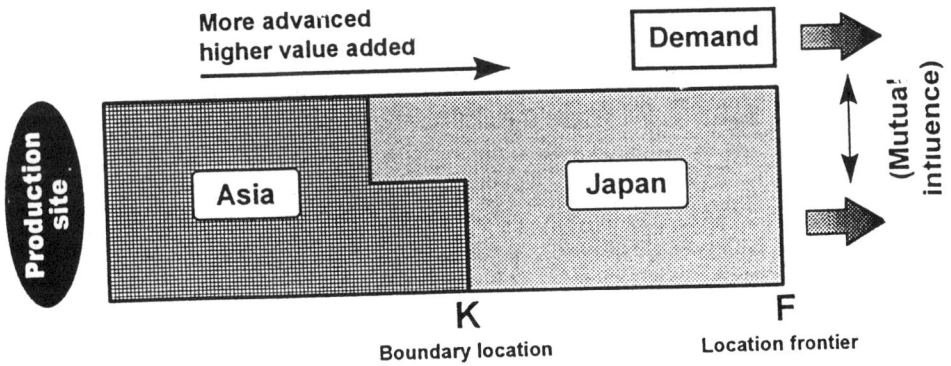

More advanced higher value added

Demand

(Mutual influence)

Production site

Asia

Japan

K

Boundary location

F

Location frontier

Figure 9-8. Determination of Production Site

Table 9-3. Case Studies of Successful Japanese Manufacturing Companies

Murata Manufacturing Company
- they have 80% of the world market share in ceramic filters, piezoelectric products;
- R&D expenditure is 7% of sales turnover;
- they are front-runners in multi-circuit parts;
- they are expanding in the business area of communications such as mobile phones.

KOA
- they have implemented a shop system that covers each product from order receipt to delivery;
- delivery time is 24 hours
- inventory has been reduced to 25% of its former level;
- overall manufacturing effectiveness has been doubled

Uniden
- 100% overseas production, with plants in the Philippines and China;
- 90% of parts purchased from overseas suppliers;
- R&D and management have been centralised in the main office in Japan and they are a lean organisation;
- they are planning to re-import their products to Japan based on sales liberalisation in the areas of automobiles, mobile phones, etc.

total productive maintenance (TPM). Moreover, one has to have a good nervous system to connect the brains with the muscles, which means just-in-time production. So, in manufacturing one needs to have TQM, JIT, and especially today TPM, because it really cuts costs. Recent examples include a chemical manufacturer who introduced TPM and reduced its primary cost index from 1.0 to 0.77—a cost reduction of 23% over a 3-year period; a semiconductor maker who reduced production costs by 45% over a 4-year period due to TPM; and an office furniture manufacturer who succeeded in reducing the man-hours needed to produce a panel by 33.8% by using TPM.

In conclusion, I would like to stress that the way the Japanese have managed in the past and the way they want to continue to manage in the future is to match manufacturing capability with market changes—the effects of the disparity in the human resources of an organisation has become more, not less, significant as the demand on manufacturing increases to match changes in the market. The general trend of organisation change has been very simple since World War II—Japan has always sought to incorporate flexibility into its systems so that change can take place to match market needs. Today the most important thing continues to be the ability to change, and to change quickly enough. There is a very serious need for Japan to cope with increasing change and very intense competition from the Asian tigers. Japanese manufacturing companies are seeking to do this through just-in-time product development and cost reduction, the two main ways of maintaining the country's current strength in manufacturing.

CHAPTER 10

THE JAPANESE MODEL - WHAT IS IT AND TO WHAT EXTENT HAS IT DIFFUSED TO THE WEST?

C.A. Voss, Centre for Operations Management, London Business School, London, England

10.1. BACKGROUND

Over the past 30 years, Japanese companies have developed a series of distinctive new approaches to the management manufacturing. The resulting changes in the level of both product and production performance in sectors such as electronics and automotives have resulted in Japan being seen as the world leader in manufacturing. These approaches have been known by many names, initially, and still in Japan as Toyota Production System, and in the West as Just-in-Time management, continuous flow production and World Class manufacturing. Most recently and probably most accurately the term *lean production* has been coined by Krafcik (1988) and Womack et al. (1990) based on their study of the global automotive industry. Approaches such as Total Quality Management have also been a central part of Japanese approaches.

Over the past 20 years, these approaches have become known in the West and have begun to diffuse. Some Japanese practices have become well-known in the West, such as Just-in-Time and Total Quality Management, whilst others such as Quality Policy Deployment and Total Productive Maintenance have not.

Lean production was developed at Toyota during the 1950s and refined during the 1970s. By the end of the 1970s it was being diffused throughout the Japanese motor industry and into other industrial sectors, but remained virtually unknown in the West. Although the Japanese benefited from contact with Western manufacturing techniques, transfer of information in the reverse direction was rare.

The first recorded transmission of details of these approaches came in 1977 when a paper by Sugimori, Kushnohi, Cho, and Uchikawa describing the Toyota approach was published. In the late 1970s, Western businessmen and academics began visiting Japan and observing the radically different manufacturing approaches being used there. Initially, companies began to develop some Just-in-Time aspects of lean production together with some of the more easily observable techniques, notably quality circles and the Kanban scheduling technique. Early attempts were made to emulate the Japanese using these techniques, but success was limited, primarily because the techniques were only components of a complex set of interlocking approaches and, used in isolation, could achieve little.

P. Lindberg et al. (eds.), International Manufacturing Strategies, 179-191.
© 1998 *Kluwer Academic Publishers. Printed in the Netherlands.*

Full adoption of these approaches outside Japan began in the early 1980s, triggered by the publication of a book that for the first time described Japanese production methods in the full breadth and in easily understandable form (Schonberger, 1982). At the same time, a number of large companies, which included Ford, Hewlett-Packard, and IBM, sent study teams to Japan to try to understand in depth the approaches being used by the Japanese. Third, some of the Japanese transplants, notably Kawasaki Motorcycles and NUMMI, a Toyota joint venture, were practising lean production in the West, providing another platform for diffusion.

Japanese approaches were adopted by an increasing number of Western companies during the 1980s. Companies traditionally at the leading edge of manufacturing practice first began to experiment with them. Adoption was at first focused in the electronics industry and in parts of the automotive industry, although General Motors was notorious for being quite late in adopting lean production in its factories. Other companies, particularly those in engineering-based industries, soon followed. The role of Japanese transplants was less important in the early stages in the UK than in the USA: the most influential of these, Nissan, did not begin production until adoption of Japanese approaches was well underway in the UK.

The characteristics of lean production as derived from Womack's study have been summarised by Oliver et al. (1994) as including:

- Team based work organisations
- *Kaizen* or continuous improvement
- Lean manufacturing operations
- Low inventories
- Management of quality by prevention
- Small batch, Just-in-Time production
- High commitment human resource policies
- Close, shared-destiny relationships with suppliers
- Cross-functional development teams
- Close links with customers to allow make-to-order strategies

Of course, development of new manufacturing practices has not been confined to Japan. There have been some distinctive practices developed in other countries. Examples of these include Materials Requirements Planning (MRP) and Manufacturing Resource Planning (MRP II), developed in the USA, and ISO 9000 quality registration, developed from the BS5750 standard in the UK. In addition, it may be wrong to attach a national origin to many concerns and practices in manufacturing such as CAD/CAM, benchmarking, value analysis, etc.

10.2. RESEARCH QUESTIONS

The main purpose of our analysis of the IMSS study data has been:

- to identify areas where there are significant differences between Japan and the rest of the world;
- to examine the practices of the best practice countries—the United States and Japan—and to compare them with the rest of the world;
- to examine certain questions in more depth.

Previous studies of Japanese firms have focused on the large manufacturing sites such as automobile production. It has been an unanswered question as to whether lean approaches and the performance obtained by companies such as Toyota would also be found in smaller companies. Our sample enables us to do this, because it was composed of a wide range of companies with an average site size of just over 1000 people. Second, despite the increasing attempts to emulate Japanese firms, it is unclear whether Western firms have been able to do this, and whether having adopted the practices they are able to reap the benefits. In the context of comparing Japan with the West we sought to answer the following specific questions:

1. Is the pattern of manufacturing practices in our sample of Japanese companies consistent with lean production?
2. To what degree has the West adopted Lean Production or maintained a traditional pattern of manufacturing?
3. To what degree are there differences in payoff between Japan and the West?

10.3. METHODOLOGY

Manufacturing Practices
The IMS survey included questions on (1) the level of use and (2) the perceived payoff from different manufacturing activities. In order to explore the differences between Japan and the West, we have classified these activities as either being associated with Japan, for example those associated with Lean Production; those associated with the West for example MRPII and ISO 9000, and general practices, those that we would not necessarily associate with either East of West. The classification of these practices is shown in Figure 10-1.

Japan	West	General
Total Productive Maintenance	MRP	Benchmarking
QPD	MRP II	Value Analysis
Kaizen	Activity Based Costs	DFA/DFM
JIT/Customer Delivery	Plant-in-Plant Redesign	Health & Safety
JIT/Lean Production	ISO 9000	CAD
TQMP		CAM
Teams		Mfg. Strategy
QFD		Environment
Zero Defects		Energy Conservation
SMED		
Kanban		
SPC		
Simultaneous Engineering		

Figure 10-1. Manufacturing Practice - By Origin

Preventive Approaches
It has been argued that preventive approaches to manufacturing are central to Japanese manufacturing. The survey data contained two sets of data that enabled this argument to be examined: first, the split between preventive and rectifying maintenance, and second the split of quality costs on the traditional cost of quality approaches. In addition, the level of adoption of Total Productive Maintenance, a manufacturing practice that focuses on prevention, was measured.

Leanness
Lastly, lean production should be reflected physically in low inventories, in particular work in process and short lead times. These inventory levels and lead times were both measured.

10.4. Discussion of Results

The mean and standard deviation for each item was calculated by country. A one-tailed t-test was used to determine whether the score for each item by country was higher or lower than the score for the rest of the world.

Japan - The Lean Model
The results of the comparison of the level of use of each individual practice against the rest of the world are shown in Table 10-1, and between UK, US and Japan in Table 10-2.

Table 10-1. Adoption of Manufacturing Practices - Comparison of Japan and the West by Approach

Approach	Japan		Sig.	Total Sample	
	Score	Rank		Score	Rank
Japanese (lean) approaches					
Total Productive Maintenance	4.59	1	.000	2.36	24
Quality Policy Deployment	4.32	2	.000	2.78	20
Kaizen (Continuous improvement)	4.23	3	.000	3.00	3
JIT Customer delivery	3.96	6	.000	2.91	11
JIT Manufacturing	3.93	7	.000	3.03	4
Total Quality Management	3.85	10	.006	3.18	9
Team approaches	3.77	12	.072	3.38	2
Quality Function Deployment	3.73	13	.000	2.56	19
Zero Defects	3.64	15	.001	2.73	16
Single Minute Exchange of Dies	3.42	16	.000	2.40	26
Kanban (pull Scheduling)	3.31	20	.013	1.47	13
Statistical Process Control	2.73	24	.580	2.89	25
Simultaneous Engineering	2.68	25	.632	2.56	21
General Approaches					
Health and Safety programs	4.12	4	.002	3.50	6
Benchmarking	4.04	5	.000	2.42	23
Computer Aided Design (CAD)	3.92	8	.164	3.60	1
Environment protection programmes	3.92	9	.000	3.16	18
Defining a manufacturing strategy	3.80	11	.012	3.31	5
Value Analysis	3.69	14	.000	2.64	15
Design for manufacture/assembly	3.39	17	.000	2.49	12
Computer Aided Manufacture (CAM)	2.75	23	.840	2.69	14
Western Approaches					
Energy conservation program	3.36	18	.001	2.58	27
MRP	3.33	19	.527	3.16	8
Activity based costing	3.19	21	.001	2.40	22
Plant within plant redesign	3.14	22	.401	2.91	7
ISO 9000	2.62	26	.128	3.14	17
MRP II	2.26	27	.151	2.66	10

These data strongly support the view that Japan has a distinctive set of production practices. The data show both the degree of use of each practice, the ranking of the practices in order of use, and whether the level of use of each practice is significantly greater than the total sample. If we examine the practices associated with Japanese approaches of lean production, we see that they both rank highly and have levels of adoption that are significantly higher than those of the Western companies. If we look at those practices associated with the West, we find a different picture. They rank lowly: all six 'Western' practices are in the bottom 10 practices out of 27 overall, and in only two are they significantly greater, and in two of the most prominent Western practices, the adoption is lower, though not significantly than the West. In the unclassified 'general' approaches, there is no pattern in the ranking, though in most the level of adoption is significantly higher than the West.

Table 10-2. Manufacturing Practices - Comparison of Japan with UK and US

Approach	Japan Score	UK Score	t-test	Sig.	USA Score	t-test	Sig.
Japanese (lean) approaches							
Total Productive Maintenance (TPM)	4.593	1.568	0.000	***	1.000	0.000	***
Quality Policy Deployment	4.320	2.143	0.000	***	3.290	0.000	***
Kaizen (Continuous Improvement)	4.227	2.500	0.000	***	1.000	0.010	**
JIT-Customer Delivery	3.962	2.179	0.000	***	3.533	0.000	***
JIT-Manufacturing	3.926	2.475	0.000	***	2.333	0.121	
Total Quality Management (TQM)	3.846	2.882	0.003	**	3.538	0.300	
Team Approaches	3.769	2.889	0.538		1.000	0.245	
Quality Function Deployment (QFD)	3.731	2.162	0.000	***	3.414	0.000	***
Zero Defects (ZD)	3.636	2.425	0.005	**	3.351	0.007	**
Single Minute Exchange of Dies (SMED)	3.417	1.853	0.000	***	2.147	0.001	**
Kanban (Pull Scheduling)	3.308	1.923	0.002	**	3.000	0.133	
Statistical Process Control (SPC)	2.727	2.460	0.354		3.697	0.382	
Simultaneous Engineering	2.680	2.077	0.260		1.000	0.017	*
General approaches							
Health and Safety Programmes	4.120	3.395	0.022	*	1.000	0.888	
Benchmarking	4.042	1.795	0.000	***	1.000	0.000	***
Computer Aided Design (CAD)	3.920	3.658	0.910		2.844	0.709	
Environmental Protection Programmes	3.920	2.889	0.002	**	1.000	0.447	
Defining a Manufacturing Strategy	3.800	3.056	0.058		3.000	0.064	
Value Analysis	3.692	2.675	0.018	*	2.955	0.000	***
Design for Manufacture/Assembly	3.391	2.179	0.002	**	2.923	0.015	*
Computer Aided Manufacture (CAM)	2.750	2.667	0.948		3.414	0.368	
Western approaches							
Energy Conservation Programmes	3.360	2.763	0.286		1.000	0.245	
Materials Requirements Planning (MRP)	3.333	2.333	0.214		2.594	0.420	
Activity Based Costing (ABC)	3.192	1.667	0.000	***	1.000	0.000	***
Plant within Plant Redesign	3.143	3.000	0.968		2.452	0.631	
ISO 9000	2.619	3.711	0.001	**	2.639	0.382	
MRP II	2.263	2.794	0.125		2.629	0.066	

We can conclude from this that the Japanese companies do have a distinctive pattern of use of manufacturing practices. This is consistent with the lean production model described earlier, but also includes high level of use of general 'good practice'.

In contrast, the adoption of practice by Western companies would seem to have no distinct pattern. Some lean production practices are ranked highly, such as Kaizen, JIT manufacturing and teamwork, but in contrast to the Japanese so are MRP I and MRP II. Overall the level of use of these practices is less than that of Japanese companies. A possible explanation for the latter is respondent bias, but there is some evidence to support the view that any impact of this is limited in this case [1].

10.4.2. JAPAN - A PREVENTIVE CULTURE?

The results of comparison data on preventive approaches between Japan and the rest of the world are shown in Table 10-3. The most striking pieces of data are the level of adoption of Total Productive Maintenance (TPM), in Japan and the contrast between Japan and the West in its adoption, and the emphasis on prevention in Maintenance. TPM ranks first in the usage in Japan, and 24th out of 27 in the total sample. The data on spending on preventive versus rectifying maintenance show an almost reverse pattern between Japan and the total sample. which is significant at the .000 level.

This emphasis on prevention is reinforced by the emphasis on prevention shown in the breakdown of quality costs. Using the standard breakdown of costs of quality into inspection, internal external and preventive costs, the preventive costs in Japanese firms was 31.8% compared with 22.1% in the total sample, significant at the .072 level. This suggests that the Japanese have an underlying philosophy of prevention in manufacturing, one that is significantly different from Western companies.

10.4.3. PAYOFF - IMPLEMENTATION OR CULTURE?

The results of the comparison of the level of payoff of manufacturing practices are shown in Table 10-2. These data show that in most areas Japanese companies report significantly higher payoffs than their Western counterparts. This is consistent with reported difficulties that Western companies have in achieving payoff from new approaches such as Total Quality Management, and difficulty dealing with multiple initiatives.

Table 10-3. Spending on Maintenance and Quality (percentage and country rank)

Country	Maintenance				Quality							
	Preventive		Rectifying		Inspection		Internal		Prevention		External	
	(%)		(%)		(%)		(%)		(%)		(%)	
Argentina	33.8	13	66.2	13	39.5	18	30.8	15	24.9	6	6.8	1
Australia	41.2	5	58.8	5	24.6	3	25.1	8	30.1	3	20.2	12
Austria	37.2	9	62.8	9	39.5	17	22.6	4	20.3	15	17.6	9
Brazil	32.8	15	67.2	15	35.5	15	22.2	3	26.9	4	16.4	7
Canada	23.5	19	76.5	19	33.8	14	30.6	14	20.9	14	14.6	6
Chile	34.3	12	65.7	12	29.5	9	45.0	19	15.5	17	9.8	2
Denmark	40.9	6	59.1	6	24.9	4	24.1	7	24.0	9	27.0	18
Finland	49.0	2	51.0	2	29.0	8	29.0	12	17.3	16	24.8	17
Great Britain	29.5	18	70.5	18	28.0	6	18.8	1	30.2	2	22.0	14
Germany	33.8	14	66.3	14	31.9	12	23.8	6	24.5	8	19.9	11
Netherlands	43.5	3	54.5	3	39.5	16	25.4	9	23.7	10	11.4	3
Italy	37.7	8	62.3	8	44.3	19	19.8	2	14.3	18	21.6	13
Japan	66.2	1	33.8	1	30.5	10	23.7	5	31.8	1	14.0	5
Mexico	34.7	11	65.3	11	28.8	7	40.1	18	12.1	19	19.1	10
Norway	30.0	17	70.0	17	17.2	1	31.8	16	22.7	11	28.2	19
Portugal	41.4	4	58.6	4	31.4	11	25.8	10	25.8	5	16.9	8
Spain	30.6	16	69.4	16	33.2	13	28.7	11	24.8	7	13.3	4
Sweden	39.5	7	60.5	7	25.6	5	29.8	13	21.4	12	23.7	16
USA	35.4	10	64.6	10	21.7	2	33.7	17	20.9	13	23.6	15
Total	37.2		62.7		31.0		28.2		22.1		18.9	

There are a number of possible explanations for this difference. First is skill in implementation. It is not enough just to adopt a new approach; its implementation must be managed properly. This may be in two dimensions. First is the management of implementation of the individual practices. Second, and more subtle, but maybe more important is the management of the total system. Lean production is more than the sum of individual practices. Successful lean production requires implementation of these in an integrated manner. Each must link clearly with the others. Unintegrated implementation, the failure to take a total approach to the manufacturing system is likely to be ineffective.

It has also been argued that Japanese approaches are dependent on Japanese culture, and that difficulty in getting payoff is due to trying to transfer culturally embedded approaches to the West. Whilst there may be some truth in this, the well known difficulties of implementing MRP II indicates that implementation problems in the West are not confined to Japanese practices. This is supported by the data in

Table 10-4. Comparison of Payoffs for Adoption of Practices in Japan, UK and US

Approach		Japan	UK			USA		
		Score	Score	t-test	Sig.	Score	t-test	Sig.
Japanese approaches								
Kaizen (Continuous Improvement)	23.000	4.600	2.852	0.000	***	1.000	0.000	***
Total Productive Maintenance (TPM)	24.000	4.407	1.942	0.000	***	1.000	0.000	***
Quality Policy Deployment	16.000	4.160	2.444	0.000	***	2.280	0.000	***
JIT-Manufacturing	6.000	4.120	2.813	0.000	***	2.938	0.060	
Total Quality Management (TQM)	1.000	4.043	2.803	0.000	***	3.025	0.016	*
JIT-Customer Delivery	7.000	4.042	2.315	0.000	***	1.944	0.001	***
Team Approaches	21.000	3.917	3.662	0.573		1.000	0.406	
Zero Defects (ZD)	10.000	3.800	2.803	0.007	**	3.053	0.015	*
Quality Function Deployment (QFD)	14.000	3.800	2.566	0.001	***	2.840	0.002	**
Kanban (Pull Scheduling)	9.000	3.762	2.370	0.031	*	3.718	0.234	
Single Minute Exchange of Dies (SMED)	8.000	3.591	2.286	0.027	*	3.316	0.004	**
Simultaneous Engineering	19.000	3.333	2.491	0.170		1.000	0.907	
Statistical Process Control (SPC)	2.000	3.222	3.104	0.790		3.167	0.622	
General Approaches								
Computer Aided Design (CAD)	12.000	4.167	3.779	0.039	*	2.556	0.341	
Health and Safety Programmes	27.000	4.167	3.429	0.001	***	1.000	0.431	
Benchmarking	22.000	4.087	2.056	0.000	***	1.000	0.000	***
Environmental Protection Programmes	26.000	4.042	2.821	0.000	***	1.000	0.010	**
Value Analysis	15.000	3.875	3.030	0.001	**	3.423	0.003	**
Defining a Manufacturing Strategy	18.000	3.760	3.206	0.056		3.333	0.178	
Design for Manufacture/Assembly	13.000	3.571	2.552	0.006	**	3.594	0.600	
Computer Aided Manufacture (CAM)	11.000	3.474	2.932	0.281		2.962	0.723	
Western approaches								
Materials Requirements Planning (MRP)	4.000	3.875	2.825	0.023	*	3.086	0.255	
MRP II	5.000	3.500	3.347	0.449		2.156	0.325	
Energy Conservation Programmes	25.000	3.480	2.969	0.041	*	1.000	0.058	
ISO 9000	3.000	3.400	4.000	0.112		2.226	0.046	*
Activity Based Costing (ABC)	20.000	3.381	2.200	0.071		1.000	0.005	**
Plant within Plant Redesign	17.000	3.368	3.214	0.930		3.528	0.891	

Table10-4 that show that the payoff for most non-Japanese practices as lower than that for Japanese companies, even when rates of use are higher. There may be some respondent bias that affects the data, but as argued in endnote 1, this is probably low. Finally, it is possible that the Japanese processes of slow incremental change has enabled them over a period of 30 years to adopt these new approaches. In contrast, the attempts by the West to catch up may have resulted in companies trying to adopt too much too fast.

10.4.4. JAPANESE MANUFACTURING PRACTICES AND PERFORMANCE

It is to be expected that companies adopting lean production will be considerably 'leaner' than those not doing so. Leanness should be reflected in lower levels of stocks, particularly raw material and WIP, higher throughput efficiency ratios, and shorter delivery lead times. Techniques by themselves are not necessarily sufficient to lead to improved performance or lean production. Hanson et al. (1994) in a study of 663 European companies found that a significant set of the sites studied who had adopted a wide range of 'best practice' had failed to realise performance improvements.

The results of performance comparisons between Japan and the rest of the world are shown in Table 10-5. These data indicate that Japanese companies are indeed significantly leaner than their Western counterparts in all of these areas. The companies also show higher levels of capacity utilisation and delivery accuracy. This is not reflected in differences in market share or return on investment. There may be two explanations for the latter. First, that the Japanese companies in the sample were predominately competing in the domestic Japanese markets (on average exporting only 21% of their output). The Japanese market is notoriously competitive and supports low levels of profitability. Second, Japanese companies are reputed to look for lower returns on capital than Western companies.

Table 10-5. Percentage Improvement (1992 = base)

	Japan		World
	Change	Rank	Change
Conformance to specification (manufacturing quality)	27.0	8	28.8
Average unit manufacturing cost	15.1	7	14.1
Inventory turnover	11.7	17	26.5
Speed of product development	15.8	12	20.2
On-time deliveries	15.7	17	26.8
Equipment changeover	7.0	18	18.3
Market share	5.5	11	11.3
Profitability	4.7	13	10.8
Customer service	18.7	7	19.6
Manufacturing lead time	13.4	19	28.3
Procurement lead time	12.1	13	18.9
Delivery lead time	9.9	19	24.2
Product variety	12.9	10	17.4

10.4.5. THE WEST - LOSING OR CATCHING UP?

Since the early spread of knowledge about Japanese manufacturing methods, Western companies have tried to identify these practices and where appropriate adopt them.

This is in effect a diffusion process. Data on the diffusion of JIT in the UK have been published by Voss and Robinson (1987) and Voss (1992). Voss conducted detailed surveys of the spread of JIT/Lean Production, collecting data in 1985 and 1990. He used two measures of adoption: (1) whether the firm had implemented Lean Production in any form, and (2) whether it had a formal programme for the introduction of Lean Production. The results are summarised in Table 10-6.

The 1985 data indicate early stages of awareness of JIT and the elements of Lean Production. Whilst half the respondents were trying something, only 16% had initiated a serious programme. Given that the first reports of JIT management to the West had been in 1977 (Sugimori et al.), and that the Toyota Production System was well-established by 1970, this does not represent a very rapid rate of diffusion. However, as described earlier, significant diffusion did not begin in the UK until the first major publication on Japanese manufacturing methods by Schonberger in 1982.

The hypothesis that the West is catching up is supported by data from various sources that indicate that the best Western companies have adopted Japanese practices. These also indicate that there is a wide spread between the best and the worst. Data from a study of 663 northern European companies (Hanson et al. 1993) show that although the average level of adoption of Japanese manufacturing practices is not high, a small number of companies (2%), have successfully adopted them and realised the performance benefits, and a further 45% are well down the line to adoption. The International Motor Vehicle Project results reported by Womack et al. (1990) indicate that the gap between Japan and the West has remained wide in this industry.

However, the catch-up view of the West must be tempered by the consideration of whether they will catch up, or whether most companies will be running hard just to maintain their position. The case of Total Productive Maintenance illustrates this. TPM is one of the more recent developments in Japan, but has relatively low adoption in the West. This indicates that there is a continuing lag, and that catch up may not be taking place, rather that Japan is continually improving and though the best Western companies may be close to the leading edge, the majority although improving, may not be any closer to Japanese than they were five years ago.

Table 10-6. Spread of Lean Production in the United Kingdom

	% of UK firms	
	1985	1990
Implementing Lean Production	50%	79%
Formal programme for JIT/Lean Production	16%	58%
n	123	87

Source: Voss (1992)

10.4.6. APPROPRIATENESS FOR THE WEST

The above data have indicated the nature of the Japanese model, and the degree to which it has led to superior performance. It has also indicated the lag in diffusion of such practices in the West. A question that arises is the degree to which such practices should be adopted in the West. Womack et al. (1990) argue strongly that this is the only model for manufacturing, and support it with data from the automobile industry. However, Japanese approaches are not a set of individual techniques to be adopted overnight. They both interlink and build on each other. Ferdows and de Meyer (1990) argue that there is a natural sequence in implementing new approaches, and we might conclude from this that adoption of Japanese approaches will depend on the starting point of a company.

We can also postulate industry effects. Just-in-Time approaches are felt to be appropriate for repetitive environments. However, the IMSS data indicate that in many countries the level of make-to-customer order is much higher that make to forecast. This may indicate a low level of repetitive manufacture.

A further issue may be the local socio-economic environment may not be conducive to full or partial adoption of Japanese practices. In the above analysis we have compared Japan with all other countries in the sample, but there is considerable variation between countries. For example, the lack of a good quality supplier infrastructure may make JIT supply impossible; local culture may not support teamworking or employee involvement; and the variability arising from small local markets may make repetitive manufacture and hence lean production more difficult. For most countries, the change trajectory is based on the local socio-economic context and history. Some countries for good reasons are following a change trajectory based on hard, technology investment with softer, organisational approaches to follow.

However, we should not ignore the evidence that our Japanese sample, which contains medium-sized companies as well as large, many with a high proportion of make to order, has superior performance in many areas. We must examine more closely what is immutable in a country and what can be changed to lead to increased performance.

10.5. Conclusions

The above results have developed a picture of Japanese industry that is still significantly different from the West. This picture is summarised in the model in Figure 10-1. At the core, Japanese companies have much higher levels of adoption of Lean Production practices. To support this they seem to be able to get more out of a given practice indicating that they are probably better at implementation. In parallel with this they demonstrate a strong attitude of prevention, which goes beyond the traditional Total Quality route, and includes maintenance and maybe is reflected in more subtle ways in manufacturing. Finally, they also have higher level of adoption of some general manufacturing practices such as design for manufacture and value analysis. Our data indicate that this set of practices leads to superior performance that is consistent with a 'lean' approach to production. Overall, this suggests that the Japanese have an distinctive underlying philosophy in manufacturing that is different from Western companies and more effective.

Looking at these results from the West's point of view, these data are very disappointing. Lean production does seem to lead to superior performance, something that other recent studies (Womack (1993), Oliver (1994), Hanson et al. (1995)) have indicated. Yet, 20 years after the start of diffusion of Japanese practices, most Western companies are still well behind. We must be wary of averages. The European study by Hanson et al (1994) quoted earlier shows that the best Western companies have adopted Japanese practices and are achieving better performance than their poorer counterparts. However, this study also showed that there is a wide spread between the best and the worst.

It would seem that even today, Western companies overall are in the early stages of adoption of the Japanese practices that are often characterised by lean production. As diffusion theory would indicate, there are innovators who are well ahead, but there are far too many late adopters. It is also likely that the Japanese have moved on to new practices such as total productive maintenance and target costing, as well as having used the last 20 years to refine and make perfect practices that Western companies are adopting.

At best the data from this conclusion present worrying facts about practice in the industries examined in the West in this study. At worst, they indicate that to be successful Western companies will have to do more than just adopt practices faster and implement them better, they will also have to change their culture in manufacturing as well.

References

De Meyer, A., Nakane, J., Miller, J., and Ferdows, K. (1989) 'Flexibility, the next competitive battle - the Manufacturing Futures Study', *Strategic Management Journal*, **10**, 135-144.

Ferdows, K., and DeMeyer, A. (1990) 'Lasting improvements in manufacturing performance', *Journal of Operations Management*, **9**, 2, 168-183.

Hanson, P., and Voss, C. (1993) *Made in Britain*, IBM (UK) and London Business School, London.

Hanson, P., Blackmon K., Oak B., and Voss, C. (1995) 'Competitiveness of European manufacturing', *Business Strategy Review*.

Krafcik, J. F. (1988)'Triumph of the Lean Production system', *Sloan Management Review*, **30**, 1, Fall.

Oliver N., Delbridge R., Jones D., and Lowe J. (1994) 'World class manufacturing: further evidence in the Lean Production debate', *British Journal of Management*, June, 53-64

Schmenner, R., and Rho, B.H. (1991) 'An international comparison of factory productivity', *International Journal of Operations and Production Management*, **10**, 4, 16-31.

Schonberger, R. (1982) *Japanese Manufacturing Techniques*, The Free Press, New York.

Schonberger, R. (1986) *World Class Manufacturing: The Lessons of Simplicity Applied*, The Free Press, New York.

Sugimori, Y., Kushnohi, K., Cho, F., and Uchikawa, S. (1977) 'Toyota Production System and Kanban system: materialization of Just-in-Time and respect-for-human system', *International Journal of Production Research*, **15**, 6, 533-569.

Voss, C.A., 'The diffusion of Lean Production from Japan to the West', Paper presented to the *Academy of Management Conference*, Las Vegas, NV, August 1992.

Voss, C.A. and Robinson, S.J. (1987) 'Application of Just-in-Time manufacturing practices in the United Kingdom', *International Journal of Operations and Production Management*, 7, 4, 46-51.

Womack J.P., Jones D.T. and Roos D. (1990) *The Machine That Changed the World*, MacMillan.

Appendix 10-1.

A criticism of cross-national survey research is that results such as those described above could have an alternate explanation—respondent bias. That is in this case Japanese respondents could have put a higher score for the same actual level of use or payoff for a practice. A potential test for this is to examine questions where Japanese might be expected to score lower or equal to the Western respondents. To test for respondent bias we examined Japanese adoption levels of 'Western' practices. If Japanese managers had tended to respond more positively to 1-5 scale questions than there Western counterparts, we would expect that this would also show up in there response to adoption levels of Western practices. (This may be less true for general practices where Japan has been adopting them for a long time). If we examine the patterns of adoption of Western practices by Japanese companies in Table 10-1, we find much lower and more varied differences between levels of adoption in Japanese and Western companies. There being no significant difference in many of the core practices such as CAD, CAM, ISO 9000, MRP, etc. with a number having slightly lower, but not significant different, means. We can conclude that the impact of respondent bias in answering scale questions is probably low and that are second explanation that there are significant differences between Japanese companies and the West can be supported. The evidence from the IMS survey thus supports the second contention, that Japanese companies are different from Western companies.

CHAPTER 11

MANUFACTURING STRATEGY OF NORWEGIAN ISIC 38 INDUSTRY: A SYSTEMATIC PERSPECTIVE

Hongyi Sun, Jan Frick and Roar Hjulstad, Department of Management, Stavanger College, Norway

11.1. Introduction and a conceptual model

Norwegian ISIC 38 companies depend heavily upon international markets for both product sales (57%) and raw materials purchases (62%). In such international markets, competitiveness is a key issue. Only by knowing how well competitors perform can companies find the right place to make improvements. In this chapter, we will look at the manufacturing strategy of Norwegian ISIC 38 companies using the IMSS data, comparing the Norwegian sample with the IMSS sample companies in order to find out the significant similarities and differences. Additionally, we will provide global benchmarks so that Norwegian companies can compare themselves with companies in the world's leading countries.

The industrial structure of Norway differs from other developed countries in this survey because the oil industry and other raw material industries such as aluminium, paper, and ferroalloys are the main pillars of the Norwegian economy, whilst engineering manufacturing industry is in the minority. There are only 843 companies in the ISIC 38 industry in Norway, a very small percentage of the total Norwegian manufacturing industry.

Whilst we cannot draw conclusions about manufacturing strategy of Norwegian industry as a whole based on the data from the 20 companies that participated in the IMSS project, we think that the sample companies are representative of the ISIC 38 industry in Norway. Therefore, although our findings and results are limited to the Norwegian ISIC 38 industry, some of the topics such as human and organisational issues may be generalised to other manufacturing industries, supported by other information and knowledge of Norwegian companies from our previous research and consulting.

The presentation of this chapter will be guided by the conceptual model shown in Figure 11-1 (Sun and Gertsen, 1994; Gertsen and Sun, 1994). This figure illustrates the scope and relevant variables of manufacturing strategy in this chapter. In this conceptual model, manufacturing is only one of the functions of a company—other functions such as marketing and R&D as well as company performance will also influence the

P. Lindberg et al. (eds.), International Manufacturing Strategies, 193-214.
© *1998 Kluwer Academic Publishers. Printed in the Netherlands.*

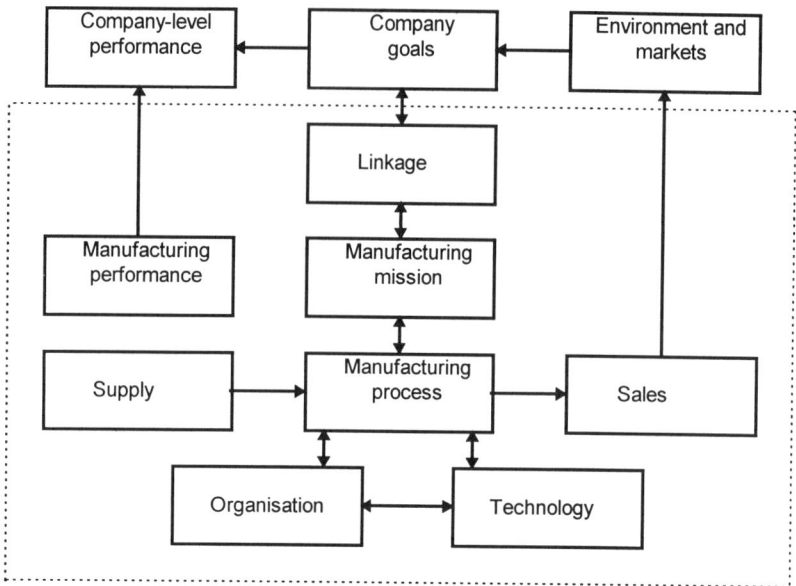

Figure 11-1. A Conceptual Model of the IMSS's Scope of Manufacturing Strategy

company goals. However, since this study is of manufacturing strategy, other functions will not be the focus.

Manufacturing strategy is concerned with the development and implementation of plans that affect the firm's choices of production resources, the development of these resources, and the design of the infrastructure to control the operation activities (Cohen and Lee, 1985). In short, manufacturing strategy will include at a minimum the elements of goals/market, practice, and performance. Goals/market/strategy include management statement of goal priority (e.g., quality, cost, delivery), market characteristics such as market coverage and customer focus, and the extent to which manufacturing strategy is linked to corporate strategy. Practice is divided into three groups of factors: programmes, technology and organisation. Performance factors concern both company performance, such as return on investment and inventory turnover, and manufacturing performance, such as cost, quality and delivery.

11.2. Manufacturing Strategy of Norwegian ISIC 38 industry

11.2.1. PRIORITY OF COMPANY GOALS

We found that the Norwegian companies placed different priorities on company goals than did companies in other survey countries. First, participants were asked to rank six business unit goals by the degree of their current importance to the company on a scale

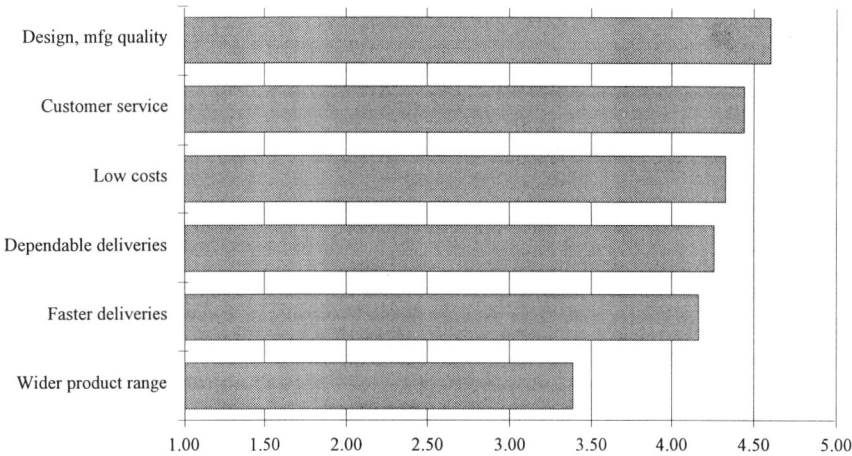

Figure 11-2. The Priority of Company Goals in 1992 - IMSS (1-5 scale from 'not important' to 'very important')

of 1 (not very important) to 5 (very important). The average for each goal is illustrated in Figure 11-2. Although the averages seem to be rather close, the difference between each pair of goals is statistically significant.

These results show that offering superior product design and manufacturing is considered to be the most important goal, followed by customer service and cost reduction. However, the Norwegian data show a rather different pattern, as shown in Figure 11-3. Most of the Norwegian companies give top priority to dependable delivery and customer service, which relate to both marketing and manufacturing.

Second, to assess the trend of goal priority in the near future, the survey also asked what goals will be important in the next two years, and whether the company has quantified corresponding goals. Sixteen different manufacturing priorities were measured. In order to highlight the main priority differences in future goals, the sixteen variables were grouped using factor analysis. Five factors were obtained:[1]

- *New product introduction*: Speed of introducing new products to market, including rapid design changes and reduce new product development cycle.
- *Cost reduction*: Reduce cost including reduce unit cost, overhead cost, material cost and improve direct labour productivity.
- *Delivery speed*: Increase delivery speed including increase delivery speed, increase delivery reliability and reduce manufacturing lead-time.
- *Quality*: Improve conformance quality and supply quality as well as delivery reliability.

---

[1] Two priorities—volume changes and white collar productivity—are not included in any of the factors.

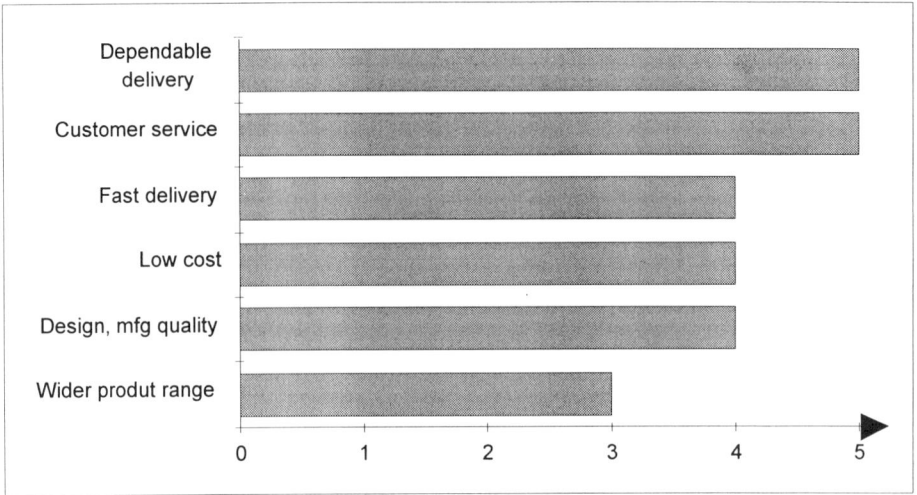

Figure 11-3. The Priority of Company Goals in 1992 - Norwegian Companies (1-5 scale from 'not important' to 'very important')

- *Supply chain management*: Improve supply including reduce number of suppliers, reduce procurement lead time and reduce inventory.

The average scores for the five factors are calculated and illustrated in Figure 11-4. The analysis shows that companies will give high priority to cost reduction in various

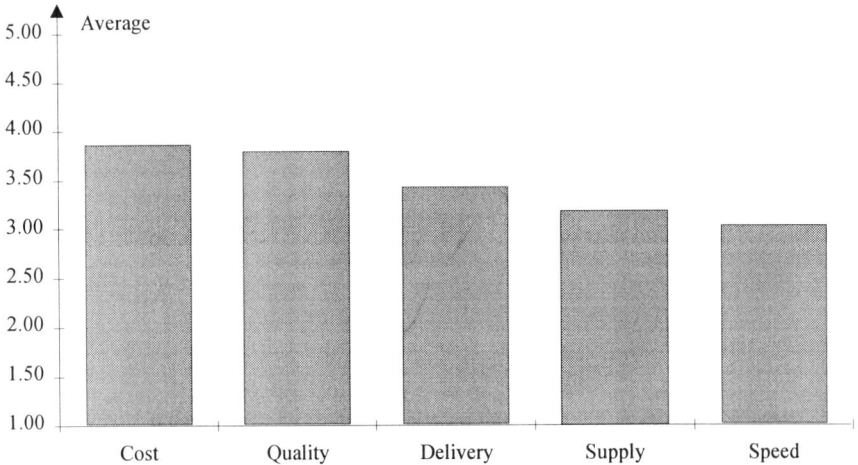

Figure 11-4. The Priority of Company goals in the Next Two Years - IMSS (1-5 scale from 'not important' to 'very important')

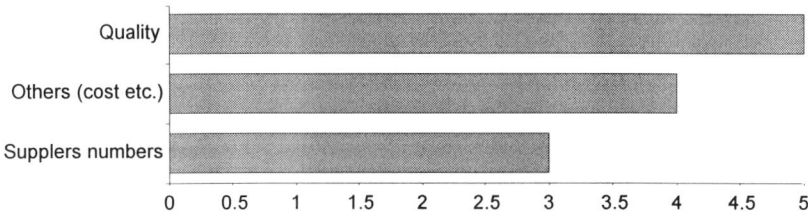

Figure 11-5. The Future Goals of Norwegian Companies (1-5 scale from 'not important' to 'very important')

areas. This suggest a focus shift from quality and delivery to cost. At least cost and quality are nearly equally important. However, the conclusion only fits the next two years. The focus may shift to other goals beyond two years. Additionally, this is only a general pattern. The possibility cannot be ruled out that a specific company or specific country may still focus on quality and delivery rather than cost reduction. For example, Norwegian data show a shift from delivery to quality as shown in Figure 11-3 and Figure 11-5. (Since the Norwegian sample is small, they cannot be reorganised by factor analysis).

11.2.2. MARKET SITUATION AND ITS INFLUENCES ON GOAL PRIORITY

The market situation includes the market share of the company's dominant product line, the market share of competitors, and the direction of market development. Each of the market dimensions will be discussed in the following subsections. The main difference found is that Norwegian companies have a higher average market share than the general IMSS sample.

Company market share
Market share ranged from 0.5% to 100% with an average of 34%, as shown in Table 11-1 below. The Norwegian average market share of 49% was the highest among the 20 countries. One reason may be that most of the Norwegian companies are small and focus on one or two specialised products in a niche market; a second is that many Norwegian companies are regular suppliers to sister companies or oil companies.

Table 11-1. The Benchmarks for Market Share of Dominant Product Line

	Min	25%	50%	75%	Max.	Avg.
Company market share						
IMSS	0.5	15	30	50	100	33
Norway	12.0	26	50	59	100	49
Competitor's market share						
IMSS	2	15	20	30	95	24
Norway	6	15	30	32	50	25

Table 11-2 Direction of Market Development (% of companies)

Market development	Rapidly growing	Growing	Stable	Declining	Declining rapidly
IMSS (n=585)	4%	26%	50%	19%	1%
Norway (n=18)	0%	28%	44%	28%	0%

Leading competitor's market share
In the IMSS sample, respondents reported that their main competitors' market share varied from 2% to 95%, with an average of 24%, as shown in Table 11-1. The average for the 20 countries was very similar. Norwegian companies have an average of 25%, which is very close to the overall IMSS average, indicating that the competitive environment of Norwegian companies is much the same as other countries.

Direction of market development
Norwegian companies followed nearly the same pattern as the world sample with regard to the direction of market development, as shown in Table 11-2. Half of the companies expected a stable market, 30% a growing market, and 20% a declining market. However, few companies expected their market to rapidly grow or decline.

11.2.3. PERFORMANCE

This section focuses on the Norwegian companies performance improvements, which showed positive results for profitability and on-time delivery, but less improvement in market share, customer service, inventory turnover and product variety. The performance improvement will be discussed at the company level and manufacturing level respectively. Each of the performance indicators was measured subjectively by the percentage of improvement (either positive or negative).

Company-level performance
Company-level performance indicators reflect the competitiveness of the company as a whole, rather than at the functional level, although company performance is related to and influenced by the functional area performance. Company-level performance indicators include market share, customer service, product variety, on-time deliveries,

Table 11-3. Company-Level Performance Improvement (%)

Benchmarks	Norway						IMSS					
	Min	25%	50%	75%	Max.	Avg.	Min	25%	50%	75%	Max.	Avg.
On-time delivery	5	5	10	45	500	57	-30	10	20	30	1000	30
Market share	-20	-7	5	10	30	2	-100	1	8	15	1000	14
Customer Service	5	5	10	15	20	11	-90	10	15	25	200	23
Product variety	-100	-20	10	10	40	-7	-200	5	12	20	1000	21
Inventory turnover	-85	10	20	20	199	11	-10	-8	10	30	100	19
Profitability	-85	10	20	30	999	29	-200	-10	5	20	600	12

Table 11-4. Manufacturing Performance Improvement (%)

Benchmarks	Norway						IMSS					
	Min	25%	50%	75%	Max.	Avg.	Min	25%	50%	75%	Max.	Avg.
Quality	5	10	20	20	300	35	-10	10	20	30	1000	30
Speed of product development	-10	5	15	29	200	38	-30	10	15	30	400	24
Equipment changeover	5	8	10	13	50	14	-50	6	15	30	300	23
Mfg lead-time	-20	10	15	30	100	24	-80	10	20	30	999	31
Cost reduction	-5	10	10	20	30	14	-50	5	10	20	300	15

profitability, and inventory turnover, as shown in Table 11-3. These indicators are reported by quartile, as well as minimum, maximum, and average.

On average, Norwegian companies reported a higher level of improvement in on-time delivery (57% versus 30%) and profitability (29% versus 12%). The results are consistent with their emphasis on delivery as a goal priority. Others indicators such as market share, customer service and inventory turnover were lower than IMSS. Market share can be explained by the fact that the absolute market share of Norwegian companies is somewhat higher, making it more expensive to increase the market share further. Product variety is due to the fact that Norwegian companies are small- and medium-sized companies that specialised in a few products.

Manufacturing Performance
Manufacturing performance is related to the following manufacturing objectives: conformance to specification (quality), equipment changeover (flexibility), manufacturing lead time (delivery/dependability), and unit cost (cost and productivity). These variables are consistent with the manufacturing objectives proposed by researchers (e.g., Skinner, 1978; Macbeth, 1989; Riis, 1992; Tunalv, 1991). Traditionally, the speed of product development has been influenced mainly by the R&D department. However, current product development practices such as concurrent engineering are very much cross-functional. Therefore, the speed of product development can also be recognised as a performance indicator of manufacturing. The results for manufacturing performance improvement are listed in Table 11-4. Norwegian companies scored neither highest nor lowest on most of the indicators.

Perhaps it is difficult to form an overall picture of the company by separate individual indicators—multiple aspects of performance need to be considered. Therefore, we have formulated a comprehensive index for the improvement of performance as the average of the percent ranks (0-100%) of all the performance indicators. The comprehensive improvement index is shown in Figure 11-6. Norwegian ISIC 38 industry as a whole is somewhere in the middle. A few individual Norwegian companies are outstanding, for example, companies 2 and 7. Company 2 will be further described later as a case.

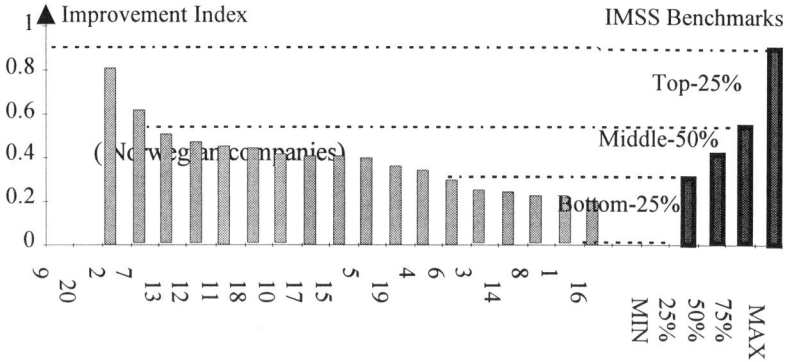

Figure 11-6. The Benchmarks for the Comprehensive Index of Performance Improvement

On the other hand, performance improvement does not necessarily reflect the competitiveness and current status of performance of a company, which are determined by two factors. The first is its current level of performance, and the second is the amount of improvement. Some companies are already rather competitive and, even if they did not improve much they would still be competitive. Other companies might have a low current level of performance and, even if they improve greatly they will still not be competitive. Thus, these data reveal more about performance improvement than competitiveness.

To understand the absolute standing of a company, other performance indicators must be considered. Figure 11-7 provides a comprehensive performance index, which is the average of the percent ranks of return on investment (ROI), sales on investment

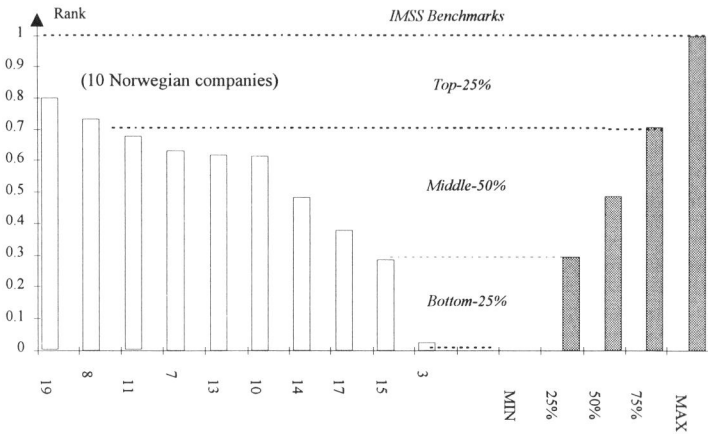

Figure 11-7. Comprehensive Performance Index - Norwegian Companies Benchmarked against IMSS Sample

Table 11-5. Corporate and Market Goals Translation into Manufacturing Strategy vs. Performance Improvement (%)

% of performance improvement	(1= none, - , 3= informal , - , 4= formal)					Significance
	1	2	3	4	5	
Quality	.	14	27	25	26	yes
Cost	.	8	15	12	11	no
On-time delivery	.	11	25	24	26	yes
Inventory turnover	.	19	22	24	30	yes
Market share	.	-1	8	12	8	yes
Profitability	.	-4	6	11	15	yes

Note: Level 1 is omitted since only a few cases existed)

(SOI), profit per person (PPP), sales per person (SPP) and return on sales (ROS). The index is an aggregate indication of capital productivity and labour productivity. In Figure 11-7, it is rather easy to find where the Norwegian companies are located against the IMSS benchmarks. A score of 100% means the company is in the first place. A score of 50% means that the company is in the middle.

Two of the ten Norwegian companies are in the top 25% group. Four of them are rather close to the top 25% group. Three of them are in the middle. Only one of the ten companies is in the bottom 25% group. The ten Norwegian companies have an average of 53%, whilst the IMSS average is 49%.

11.2.4. COMPANY GOALS AND MANUFACTURING STRATEGY

The linkage between corporate goals and manufacturing strategy (both top-down and bottom-up) correlates significantly with manufacturing and economic performance. However, analysis of variance (ANOVA) reveals that the more the manufacturing is involved in the formulation of company goals—and *vice versa*—the higher the performance, as shown in Table 11-5 and Table 11-6, respectively.

The pattern for Norway, as shown in Table 11-7, is more or less the same as that for the IMSS. A general suggestion can be made that a formal process should be established to link corporate strategy and manufacturing strategy. What is more, the linkage should be mutual, rather than either top-down or bottom-up. However, the

Table 11-6. The Influence of Manufacturing Strategy on of Company Strategy Development vs. Performance Improvement (%)

% of Performance improvement	(1= none, - , 2= some, - , 5=a lot)					Significance
	1	2	3	4	5	
Quality	.	22	23	28	35	yes
Cost	.	8	13	14	16	yes
On-time delivery	.	21	20	27	29	yes
Inventory turnover	.	21	23	25	34	yes
Market share	.	8	7	11	12	yes
Profitability	.	7	2	13	23	yes

Note: Level 1 is omitted since few cases existed

Table 11-7. Benchmarks for Top-down and Bottom-up Linkages (%)

	Top-down linkage					Bottom-up linkage			
	None	----	Informal	-----	Formal	Not at all	Partially		A lot
IMSS	7%	8%	33%	36%	17%	3%	15%	39% 30%	13%
Norway	5%	15%	30%	30%	20%	0%	10%	45% 20%	25%

IMSS data suggest that companies pay more attention to top-down linkage (36%+17%), than to bottom-up approaches (30%+17%). In the Norwegian companies, bottom-up and top-down linkages were more balanced. In about 25% of the Norwegian companies, manufacturing has a lot influence on the development of corporate strategy, whilst the percentage for the total IMSS data is only 13%, as shown in Table 11-7.

11.2.5. THE SUPPLY CHAIN

Sources of supply
The main difference in the supply chain between Norway and IMSS is the geographic location of suppliers. Other dimensions such as number of suppliers, relationship with suppliers, and the percentage of JIT materials/components are more or less the same. The Norwegian companies depend upon international markets of raw materials, as listed in Table 11-8. About 62% of raw materials were purchased from abroad, mainly from the European Union (EU), compared with about 42% for the IMSS.

International markets
Among the Norwegian ISIC 38 companies, 57% of sales is exported to other countries, higher than the IMSS average of 37%, and the highest among the 20 countries (see Table 11-9). 25% of Norwegian companies export more than 80% of their products, whilst the same figure for IMSS is only 66%. The finding is consistent with the Norwegian annual statistics. Svendsen (1992, p.15) reports that, for all the Norwegian manufacturing industries (excluding oil and gas), 45% of their revenue is derived from sales abroad, and 16% of the revenue represents the sale of products produced and sold abroad. In sum, 61% of the Norwegian manufacturing revenue is from abroad. Since the Norwegian domestic market is limited, Norwegian companies have to export and would like to export. In fact, the country has a long tradition of exporting, from sardines to salmons, from petroleum to platforms.

Most of the exports go to the EU countries, about 85% of total exports. In 1992 Germany became the most important export market of Norway; this marked the first time that Sweden was not the most important export market for Norway, mainly due to

Table 11-8. Sources of Purchase of Norwegian Companies

Area	Norway	EU	Others
Average % of purchase	38%	52%	10%

Table 11-9. Exports as a Percentage of Sales

Benchmark	Min	25%	50%	75%	Max	Avg
IMSS	0	6	25	66	100	37%
Norway	0	5	30	80	100	52%

revealed that 37% of companies voted Germany as the most important export market of Norway through 1997. It is forecast that Sweden will hold a strong second place. However, it looks as if Sweden's EU membership and Norway's rejection may negatively influence the economic relation of the two countries. Additionally, Norway has been actively exploring the export markets of developing countries in Asia, especially South-East Asia. The Norwegian government initiated an Asian Strategy for Export at the end of 1993, and its premier visited China with a large business delegation the next November to further promote the Norwegian exports.

11.2.6. ORGANISATION AND PEOPLE MANAGEMENT

Structure
Most of the companies in the IMSS sample were small or medium-sized enterprises (SMEs), with an average number of about 770 employees. Only one Norwegian company had more than 1000 employees, whilst the rest were SMEs, who account for a large percentage of manufacturing industry in Norway. Among the 843 ISIC 38 Norwegian companies, 22 (3%) companies had over 700 employees, 229 (27%) between 50 and 700, and 592 (70%) fewer than 50.

Although the number of organisational levels varied from 2 to 9 in the IMSS sample, most companies, about 83%, reported having from 3 to 5 levels. 2 to 6-level companies occupy 5% respectively. Very tall or very flat organisation are very few. Norwegian companies are all from 3 to 5 levels, which is much flatter than most of the IMSS and most of the countries. The organisational level is slightly correlated with the size of the company (r=0.27, p=0.001). However, another factor may be the influence of lean production on organisational structure, and partly only to the smaller size of Norwegian companies.

The span of control varies from a few to one hundred for the IMSS sample, whilst the span of control in Norwegian companies was quite moderate, being neither too narrow nor too wide. Traditional organisation theory suggests that the span of control is related to the size of the organisation; however, the empirical data did not show any relationship between the span of control and the number of organisational levels, which may be the result of moving towards leaner and flatter organisations.

Table 11-10. Companies Using Different Payment Systems (%)

	Group incentive	Individual incentive	Fixed salary
% of IMSS (n=449)	33%	34.5%	34.5%
No. of Norwegian (n=18)	27%	6%	67%

Table 11-11. Percentage of Companies using Different Incentive Systems

	Productivity	Quality	Output	Profit
IMSS (n=435)	24%	44%	12%	22%
Norway (n=6)	1	5	2	0

Motivation

Payment system and Incentive methods. The percentages of companies using different payment systems are illustrated in Table 11-10. Within the IMSS sample, about one-third of respondents reported using each of the three payment systems. However, about 67% percent of Norwegian companies used a fixed salary, and 27% group incentive systems, whilst only 5% used individual incentive systems. This may reflect the social welfare heritage of the Scandinavian countries, the fact that most of Norwegian labour unions require fixed salaries, and the long history of the Social-Technical movement in Nordic countries. It may also indicate payment system changes on the shop floor after the introduction of new technologies. Our previous research (Sun, 1993) found that in one company, workers working around FMS get a fixed monthly salary, similar to office clerks.

The adoption of different incentive payment systems is shown in Table 11-11. 67% of companies replied that they use an incentive system, whilst 34.5% said they use fixed salary. However, that 435 companies replied that they used certain incentive methods suggest that most (132) of the companies using fixed salary (151) also used incentive systems. To our knowledge, both fixed salary and incentive methods are used together in Norway. The main incentive system was based on quality, used by about 44% of the companies. The second most popular incentive method was productivity, used by 24% of the sample. About 22% of the companies used profit as an incentive method. The gain-share programmes proposed recently suggest that, as the final performance measure of the company is profit, profits should be shared by all the people in the company. It is also worth mentioning that the traditional workshop incentive methods, output or piece payments, have not been that popular. The Norwegian data are not sufficient to give further comments based on data. However, according to our knowledge, productivity is the most commonly used incentive or bonus method.

Personnel development. Payment systems and incentive methods are only the way to allocate the gains of a company. However, if nobody is able to fulfil the goal, nobody will gain anything. How to capture this aspect? In this section, we will look at the factors that are related to the development of personnel as well as groups. The specific dimensions as well as the benchmarks are shown in Table 11-12. Norwegian companies have fewer job classifications, thus, each worker will be responsible for more tasks. This may be due to the small company size.

Table 11-12. Benchmark for Dimensions of Personnel Development

Dimensions	Measurement	IMSS Benchmarks					Average	
		Min	25%	50%	75%	Max	IMSS	NOR
Job classifications	No. of job classifications	1	5	9	15	120	14	11
Suggestions	No. of suggestions per employee per year	0.02	0.5	1	3	90	4	2
Teams	% of employee in teams	0.5	10	30	75	100	42	38
New worker training	Hours per year per person	3	30	50	100	750	90	145
Regular worker training	Hours per year per person	1	11.5	21	42	225	32	21
Skilled workers	% of skilled workers	1	20	46	70	100	46	45
Job rotation-1	1-5 scale from never to frequently	1	3	3	4	5	3	3
Job rotation-2	Average number of jobs in rotation	2	2	3	4	9	3	4
Personnel turnover	% change of personnel per years	0.01	2	4	10	43	9	3
Absent rate	% of absentee	0.03	3	4.3	6.5	15	5	4

The average number of suggestions per employee per year in Norway was not only lower than the IMSS average, but also lower than in many other countries. Japan has an average of 34 suggestions per person per year, the highest in the survey. On average, Japanese employees provide 10 times more suggestions than employees in other countries. The difference is significant ($F=10$, $p<0.001$). One explanation is that Norway has a different recording system, where not all the suggestions are recorded, whilst in other countries, for example, in Japan, all the suggestions are written down.

The Norwegian use of teams is less than the IMSS average, which is a little strange, as, since the 1970s, Scandinavian countries have implemented the Social-Technical System as well as teams. The average may be diluted by some companies that have few employees working in teams. In at least one-fourth of the Norwegian companies more than 66% of employees work in teams. The average percentage of employees working in teams in Japan is 56%, the highest among the 20 countries.

The training time for new workers in Norway is much longer than the IMSS average. In fact, the Norwegian average hours for new worker training is in second place, next to Portugal. This is because many companies related to oil companies are influenced by the complicated features of offshore operations. The Norwegian new workers are very well prepared. On the other hand, the training hours for the regular workers are less than the IMSS average, which can be explained by the longer training hours for new workers.

The percentage of skilled workers and the degree of job rotation in Norway are close to the IMSS average. Both the percentage of personal turnover percentage and the absenteeism rate in Norway are lower than the IMSS average. This is an indication of stable employees and stable employment in Norway. Historically, due to geographical reason and available job opportunities, Norwegians have traditionally stayed in their hometowns instead of going to other places for jobs.

11.2.7. TECHNOLOGY

Integration level of manufacturing technology
A technology issue that is related to both process technology, engineering technology and information technology is the level of integration. All kinds of cases can be found along the scale, from no computerised integration to fully computerised integration. However, 44.4% of companies did not implement any advanced manufacturing technologies. This is consistent with the percentage of companies that adopted NC machines (45%). At the other extreme, only 0.33% of the sample companies, 2 out of the 600 sample companies, have reached the full integrated level of integration. Most are somewhere in the middle, ranging from stand-alone to 'islands of automation'. The integration patterns are more or less the same for all the participating countries, including Norway.

Technology diffusion
The adoption of new process technology is one of the major innovations in manufacturing companies. Many companies have adopted some kind of advanced manufacturing technology, such as CNC, MC, FMS, robots, etc. The benchmarks for these AMT items are illustrated in Table 11-13. Since the adopters of computer-aided process technologies are still rather few, a separate measure of adoption rate is also provided.

The use of AMT process technologies is still very low compared with conventional machines. For example, only 18% of the companies have adopted FMS. NC machines account for the largest percentage (55%)—less than we expected. On average, Norwegian companies are quite close to the world average in percentage of companies adopting advanced process technologies except for the application of robots in assembly. This is because many Norwegian ISIC 38 companies do not have assembly operations; for example, robotics would not be relevant to a cable company.

No significant statistical relationship was found between technology and performance improvement for various measurements such as the number of equipment, technology intensity, or the percentage of adoption. However, it was found through ANOVA (Analysis of Variance) that there are differences in performance improvement

Table 11-13. Benchmarks for the Diffusion of Advanced Process Technologies and Adoption Rate (Percentage of Companies Adopting)

Area	Advanced Technology	IMSS Benchmarks					Adoption Rate	
		Min	25%	50%	75%	Max.	World	Norway
Manufacturing	FMS	0	0	0	0	150	18%	15%
	NC	0	0	2	10	500	55%	65%
	M-centre	0	0	0	2	170	34%	45%
	Robots	0	0	0	1	600	29%	20%
Assembly	Robots	0	0	0	0	398	15%	0
	FMS	0	0	0	0	300	18%	15%

between those companies that adopted advanced process technologies and those did not. However, the differences do not mean that the more new technology is adopted, the more performance will be improved. According to our previous research (Frick and Riis, 1991a&b; Sun, 1993; Sun, 1994), those companies that combined organisational and technological development tended to improve performance more than others. Based on these studies, we have proposed a balanced development approach.

11.2.8. MANUFACTURING PROCESS

It is the manufacturing process (or production process) that transforms the raw materials into the final products. If the manufacturing process cannot meet the goals or market requirements, companies will still not make any profits, even if there is a huge demand.

Percentage of forecast versus customer orders
A question at the beginning of the production process is: how is the production triggered? Production can be triggered by forecast order and/or customer order. The results for customer order for the IMSS and Norwegian samples are in Table 11-14. The data indicate that both Norwegian and other countries' companies produce more to customer order than forecast—nearly 34% of the sample companies produce totally (100%) according to customer order; in Norway, the corresponding figure is 42%. Nearly 62% of the IMSS sample and 72% of Norwegian production is to customer order when measured by number of orders.

In fact, very few companies—only 2% of the sample companies—produce totally to forecast order. The pattern is even more obvious for the Norwegian companies. From this, we infer that customer satisfaction, customer quality requirements, customer service, etc. will become more and more important for Norwegian companies.

Percent of capacity utilised
The Norwegian average capacity utilisation is rather high among the 20 countries, and higher than the IMSS average, as shown in Table 11-14. Additionally, it was found that the Norwegian minimum capacity utilisation is 65%, whilst the IMSS average is 10%. This can be explained by (1) the stable market of Norwegian companies, and (2) manufacturing capacity policy. About one-third (30%) of Norwegian companies claimed that capacity should be less than demand, whilst only 12% of IMSS companies adopted such a policy.

Degree of product-process variety
According to averages, there is no obvious difference between Norway and IMSS in terms of process variety. However, the detailed benchmarks revealed some differences, as shown in Table 11-14. The IMSS data are normally distributed along the scale from

Table 11-14. Benchmarks for Manufacturing Process

	Max	25%	50%	75%	Min	Average
Customer orders (%)						
IMSS	0%	30%	70%	100%	100%	64%
Norway	0%	30%	91%	100%	100%	72%
Capacity utilisation (%)						
IMSS	10	70	80	90	130	81
Norway	65	80	89	97	110	87
Process variety[1]						
IMSS (%)	12%	21%	24%	30%	13%	3.2
Norway (No.)	3	2	2	8	0	3.1

Note:
1. 1 = many different processes for different products; 5 = single process routine for all products.

many processes to a single routine. However, more than half of the Norwegian companies are rather close to a single routine. This may also be explained by their smaller size and limited scope of products.

Inventory in production days
Inventory holdings measured as days of production varied from a few days to several hundred days. Perhaps it is difficult to compare the days of inventory of different companies, even they are in the same industry. For example, a car manufacturer and a ship manufacturer will differ a lot in inventory days. This is exactly what we found out from the Norwegian data. If we look at the normal average of Norway and IMSS, we will find out Norway has longer inventories in raw material/component, in-process as well as finished goods. However, the differences are not statistically significant.[2]

Order-to-delivery lead-time
The Norwegian average (49 days) order-to-delivery time is 9 days longer than the IMSS average (40 days). However, the F-test is not significant (F=0.5, p=0.48). The explanation is that the average was inflated by one or two companies.

11.2.9. SUMMARY

We found that the Norwegian ISIC 38 companies are different from the IMSS as a whole in many aspects. Their special features are summarised below.

1. *Company goals and manufacturing objectives* - The IMSS data revealed a shift from quality to cost, in line with the 'sand cone' model. However, Norwegian sample companies do not follow the same pattern, instead emphasising dependable

---

[2]   In reviewing the Norwegian data, we found that a Norwegian ship-building company has a 300-day in-process inventory and there are only six Norwegian companies answering this question. It is not difficult to conclude the higher Norwegian averages were inflated by this ship-building company. If this company is excluded, the Norwegian average will be very close to the IMSS average.

delivery and customer service. This may be because the Norwegian companies were small suppliers to oil companies or sister companies.

2. *Market environment* - Although Norwegian companies faced the same competitive environment in terms of competitors' market share and the stability of market development, they reported higher market shares.

3. *Performance improvement* - Norwegian companies had higher average improvement in on-time delivery and profitability, consistent with their company goals. Norwegian companies emphasise on-time delivery more than others. There are no obvious differences in the improvement of manufacturing performances between Norwegian companies and IMSS companies. On average, the performance improvement of Norwegian companies was above the IMSS sample.

4. *Linkage between company goals and manufacturing* - The linkage or integration between company goal and manufacturing was positively related to both the manufacturing and company performance. However, more companies in the IMSS sample had implemented top-down linkages, whilst fewer had implemented bottom-up linkages. The Norwegian companies had more balance between bottom-up and top-down linkages.

5. *Supply chain* - Norwegian companies import around 60% of raw materials and components, higher than average. Norwegian companies heavily depend on international markets, with about 57% of their sales going to the international market, especially Germany, Sweden, Denmark and other EU countries. The government is working hard to explore the Asian market

6. *Organisation* - Norwegian companies really have some unique features in organisation and personnel aspects, including small size, flat structure, fewer job classifications, more employee training, and stable employment. The payment system is also quite different from other countries, characterised by more fixed salaries plus productivity-based bonuses.

7. *Technology* - The most distinctive finding about the level of adoption of advanced manufacturing technology (AMT) by Norwegian companies was the combination of technological and organisational development, as well as the integration between technology, organisation and the manufacturing missions and/or company goals.

8. *Manufacturing process* - Most Norwegian companies reported: (1) a single routine process for all or most products, (2) higher utilisation of their capacities, (3) customer-order-driven production, and (4) short delivery times.

In conclusion, the manufacturing strategy of Norwegian companies is characterised by (1) emphasising delivery and customer service, yet not ignoring other manufacturing missions such as cost reduction and quality; (2) proper inter-linkage of company goals and manufacturing strategy; (3) balanced development of advanced manufacturing technologies and organisation/personnel; and (4) focused/specialised production process. Such a manufacturing strategy does not focus only on one or two elements of manufacturing, but all of them as well as their integration. Under such a manufacturing strategy, the goals and the means, the organisation and technology, the internal and the external, are well linked and interrelated. In one word, it is a systematic manufacturing strategy as illustrated in Figure 11-1. In next section, a case company will be described to provide the details of such a systematic manufacturing strategy.

11.3. Case Study - Company K

Company K, one of the Norwegian sample companies in IMSS, has been studied and followed for about ten years in our research (Frick and Riis, 1991a&b; Frick, Gertsen, Hansen, Riis and Sun, 1992; Sun, 1993). Both quantitative data from IMSS and qualitative data from our previous research will be used in this section. Company K is a manufacturer of agricultural equipment. It had 560 employees in 1992. By IMSS standards, it is a medium-sized company.

Market environment and company goals
According to the forecast by a consulting company, the world-wide market for agricultural equipment will drop, and in the future the market will be dominated by a few large manufacturers rather than many small manufacturers. In fact, Company K had already experienced the market change by 1989, when it developed a strategy based on the market forecasts. The strategy specified three main company goals:

1. Increase market share by 50%
2. Increase profit by 10%
3. Shorten lead times

Manufacturing Strategy
The company's manufacturing strategy was formulated to support the company's goals. Additionally, it covered manufacturing missions, technology, organisation, and process as well as their integration. Other functional strategies such as marketing strategy and R&D strategy were also formulated. The basic principle of the manufacturing strategy was to make sure that the company can produce proper volume with proper quality at proper time. The detailed elements included:

1. Shorten lead time by 30% by improving material flow.
2. Reduce set-up time
3. Increase inventory turnover speed from 2.7 to 3.0 times
4. Flatten the organisation
5. Introduce new technology in order to develop competence.

As in most other Norwegian companies, the company highly emphasises delivery. However, it does not ignore other manufacturing missions such as quality and cost.

IMSS data revealed that in 1992 Company K equally emphasised delivery, cost and quality, as well as customer service.

Smooth production process
Under the manufacturing strategy of the company, the production process should have a smooth and simple material flow. Using the production management concept (Riis, 1990) and activity chain, the production flow was analysed in the fall of 1989 as the first step of implementing computer-integrated manufacturing (CIM). The activity chain concept made the production management aware of the need to form a coherent picture of the development of its technology. Following the production management concept, a rough model of an idealised production flow was established. Under this production flow, high-volume parts were separated from low-volume ones. Further, the plant was divided into homogeneous planning units, that is, production groups.

As a result of this project, a production process containing 200 components was simplified from 140 or 200 operations to 40. The simplification of the process made automation and integration easier. The internal inventory and work-in-process were reduced.

We can see that the practices at the operational level are consistent with the manufacturing mission as well as the company goals. If the strategy has not clearly specified the goal and the related operational practices, the activities on the shop floor will be misdirected.

Development of Technology and Organisation
To improve the manufacturing process further and to support the company goals, Company K has heavily invested in new manufacturing technologies. By 1990, the company had installed the first part of a token ring network, several CAD workstations, FMS for welding, and real-time work-in-process data collection systems in all production departments. In terms of integration, it fell somewhere between islands of automation and partial functional integration. What is more, this technological innovation was accompanied by many organisational changes.

Company K began to reorganise its organisation in 1992. Several production departments were reorganised into only two. As a result, most of the staff was removed. At the end of the reorganisation, the whole organisation had changed from a traditional hierarchical structure to a flat, group-based structure with fewer staff functions. Currently, there are three levels in its organisation structure, lower than most of the IMSS companies.

The company also encouraged employee participation and involvement. For example, one of its development project involved about 100 persons. Only five of them were outside consultants. The principle of running the project was to involve all people who would be affected by the project. Involvement was also taken as an opportunity for education, training and learning. About 40% of the employees were working in teams, and 70% of the direct workers were skilled. Both figures are above the IMSS averages.

The systematic manufacturing strategy emphasised not only the balance of technology and organisation, but also the linkage among technology, organisation and external factor such as market changes. For example, the reorganisation mentioned above was planned to be conducted in 1993. However, since the market started to fall in 1992, the company decided to reorganise the company earlier than planned in order to face the market change quickly.

Company Performance

In the period from 1989 to 1995, there were dramatic changes in its market, especially during 1992, when the market of agricultural equipment dropped dramatically all over the world. In 1988, Company K's exports accounted for 80% of its total sales, with its main market Scandinavia. In 1992, the company expanded its market to all of Europe and the share of exports increased to 90%. The company managed to increase export and market share in a declining market. In this period, the market of this company kept growing. By IMSS benchmarks, the average improvement of performance was in the top 25% group. As a result, the company managed to make money during the period.

We believe that market share increase and profit increase came from the accomplishment of the manufacturing objectives. The lead time for cutting and shipping was planned to be cut by 30%. In fact, it was reduced from 2 to 3 months to 10 to 15 days, nearly by 50%. The inventory turnover speed was planned to be increased from 2.7 to 3. In 1992, the speed had reached 5.4. Organisation and technology were also developed as planned. Suppose the related manufacturing objectives under the company goals had not been achieved, would the company have increased its market share and profitability? Additionally, suppose the manufacturing objectives were not in line with the company goals, would the company-level performance have been higher even the manufacturing performance were significantly improved? It can be concluded a systematic manufacturing strategy is one of the necessary conditions for its success.

11.4. Conclusions

In this chapter, we have revealed a systematic manufacturing strategy from the empirical evidences of Norwegian companies. The systematic manufacturing strategy is characterised by mutual, dynamic and dialectic linkage of company goal, technology and organisation of the process. The principle of such a systematic strategy is:

1. A manufacturing strategy must represent the external market requirements and the context. More precisely, the external market requirement will be reflected first by the company goals, which will be translated into manufacturing objectives or missions then.
2. A manufacturing strategy should be based on the internal advantages in the organisation and technology of the manufacturing process. The direct linkage with

technology and organisation of the manufacturing process is the manufacturing objectives derived from company goals, rather than the market requirement directly.

A manufacturing strategy fulfilling the above two requirements will function as a link between the external market requirement and the internal organisational and technological capacity. Without a proper strategy, the technological and organisational resources and capacities will be unfocused and very often misdirected.

A good manufacturing strategy should address two questions: 'What business are we in?', and 'What is our basis of competitive advantage?'. It has to be emphasised that both of these questions have to be properly and systematically answered. If only the first question is answered and the second is not, the answer to the first must be reconsidered. For example, the identified market opportunity has to be redefined until there is a sufficient basis for the competitive advantages. If the basis cannot be found or built, the market opportunity has to be abandoned! On the other hand, if the planned technological and organisational development will not support the manufacturing objectives as well as the company goals, the plan should not implemented. Finally, the mutual and dialectic interrelations of objective, technology and organisation is the core of Norwegian systematic manufacturing strategy, and holds the key of success.

References

Cohen, M.A. and Lee, H.L. (1985) 'Manufacturing strategy: concepts and methods', in P.R. Kleindorfer (ed.), *The Management of Productivity and Technology in Manufacturing*, Plenum Press, N.Y.

Frick, J. and Riis, J.O. (1991a) 'Activity chain as a tool for integrating industrial enterprises', in Eloranta (ed.), *Advances in Production System*, Elsevier Science Publishers.

Frick J. and Riis, J.O. (1991b) 'Organizational learning as a means for achieving both integrated and decentralized production systems', in G. Doumeingts, et al., *Computer Application in Production and Engineering*, Elsevier Science Publishers.

Frick, J., Gertsen, F., Hansen, P.H.K., Riis, J.O., & Sun, H. (1992) 'Evolutionary CIM implementation - an empirical study of technological-organizational development and market dynamics', in the proceedings of CIM-Europe 8th annual conference, Birmingham, UK, May, 38-49.

Gertsen, F. and Sun, H. (1994) 'Manufacturing strategy and performance: a survey of Danish industry', In: K.W. Platts, M.J. Gregory, and A.D. Neely, (eds.) *Operations Strategy and Performance, The Proceedings of the 1st International Conference of the European Operations Management Association*, Cambridge, UK, 27-29 June 1994.

Macbeth, D.K. (1989) *Advanced Manufacturing, Strategy & Management*, IFS Publications, UK.

Riis, J.O. (1990) 'The use of production management concepts in the design of production management systems', *Production Planning & Control*, 1, 1, 45-52.

Riis, J.O. (1992) 'Integration and manufacturing strategy', *Computer in Industry*, 19, 37-50.

Skinner, W. (1978) *Manufacturing in the Corporate Strategy*, John Wiley, NY.

Sun, Hongyi (1993) *Patterns of Organization and Technology Development with Strategic Considerations: Managerial Implications for Advanced Manufacturing Technologies*, Ph.D. thesis, Department of Production, Aalborg University, Aalborg, Denmark.

Sun, H. and Gertsen, F. (1994) 'The holistic perspective for the management of technology: a conceptual model and empirical evidences', in: P.T. Kidd, and W. Karwowski, (eds.) *Advances in Agile Manufacturing, Proceedings of The 4th International Conference on the Human Aspects of Advanced Manufacturing and Hybrid Automation*, Manchester, UK, 6-8 June 1994, OIS Press, 149-152.

Sun, H. and Riis, J.O. (1994) 'Organizational, technological, strategic, and managerial issues along the implementation process of advanced manufacturing technology: a general framework of implementation guide', *International Journal of Human Factors in Manufacturing*, **4**, 1, 23-36.

Sun, Hongyi, (1994) 'Patterns of Organizational Changes and Technological Innovations', *International Journal of Technology Management*, **9**, 2, 213-226.

Svendsen, B. (1994) 'Stagnation in 1992 - but some positive signs for the export', in *Norway's Biggest 10000 Companies*, Økonomisk Literatur, Oslo, Norway.

Tunälv, C. (1991) *Manufacturing Strategies in the Swedish Engineering Industry*, Ph.D. thesis, Chalmers University of Technology, Gothenburg, Sweden.

CHAPTER 12

THE NEW STRATEGY FOR THE PORTUGUESE INDUSTRY

J. B. Gouveia, Universidade Católica Portuguesa, C.R. Porto, DEEC, FEUP,
Universidade do Porto, and R. Sousa, , Universidade Católica Portuguesa, Porto,
Portugal

12.1. Introduction

Portugal's entry into the European Union in 1985 stimulated the growth of the country's economy. Real GDP growth was very strong from 1985 to 1990, averaging 4.4% per year, although the recession beginning in the early 1990s has restrained this growth. Part of the reason for this growth is that Portugal has an important industrial history. Industry is and will remain the engine of Portugal's economic and social development. The country's strongest industry sectors have been textile and apparel, including hosiery and knitwear, carpets and textiles, clothing, footwear and leather, and yarn and cloth. Other strong sectors are related to the forestry segment, including pulp and paper and other wood products, beverages, and metal and mechanical products.

Until recently, Portugal has been protected from the full brunt of competition within the European Union, but this protected status is soon to change. This means that strong actions must be taken to catch up with the rest of the European Union and outdistance new competitors from other economic regions. Portugal has traditionally ranked in the bottom quartile of the developed countries in the Organisation for Economic Co-operation and Development, along with other Southern European countries Greece, Italy, Spain, and Turkey. These countries have traditionally been able to compete based on a lower wage rate, but the opening up of the formerly non-market economies of Central and Eastern Europe, together with the rise of the newly-industrialising countries in the Far East and Pacific Rim, have largely made this an impractical strategy to pursue for further economic development.

The climate of change in this decade and the conditions created by these changes will bring an unprecedented pressure to bear on Portugal's economic structure. Although much has been achieved in the last few years in terms of the resurgence of a new, more dynamic and modern enterprise culture that is better equipped to meet the challenges of the future, some of the competitive disadvantages facing Portuguese companies in the European context remain to be overcome. They will have to be minimised during this decade by virtue of industrial policy—whose success greatly depends on three main conditions:

P. Lindberg et al. (eds.), International Manufacturing Strategies, 215-233.
© *1998 Kluwer Academic Publishers. Printed in the Netherlands.*

1. macroeconomic stabilisation
2. completion of the Single Market and construction of the European Union
3. strengthening of the economic and technological base

Comparative disadvantages that must be overcome include:

1. a shortage of both human and natural resources
2. an entrepreneurial approach that is still individualistic, not very receptive to initiatives involving business co-operation
3. a very negative technological and energy balance
4. insufficient basic infrastructures and educational infrastructures, despite the benefits of rapid modernisation
5. a shortage of strong and sizeable business groups
6. insufficient critical mass for companies to be able to develop more aggressive internationalisation strategies
7. insufficient attention given to qualitative development factors, called dynamic factors of competitiveness
8. lack of a quality image in external markets
9. lack of a delivery reliability image in external markets
10. shortcomings in the link between industry and services

With the construction of a European Union, Portuguese businesses, in particular the small and medium-sized enterprises (SMEs), are going to be faced with various challenges, including:

1. wider and more demanding markets
2. diversified products with a greater technological component and fiercer competition both from the Community producers and third-country producers
3. better prepared human resources in other countries
4. more developed technology, and more flexible forms of management, organisation and production

On the other hand, Portugal's presence in the European Union and in the Economic and Monetary Union offers the following potential benefits:

1. the creation of positive expectations for the development and stability of the country's economy, in both national and foreign organisations, thereby creating a favourable climate for investment and for securing the substantial flows of savings on the international markets
2. the reduction of national boundaries and an expansion from national scale in markets and resources, which means that national companies will have better prospects for penetrating external markets (and, in particular, the Community

markets), thus operating on a more favourable scale and benefiting from changes in the supplier and consumer markets, as well as from the potential created by the mobility of human resources and the organisation of economic activity.

12.2. Portuguese Industry in the 1990s: Overcoming Competitive Disadvantages

At the end of the 1980's and beginning of the 1990s, Portuguese companies began to think in terms of the dynamic factors of competitiveness. Their strategy was to abandon low-cost and low-value-added products and to rationalise production.

It is now time for the manufacturing sector to move on to the next stage of competition, which involves implementing differentiation strategies at product level, R&D and distribution, and identifying new specialisation niches. There should also be an increase in business co-operation, above all at SME level, and in mergers and concentrations: a vital reshaping given the challenges of the Single Market. Companies must begin their process of internationalisation, not only in Europe but in other areas.

For the Portuguese economy, accession to the European Rate Mechanism (ERM) of the European Monetary System (EMS) put a definitive end to a model of development based solely on the price factor and a cheap labour force. In the past, the exchange rate has been a fundamental tool for restoring the competitiveness of the Portuguese export sectors. In the future, qualitative development factors will become increasingly important for Portuguese industry, provoking a more selective and qualitative growth, and directed at the modernisation of management and of technological processes and where internationalisation plays a decisive role.

In order to compete internationally, Portuguese businesses will have to change their productive specialisations, identify market niches and pursue appropriate strategies in terms of their location and size. Scale and scope economies, technological innovation, sophisticated techniques, workforce qualifications and the ambitious efforts towards internationalisation are determining factors in this new model of economic development centred around entrepreneurial efficiency and exchange-rate stability.

With the support of the Community funds, Portuguese industry is expected to move closer to the standards of the most developed countries in Europe whilst not overlooking the traditional sectors of textiles, clothing and footwear—which are being upgraded—and mineral and forestry natural resources.

In the next century, the new technological paradigm will lead to changes in the industrial structure. Elements of this new paradigm are: a more diversified industrial structure—less specialised in mass-produced products, and more compatible with the consumer patterns of demanding markets; industries that are more energy-efficient and that increasingly turn to environmentally-clean technologies. At the business level, businesses will compete in an open economy, taking on board the idea of total quality; they will be innovative, intelligent and dynamic; and will build on the talent of their entrepreneurs and on the motivation and qualifications of their workers.

12.3. Industrial Policy for the 1990s: Competitiveness and Internationalisation

It is important to pursue the adoption of strategies that encourage adjustments in Portugal's industrial structure to encourage a strong productive infrastructure, a dynamic export sector, increased distribution capacity and greater commercial aggressiveness, ensuring the continuation of both rapid and sustained economic growth. The essential aims of sector-based economic policies will be to:

1. encourage and channel investment into those productive areas that are likely to be the most profitable in the future, especially new activities
2. promote modernisation, rationalisation and increased business efficiency in those activities underpinning Portugal's entrepreneurial base
3. create a favourable climate for the re-orientation of the productive infrastructure by encouraging a competitive supply structure, based on the production of goods that are innovative, top quality and with a high technological content

Furthermore, a constant concern over the last few years in Portuguese economic development policy has been the need to overcome the structural bottlenecks of the Portuguese economy via the process of European integration. This has been an important factor in mobilising the resources and aspirations needed for rapid modernisation. This is particularly applicable to traditional industrial sectors such as textiles, clothing and footwear. Industrial policy, which concerns the broad range of industry sectors, is concerned with:

1. the creation of infrastructures and services to support industry
2. training human resources
3. greater use of the dynamic factors of competitiveness
4. introduction of clean technologies, respectful of the ecological balance, in productive activities
5. encouraging business investment to promote both a desirable increase in productivity and business efficiency as well as the move to industrial diversification and internationalisation

Three major priority areas within this industrial policy context that will be part of the new industry support program are:

1. *business restructuring and re-scaling*, involving both productive aspects and the dynamic factors of competitiveness, as well as commercial aspects and the question of business concentration and the creation of economic groups that are of an appropriate size in terms of productivity and financial capacity.
2. *internationalisation*, putting operations in the domestic market on the same footing as operations in other markets where the company is present; the idea of export is

thus replaced with the concept of sales to the market. This will mean having the conditions necessary to compete in productive and commercial terms in industrial markets, in particular, those of the European economic area.

3. *the financial framework*, so as to minimise the traditional financial disadvantages facing national companies compared with their European counterparts.

12.4. Current Issues for Manufacturing in Portugal

Having described the background and context of Portuguese industry, we will now focus on the current issues for manufacturing in Portugal. Based on the IMSS data, we will analyse two issues we consider of particular importance for Portugal: delivery dependability and the adoption of the Just-in-Time philosophy.

The strategic goal of any company's manufacturing function is to support, better than competitors, those criteria that enable products to win orders in the market place. One such criterion that is gaining an increased importance is *delivery dependability* (or delivery reliability). It is thus essential that companies deliver their products on time in order to compete. In fact, in many businesses it now constitutes a qualifier and, very often, an order-losing qualifier[1].

Delivery dependability is one of the most serious handicaps of Portuguese industry. This handicap is critical, as a general image of Portuguese companies as unreliable has gradually been created in external markets. We will try to understand why Portuguese companies are performing poorly in terms of delivery dependability and what might be done to improve the present situation.

The second issue concerns the degree of adoption of *JIT practices* by Portuguese companies. It is important to investigate to what extent traditional practices are being replaced by JIT-related practices, since this would be an indication of both the degree of modernisation of the Portuguese industry and also a way to improve manufacturing performance. The implementation of JIT-related practices will of course also contribute to the improvement of delivery dependability, but we will analyse JIT practices independently, and not as a means to improve the delivery dependability problems of Portuguese companies.

Where appropriate, we will make comparisons between Portugal, the best practice leaders Japan and US, and the total sample (all 19 countries in the survey).

12.4.1. DELIVERY DEPENDABILITY IN PORTUGAL: EVIDENCE FROM THE IMSS SURVEY

The general perception of the poor performance of Portuguese companies in terms of delivery dependability is substantiated by the survey data.

---

[1]   Hill, T. *Manufacturing Strategy - The Strategic Management of the Manufacturing Function*, 2nd edition, The Macmillan Press Ltd., 1985, p.72.

Table 12-1. The Degree of Importance of Goals to the Companies

Goal	Portugal		Japan		US		Total	
	Avg.	Rank	Avg.	Rank	Avg.	Rank	Avg.	Rank
Product design and manufacturing quality	4.76	1	4.93	1	4.83	1	4.60	1
Dependable deliveries	4.60	2	3.96	5	4.25	4	4.23	4
Superior customer service	4.58	3	4.46	3	4.78	2	4.44	2
Lower manufacturing costs	4.56	4	4.89	2	4.32	3	4.33	3
Faster deliveries	4.22	5	4.19	4	4.08	5	4.16	5
Wider product range	3.58	6	3.42	6	3.50	6	3.39	6

Notes:
1. 1= not important, 5 = very important

Looking at the degree of importance of goals to companies (Table 12-1) we observe that Portugal agrees with the best practice leaders, Japan and the US, and the world, on what are the most and less important goals: product design & manufacturing quality, and wider product range respectively. However, Portuguese companies ranked lower manufacturing costs as less important than their counterparts, behind dependable deliveries. This suggests that Portugal is indeed still trying to achieve a sufficient level of delivery dependability whilst other countries, having achieved that, are now shifting their focus to reducing manufacturing costs.

In fact, in terms of the actual percentages of orders delivered late, Portugal does not perform well, with an average of 11.1% of orders delivered late, which ranks 10th among all countries, well below the leaders, Japan (4.8%) and US (9.8%). Figure 12-1 shows that the most frequently cited reasons for late deliveries are production bottlenecks, material shortages, due date changes and design changes.

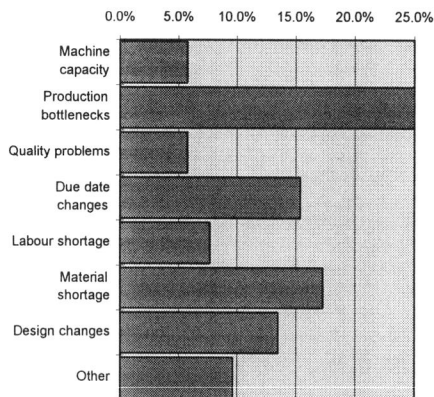

Figure 12-1. Portugal - General Reason for Late Deliveries

Business unit goal	Objectives
Lower manufacturing costs	Reduce unit cost; reduce materials costs; reduce overhead costs; improve white collar productivity; improve direct labour productivity.
Faster deliveries	Reduce manufacturing lead times; reduce procurement lead time; reduce number of suppliers; reduce inventories; increase delivery speed.
Dependable deliveries	Increase delivery dependability; improve ability to make rapid volume changes.
Product design and manufacturing quality	Improve conformance quality; reduce new product development cycle; improve ability to make rapid design changes.
Wider product range	Reduce new product development cycle; improve ability to make rapid design changes; improve ability to make rapid volume changes.

Figure 12-2. Relationship between Business Unit Goals and Objectives

In the next sections we will try to understand why Portuguese companies are performing poorly in delivery dependability. We will also point out some possible ways of improving the present situation.

Delivery Dependability Issues in Portugal

1. Communicating company goals to the business unit's manufacturing functional area. We have seen that delivery dependability is the second most important goal for Portuguese companies. However, this goal can only be achieved if it is also endorsed by the manufacturing function, that is, if there is an adequate translation of the importance of this goal from the corporate level to the business unit manufacturing function. The IMSS survey includes data concerning the degree of importance in the next two years of several objectives for the business unit manufacturing function (1= Low, 5= High). We grouped these in the following five categories of business unit goals:

We then computed an importance index for each business unit goal by averaging the degrees of importance of the corresponding objectives. Table 12-2 shows the computed business unit goals indexes together with the company goals indexes of Table 12-1.

We note that whilst dependable deliveries is the second-ranked company goal, it is only ranked fourth as a business unit goal. On the other hand, lower manufacturing costs is only ranked fourth as a company goal but first as a business unit goal. There are clearly some problems in communicating company goals to the business units manufacturing function. In particular, the manufacturing function can only significantly improve delivery dependability if it takes on board the importance of this goal.

2. Type of processes used. It is useful to analyse the type of processes used by Portuguese companies, since this significantly influences delivery performance, and

Table 12-2. Consistency between Company Goals and Business Unit Goals for Portugal

	Company goals		Business unit goals	
	Rank	Avg.	Rank	Avg.
Product design and manufacturing quality	1	4.76	2	3.88
Dependable deliveries	2	4.60	4	3.65
Superior customer service	3	4.58	-	-
Lower manufacturing costs	4	4.56	1	4.10
Faster deliveries	5	4.22	3	3.71
Wider product range	6	3.58	5	3.46

thus should contribute to a better understanding of the Portuguese situation. Table 12-3 shows the percentage of companies using each one of three types of processes in fabrication and assembly: one-off (unique products), batch and line.

Portuguese companies seem to use more one-off processes in fabrication than the rest of the countries (29.4%, ranked 3rd). The use of this type of process in assembly is similar to the world average (21.4%, ranked 12th). Both in fabrication and assembly, Portuguese companies report little use of batch processes (ranked 18th and 17th for fabrication and assembly, respectively). There is a surprisingly high proportion of Portuguese companies using line processes both in fabrication and assembly.

This finding may be explained by the small product diversity and high production volumes observed in the sampled Portuguese companies. In fact, in 1991, the average production volume for Portugal was 32.3 million units and the average number of different products was 68, compared with a total sample average of 6.65 million units and 706 different products. Since the expected changes in the following 5 years in the latter attributes that were reported by Portuguese companies in the survey were similar to or less than the total sample expected changes, one might conclude that the 1991 situation may still hold at present.

It is interesting to note that despite this high volume/low diversity context, there is still, as we have seen, a high percentage (29.4%) of companies using a one-off process in fabrication. As we know, lead time control is a key feature and a difficult task in one-off processes. This may contribute to the poor performance of Portuguese companies in delivery dependability.

Table 12-3. Type of Process used in Fabrication and Assembly

	Fabrication						Assembly					
	One-off, unique		Batch		Line		One-off, unique		Batch		Line	
Country	% R	ank	% R	ank	% R	ank	% R	ank	% R	ank	% R	ank
Portugal	29.4%	3	32.4%	18	38.2%	4	21.4%	12	17.9%	17	60.7%	2
Japan	6.5%	18	29.0%	19	64.5%	1	5.6%	17	5.6%	19	88.9%	1
US	23.3%	7	50.0%	13	26.7%	10	17.9%	7	28.6%	13	53.6%	10
Total	15.7%		53.4%		30.9%		23.2%		33.7%		43.0%	

Table 12-4. Market Demand Variation, Stability of Production Schedules, and Lead Times

Country	Variation in customer/market demand concerning order types, technologies, quality requirements, etc.		Average length of frozen period in production schedule and average lead time (from customer order to delivery)			
	Average index[1]	Rank (descending)	Length of frozen period		Lead time	
			Weeks	Rank	Days	Rank
Portugal	2.90	5	0.79	19	56.4	11
Japan	2.96	7	1.65	17	32.5	4
US	2.98	9	5.26	5	41.8	9
Total	3.00		4.49		54.9	

1. 1= great difference between orders/customers, 5 = little difference

3. Market demand variation and stability of production schedules. Another factor influencing delivery performance is the degree to which customer/market demands vary in order types, technologies, quality requirements, etc. Table 12-4 shows that Portuguese companies experience a high degree of variation in customer/market demands, with Portugal exhibiting the 5th highest average variation index. This is in turn reflected in the split of production orders between forecast and customer orders, with Portugal having the fifth highest average percentage of customer orders (73.1%), higher than Japan (69.3%), the US (59.5%) and the total sample (65.6%).

Besides this customer/market variability, Portuguese companies also seem to operate in a somewhat unstable environment. There are no direct measures of this in the survey, but we can use an indirect measure: the length of the production schedule's frozen period (the time period in the schedule during which no changes are allowed to the schedule). Table 12-4 shows that Portugal is the country with the shortest average frozen period in the production schedule, despite reasonably long lead times. This lack of longer term production plans (or difficulty in keeping to existing ones) is consistent with some of the most frequently quoted reasons for late deliveries: production bottlenecks, material shortage and due date changes (Figure 12-1).

The instability in the production schedules certainly contributes to poor shop floor scheduling, which might explain the problems that Portuguese companies experience with bottlenecks. It is also more difficult to predict materials usage, which could explain why material shortage is a frequent cause of lateness. Finally, due date changes may be a cause in itself for the short frozen periods.

Overall, the unstable environment faced by Portuguese companies clearly makes reliable deliveries a difficult goal to achieve.

4. Capacity issues. A company's approach to capacity is another relevant factor for delivery performance. Looking at the business unit's policy in terms of overall manufacturing capacity, Table 12-5 shows that 48.8% of the Portuguese units state that their policy is to keep capacity higher than market demand ranking 5th against

Table 12-5. Business Unit Policy for Overall Manufacturing Capacity

Percentage of companies with policies aiming to keep capacity
higher than, equal to, or lower than market demand

Country	Capacity higher than demand		Capacity equal to demand		Capacity less than demand	
	(%)	Rank	(%)	Rank	(%)	Rank
Portugal	48.8%	5	36.6%	15	14.6%	6
Japan	23.1%	16	73.1%	1	3.8%	18
US	48.7%	6	43.6%	12	7.7%	14
Total	43.7%		44.0%		12.3%	

the other countries. This policy may be related to dependability problems; for example over-capacity may be used to create flexibility through having slack capacity to deal with unplanned events.

However, Portuguese units display a high percentage of utilisation of the planned capacity of the main processes (an average of 86.5%), ranking fifth among all countries. This may indicate that the capacity policies declared by respondents are not strictly followed and are perhaps overridden by an operational emphasis on capacity utilisation (which is traditionally used as a key shop-floor performance measure),rather than on flexibility and due date compliance. This is consistent with the low degree of use of pull-scheduling and JIT Manufacturing/Lean Production activities (which will be discussed below).

The importance of keeping capacity higher than demand in order to create flexibility must be stressed in the Portuguese context of a higher degree of variability in customer orders and instability, and of a high use of one-off processes in fabrication. Failure to provide slack capacity in this context no doubt contributes to a reduced delivery performance.

It was also found that most Portuguese companies (82.1%) run products with different order sizes together on the same equipment, which supports the emphasis placed on capacity utilisation.

5. Lead time. Lead time is another important factor to bear in mind in terms of keeping to due dates. As Table 12-4 shows, Portuguese lead times rank near the middle, but above the overall average and still far longer than the leaders'.

The higher the lead time the poorer manufacturing's reaction is to changes in customer orders, including due date and design changes. The fact that Portugal is still far from the leaders in shortening lead times is consistent with the high proportions of Portuguese companies that indicated due date and design changes as one of the most frequent reasons for lateness in deliveries (Figure 12-1). This is of increased importance for Portuguese companies, since they have a high percentage of customer orders, as opposed to forecast orders.

6. Inventory. Closely related to lead times and delivery dependability are the levels of inventory held by companies at various stages: raw material/components inventory,

WIP (work in process) inventory, and finished goods inventory. Table 12-6 presents inventories at these stages:

Portuguese companies hold a high level of raw material/components inventory, a reasonably high level of WIP, and a relatively low level of finished goods inventory. We next analyse these separately.

a.   Raw   material/components   inventory. Despite   the   high   levels   of   raw material/components inventory, material shortages were the second most frequently reported reason for late deliveries (reported by 17% of the companies; Figure 12-1). One possible explanation of this apparent contradiction may be the low degree of use of MRP. Portugal is ranked only 15th in the degree of use of Materials Requirement Planning (MRP) in the last two years, with an index of 2.70 compared with a total sample index of 3.16 (1= No use; 5= High use). At the same time, the relative payoffs reported by companies that do use MRP is quite high, with Portugal ranking sixth among all countries, with an index of 3.71 compared with a 3.47 total sample index (1= Low payoff; 5= High payoff). This strongly suggests that Portuguese companies would benefit from the use of MRP systems. This conclusion is consistent with the -0.35 correlation coefficient found between raw material/components inventory and degree of use of MRP for Portuguese companies. From the survey data, it was found that a moderate percentage (38%) of the companies that are not currently using an MRP system, are planning to adopt one in the next two years.

A second possible reason to explain the high level of raw material/components inventory may be that companies use high safety stocks for protection against unreliable supply. Apparently, this hypothetical insecurity of supply bears no relationship with the sourcing policy adopted by the companies. In fact, Portuguese companies have considerably fewer suppliers than the best practice leaders. Also, the strength of relationships between companies and suppliers are close to the overall average and stronger than Japan's, although weaker than the US's (Table 12-7).

Instead, the geographical location of suppliers may be the key factor. Portuguese companies have the second lowest percentage of purchases made in their own country, most of the purchases being from other European Union (EU) suppliers (Figure 12-3). Consequently, supply lead times are bound to be longer and less reliable, thus forcing companies to hold higher levels of raw material/components inventory. A third explanation for the high level of raw material/components inventory, although

Table 12-6. Inventory Holdings (days)

	Raw materials & components		Work-in-progress		Finished goods	
	(days)	Rank	(days)	Rank	(days)	Rank
Portugal	39.5	14	24.6	12	19	8
Japan	10.9	1	12.1	1	20.5	10
US	24.8	3	26.3	15	21.5	11
Total	33.1		24.4		20.9	

Table 12-7. Number of Parts/Materials Suppliers and Strength of Supplier Relationship

(1= arm's length relationship, 5= close relationship)

Country	Number of suppliers		Strength of relationship	
	Average	Rank	Average	Rank
Portugal	363	12	3.24	12
Japan	412	10	3.11	16
US	1718	1	3.31	1
Total	469		3.32	

not possible to substantiate with the survey data, is the traditional way the purchasing function is run by Portuguese companies. Companies have traditionally emphasised making large enough purchases to receive quantity discounts, with insufficient concern for the associated costs of holding this excess inventory.

b. WIP inventory. This type of inventory directly affects lead times, and as a consequence, delivery. The reasonably high level of WIP inventory for Portuguese companies may result from little use of Lean Production practices (see the next section). Surprisingly enough, on average Portuguese companies display a very good throughput efficiency (66.2%, the fourth highest), the time products are worked on as a percentage of the total manufacturing lead time (start of first operation to finish of last operation).

c. Finished goods inventory. The low level of finished goods inventory observed for Portuguese companies may be explained by the high percentage of production orders that are customer orders (as opposing to forecast orders). This leads to reduced production to stock. The fact that demand is not, in its majority, satisfied from stock, makes lead time control even more important.

7. Product Development. Since design changes has been one of the most frequently quoted reasons for late deliveries, it is worth investigating whether this has to do with deficiencies at the product development level, or instead, an inability to react to a

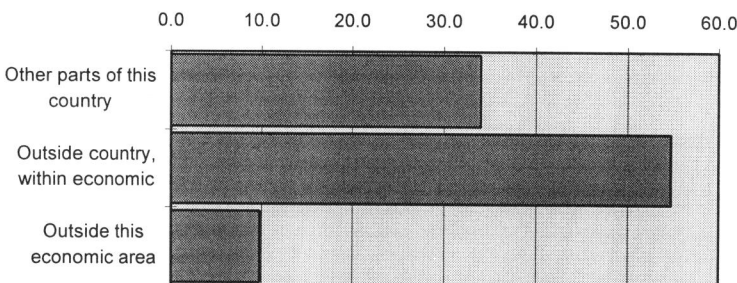

Figure 12-3. Percentage of Total Purchases by Origin (Average)

Table 12-8. Co-ordination between Design and Manufacturing (Country Averages and Ranks)

Country	Information transfer from design to manufacturing		Frequency of rotation of people between design and manufacturing		% of blueprints subject to ECOs	
	Avg.[1]	Rank	Avg.[2]	Rank	Avg.	Rank
Portugal	3.34	13	2.11	5	33.1%	12
Japan	3.08	17	1.70	1	13.8%	3
US	3.80	4	20.3	2	60.2%	18
Total	3.38		2.21		29.5%	

Notes:

1. 1 = one-way communication of specification; 5 = active contribution of manufacturing in design process.
2. 1 = continuously, 5 = never.

normal and acceptable degree of design changes. The survey data point towards the latter, with no outstanding differences between Portugal and the other countries in product development.

In fact, in the organisational co-ordination between design and manufacturing, Portugal occupies an average position, with 25.5% of the respondents stating that co-ordination is made through rules and standards, 18.2% through formal meetings, 14.5% through informal meetings, 27.3% through cross-functional task forces, 10.9% through personal contacts, and 3.6% via other means. The results for the information transfer between design and manufacturing, rotation of people, and percentage of blueprints subject to Engineering Change Orders (ECOs) are summarised in Table 12-8.

Portugal is slightly below average position in the contribution of manufacturing in the design process. It is interesting to note the high degree of rotation of people between design and manufacturing. As far as the percentage of blueprints subject to ECOs is concerned, Portugal is moderately above average.

In conclusion, there is no strong evidence that late deliveries attributed to design changes are mainly due to problems at the product development level and its co-ordination with manufacturing. As a consequence, Portuguese companies should primarily concentrate on improving their ability to react to design changes.

Recent trends related to delivery dependability

Table 12-9 presents the percentage improvements achieved, on average, by Portuguese companies from 1990 to 1992 in terms of on-time deliveries, and three associated factors: manufacturing lead time, procurement lead time and delivery lead time.

We can see that despite the need for improvement in the above performance indicators, the gap has been widening between Portuguese companies and their international counterparts, except in procurement lead time, which has undergone one of the largest improvements among all countries. Swift action is recommended in order for Portuguese companies to improve their competitiveness, especially now that competition with other EU countries will increase due to the expiration of the EU transition period granted to Portugal.

12.4.2. THE ADOPTION OF JUST-IN-TIME CONCEPTS IN PORTUGAL

JIT manufacturing is a broad philosophy of continuous improvement that includes several categories of effort that are mutually supportive. Using survey data, we will study the adoption of JIT concepts in three areas: quality, JIT production (in what concerns the flow of goods through different production phases) and JIT culture. We are interested in practices broadly associated with JIT manufacturing, not all of them necessarily associated with traditional JIT practices as first introduced by the Japanese.

Quality

In JIT manufacturing the approach to quality should involve every department and every employee in the company. Quality is viewed as a company culture, seeking to continually improve products and processes. In this section, we will analyse the use of the following quality-related activities by Portuguese companies: QFD (Quality Function Deployment), QPD (Quality Policy Deployment), Total Productive Maintenance, Zero Defects, SPC (Statistical Process Control), TQMP (Total Quality Management Program) and ISO 9000. Table 12-10 and Table 12-11 show the degree of use and relative payoffs of these and other activities.

We observe that the degree of use of quality related activities is quite good for Portugal, with a high use of QFD, QPD, Total Productive Maintenance, Zero Defects, and a fair degree of use of SPC, TQMP and ISO 9000. In fact, except for TQMP and SPC, the use of quality-related activities is always higher than in the US and total sample. The implementation of these activities seems to have been quite successful, since the average values of the relative payoffs reported are quite high, and again, with only the exception of ISO 9000 certification, higher than those reported by the US and total sample. It is also worth reporting that for the activities with lower usage ranks (SPC, TQMP and ISO), a high percentage (close to 70%) of the companies not using them at the present plan to do so within the next two years.

As far as how maintenance and quality spending are split, Portugal follows a spending pattern similar to Japan, clearly the leader in this respect. This pattern is characterised by an emphasis on prevention, as shown in Table 12-12.

Table 12-9. Average Percentage Improvement from 1990 to 1992

	% improvement	Rank
On-time deliveries	20.3%	14
Manufacturing lead time	14.9%	18
Procurement lead time	27.4%	3
Delivery lead time	17.4%	16

Table 12-10. Degree of Use of Activities in the Last 2 Years (1 = no use, 5 = high use)

Activity	Portugal			Japan			US			Total
	Degree of use	Rank	Var. coeff. (%)	Degree of use	Rank	Var. coeff. (%)	Degree of use	Rank	Var. coeff. (%)	Degree of use
Quality										
QFD	3.71	2	35	3.73	1	28	2.16	12	55	2.56
QPD	3.83	2	31	4.32	1	19	2.33	14	62	2.78
TPM	3.04	2	34	4.59	1	16	2.12	11	52	2.36
ZD	3.00	5	49	3.64	2	32	2.59	10	55	2.73
SPC	2.87	8	52	2.73	10	53	3.03	6	39	2.89
TQMP	3.07	10	48	3.85	2	31	3.54	4	32	3.18
ISO9000	2.93	12	57	2.62	15	63	2.25	17	67	3.14
JIT Production										
SMED	2.50	4	47	3.42	2	35	2.23	7	59	2.40
JIT/cust	3.04	5	50	3.96	1	31	2.64	12	54	2.91
Kanban	2.46	11	63	3.31	3	48	2.69	6	57	2.55
JIT/LP	2.83	12	58	3.93	1	32	3.36	4	46	3.03
JIT culture										
Bench-marking	2.78	4	51	4.04	1	27	2.15	11	53	2.42
Kaizen	3.00	9	45	4.23	1	30	3.32	6	38	3.00
Teams	3.30	13	42	3.77	2	29	3.41	8	37	3.38
Averages			47			34			50	

Variation coefficient = 100 x (standard deviation / mean)

Table 12-11. Relative Payoff of Activities in the Last 2 Years (1 = low, 5 = high)

Activity	Portugal			Japan			US			Tot.
	Rel. Payoff	Rank	Var. coeff. (%)	Rel. Payoff	Rank	Var. coeff. (%)	Rel. Payoff	Rank	Var. coeff. (%)	Rel. Payoff
Quality										
QFD	3.75	2	29	4.16	1	19	2.95	13	44	3.19
QPD	3.64	3	32	3.80	1	23	2.84	14	43	3.24
TPM	3.65	4	34	4.04	1	23	3.35	7	33	3.38
ZD	3.27	5	41	3.22	6	38	3.05	10	39	3.08
SPC	3.38	7	39	3.80	2	18	2.92	14	48	3.28
TQMP	3.37	7	36	4.41	2	18	2.53	14	47	3.08
ISO9000	3.20	13	49	3.40	7	44	2.46	19	56	3.28
JIT Production										
SMED	3.57	8	32	4.04	1	26	2.84	16	49	3.35
JIT/cust	3.00	8	40	3.59	3	28	2.56	12	52	3.07
Kanban	3.07	13	50	3.76	5	31	3.31	10	42	3.31
JIT/LP	3.28	16	51	4.12	2	23	3.54	10	37	3.60
JIT culture										
Bench-marking	3.18	6	38	4.09	1	21	2.45	18	55	3.08
Kaizen	3.53	10	34	4.60	1	15	3.53	11	32	3.60
Teams	3.29	17	36	3.92	3	22	3.68	10	32	3.65
Averages			39			26			41	

Variation coefficient = 100 x (standard deviation / mean)

Table 12-12. Quality and Maintenance Spending (country averages and ranks)

Country	Proportion of maintenance spending on preventive maintenance		Proportion of quality spending on							
			Inspection/ control		Internal quality		Prevention		External quality	
	%	Rank	%	Rank	%	Rank	%	Rank	%	Rank
Portugal	41.4	4	31.4	11	25.8	10	25.8	5	16.9	8
Japan	66.2	1	30.5	10	23.7	5	31.8	1	14.0	5
US	35.4	10	21.7	2	33.7	17	20.9	13	23.6	15
Total	37.2		31.0		28.2		22.1		18.9	

The survey data seem to indicate a healthy commitment of Portuguese companies to quality. The goal of production design and manufacturing quality (considered, on average, the most important company goal for Portuguese companies) seems to be adequately being pursued. This is confirmed by the fact that quality is the second-ranked business unit goal (Table 12-2) and by the average 24.1% improvement in manufacturing quality (conformance to specification) reported by Portuguese companies from 1990 to 1992. It thus comes as no surprise that one of the least frequently reported reasons for lateness is quality problems.

JIT Production

JIT production seeks to minimise WIP inventory through producing in small lots and synchronising work centres so that the right items are at the right place at the right time. The survey data enable us to analyse the use of four activities related to JIT production: SMED (Single Minute Exchange of Die), Pull-Scheduling (e.g., *Kanban*), JIT Manufacturing/Lean Production, and JIT (frequent) deliveries to customers.

Table 12-10 shows that there is a high degree of use of JIT deliveries to customers and SMED (higher than in the US and total sample), but a relatively low degree of use of Pull Scheduling and JIT Manufacturing/Lean Production (lower than in the US and total sample). All these activities are being used less in Portugal than in Japan. The low use of Pull Scheduling and JIT Manufacturing/Lean Production may be related to the high percentage of companies that use one-off processes in fabrication. This may also contribute to the relatively high levels of WIP inventory experienced by Portuguese companies, and to the high percentage of companies reporting production bottlenecks as a major reason for late deliveries.

As far as relative payoffs stemming from the use of activities are concerned, again there is a marked difference between those of JIT deliveries and SMED and those of Pull Scheduling and JIT Manufacturing/Lean Production (Table 12-11). The former are reasonably good (both rank 8th), whilst the latter are poorer (JIT Manufacturing ranks 16th and Pull Scheduling 13th). The low degrees of use and associated low payoffs for the latter could suggest that companies haven't adopted these activities long enough to achieve significant payoffs and/or that they are not implementing them properly. The moderate adoption percentages for Pull Scheduling (43%) and JIT Manufacturing/Lean

Production (50%) for the next 2 years reveal that Portuguese companies are indeed still réluctant to adopt these practices.

JIT Culture

The implementation of JIT concepts must be accompanied by a change in company culture towards a philosophy of continuous improvement and strong people involvement. In the IMSS survey, the level of use of three activities can be used to measure how JIT culture is being assimilated by Portuguese companies: Implementation of a team approach (work groups), *Kaizen* (continuous improvement) and benchmarking.

Teamwork. Portuguese companies on average have a reasonably low degree of use of teams (lower than Japan, the US and the total sample averages) and the average payoff reported for Portugal is one of the lowest (ranked 17th). Despite this, it seems that more companies are adopting team approach, as 75% of the companies not currently using it are planning to do so in the next two years.

Continuous Improvement. In the use of continuous improvement (*Kaizen*), Portugal ranks around average, although still behind Japan and US. Relative payoffs and adoption percentages for the next two years are moderate.

Benchmarking. The degree of use of benchmarking activities is associated with continuous improvement, since its purpose is to gauge a firm's competitive position with a view to improving it. Surprisingly enough, Portuguese companies seem to be doing very well here, with an average degree of use higher than US and the total sample, but still below Japan. Relative payoffs are also good, being on average higher than those reported for the US and the total sample, but again below those reported for Japan.

JIT in Portugal

The survey data seem to indicate that Portuguese companies are not adopting the JIT philosophy in a global and integrated manner, but are instead adopting some of its components, at a moderate rate. This is confirmed by the wide gap between Portugal and Japan (the country where these practices are most disseminated), with Portuguese average use and average payoff indexes systematically below Japan's, with the exception of SPC (for which use and payoff are higher for Portugal) and ISO 9000 certification (for which use but not payoff is higher for Portugal). This conviction is strengthened if we compare the variation coefficients for the degree of use and relative payoff of activities between Portugal, Japan, and the US (Tables 13-10 and 13-11). The variability of the answers among the Portuguese companies are for most activities higher than the Japanese, but similar to the US. The average (taken over all the activities) of the variation coefficients for the degree of use is 47% for Portugal and

50% for the US, compared with only 34% for Japan. For payoffs, that average is 39% and 51% for Portugal and the US respectively, compared with only 26% for Japan. This suggests a more consistent, widespread and longer use of JIT practices in Japan, when compared with Portugal, which comes as no surprise.

12.5. Conclusions

This chapter began with a description of the background and context of the Portuguese industry. Until recently, Portugal has been protected from the full brunt of competition within the European Union, but this protected status is soon to change. As a result, two main issues that concern manufacturing competitiveness in Portugal must be addressed.

First, delivery dependability, which indicated a major competitive disadvantage for Portugal. The reasons for lateness most frequently reported by Portuguese companies were production bottlenecks, material shortages, due date changes, and design changes. Several factors possibly leading to the poor performance of Portuguese companies in terms of delivery dependability were identified:

- Inadequate communication of the importance of delivery dependability from the corporate level to the business unit's manufacturing function.
- High degree of use of one-off processes in fabrication when compared with the other countries and given the low product diversity and high production volumes observed in the sampled Portuguese companies.
- High degree of variation in customer/market demand concerning order types, technologies, quality requirements, etc., leading to a high percentage of customer orders.
- Instability in the production schedules, making difficult shop-floor scheduling and materials planning.
- Geographical distance of suppliers.
- Low degree of use of MRP systems.
- Apparent excessive emphasis on capacity utilisation. More slack capacity would help, especially in a context of high use of one-off processes, high variation in customer orders and unstable production schedules.
- Reasonably high WIP inventory, possibly due to low use of pull-scheduling and JIT manufacturing/lean production practices.
- Reasonably high lead times, limiting the capability to react to changes rapidly.
- Possible poor reaction to design changes.

It was observed that delivery dependability has been improving in the last two years, but at a slower rate than other countries, which means that Portuguese companies do not seem to be closing the gap to their international competitors.

The second issue that must be addressed concerns the degree of adoption of JIT-related practices by Portuguese companies. It was deemed important to investigate to which extent the traditional practices were being replaced by JIT-related practices, since this would be an indicator of the degree of modernisation of the Portuguese industry and also a way to improve manufacturing performance. The use of JIT practices was investigated at three levels: Quality, JIT Production and JIT Culture.

Portugal seems to have a good degree of use of quality related activities, with good payoffs. This confirms that Portuguese companies are no longer competing primarily on cost and with little concern for quality.

In what concerns JIT production, there is a reasonably low degree of use and payoffs of pull-scheduling and JIT manufacturing/lean production, but a good degree of use and payoffs of SMED and JIT deliveries to customers.

In terms of assimilation of a JIT company culture, there is a reasonably low degree of use of team approach, a moderate use of a continuous improvement approach, and a high use of benchmarking.

Overall, it seems that Portuguese companies are not adopting the JIT philosophy in a global and integrated manner, but are instead punctually adopting some of its components. There is a high degree of variation between companies in what concerns the adoption of JIT concepts, which suggests that its use is not yet widespread and universally accepted.

In conclusion, two critical issues for Portugal, delivery reliability and adoption of JIT-related practices, have much scope indeed for improvement, and constitute two promising avenues for increasing the competitiveness of manufacturing in Portugal.

CHAPTER 13

MANAGING MANUFACTURING IN AN ECONOMY IN TRANSITION:
SPAIN'S CHALLENGES AND RESPONSES

*Gustavo A. Vargas, Instituto de Empresa, Madrid, Spain, and California State
University, Fullerton, CA, USA*

13.1. Introduction

A number of European firms find themselves at the vortex of economic, market and
technological forces that pull at them in a variety of directions. The twin phenomena of
Western European economic integration and Eastern European socioeconomic
transformation are changing the economic landscape within which firms must compete
and succeed. These creative and regenerative processes are also painful and, at least
initially, quite upsetting and destructive. The 20-50% plummeting of Eastern European
per capita GDPs and the rise of Western European unemployment to the 10-20% level
are clear indications of the current upheaval. Manufacturing is, perhaps, the economic
sector most seriously affected by the forces of globalisation and integration, as shown
by the ongoing dramatic downsizing of 'traditional' industries such as steel, textiles,
and automotives. Manufacturing is also of enormous significance since it lies at the core
and foundation of any industrialised and/or post-industrial society (Dertouzos, 1989).
Clearly, globalisation and integration offer manufacturing both challenges and
opportunities, which must be diagnosed and addressed (Kennedy, 1993).

Perhaps nowhere is such task more daunting than in a mid-sized economy such as
Spain's, which has to overcome the hurdles of a late incorporation into the European
Union (EU) and of a less developed infrastructure (Allard & Bolarinos, 1992). Thus,
learning how leading manufacturing firms manage to cope with the global challenges
whilst based within Spain's particular setting is of paramount importance if Spanish
manufacturing is to have any future at all. 'Best practices', however, can hardly claim to
be neutral to and unaffected by the specific macroeconomic and sociocultural
environments within which specific firms operate (Thurow, 1992; Vargas & Johnson,
1993; Whybark & Vastag, 1993). As Figure 13-1 indicates, government policies, work
rules, social attitudes, factor costs and competition intensity all play a vital role in
shaping what constitute 'best practices'. A brief review of the characteristics of this
study and of its relevant environmental aspects is thus called for to understand the
Spanish experience.

P. Lindberg et al. (eds.), International Manufacturing Strategies, 235-258.
© 1998 *Kluwer Academic Publishers. Printed in the Netherlands.*

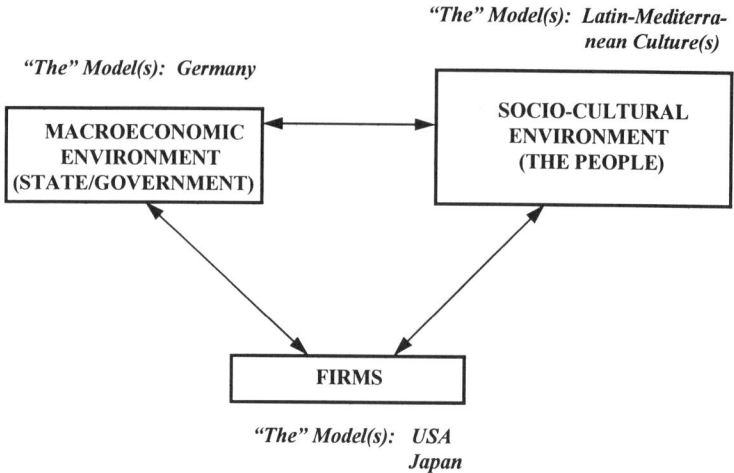

Figure 13-1. The Three Way Interplay: Spain's Participants and Foci

13.1.1. POSITIONING AND CHARACTERISTICS OF THE STUDY.

Within the assembly manufacturing industries covered in this survey, a number of critical economic and managerial issues tend to converge. Complex strategic issues such as downsizing, time-based competition, outsourcing, flexible/intelligent automation, and sociotechnical integration are but a few of the challenges present in these settings (Dertouzos, 1989).[1]

For Spain itself, the economic importance of these industries can hardly be overemphasised. In the transportation equipment sector, for instance, Spain's share of global car exports increased between 1987 and 1990 from 6.36% to 8.84%, with over 95% of Spain's car exports going to other EU nations. A similar assessment can be made of the telecommunication equipment industry within the electrical machinery and equipment sector (Allard & Bolarinos, 1992; Economist Intelligence Unit, 1994).

In addition, these industries significantly affect and are affected by their social and economic environments due to the preconditions required for their effective operation, such as the existence of a network of suppliers, a critical mass of professional and skilled personnel, adequate regulations of the labour markets, a working physical/communications infrastructure, and sophisticated distribution channels. For Spanish firms, there are many valuable lessons to be learned from these industries, since although Spain's productivity (e.g., value added/person employed) is lower than that of

---

[1]   Fabricated/assembled metal products, electrical/non-electrical machinery and equipment, transportation equipment and scientific/control instruments

the leading EU economies, it is certainly much closer to them in manufacturing productivity as a whole and stands at a par with them in the transportation and electrical-electronic industries.

13.1.2. THE MACROECONOMIC ENVIRONMENT OF SPAIN.

The importance of Spain's achievements in the last decade is frequently lost in the ensuing social and political turmoil. During the 1980s the Spanish economy experienced a profound transformation that turned it into one of the 'success stories' of Europe. The magnitude of the positive changes experienced by the Spanish economy is clearly displayed by the fact that today Spain is the eighth largest industrialised economy of the world, as measured by its Gross Domestic Product (GDP) (World Competitiveness Report, 1994).

The incorporation of Spain in the EU has been enormously beneficial for the Spanish economy, and not only because of the 'cohesion funds'. Accession to the EU fuelled the expansion of the Spanish economy in the 1980's and is today the engine of its industrial reconversion, by virtue of the reorientation of the economy towards exports and productive investments. In 1992, exports represented 15.2 % of GDP, and in 1993 exports reached 19.8% of GDP. Between 1986 and 1991, the Spanish GDP growth rate was 1% higher than the EU average. Key manufacturing sectors experienced growth by leaps and bounds. The automotive sector, for instance, became the third largest one in Europe, with an annual growth rate of 4% and with an output of 1.8 million vehicles in 1992, of which 80% were exported. Investment in education also grew dramatically. In the 1981–91 period, expenditures in public education rose at an annual rate of 14.5%, and college enrolments increased twofold. In the same time span, the annual growth rate in fixed assets was 5.5%, the highest in the EU and barely below Japan's.

The brutal 1992–94 recession exposed some lingering critical economic weaknesses. Their roots could be found in the steady rise in labour wages, in a fairly rigid labour market in spite of its recent liberalisation, in the still low investment in training, and in the relatively low investment in R&D. The business environment characteristics presented in Figure 13-2 show Spain lagging behind in a number of significant aspects, with labour productivity being the only encouraging aspect.

Together, this combination of factors makes for some unpleasant consequences. The Spanish economy is heavily dependent on imported technology and foreign patents, yet the future of Spain's economy and of its manufacturing sector is contingent on the generation and exportation of goods and services with a high component of technological sophistication. Since in an economic environment with free flows of capital, the location of medium- to high-technology firms is determined by the costs of inputs (labour, capital) or the accessibility of relevant know-how (products, services, processes), higher labour costs and lower innovation ability constitute a serious source of structural instability for the Spanish economy.

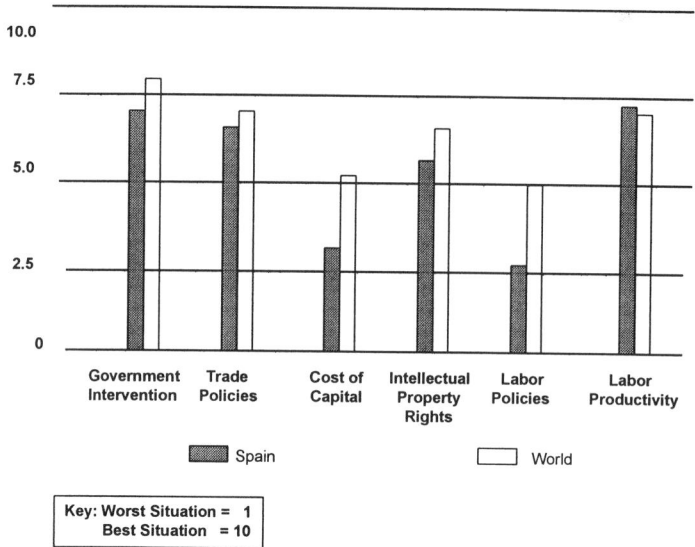

Figure 13-2. Business Environment Characteristics

During the last half of 1994 and first half of 1995, signs of a strong economic recovery cropped up. The Industrial Production Index (IPI) grew at an annual 4.6% growth rate, and the GDP grew at a annual 4.5% growth rate in the last quarter of 1994 . Furthermore, in the first quarter of 1997, the IPI and the GPD grew at an annual rate of 12.5% and 3.0%, respectively. The devaluation of the Spanish peseta has helped restore a measure of international competitiveness, and with the productive sectors working at 70% of capacity, both costs and inflation can be kept down (Economist Intelligence Unit, 1994). Yet, the long-term structural problems persist.

Overall, Spain's 'European gamble' at the macroeconomic level still has to translate in a parallel, effective gamble at the microeconomic level at a national scale. The need for the existence of a robust manufacturing sector is not a subject of discussion in Spain: it is agreed that a highly-developed society requires a successful 'manufacturing hard-core' for otherwise the nation becomes regressive and *externally dependent*. But, ultimately, it is the actions and practices adopted by the firms in response to their microeconomic challenges that determine a nation's success or failure. Spain's economic history and future would thus be incomplete without learning from the live experience of firms able to thrive within their specific national and international circumstances.

13.1.3. THE SOCIOCULTURAL ENVIRONMENT OF SPAIN

Spain clearly differs from other countries in a number of key sociocultural dimensions. However, even when countries and environments are different, the relevant issue is the degree of congruency between a country's sociocultural traits and the implementation of specific business practices. A key question, thus, is whether the Spanish traits may constitute an impediment to the achievement of world-class manufacturing.

This is no trivial matter, considering that in the last two decades successive paradigms such as Material Requirements Planning (MRP), Just-in-Time (JIT), and Total Quality Management (TQM) have been propounded as 'the' way to manage processes and operations and of competing successfully in increasingly global and complex markets. Even more recently, other paradigms such as TBC (Time-Based Competition) and BPR (Business Process Re-engineering) have been propounded. However, many firms have encountered serious difficulties in implementing the paradigms in spite of the efforts and resources invested. Real-life experience has demonstrated that these paradigms implicitly assume the existence of certain environmental features upon which the various paradigm components rest.

To wit, the full power of MRP's integrative capability rests upon the assumption of input data reliability and exact schedule execution, assumptions that become feasible only if data are gathered and processed promptly, shop floor priorities are observed, and any deviations are reported as and when they occurred. In other words, MRP requires that everybody tell the truth and be willing to do as commanded; otherwise, the system simply breaks down. Likewise, the requirements of participative management and multiskilled workers present in the TQM paradigm become viable only when a highly educated work force, a tradition of democratic practices, and a habit of teamwork are present. Normally, firms operating in a highly industrialised environment can count on these environmental attributes, whilst firms operating in the Third World and former non-market economy countries rarely, if ever, have access to any one of such attributes.

In Spain's case, Hofstede's dimensions shown in Figure 13-3 (Hofstede, 1991) do not seem to be a cause of concern in implementing world-class 'best practices', with one major exception. Spanish society at large is significantly more conservative (e.g., higher uncertainty avoidance) than other industrialised societies, a fact that may seriously slow down the adoption of necessary changes in business practices and cause poor economic and competitive performance.

Perhaps a more worrisome aspect is the lack of a common focus among the Spanish participants in the industrial reconversion and European integration process. For almost two decades now, successive Spanish governments have progressively steered away from protectionism and *dirigisme*, privatising and rationalising numerous economic sectors, and promoting open-door trade policies whilst preserving an extensive 'social net' for the less fortunate members of society. Effectively, the Spanish national government increasingly concentrates on macroeconomic (e.g., fiscal policies,

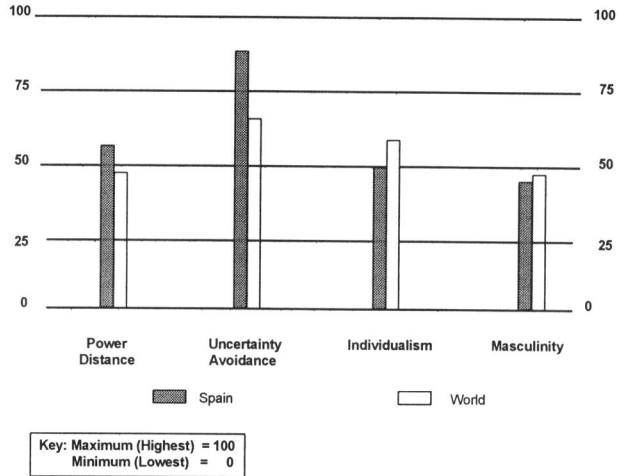

Figure 13-3. Key Sociocultural Dimensions (Hofstede's Dimensions/Indices)

infrastructure) and social (e.g., education, housing, health) issues; implicitly or explicitly, the national government determinedly attempts to follow the German model (most successful in Europe) (Allard & Bolarinos, 1992; Economist Intelligence Unit, 1994; World Competitiveness Report, 1994). Firms, in turn, try to follow the American or Japanese models (most successful in business) and the workers, as a people, maintain their Latin-Mediterranean identity (longest lasting of Western cultures). It is this three-way interaction (or lack of it), as depicted in Figure 13-1, that must be kept in mind when looking at the actions and practices of firms in Spain.

13.2. Results and Discussion

Only some of the most important aggregate survey results are shown here and detailed in the following sections. For benchmarking purposes, Spanish firms are selectively compared against those from the United States, Japan, Germany, Italy and Argentina. These countries were selected because of world-wide economic leadership (USA, Japan, Germany), comparable degree of economic development and sociocultural traits (Italy), and membership among industrialising nations (Argentina).

13.2.1. METHODOLOGY AND PROFILES

The profile of survey respondents in terms of their organisational and technical dimensions is shown in Table 13-1.[2] The survey profile closely approximates the overall census profile of Spanish firms within the manufacturing sectors that were the subject of this study. As shown in Table 13-1, about two-thirds of the firms had foreign home bases and were not Spanish-owned (e.g., foreign subsidiaries, joint ventures, etc.), with more than one third of them directly linked to other EU nations. Most firms were of middle or large size (about 85%), and were located in the major industrial heartlands of Madrid and Catalonia (about 67%). The firms were almost evenly split between high-

Table 13-1. Profile of Participating Spanish Firms: Organisational and Technical Dimensions

Dimension	% Responses
Origin	
Spain	33
Within European Union (EU)	
Other (non-EU)	31
Ownership	36
Wholly-owned, domestic	33
Foreign subsidiary	49
Joint Venture	12
Others	6
Size of the Organisation (number of employees, Spain)	
up to 100	15
101 - 500	52
501 - 1,000	19
over 1,000	14
Location of Plants- Spain	
Madrid region	40
Catalonia region	27
Other regions	33
Main Lines - Product/Industry	
Metal products, except machinery and equipment (381)	33
Machinery and equipment, except electrical (382)	19
Electrical machinery and equipment (383)	19
Transportation equipment (384)	7
Professional/scientific/measurement/ control equipment (385)	22
Volume of Operations - Spain (N° of Units/year - Dominant product lines)	
up to 100,000	59
101,000 - 500,000	5
more than 500,000	36

Note:
381, 382, 383, 384 and 385 correspond to *ISIC* (Industrial Standard Classification Code)

---

[2]   Both primary and secondary mailings were addressed to a selected group of 80 leading firms involved in national and transnational activities, operating in Spain and with corporate headquarters both inside and outside the country. Spanish and English versions of the survey questionnaire were made available to the survey participants. The overall response rate was 36%, or 29 firms in total.

volume operators (41% produce more than 100,000 units/year) and low-volume operators (59% produce less than 100,000 units/year). The middle/high-technology subsectors (e.g., electrical, non-electrical, and transportation machinery and equipment) were heavily represented (45%), with the low-technology and high-technology subsectors also being well represented (33% and 22%, respectively).

There exists a deliberate bias in the profile of respondents, since the goal is to identify the practices of successful firms that possess considerable competitive and economic strength. Within this bias, the set of participating firms provides a well-balanced sample of Spain's manufacturing in the sectors of interest.

13.2.2. STATUS, CHALLENGES AND GOALS

The benchmarking of Spain against the reference countries displays some important results, as shown in Figure 13-4 and in Table 14.2. It is clear that the Spain constitutes an excellent economic risk, judging by its ROI (Return on Investment), since Spain is bested in this category only by Germany and the United States (and not by much in the latter case). However, there also exist serious performance difficulties and potential erosion of competitive advantages. Inventory turnover is relatively low, order-to-delivery lead time is exceedingly long, and process efficiency has much to improve.

There is no surprise in the relative ranking of the major end-item cost components. The cost structure is basically 'neutral', which may indicate a long-term erosion of structural competitive advantages. This is particularly worrisome in regards to salaries/wages, which (either because of low wages or high productivity) have constituted a traditional advantage of the Spanish assembly manufacturing sector . Although there already exists a whole spectrum of solutions, from labour market reforms to more investment in training and education and to greater automation of high-value-added operations, their implementation is just starting up.

Table 13-2. Economic Results, Management Challenges, and Cost Structures

Attribute	Country						
	Spain	World	USA	Japan	Germany	Italy	Argentina
Economic results							
Return on Investment (ROI)	10.7	13.6	11.6	9.9	15.2	5.8	9.4
Management Challenges							
Inventory turnover	7.5	8.1	9.1	16.8	7.2	8.3	6.8
Order-to-delivery lead time (days)	40.2	54.7	41.8	32.5	34.8	85.1	35.7
Process efficiency (% of work time/process time)	53.3	32.7	n.a.	74.2	33.3	26.2	49.2
Production cost structure (% of end item costs)							
Direct materials	56.1	54.6	53.6	64.4	44.2	60.0	48.1
Direct labour	21.7	20.3	15.9	14.6	25.8	18.2	26.2
Overhead	22.1	25.1	30.5	21.0	30.0	22.4	25.7

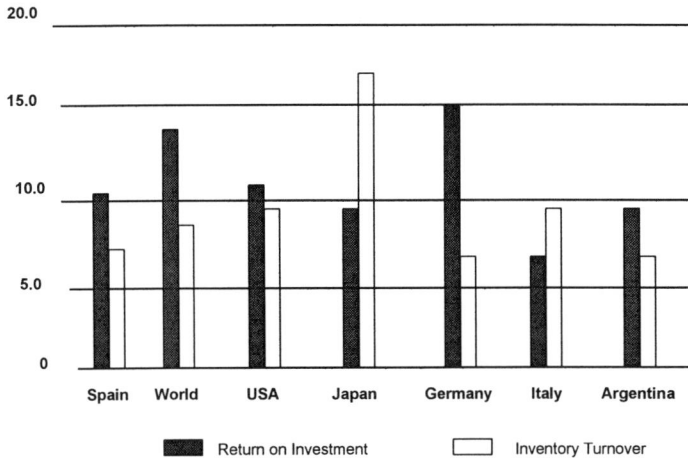

Figure 13-4. Economic and Managerial Performance

Table 13-3 shows the competitive position and challenges faced by Spain's assembly manufacturing firms, vis-à-vis their dominant and competing product lines. The participating firms enjoy, on average, a market share of 36% versus a 27% market share

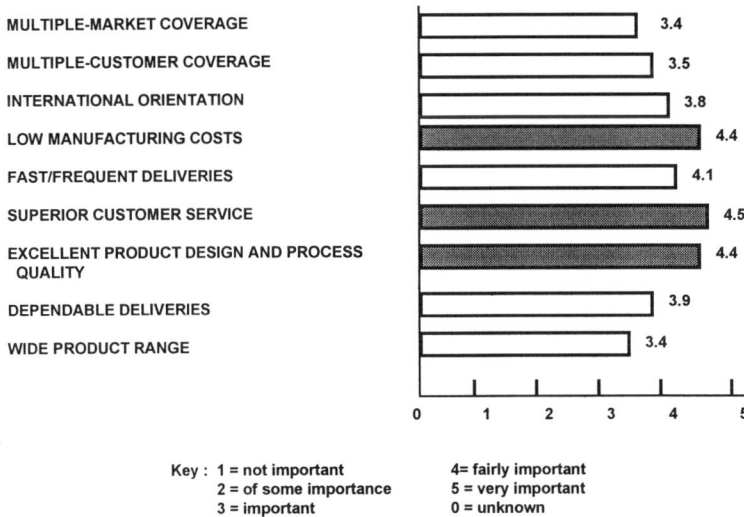

Figure 13-5. Long-term Drivers and Objectives of Spanish Firms

Table 13-3. Competitive Positioning and Challenges of Spanish Firms: Markets, Competitors and Customers

Dimension	% Responses
Dominant Product Line / Own Firm	
Mean	36
Maximum	80
Minimum	10
Dominant Product Line / Competing Firms	
Mean	27
Maximum	55
Minimum	10
Market Growth and Changes	
Rapid growth	7
Growing	10
Stable/stagnant	53
Declining	27
Declining rapidly	3
Variations in Customer Needs / Requirements	
Many highly marked differences	14
Many marked differences	28
Some marked differences	10
Few significant differences	38
Little difference	10

going to their most important competitors. The spread of the market shares shows that competition is, nevertheless, quite tough. The firms' market shares range from 80% to 10% whilst the competitors' market shares go from 55% to 10%, indicating the existence of a substantial overlap with little room for comfort. About half of the firms face changing markets, a fact that makes for quite a fractured view. Whilst most firms (53%) consider that their main product lines are stable, a significant number of firms (17%) consider that their markets are growing (sometimes very rapidly), or declining (30%). Where the real crunch comes is in the variations in customer requirements, with the majority of the firms (52%) faced with increasingly high variations in requirements whilst most of their markets (83%) are stagnant or declining. The implication here is that achievement of productivity gains and cost reductions through economies of scale is largely precluded. Size, on the other hand, is not necessarily a competitive advantage, which bodes well for the many small- and mid-sized firms.

The long-term goals and drivers of Spanish firms are presented in Figure 13-5. On a 5-point scale, the four most important drivers are: customer service (4.50), low costs (4.4), design and quality (4.4), and delivery speed (4.1). Clearly, the previous insularity of Spain's economy and firms has already been largely removed, since they are affected by the same drivers operating at a global scale and are competing along similar lines. At the same time, there is still a heavy dependence on low costs to achieve significant competitive advantages. Most of these four drivers can be handled through good internal logistics, but the most important one (i.e., customer service) requires flawless, highly integrated interfaces among a firm's major functions (e.g., marketing, engineering, operations, information systems).

Within the remaining five drivers, the higher-ranked drivers of delivery dependability (3.9) and international focus (3.8) are consistent with the four top-ranked ones. However, the fact that the drivers of multiple customers (3.5), multiple markets (3.4), and product variety (3.4) are the three lower-ranked ones implies a key long-term strategic thrust oriented to the pursuit of a geographical expansion of markets for a narrow range of products and customers served by excellent internal and external logistics. Specialisation, then, at a global/regional scale seems to be the strategy of choice, since it will result in superior customer service, lower costs and excellent quality.

13.3. Responses: Actions and Programs

13.3.1. RESOURCE ALLOCATIONS

The pattern of allocations of resources chosen by Spain's assembly manufacturers to deal with the challenges they face is shown in Table 13-4. The typical composite expenditures in the critical areas of research and development (R&D), process equipment, and training are fairly robust (13% of sales), although still more oriented towards the medium-term (equipment: 7.7%; training: 2%) than towards the long-term (R&D: 3.7%). Compared against the typical R&D expenditures in the American automotive (about 4% of sales) and electrical/electronics (about 6% of sales) industries, the Spanish mix of expenditures is well aimed, provided that it can be sustained over time. Better equipment and educated personnel will ultimately enhance productivity and competitiveness, and a well-known requirement for successful R&D implementation is the existence of sufficient numbers of well-trained personnel.

In the long term, however, the R&D effort must be increased, so that new technologies can be accessed to in timely fashion, either by the firms alone or by means of joint ventures and strategic alliances. Otherwise, the Spanish firms would acquire a dependent character, become far removed from the own organisations' 'hard core', and be more exposed to boom-and-bust cyclical swings.

A serious management deficiency in Spain's firms is apparent in the lopsided expenditures in corrective/emergency maintenance (about 69.4%) as against preventive maintenance (30.6%). To invest 7.7% of the firm's revenues in new process equipment acquisition, and not to assign enough resources for equipment preservation is counterproductive. The preventive maintenance effort, although substantial today, can afford some significant improvement; unless and until this pattern is reversed the goals of customer service, product design and process quality can be seriously compromised. On the other hand, Table 13-4 does show that the quality expenditures are aimed in the

Table 13-4. Patterns of Key Expenditures in Spanish Firms

Dimension	Attribute	Mean Responses
Research and Development	Average proportion of total sales	3.70
Process Equipment Acquisition	Average proportion of total sales	7.70
Training and Education	Average proportion of total sales	2.00
Maintenance Management	Average proportion of total maintenance expenditures	
	Preventive Maintenance	30.6
	Corrective/Emergency Maintenance	69.4
Quality Assurance	Average proportion of total quality costs	
	Prevention	24.8
	Appraisal	33.2
	Internal failures	28.7
	External failures	13.3

right direction, with the bulk of them (58%) going into prevention and appraisal, although the high costs associated with internal and external failures (42%) still need improvement.

In general, Spain's assembly manufacturing industry seems to be allocating the resources needed to support its goals in an effective fashion. The caveat here is that the effort be sustained over time, since their top-ranked goals (service, quality, costs) can be simultaneously attained in the long haul only.

13.3.2. EXTERNAL LOGISTICS

The main characteristics of the external logistics of Spain's assembly manufacturers are presented in Table 13-5. Clearly, the paramount national goal of greater participation in the EU's economic activities has been pursued and largely achieved by these Spanish firms, since, by large numbers, they tend to source inputs (30% of total purchases), place products (33.2% of total sales), and become suppliers for their companies' customers (50% of all plant locations) within the EU region. Coupled with the fact that 50% of all purchases and 43.2% of all sales are of international (i.e., non-Spanish) nature, a 'trading-nation' status seems to have been already attained here. A counterpoint to the good news may be found in the destabilising effects of international currency exchange rates, which could negatively affect a company's profitability whenever inputs (and thus processes) are too dependent on foreign sources; the cautionary need thus exists for a careful allocation of inputs to foreign sources.

On the procurement side, Table 13-5 shows significant achievements, since 45% of firms maintain co-operative relations with vendors/suppliers, and 33% of firms operate in a JIT mode. These figures are quite encouraging, given the large amounts of international purchases and sales engaged in. They also seem to demonstrate that, at least at the national level, both co-operative vendor/supplier relations and JIT deliveries are firmly established. The Achilles' heel of these accomplishments lies in the large

Table 13-5. Characteristics of External Logistics:. Participating Spanish Firms

Dimension	% Responses
Sourcing	
Average proportion of total purchases	
from within Spain	50.0
from within EU (excluding Spain)	30.0
from outside the EU	20.0
Sales	
Average proportion of total sales	
within Spain	56.8
within EU (excluding Spain)	33.2
outside the EU	10.0
Location of Plants manufacturing dominant product lines	
only plant in company in Spain	42.3
only plant in company in EU (excluding Spain)	15.4
one of several plants in EU (excluding Spain)	34.6
one of several plants in Spain	7.7
Procurement	
Relations with Vendors/Suppliers	
mainly 'arms-length' type	55.0
mainly co-operative type	45.0
Vendor/Supplier deliveries of raw materials/components	
JIT mode	33.0
other modes	67.0
Number of Vendors/Suppliers per firm[1]	477 (*)

Notes: 1. Given in absolute numbers, *not* in % responses

number of vendors/suppliers per firm (477), where 'cascade-type' systems or more JIT deliveries may be of help.

13.3.3. INTERNAL LOGISTICS

Table 13-6 displays the main characteristics of the internal logistics of the firms. It is not surprising that, in terms of demand management, firm customer orders tend to prevail (74.1%) over forecast orders (25.9%), since Table 13-3 already points to the increasing variations in customer requirements. The impacts of the prevailing pattern of customised orders are reflected in a dominant multiple-routing process structure (about 54%), in relatively low levels of capacity utilisation (82.2% of planned capacity) and in long delivery times (about 40.2 days). Capacity is clearly used as a buffer to cope with uncertainties of demand and production. In addition, inventories are also used as buffers to cope with uncertainties of supply (30 days in raw materials/components), demand (23 days of finished goods) and production (23 days of work-in-process). Inventories, however, seem to be fairly well managed with rates of 7-11 turns/year (7.5 turns/year as

Table 13-6. Characteristics of Internal Logistics: Participating Spanish Firms

Dimension	Mean Response
Demand Management	
Sources/releases of production orders	
sales forecasts (in %)	25.9
firm customer orders (in %)	74.1
Capacity Management	
Capacity utilisation in main processes	
number of hours/day (in hours)	15.7
fraction of scheduled capacity (in %)	82.2
Production Management	
Typical length of order-to-delivery lead time (in days)	40.2
Process structure and complexity	
multiple routings, mainly (in %)	54.0
single routing, mainly (in %)	46.0
Inventory Management	
Typical inventory levels held in stock (in production days)	
raw materials / components	30.2
work-in-process	23.3
finished products / end-items	22.8
Product Design and Development	
Mechanics of Marketing-Engineering-Manufacturing Interfaces (in %)	
use of rules / standards / norms	51.7
formal meetings	13.8
informal meetings	6.9
cross-functional task forces	17.2
personal contacts	10.4

an average, better than Germany) although below other benchmark countries (Japan, USA, Italy). In general, the internal logistics of the firms seems to be highly supportive of their external logistics and competitive requirements.

Perhaps the most troubling aspect of the internal logistics is the mechanics of product design and development used to respond to the increasing variations in customer requirements. The dominant mechanics still seems to be of the sequential/arms-length type (51.7% use rules and standards) rather than of the concurrent/interactive type (17.2% use cross-functional teams, and 13.8% use formal meetings). The resulting long product development cycles may be detrimental to the firms' fast access to new markets and to the strategic emphasis on customer service. If part of the strategic thrust is directed towards regional specialisation, with geographical market expansion and without high product variety, as pointed out by Figure 13-5, sequential product design and development may well be a viable medium-term approach. In the long term, however, it will not do. The international benchmarking here is hardly encouraging since it shows that Spain is much more tilted towards norms and standards than most other countries. US firms, by way of contrast, have largely implemented the concepts of cross-functionality and simultaneous engineering.

13.3.4. PROCESSES AND TECHNOLOGIES

Table 13-7 segments production processes into two fundamental stages: fabrication (part/component production) and assembly (end-item production). Table 13-7 shows that batch production, both in fabrication and in assembly, is the dominant mode among Spanish firms. This is consistent with the realities of market segmentation, requirement variations and multiple routings. However, this also points out two serious challenges.

The first major challenge refers to the difficulty of attaining economies of scale and efficient capacity utilisation, resulting in more expensive end-items. From Table 13-7 it follows that Japanese firms, for example, are achieving economies of scale by fabricating a large number of components common to a variety of end-items. American firms, on their part, seem to have opted both for batch production and for flexible automation (that is, the 'technological' option). A second major challenge that comes out of Table 13-7 is the economic difficulty of acceding inexpensive flexible automation ('hard' automation is fairly inexpensive but rigid). Requirement variations and multiple batches make flexible, but expensive, automation necessary precisely when profit margins are reduced by loss of economies of scale and inefficient capacity utilisation. In a nutshell, a truly serious technological and economic 'flexibility dilemma', that perhaps can only be solved by the adequate use of human resources.

It is not surprising, then, that the survey has revealed such an enormous variety in levels of automation among Spanish firms. For the time being, basically all benchmark countries are into mostly light automation, since their firms face the same dilemma as the Spanish firms. Both the United States and Japan show the greatest usage of automation by combining levels 1 through 4 (58.5% for USA, 44.4% for Japan).

Table 13-7. Types of Processes

Country	(% responses in each category)			
	One off	Batch	Line	Others
Fabrication				
Spain	6.9	65.5	17.2	10.4
World	14.5	49.7	28.5	7.3
USA	17.1	36.6	19.5	26.8
Japan	6.4	29.0	64.5	0.1
Germany	18.7	37.5	43.7	0.1
Italy	7.3	48.8	29.3	14.9
Argentina	9.8	56.1	26.8	7.3
Assembly				
Spain	17.2	37.9	34.5	10.4
World	19.2	27.8	35.5	17.5
USA	12.2	19.5	36.6	31.7
Japan	3.7	3.7	59.3	33.3
Germany	29.2	29.2	41.6	0.0
Italy	19.5	14.6	51.2	14.7
Argentina	9.8	19.5	36.6	34.1

Note: Data have been adjusted to eliminate deviations in percentages

The other benchmark countries use mostly 'other' automation modes (48.6% for Spain). In Spain's case, this means using people-based processes (based on a combination of both specialised and flexible workers). Today, the Spanish 'flexibility dilemma' in production processes seems to be dealt with through human resources.

13.3.5. SOCIO-TECHNICAL DIMENSIONS

Adequate program and process implementation requires adequate organisation and personnel. Table 13-8 provides an international benchmark of human resources for Spanish firms, and the signals are mixed but mostly positive. The most encouraging aspect is that the number of organisational levels is among the lowest in the world (3.9), a fact that is fostered by the massive training effort for new personnel (206.8 hours), the high number of multiskilled workers (51.8% of personnel), and the high use of incentives (69% together for group and individual incentives). The presence of these aspects will tend to provide the Spanish firms with a 'social/human' response to the 'flexibility dilemma' present in the process structures.

The least encouraging aspects are in the low employee/supervisor ratios (21 - 23), the large number of job classifications (12.6) and the sparse training for regular employees (28.1 hours/person/year). All these aspects will tend to reinforce the sociocultural trait of risk avoidance, and to hamper the flexibility required from the work force to cope with the 'flexibility dilemma'. Use of teams, which may overcome risk avoidance and allow for flexibility, is at a mid-level usage (32% of personnel involved).

The figures seem to indicate a transition stage towards a more decentralised and delegative mode of personnel management, that would eventually increase process flexibility and responsiveness. Clearly, some of the key fundamentals are in place (new-employee training, incentives, organisational levels, teams) but the overall

Table 13-8. Job Classifications, Training and Multi-Skilling

Country	Number of Job Classifications	Work Force in Teams[1] (%)	Training of new Production Workers (hours)	Training of Regular Work Force (hours/year)	Multiskilled Workers[2] (%)
Spain	12.59	32.04	206.8	28.1	51.8
World	16.17	36.56	109.9	35.5	45.7
USA	25.73	47.58	66.4	30.3	54.9
Japan	8.57	62.81	136.8	19.4	27.4
Germany	6.33	28.91	225.9	107.7	26.1
Italy	7.57	30.04	106.2	27.0	37.4
Argentina	8.11	31.21	112.3	33.5	41.1

Notes:
1. A team is a group of employees performing a task, with high degree of responsibility for planning, execution and control.
2. A multiskilled worker is able to perform several operational tasks.

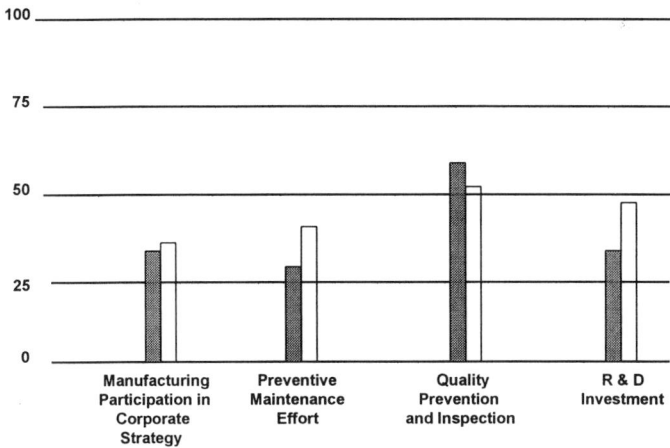

Figure 13-6. Key Manufacturing Management Mechanisms and Decisions (Proactive Dimensions)

organisational structure has not fully evolved yet in that direction. At one point in time, a choice of continuous training mode will have to be made by Spain's firms between teams (Japanese) or in-class work (German) or somewhere in between (American). It is here where the sociocultural environment will become critical for selection and fine-tuning of programs

13.3.6. KEY MANAGEMENT MECHANISMS AND DECISIONS

Taken together, Table 13-9 and Figure 13-6 provide a revealing and symptomatic view of key manufacturing management aspects. The international benchmarking comparison shows that Spain's firms exhibit one of the world's lowest degrees of influence and participation by the manufacturing function in the setting of corporate goals and in the development of corporate strategy. If to this the sequential and dilatory mechanics used for new-product development is added, a serious situation of lack of 'fit' between formulation and implementation of corporate strategy becomes evident in Spanish firms. The existence of such a situation is symptomatic of a state of internal disarticulation, which may be conducive to the grave erosion of positive internal and external logistics aspects, and to the loss of competitiveness.

Table 13-9 and Figure 13-6 also reveal that Spanish firms are deficient in preventive maintenance, active in quality prevention and inspection, and modest in R&D. The overall reading is, thus, that Spain's firms do not possess enough internal articulation and do not emphasise enough long-term, proactive efforts. As can be seen in Table 13-9 and, further ahead in Table 13-10, the effort is there; what is questionable today is its intensity, its orientation, and its effective articulation.

Table 13-9. Key Manufacturing Management Mechanisms and Decisions

COUNTRY	Participation in Corporate Strategy Development[1]	Preventive Maintenance Effort[2]	Quality Prevention and Inspection[3]	R & D Investment[4]
Spain	3.2	30.6	58.0	3.7
World	3.4	37.2	53.1	4.8
USA	3.5	35.4	42.6	5.2
Japan	4.1	66.2	52.2	3.4
Germany	2.9	33.8	56.4	2.8
Italy	3.2	37.7	58.6	4.5
Argentina	3.6	33.8	64.4	11.1

Notes:
1. scale is from 1 (= none) to 5 (= a lot)   3. % total quality costs
2. % total maintenance expenditures   4. % total sales

13.3.7. PROGRAMMES AND PRACTICES: THE 'QUALITY THRUST'

One of the 'grand questions' of manufacturing management ·is the degree of transferability and usefulness of world-class 'best practices' among different sociocultural and macroeconomic environments. This 'grand question' was, in fact, 'the' paramount concern in the minds of the participating Spanish managers.

Table 13-10 provides a picture of the degree of usage of specific programs/actions both within Spain and in other countries from the viewpoint of alternative strategic thrusts (e.g., quality, leanness/costs, flexibility), and it allows for two very important readings:

a. The most important strategic emphasis for Spanish manufacturers is quality, whose importance was already highlighted in Figure 13-5 (e.g., customer service, design and process quality). Table 13-10 shows that the Spanish assembly manufacturers are definitely serious about quality, given the variety and intensity of programs/actions used, putting Spanish firms above or at a par with world-wide indicators. Further, the Spanish profile is very similar to that of the American firms, a fact highly encouraging considering that the US firms have been able to stop and significantly arrest the 'Japanese onslaught' in the last 5 years, and in precisely the automotive and electronic sectors. Given that Spain's firms started their quality programs/actions later than the American firms, additional time will be needed to get to where their American counterparts stand today. It can also be seen in Table 13-10 that in the strategic quality struggle Spain's assembly manufacturers are well positioned vis-à-vis First World (Germany, Italy, Japan, USA) and industrialising (Argentina) competitors. The direction is right, but constancy of purpose is needed.

b. Beyond quality, there also exists a high degree of adoption of 'best practices' with emphases on leanness/cost and flexibility, not only in Spain but throughout

Table 13-10. Manufacturing Management Practices and Programmes: Degree of Usage

Practices and programmes	Country							Emphasis		
	ESP	IMSS	USA	JAP	DEU	ITA	ARG	Q	LP	FM
Total Quality Management (TQM)	3.3	3.2	3.5	3.8	N/A.	2.7	2.8	X		
Statistical Process Control (SPC)	3.4	2.9	3.0	2.7	3.0	3.0	2.8	X		
ISO 9000 Norms	3.4	3.1	2.2	2.6	3.3	2.7	2.0	X		
Material Requirements Planning (MRP)	3.8	3.2	3.7	3.3	2.3	3.2	2.3			X
Manufacturing Requirements Planning (MRP II)	3.2	2.7	3.2	2.3	1.9	1.9	1.5			X
Just-in-Time (JIT) / Lean Production	3.0	3.0	3.4	3.9	3.0	2.3	2.2		X	
Just-in-Time (JIT) / Deliveries	3.0	2.9	2.6	4.0	2.5	2.4	2.4		X	
Single-Minute Exchange of Dies (SMED)	2.2	2.4	2.2	3.4	1.0	2.8	1.8		X	X
Pull/Kanban Scheduling	2.7	2.5	2.7	3.3	2.5	2.2	1.7		X	
Zero-defects programs	3.2	2.7	2.6	3,6	2.7	2.0	1.8	X		
Computer-Aided Manufacturing (CAM)	3.1	2.7	3.1	2.7	2.6	2.2	1.5			X
Computer-Aided Design (CAD)	3.8	3.6	4.0	3.9	3.6	3.8	2.2			X
Design for Assembly / Manufacturability (DFA/DFM)	2.5	2.5	2.6	3.4	1.9	1.8	1.3	X	X	X
Quality Function Deployment (QFD)	3.1	2.6	2.2	3.7	1.3	1.4	2.5	X		
Product value analysis's/redesign	2.5	2.6	2.4	3.7	2.7	2.2	2.2		X	
Quality Policy Deployment (QPD)	3.6	2.8	2.3	4.4	3.0	1.8	2.3	X		
Plant-within-a-plant reorganisation	2.8	2.9	2.9	3.1	2.8	2.8	2.0		X	X
Defining a Manufacturing Strategy	3.7	3.3	3.2	3.8	3.2	2.8	3.0	X	X	X
Simultaneous engineering	2.1	2.6	3.5	2.7	2.5	2.1	1.6			X
ABC costing	2.0	2.4	1.9	3.2	2.5	1.8	2.4			X
Implementing work teams	3.2	3.4	3.4	3.8	2.9	3.5	2.5	X	X	X
Benchmarking	2.7	2.4	2.1	4.0	2.3	2.1	2.1	X		X
Kaizen	3.1	3.0	3.3	4.2	1.7	2.2	2.2	X	X	
Total Productive Maintenance	2.8	2.4	2.1	4.6	2.2	1.8	2.0	X	X	X
Energy conservation programs	2.6	2.6	3.0	3.4	3.1	2.4	1.9	X	X	X
Environmental protection programs	2.9	3.1	3.7	3.9	2.8	3.2	2.5	X	X	X
Health and safety programs	3.7	3.5	4.1	4.1	1.8	3.5	3.1	X	X	X

Notes :

1. Degree of Usage: 1 = None to 5 = High

2. Q = Quality; LP = Lean Production; FM = Flexible Manufacturing

the world. Encouraging as these results may be in their implications about the 'environment-neutrality' of the 'best practices', it must be remembered that the industries and firms surveyed here (e.g., machine tools, automotive, electronics, instruments) are of necessity based on a high degree of globalised know-how and sophistication in their operations. Nevertheless, the reading is highly encouraging, and particularly so for Spain given its national goals of pan-European convergence and world-wide competitiveness.

13.4. Conclusions

Overall, the composite picture presented by Spain's key assembly manufacturing sector is one of an intense, eclectic struggle, focused on strategies with multiple objectives, and accompanied by a process of industrial restructuring and regional integration.

A key outcome of the analysis presented here is that, within their specific sociocultural and economic setting, the leading Spanish firms have been able to effect the successful implementation of a variety of advanced manufacturing's 'best practices', regardless of the origin of those practices. Further, Spain's leading manufacturers appear to be on the right path, as confirmed by their results shown in Table 13-11 and by their effective attainment of a 'trading-nation' status.

Another key outcome of this analysis is the need for urgent attention to some critical issues and priorities. Chief among them is the need for a greater intra-organisational articulation and a greater (and better) proactive management capability. The advances achieved to this date appear as an archipelago of 'islands of excellence', which are not organically interconnected in a consistent fashion yet. Just as critical as intra-organisational articulation, a need to implement well-balanced advances in the sociotechnical aspects within the firms also exists.

Figure 13-7 and Figure 13-8 point out yet another key outcome of this analysis. It appears that although Spain's leading manufacturers have improved performance significantly, selection of the performance improvement programs is far more complex than anticipated. In Figure 13-7, it can be seen that although most firms improved their speed of product development, this was not necessarily the result of greater R&D expenditures.

Table 13-11. Performance of Participating Spanish Firms: Economic and Competitive Indicators

Dimension/Indicator	Performance Evaluation (% Improvement in Last Two years)	
	Spain	World
Competitive		
Market share	-4.8	11.3
Customer service	21.9	19.6
Product variety	8.4	17.4
On-time deliveries	75.2	26.8
Delivery lead time	51.8	24.2
Economic		
Average unit manufacturing cost	15.4	14.1
Profitability	-3.2	10.8
Inventory turnover	36.3	26.5
Operations/Technology/Manufacturing		
Speed of product development	26.3	20.2
Conformance to specifications	66.3	28.8
Equipment changeover time	27.8	18.2
Manufacturing lead time	30.3	28.4
Procurement lead time	32.5	18.9

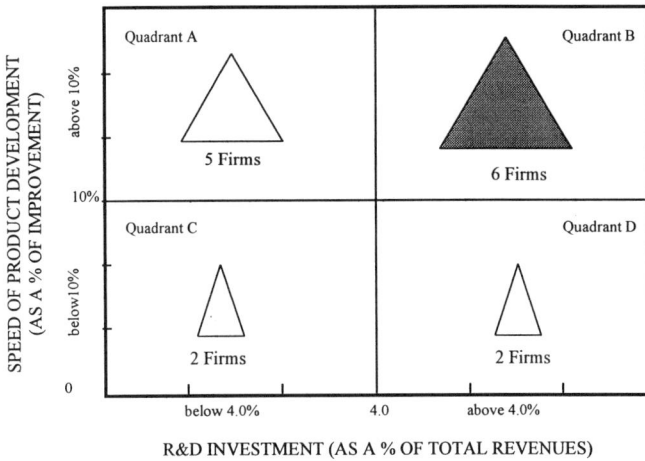

Figure 13-7. The R&D Payoff

Similarly, Figure 13-8 shows that whilst manufacturing quality went up markedly for most firms, fewer than a third of them incurred high training and development expenditures. In a nutshell, 'best practices' may be transferable among different environments regardless of origin, but the transfer mechanisms seem to be quite environment-specific.

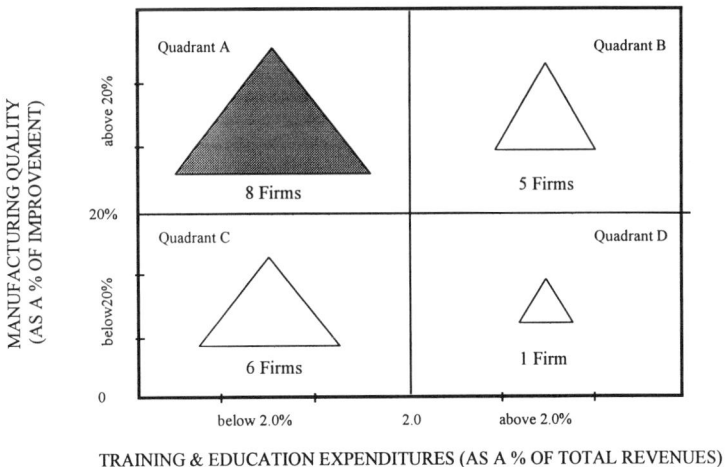

TRAINING & EDUCATION EXPENDITURES (AS A % OF TOTAL REVENUES)

Figure 13-8. The Training and Education Payoff

Spain's macroeconomic profile is worrisome, since there exist quite a few gaps and weak spots vs. both 'large economies' (USA, Germany, Japan) and 'middle/small economies' (Belgium, Denmark, Netherlands). Those gaps appear too regularly and consistently as to ignore the pattern they conform. The macroeconomic environment must be managed (e.g., economic pacts, social contracts) for the attainment and preservation of advantages and economic results of national interest.

On the other hand, a positive fact of major importance is that the Spanish sociocultural profile does not appear as an obstacle to implementation of world-class 'best practices', since this profile is close enough to the profiles of the 'large economies' in important aspects. In other words, Spain's sociocultural profile is quite flexible, with a sole major exception: its high degree of uncertainty avoidance (or, high conservatism). But, then, 'Spain is different', and so it is in its numerous achievements, its economic miracles, and its headlong charge towards the future.

13.5.   Case Study: Seat/Volkswagen

The 'Picos' System is one of the enduring legacies left in Seat/Volkswagen of Spain by Ignacio Lopez de Arriortua (a.k.a. Iñaki), the former No 2. person in the Germany-based multinational manufacturer of automotive equipment. The system's name ('Picos' means 'Peaks' in Spanish) was coined by Mr. Lopez to emphasise the need for looking into the factors that foster peak-level performance in manufacturing, and then carefully re-engineer both the social and physical environments of the organisation to consistently maintain such factors in place.

The whole idea in Picos is not to accept average performance as a fact of life, but, rather, to strive for continuous improvement in both evolutionary and revolutionary ways. The system is predicated on three key elements: suppliers, workers, and managers. By transferring to suppliers all the non-core activities of Seat, and by selecting only those suppliers that can maintain peak performance, excellence and co-operation are created at the source of the major contributor to end-item costs that is, materials. In turn, by providing protagonism to work teams, acceptance to change and innovation are fired up.

The big punch, however, comes with the manager's role. Management, under the Picos System, must ensure that new manufacturing systems are created within the organisation, and never copy the existing ones, since this will never make the company a market leader, loosing also the potential 'innovator's profits' from being first to market. To copy is to loose, recites Mr. Lopez, since to copy means being late for industrial changes, missing the train of the 'fourth industrial revolution' based on knowledge power.

Experimenting with the Picos System has been an unqualified success for Seat/Volkswagen. Productivity has increased in 28%, downtimes have been reduced in 37%, inventories have shrunk 34% and working space in 29%. With these results, is there then a threat of layoffs for the company's work force? No and yes. There is no threat because the company will simply produce more rather than reducing personnel, in principle. Economic downturns will pose a threat, but they can be dealt with to some extent by turning workers into suppliers, or, alternatively, by shifting workers into other profit-enhancing activities such as maintenance, training, or process improvements.

A heavy emphasis is placed under Picos on training as an agent of change, as a means of persuasion, as a deflector of labour conflicts, and as a linkage to suppliers. The mind-set of all participants in the value-producing chain must be adjusted first for the specific techniques and programs to work adequately. If such an adjustment is achieved, then techniques and programs will bear fruit. Effectively, a 'social-energising' phase must precede a 'technical' phase for implementation to succeed. And success, of course, brings job security and career advancements for all contributors.

The Picos System clearly works well, and not only in Spain. It has also been applied in Volkswagen plants in Germany and across Europe with equally stunning results. This does not necessarily mean that Picos is environment-neutral, but, rather, the other way around. In charting its implementation through a careful sequence of social and technical steps, Picos becomes transferable into a variety of environments.

13.6. Case Study: Rank Xerox

The 'quality struggle' is no stranger to Rank Xerox of Spain, which has been in operation for over 20 years and has consistently been a moving force for innovation in Spanish manufacturing. But the 1970s were rough on Rank Xerox, witnessing the market penetration of Japanese competitors and the loss of market share for Xerox products. Benchmarking was developed in Xerox U.S. and from there spread throughout Europe, including Spain.

The dramatic organisational and operational re-engineering set forth by benchmarking turned Xerox into a wholly quality-centred organisation. Fuji Xerox was awarded the Japanese Deming Quality award in 1980, Rank Xerox won the corporate Crystal Apple award in 1987, Xerox U.S won the American Malcolm Baldrige award in 1989, and Rank Xerox won the European Quality Award in 1992.

So, what happened at Xerox? Basically, Xerox decided to become a market leader through consistent achievement of quality, defined as 'quality is what the customer wants'. The implementation of this concept in Rank Xerox of Spain was done through two types of deployment: geographical deployment, and policy deployment.

The geographical deployment was done by breaking up the Spanish market into seven regions, and by entrusting the Regional Directors with marketing, sales, customer service, and after-sales support. Effectively, the regions were made responsible for the 'soft' non-technical elements of quality. The 'hard' technical elements of process conformance quality were concentrated in a single manufacturing facility in Coslada, where the line workers were empowered to engage in continuous process improvements, and the design effort was carried jointly by Marketing and R&D personnel.

The policy deployment was carried out through objective-setting, training, compensation, and information dispersion. Effectively, the four elements were tied together in a 'package' that in turn was linked up with every worker or work team. Management's role was set to facilitate the proper working of these 'packages' throughout the organisation.

Quality strategies and efforts generated excellent results. Rank Xerox is today the leader in Spain with over 35% of market share, productivity went up in the last 5 years by as much as 20%, and profitability increased by 25%. The acknowledged key to Rank Xerox quality success story is the deconcentration and empowerment built into the regional and policy deployments. In a nutshell, both internal and external customers were cared for, and the quality actions and programs were carefully chosen and fine-tuned to the specific customer needs and strengths.

Acknowledgements

With the contributions and co-operation of David B. Allen, Ph.D and Maria Teresa Brogeras, J.D.

References

Allard, G. and Bolarinos, J. (1992) *Spain to 2000: A Question of Convergence*, The Economist Intelligence Unit, Special Report N° M207, August.

Dertouzos, M. et al. (1989) *Made in America*, Harper Perennial.

Hofstede, G. (1991) *Cultures and Organizations*, McGraw-Hill.

Kennedy, P. (1993) *Preparing for the Twenty-First Century*, Random House.

The Economist Intelligence Unit (1994) *Spain: Country Profile 1994-95*, December.

IMD/WEF (1994) *The World Competitiveness Report 1994*, 14th edition.

Thurow, L, (1992) *Head to Head*, W. Morror & Co., Inc.

Vargas, G. and Johnson, T. (1993) 'An analysis of operational experience in the US/Mexico production sharing program', *Journal of Operations Management*.

Whybark, D.C. and Vastag, G. (1993) *Global Manufacturing Practices*, Elsevier.

CHAPTER 14

LEAD TIME REDUCTION - MANUFACTURING STRATEGY IN SWEDEN

Per Lindberg, Chalmers University of Technology, Sweden

14.1. Introduction

14.1.1. GENERAL BACKGROUND

The Swedish economy has entered a state of increasing turbulence and decreased certainty over the last five years. The economic outlook for the Swedish government is increasingly gloomy, although exports are reaching record volumes. Politically, the previously unaffiliated Swedes joined the European Union in 1995, which favoured parts of the economy (mainly those relying on exports), while other parts experienced increased competition (e.g., the agricultural sector). In the midst of the turbulence of the early 1990s, a majority of Swedish industrial manufacturers have embarked on a change route, aiming towards increased competitiveness and flexibility. The main driver for the transformation is time-based competition, the reduction of lead-times together with organisational redesign.

This chapter draws on the example of the Swedish engineering industry to illustrate these changes, using data from both external sources and the IMSS survey.

Swedish economic development in general
As stated previously, the Swedish economic and political situation has created a turbulent business environment. The government budget is in crisis. Unemployment has increased to levels previously unheard of in Sweden: from levels of 2-3% unemployment at the beginning of the 1990s to around 8% in 1995, with the total unemployment level around 14% when everyone in different unemployment programs is included. Even if the unemployment levels in Sweden become "adjusted" to EU levels through this process, the Swedish welfare system has been designed to compensate to a large extent for income losses such as those due to unemployment. This has placed tremendous stress on public finances, exploding public expenditure and creating a huge national debt. Consequently, long-term interest rates increased to almost 11% by the first quarter of 1995, and, since inflation continued to be low (2.9% in February 1995), the high real interest rate has demanded effective utilisation of capital and capacity in Swedish industry.

Beginning in November 1992, the release of the fixed exchange rate of the Swedish currency against other major currencies resulted in a dramatic devaluation of the

P. Lindberg et al. (eds.), International Manufacturing Strategies, 259-273.
© *1998 Kluwer Academic Publishers. Printed in the Netherlands.*

Swedish kroner (SEK). The fall of the kroner was around 20% against the US dollar, 37% against the German D-mark, and even more against the Japanese Yen. This threw large parts of the economy into an export boom, with a dramatic increase in capacity utilisation. As a result, Sweden continued to enjoy a positive balance of trade: in 1994, exports amounted to 471 billion SEK, and imports amounted to 398 billion SEK.

In 1994, 65% of all exports went to the EU region, with Germany and Great Britain Sweden's main trade partners (accounting for 23% of all exports). The engineering industry accounted for 50.7% of exports, with forestry-related industries in second place (16.7%). Engineering products accounted for 44.7% of imports, nearly double miscellaneous (food, textiles, etc.), which accounted for 22.6% of imports. From these figures, it can be seen that the competitiveness of the engineering industry is of major interest for the nation.

Exports had reached an all-time high in output by early 1995, but at the same time some 25% of all jobs have been shed in industry since 1991. Consequently, profits for 1994 in the major Swedish export firms were at record levels, a significant improvement from years of poor financial performance in the beginning of the 1990s. Thus, major improvements have been created in terms of overall industrial productivity and profitability over the last few years.

However, the main volume and profit increases had been created in export-oriented industries, due to the situation described above, while industries producing for domestic markets continued to be sluggish. Not only are the economic prospects of the Swedish state economy and export-oriented industries in very different situations, export industry and domestic industry also faced very different outlooks.

Change programs in Swedish industry
The improved competitiveness of Swedish exports (due to the devalued currency) is of course improving possibilities for competing in global and regional markets. However, competition is not established by factor prices alone. Internal company performance on dimensions such as productivity, conformance to quality standards, delivery speed and performance also determine the overall competitiveness of companies. Over the last 4-5 years, Swedish companies in general – and companies in the engineering industry in particular – have engaged in change efforts in order to increase competitiveness. The most significant driver has been lead-time reduction, or *Time-Based Management (TBM)* (e.g., Stalk and Hout, 1990). One pioneering company in this respect is ABB Sweden, which has inspired several followers with their T50 program.[1]

The reasons for selecting lead time as the driver for change are several (Järneteg, 1995). First, time is easy to grasp, since it is a clearly defined concept known to everyone. Secondly, it is relatively easy to measure levels of performance. Thirdly,

---

[1]   T stands for cycle times, 50 for a reduction of 50%. Thus T50 means a reduction of all cycle times with 50%.

```
┌─────────────────────────────────────────────────────┐
│          Halved lead-times are associated with...     │
│                                                       │
│              -      8.5 % production costs            │
│              -      47 % Work-In-Progress             │
│              +      9.5 % profitability               │
│                     15 % fixed assets                 │
│                                                       │
│          Source: PIMS/INDEVO, 1992                    │
└─────────────────────────────────────────────────────┘
```

Figure 14-1. Reduction of Lead time and Associated Performance Measures

there is a clear interdependency between time factors and quality issues and/or cost issues. The three are inter-linked in several ways, which is also shown by data from PIMS (Profit Impact of Market Strategy), as shown in Figure 14-1.

The interdependencies between quality, cost and lead times have been recognised in several industrial change programs,[2] together with the acknowledgement of organisational development as a necessary measure for increasing speed of operations and corresponding reduction of lead-times. Thus, the second major characteristic besides lead time focus, with all its implications for process thinking (e.g., Rummler & Brache, 1993), small-batch production, Just-In-Time principles, etc., is an emphasis on major organisational redesign. This means increased decentralisation, and increased reliance on operator-level competence development and on team-work.

Competitive development - an OECD comparison
Industrial statistics recently compiled by the OECD allow a comparison of the industrialised countries. (See Table 14-1). These figures show that Sweden is in the bottom when it comes to average GNP growth, but in the very top when it comes to industrial productivity growth.

Table 14-1. OECD Statistics For Selected Countries (Source: OECD, 1994)

Country	Average GNP growth 1983-1993 (%)	Industry % of GNP (1994)	Foreign direct investments 1994 (% GNP)	Average industry productivity growth (%) (1990–1994)
Sweden	1.2	27.4	2.0	33
Japan	3.7	41.1	0.0	-9
USA	2.8	27.0	0.3	10
Great Britain	2.2	28.7	1.5	10
Germany	2.8	36.6	0.0	-5
Australia	3.3	24.9	0.9	N/A
Finland	1.0	27.6	0.7	28

---

[2]   For example,   SAAB Automobiles change concept QLE / H (Quality, Lead-time, Economy and Human), and Volvos change concept KLE / P (which basically stands for the same concepts as SAAB)

These statistics show that Swedish industry improved dramatically between 1990 and 1994 in terms of productivity development and in terms of attracting direct investments. However, the economy as a whole is still sluggish, as indicated by poor economic development (GNP growth), and further enhanced by the low reliance on the once again very competitive industry (only 27.4% of GNP from industrial activity).

14.2. The Swedish Engineering Industry - Main Characteristics

14.2.1. COMPETITIVENESS AND STRATEGIC DRIVERS

The Swedish engineering industry has historically been known for the production of relatively specialised products, with producers placing little emphasis on volumes and cost-based competition (see Lindberg and Hörte, 1994). The main dimensions on which companies compete have been customisation, quality and delivery performance.

Table 14-2. Strategic Variables in Swedish Engineering Industry (Source: IMSS)

	Sweden		IMSS average		2-tailed t-test	
	Mean	Rank	Mean	Rank	P	Sig.
Company goals (average ranking of importance)						
Customer service	4.627	1	4.420	2	0.061	-
Design & mfg quality	4.550	2	4.601	1	0.575	-
Delivery reliability	4.433	3	4.234	4	0.093	-
Delivery speed	4.150	4	4.157	5	.0952	-
Low costs	4.117	5	4.351	3	0.040	*
Wider product range	3.186	6	3.414	6	0.144	-
Market shares						
Own market share	-	28%	-	34%	0.044	*
Main competitors market share	-	20%	-	24%	0.023	*
R&D spending						
R & D spending	-	5.7%	-	4.9%	0.334	-
Product range Number of different products						
- Current number of products	-	265	-	706	0.447	-
- Expected change next 5 years	-	+290%	-	+51%	0.004	**
Revenues from new products						
- Current revenue	-	16%	-	19%	0.372	-
- Expected change next 5 years	-	+27%	-	+32%	0.737	-

* p < 0.05

Table 14-2 supports our previous discussion of the competitive priorities of Swe͞ sh manufacturers, and suggests that they are still valid for Swedish engineering companies. It is evident from the company goals that the emphasis on lead time reduction previously stated is not just a function of improving capabilities of delivery speed, but also the general driver for improving customer satisfaction (responsiveness), quality and delivery performance in general.

The data also suggest that the Swedish manufacturers in the engineering industry *in general* are niche players, and that their main strategy is to refine their position in niche markets. The high emphasis on customer service, coupled with relatively small market shares, suggests that they operate in relatively fragmented industries.

They have, in comparison with the total IMSS sample, high R&D expenditures, indicating an emphasis on advanced products rather than on commodities. However, the Swedish manufacturers expect to rely on revenues from their existing product ranges the coming few years. The relatively narrow product range is expected to grow significantly in coming years, but the revenue from the expanded product range is expected to be modest: only 27% of revenues are expected to come from product range expansion. The wider product range puts demands on increased responsiveness, flexibility and decreased time-to-market, but the bulk of revenue is expected to be from existing product ranges. This also indicates that overall, the companies have strong existing product lines.

Economics of the firms

The Swedish companies grew a little more than average in terms of production volume the 5 years before 1992 (Table 14-3), but they did not at all expect to achieve the same level of growth as their counterparts in other countries. In retrospect, it can be concluded that the expected future growth most likely was underestimated by the Swedish manufacturers. Since 1992 the growth in volumes for export companies has been substantial, as described previously.

Table 14-3. Economics of the Swedish Firms (Source: IMSS)

	Average		2-tailed t-test	
	Sweden	IMSS	p	Sig.
Size and structure				
Production volume changes				
- past 5 yrs	43%	38%	0.757	-
- next 5 yrs	18%	37%	0.121	-
Economic results				
Return on Investment	10.3%	13.6%	0.667	-
(ROI)				
Cost structure				
Direct materials	52.6%	54.6%	0.340	-
Direct salaries	18.2%	20.3%	0.190	-
Manufacturing overhead	29.2%	25.1%	0.017	*

$* p < 0.05$

Table 14-4. External Logistics of the Swedish Firms (Source: IMSS)

	Average		2-tailed t-test	
	Sweden	IMSS	p	Sig.
Sourcing				
External / international sourcing	48%	43%	0.185	-
Number of suppliers				
- Current supplier base	219	469	0.299	-
- past 5 years	-4.1%	4.2%	0.440	-
- next 5 years	-4.5%	-5.1%	0.932	-
JIT deliveries	31%	29%		
Sales				
International sales (exports)	63%	42%	0.000	***

* p < 0.05; ** p < 0.01; *** p < .001

The financial performance of the Swedish companies in terms of ROI was, on average and in relation to other companies, relatively weak, even though in absolute terms the figures for 1992 were acceptable. The most significant feature of the cost structures is the high overhead cost, indicating a general need for reduction of these costs through a focus on value-adding activities and decrease of overhead functions.

14.3. Manufacturing strategy of Swedish manufacturers

14.3.1. SUPPLY CHAIN MANAGEMENT - EXTERNAL AND INTERNAL LOGISTICS

As previously described, Sweden is dependent on international trade, both for imports as well as for exports. The engineering industry accounts for approximately 50% of all exports, but also for a significant part of the imports. Products and components are bought outside the country for final manufacturing and assembly, and then re-exported.

The supplier base of Swedish engineering companies is relatively small and decreasing. Still, the majority of parts and components are sourced within the country. Some 37% of parts and components are sourced from within the EU region. The proportion of materials and components delivered Just-in-Time is roughly equal to the IMSS data in average. Exports dominate sales, with 48% of production sold to the EU region, and 37% sold domestically.

Table 14-5 shows that, on average, Swedish manufacturers produce roughly similar amounts to customer order (vs. forecast orders) as other corresponding companies in the IMSS sample. With this in mind, the lead times from order to delivery reported by the Swedish companies are good; some 20% below average. Since a high degree of production to customer orders (vs. forecast order production) may create longer lead

Table 14-5. Internal Logistics of the Swedish Firms (Source: IMSS)

	Average		2-tailed t-test	
	Sweden	IMSS	p	Sig.
Demand, capacity & inventory management				
Customer orders	71%	66%	0.280	-
Capacity utilisation	81%	81%	0.789	-
Inventory levels				
- Raw material	29 days	33 days	0.423	-
- Work-In-Progress	14 days	24 days	0.018	*
- Finished goods	14 days	21 days	0.072	-
Lead time - order to delivery	45 days	55 days	0.439	-
Production process (sum >100%)				
- One-off production	22%	18%		
- Batch production	67%	60%		
- Line production	26%	35%		

* p < 0.05

times, this is a qualitatively good measure of performance. It should be noted, however, that the degree of value-adding and the number and complexity of operation steps may significantly influence lead times and potentials for lead time reduction. But since the proportion of production processes is roughly equal to the comparative sample, and even a tendency for more complex and normally time-consuming one-off production, it still indicates a relatively good lead-time performance. Corresponding inventory figures are also good, especially in terms of WIP and finished inventories, where the average amount of inventories are 30-40% below the IMSS average.

In regard to time-to-market issues, co-ordination of the design/manufacturing interface is of crucial importance. Given that the majority of contacts between design

Table 14-6. Organisation and People Management

	Average		2-tailed t-test	
	Sweden	IMSS	p	Sig.
Structure				
Headcount	634	869	0.297	-
Number of organisational layers	3.8	4.1	0.061	-
Control span of first line supervisor	23.1	23.7	0.835	-
Motivation / Facilitation / Support				
Payment system / incentives (sum > 100%)				
- Fixed bonus	33%	63%		
- Group based incentives	50%	30%		
- Individual incentives	27%	31%		
Teamwork	51%	37%	0.000	***
Multiskilling (Work-force flexibility)	55%	46%	0.007	**
Decentralisation				
Scheduling and control responsibility (>100)				
- Planning dept. or foreman	79%	97%		
- Operators	28%	9%		

* p < 0.05; ** p < 0.01; *** p < 0.001

and manufacturing in the Swedish companies are taken and co-ordinated through task-forces rather than by formal rules and impersonal regulations, this seems to be a conscious means of co-ordination. The degree of manufacturing involvement in information transfer is in new product development can also be regarded as high.[3]

14.3.2.   ORGANISATION AND PEOPLE MANAGEMENT

One of the more significant features of Swedish manufacturing industry has been a long standing practice of experimentation with new forms of work organisation and decentralised organisational structures. The Volvo experiments in Kalmar and Uddevalla are probably the most widely known and publicised over the last few years (e.g., Engström et al., 1993).

The trend towards decentralised structures, teamwork and multiskilling of (primarily) operators is continuing. The Swedish companies reported significantly higher use of teamwork (the proportion of employees working in teams) and multiskilling (the proportion of employees that are skilled in several operational tasks). Decisions on operative issues (e.g., planning and control) were more decentralised, with a much higher proportion of operators responsible for scheduling and control. Payment systems promote group-based activities through a higher proportion of group-based incentive systems.

Lead-times, organisational and manufacturing performance
A high emphasis is placed on lead-time reduction, in all aspects, among the Swedish companies. Of the IMSS respondents (n=61), some 77% stated that *manufacturing lead-time* reduction is of high or very high importance as a goal for future action in manufacturing (while only 5% stated that it was of low importance). Similarly, some 52% stated that *procurement lead-time* is of high or very high importance, and 53% stated that *new product development cycle times* is of high or very high importance. 36% of the companies stated that all three lead-time indicators above are of high or very high importance as goals for future manufacturing action.[4]

A comparative analysis of the performance of the Swedish manufacturers and the average IMSS sample is shown in Table 14-7. This shows that the improvement rates for the time period 1990–1992 for Swedish companies was comparable with that of the IMSS sample, with a few exceptions.[5]

The Swedish manufacturers have substantially better improvements in product development cycle times. In all other measures, the performance improvement rates are more or less on average or below average. The technical and organisational arrangements therefore seems to have substantial effects on lead time reduction, but not

---

[3]   3.8 on a scale from 1 to 5, vs. 3.4 in the total IMSS sample
[4]   Indications are made on a scale from 1 to 5. Average score for e.g. manufacturing lead time reduction was in Sweden 4.21, and in total IMSS sample 3.96.
[5]   Indicated as a change from 1990 (index =100) to 1992.

enough on factors such as cost reduction and quality improvement during the period 1990–1992. However, productivity gains after 1992 have been exceptionally strong, thus "lagging" cycle time reductions and organisational developments, and due to dramatically increased volumes after depreciation of Swedish SEK. Volumes and profitability levels reached record heights during 1994–95, due to the devalued SEK and leaner and faster companies.

Conclusion

It is evident from the above presentation that Swedish manufacturers in general emphasise a specific set of issues, creating a particular manufacturing strategy profile. The issues that stand out are:

- A high emphasis on lead-time reduction and quality improvement in manufacturing
- A premium emphasis on decentralised, team-based organisations
- Significantly better-than-average lead-time reductions

Table 14-7. Organisational Performance Improvements 1990–1992

Performance area	Average		2-tailed t-test	
	Sweden	IMSS	p	Sig.
Competitiveness development				
Market share	+6%	+11%	0.494	-
Customer service	+17%	+19%	0.537	-
On-time deliveries	+26%	+27%	0.930	-
Unit cost	-14%	-14%	0.957	-
Economic development				
Profitability	1%	+11%	0.181	-
Inventory turns / year	+29%	+26%	0.771	-
Manufacturing development				
Product development cycle	-31%	-20%	0.024	*
Mfg lead times	-35%	-28%	0.435	-
Procurement lead times	-19%	-19%	0.986	-
Mfg. quality	+25%	+28%	0.642	-

* $p < 0.05$

14.4.   Explaining time-based improvements - two case studies

We have seen that Swedish manufacturers, *in general*, have based their manufacturing strategies on decentralisation, lead-time and inventory reductions, in support of overall strategies aimed at production of high-quality customised products in relatively small volumes and in fragmented industries. In order to provide a more substantial description of the indicated manufacturing strategies, two significant case studies will be presented; ABB Sweden[6] and Ericsson Telecom, Sweden. These cases are representative of the time-based manufacturing strategies indicated above.

14.4.1.   MANUFACTURING CHANGE PROGRAMMES - TWO EXAMPLES

ABB Sweden - T50 programme
ABB Sweden employs around 30,000 employees in Sweden, and since 1990 has been a pioneer in organisational renewal efforts with their T50-programme. This programme is focused on reduction of total cycle-times, including marketing, design and manufacturing, and aimed at a total reduction of these cycle times with 50% between 1990 and 1993. The programme is fundamentally a customer focus programme, aimed at reducing cycle times in all processes that have an impact on perceived customer value and customer satisfaction. Three central themes have been put forward in the T50 change programme:

- Cycle times
- Decentralisation
- Competence development

The underlying logic of the programme is that cycle time reduction is the key to generate overall organisational improvement (the relation between lead-time reduction and other performance measures are shown in Figure 14-1). A focus on time also serves as a simple and clear key for focusing the organisation in its change efforts. Decentralisation and competence development are necessary ingredients of the overall change, in order to speed up decision-making in crucial processes, and in order to improve productivity, commitment and morale in the organisation.

Ericsson Telecom - FOCUS
The production division of Ericsson Telecom employs around 7,000 people in several production facilities throughout Sweden. During the late 1980s and in the beginning of the 1990s, a radical change in customer profile took place in public

---

[6]   The ABB case is partly based on Berger and Hart (1994)

telecommunications, as state-owned monopolised PTTs were privatised, and telecom markets were deregulated throughout the world. As a consequence, buying behaviour by PTTs changed, and Ericsson Telecom experienced the need for a major change in their manufacturing strategy in order to respond to customer demands. In 1992, the management of the production division at Ericsson Telecom launched a new strategy for manufacturing, which was to implemented by the change programme FOCUS (For Our CUStomers). The specific aims of the change programme were to achieve:

- Lead-time reduction of 67% by 1995 (from 12 to 4 weeks for hardware production)
- Reduction of total costs by 30%
- One of the most attractive workplaces in Sweden
- Continuous improvements as a part of daily work

This combination of hard and soft measures of the change effort was the result of an ambition to generate highly motivated personnel, responsibility and commitment, in necessary combination with radical performance improvement. The content of FOCUS is a radically changed and simplified structure, with establishment of several product-focused production units. The central themes of the FOCUS programme are:

- Full responsibility in line organisation of production and product management
- Simplicity in structure, responsibilities and processes
- Product orientation (decentralised units focused on products)
- Flow-orientation
- Commitment

The logic of the FOCUS programme is thus to decentralise responsibilities for performance to lower levels, and a simplification of structures and processes in order to gain the specified performance improvements.

14.4.2. CREATING THE FOUNDATION FOR CHANGE - DECENTRALISATION

ABB - From few to many companies
In the beginning of the 1980s, ASEA consisted of a handful of diversified divisions and companies. The major decentralisation of ASEA[7] was undertaken beginning in the early 1980s, through a decentralisation from the corporate level to company levels. After decentralisation, the structure of ABB Sweden consisted of some 150 companies, each focused on specific products and markets. The structure is co-ordinated through a global matrix organisation, where each individual company has a regional/country manager as well as a manager in the business area in which the company is a part.

---

[7]   ASEA of Sweden and Brown Boveri of Switzerland merged to become ABB in the 1980s.

After the decentralisation of product and market responsibilities from corporate levels to company levels, the T50 programme introduced a further decentralisation, from the company levels down to multi-functional team levels. The guiding principle is that ABB companies should consist of several multi-functional teams with full responsibilities for the delivery of specific products to customers.

Ericsson Telecom - from a few factories to many product units
Up until 1992 the production division of Ericsson Telecom was organised into seven factories at seven locations in Sweden. This structure was complicated, since factories produced both components and subsystems, and the products were shipped between factories in a complex and non-rational way. Additionally, the organisation was functionally divided, which created further complications in handovers of products, orders and responsibilities in the organisation.

The new structure created under FOCUS consists of 25 product units (still in 7 locations), each one focused on the production of a particular component (or a set of related components), or a subsystem (or a set of related subsystems). The overall structure was simplified, in that no deliveries were made between component units or between subsystems units, thus creating a better overall flow pattern. In addition, the product units have full responsibility for the delivery of their respective product, responsibility for product revisions and updates, and overall responsibility for their respective production and economic performance. Furthermore, a decentralisation of responsibilities within the product units has been made to production teams.

The Russian doll - but with structural differences
There are a set of unifying themes in the structural changes at ABB and Ericsson: a break-up of traditionally strong hierarchies into smaller, more autonomous units. Furthermore, the decentralisation is continued from the unit level down to work-group levels. The principles for decentralisation is similar to the Russian doll; within the largest doll, you find another doll, inside which there is another doll, inside which there is another....

But there are also differences. ABB is organised as companies within the company. These companies are fully owned by ABB, but they are legally separate entities and they operate autonomously from other ABB companies, with their own balance sheets, boards, etc. The reason for this structure is twofold:

1. It gives ABB a possibility to operate flexibly in various markets and in various product areas. Each company defines its own product, market and operative structure, with the result that very specific customer segments in different markets may be satisfied by a specialised ABB company.
2. The products produced within ABB Sweden does not require too much vertical integration. This means that ABB can concentrate on the "front end" of the value chain, and create market-oriented companies.

Ericsson, on the other hand, has defined a plant-within-a-plant structure, and not a decentralisation into separate companies. The reason is that the products produced by Ericsson—telephone exchange systems—are complex integrated systems, produced in several operative stages (components, subsystems, systems, etc.) and factories. The inter-relatedness of the Ericsson factories created a need for a structural simplification, but at the same time also co-ordination in order to decrease overall lead-times and costs. The overall change and lead-time reduction efforts requires a concerted and co-ordinated change of all parts of the system, thereby avoiding a situation where one part of the entire production system lags all other parts in lead-time reduction and thus hampers overall lead-time performance.

14.4.3. ORGANISATION AROUND TEAMS AND PROCESSES

ABB - Process management through team-based organisations
At ABB, one fundamental starting point for lead-time reduction at the company level was to map processes in order to find out current practice, potentials and actions for change. The process maps also formed the basis for organising the work. The customer order processes from order receipt to production and to shipment are actually managed through authority and decision-making in team-based organisations.

Process management is a question of organising work along processes rather than anything else. Process identification and process mapping, in which the process activities and overall structure is analysed for effectiveness and efficiency are widely used tools for this purpose. This is a key feature of a time-based rationalisation program, focusing on processes rather than on individual activities.

Ericsson Telecom - flow orientation through structural simplification of processes
Many of the flow features described above are also applicable to Ericsson Telecom production. A focus on order processes reveals that the complexity of the product (and corresponding production system) created a large number of handovers, not only between different production units, but also between departments in the same production unit, for example, between design departments and production units.

The new production and organisational structure decreases the number of hand-overs through increased spans of responsibility within the product units, and is was combined with a clarification of responsibilities of product ownership, thus also simplifying communication between design and production.

14.4.4. ORGANISATIONAL DEVELOPMENT

Both ABB and Ericsson Telecom—as well as other Swedish manufacturers as was previously shown—rely heavily on group/team-based organisations. The significance of teams from a lead-time reduction perspective is (at least) twofold:

1. *Teams must be created around a natural group task.* This, in turn, means that the physical flow needs to be arranged in such a way that operations and tasks for a product or set of products are physically brought together (e.g., through group technology) into cells. This change of flow structure, often from functional layouts, creates possibilities for lead-time reduction through for example simplified planning, reduced transportation and reduction of buffers.
2. *Teams are in themselves a planning unit.* When teams are allocated responsibility for a process, decisions are taken rapidly and executed with far less slow-downs than in functional environments.

Competence development is another key area for these companies. The significance is that a competence development is primarily seen as a movement from education to on-the-job training. As an example, one multi-functional team at ABB engaged in an extensive training and education programme on assembly, scheduling and team co-ordination which amounted to 475 hours per member.

14.4.5. RESULTS FROM CHANGE EFFORTS

The ABB T50 program has resulted in substantial performance improvements.[8] Cycle times were on average reduced by 47% between 1990 and late 1993, and delivery performance increased from 85% to 94% between 1991 and late 1993. Productivity increased around 9% the first two years of the change programme, and work-in progress levels decreased some 20%.

14.4.6. CONCLUSION - DIFFERENCE AND SIMILARITIES: A SWEDISH MODEL OF CHANGE

It may be argued that ABB Sweden and Ericsson Telecom are different, and of course they are. The change programmes are not defined in similar terms, since ABB focused on lead-time only, and Ericsson also on cost. They also have some difference in structure (Company within a company rather vs. factory within a factory), and differences in flow structures due to differences in product integration. However, the similarities are larger than the differences.

These two fundamental change efforts share a set of common principles:

*   An emphasis on radical improvement in lead-time reductions. As has been seen previously,
*   this emphasis is shared with several other Swedish companies.
*   Decentralisation is a key driver in organisational change and performance improvement.

[8]   Berger and Hart (1994)

> - A simplification of flow structures and organisational processes.
> - Organisational and people development through competence development and new work organisations.
>
> As shown in Section 14-3, all of these features are commonplace in manufacturing strategy changes in Sweden.

14.5. Conclusion: The Swedish manufacturing strategy

Swedish manufacturers in the engineering industry are, in general, niche players, highly internationalised and competing through customer service, quality and delivery.

The manufacturing strategy for these organisations is primarily based on faster response times and leaner organisations in manufacturing and product development. The means for achieving this strategy is through shortened manufacturing cycles and reduced waste primarily through work organisation development and flow orientation. This is supported by case data as well as IMSS data. In product development, IMSS data indicate that the means primarily are based on closer contacts between product development and manufacturing in combination with investments in Computer-Aided Design tools.

The results from these changes are a strong cycle time development, but up until 1992–93 "normal" cost and quality developments. However, since 1992, when production volumes increased substantially, productivity has also risen dramatically.

References

Dagens Industri (1995) 'Moder Svea byter kostym', *Dagens Industri*, Wednesday, May 31.

Ellegård, K., Engström, T., Johansson, M., Nilsson, L., and Medbo, L. (1992) *Reflektive produktion - Industriell verksamhet I förändring*, AB Volvo, Goteborg.

Järneteg, B. (1995) *Time-based transformation of manufacturing systems, Ph.D. dissertation,* Department of Sociology, University of Göteborg, Göteborg, Sweden.

Lindberg, P. and Hörte, S.-Å. (1994) *Produktion Jorden Runt - strategier I 600 företag från 20 länder,* Chalmers Tekniska Högskola/IMIT, Göteborg, Sweden.

PIMS/INDEVO (1992) Internal Report, Göteborg, Sweden.

Rummler, G.A. and Brache, A.P. (1990) *Improving Performance - How to Manage the White Space on the Organisation Chart,* Jossey-Bass, San Francisco, USA.

Stalk Jr., G., Hout, T. M., (1990) *Competing Against Time - How Time-Based Competition is Reshaping Global Markets,* The Free Press, New York, NY.

CHAPTER 15

MANUFACTURING COMPETITIVENESS IN BRITAIN- FROM DECLINE TO RENEWAL?

Chris Voss and Kate Blackmon, Centre for Operations Management, London Business School, London

15.1. Introduction

In 1890, Britain was the largest and most powerful economy in the world. Now, a century later, it is not. Since 1960, other European countries such as Germany and France have not only caught up with Britain but overtaken her, as have other countries such as Japan. In particular, Britain has lost its status as a manufacturing economy. In the last 100 years or so, Britain's share of world manufacturing output has fallen from 25 per cent or more to about 4 per cent; its share of world trade in manufactures, from about 40 to less than 9 per cent (Supple, 1994). Manufacturing output grew far more slowly in the United Kingdom in the 1970s and 1980s than in other large industrial economies. Since 1979 GDP has grown as much in the United Kingdom as in France and Germany but manufacturing has grown more slowly. The percentage of the British labour force employed in manufacturing, mining, construction, and public utilities began to decline in the early 1970s from its historic level since the middle of the nineteenth century of about 43 per cent, falling to 38% by 1981 and below 30 per cent by the end of the 1980s (Supple, 1994).

On the other hand, the British have recently begun to reverse this trend. Labour productivity has grown at an annual rate of more than 4% since 1979, far faster than in France and Germany (Eltis, 1995).

The goal of this chapter is to use economic history, empirical evidence, and economic theory to describe the evolutionary patterns of manufacturing competitiveness in the United Kingdom and to draw out some lessons that are relevant to other countries. To begin, some alternate definitions are provided. Next, the evolutionary path of British competitiveness is briefly summarised in a historical context. Some explanations for the rise and fall of British dominance are then presented. Finally, some conclusions and some suggestions for increasing competitiveness are given.

15.1.1. DECLINE IN MANUFACTURING COMPETITIVENESS - A DEFINITION

The change in Britain's relative status has been accompanied by a general perception of decline by economist, historians, politicians, and the general public (Supple, 1994;

P. Lindberg et al. (eds.), International Manufacturing Strategies, 275-292.
© 1998 *Kluwer Academic Publishers. Printed in the Netherlands.*

Tomlinson, 1996). This perception has been exacerbated by the spread of global communication, enhanced social aspirations, and the visible decay of facilities which are increasingly costly to update, and by objective measures of national income, industrial production, productivity, and shares of world trade (Supple, 1994). Nevertheless, most people in Britain enjoy a higher standard of living than ever before, with widespread availability of telephones, televisions, automobiles, foreign holidays, etc. Has Britain indeed declined in actual terms, or only in the frame of league tables such as *The World Competitiveness Report*?

In the context of national economic context, decline can take on a number of meanings:

1. A decline in living standards.
2. A reduction in relative economic power on the world stage, as the costs of meeting Britain's military, imperial and political position in the world have outstripped the growth of the economy, leading to a change in aspirations and a withdrawal from the international role.
3. the inadequacy of the growth GDP to meet public and private aspirations for goods and services, most commonly found in the political arena.

We may thus take it as generally accepted that judged according to a number of different criteria, that Britain has suffered a relative, if not absolute, decline, over the past century, including Britain's relative importance on the world stage, in its ability to meet the public's aspirations for improvement, and in the performance of its manufacturing sector. Whilst the first kind of decline has clearly not occurred, one of the major reasons for the perception that the second and possibly the third types of decline have occurred in Britain is the loss of competitiveness of Britain's manufacturing sector, especially the reduction in the relative significance of manufacturing.

First, the structure of the economy has changed, in particular the balance between manufacturing and services (especially financial services). Deindustrialisation, the shrinkage or even disappearance of the manufacturing sector that has resulted from Britain's poor industrial performance since 1960, is perceived to be a serious problem for the whole economy, not just the industrial sector itself. Deindustrialisation can damage a nation's wealth because of the continued importance of world trade in manufactured goods and the symbiotic relationship between the manufacturing and service sectors. It is generally considered that service industries are themselves reliant on manufacturing (e.g., Cohen and Zysman, 1985). Despite Britain's continued dominance of the world's financial and commercial systems, deindustrialisation would thus put Britain in the position of a rentier economy, like Holland.

Second, the British economy has undergone significant structural change in the relative strength of different manufacturing sectors. Here, the fear is that '... structural change would dilute the quality of industry and work and social life, and undermine heavy and sophisticated industry, rather than shifting toward new technologies and higher value-added sectors' (Kitson and Michie, 1995, p. 196).

15.2. The Evolution of Manufacturing Competitiveness in the UK

In examining the changes in the competitiveness of the British manufacturing sector, one must do three things (Eichengreen, 1996):

1. place Britain's post-war relative decline in its historical context.
2. provide an international comparative perspective.
3. distinguish proximate from fundamental determinants.

15.2.1. HISTORICAL CONTEXT

During the 18th century and the early part of the 19th century, Britain was the world's leading economic power. The UK was the first industrial nation, being the cradle of the Industrial Revolution, and dominated world markets by the time of the Great Exhibition in 1851. This was largely due to the early start of the Industrial Revolution there. During the 1700s and early 1800s, most industries made the transition from mercantile to industrial capitalism by concentrating production in small workshops, then mills and factories. The first part of this transformation saw the rise of industrial organisations with large investments in buildings and machinery, and the employment of a steady workforce and professional managers began to emerge. Cost accounting began to develop as an aid to managerial decision-making. Over nearly a century, the adaptation of the workforce to factory life began to take hold. The major industries in the UK during this period were the railways, iron, cotton, and engineering. In particular, the railways made many contributions to industrial management. The rise of British industrial capitalism was also facilitated by access to financial capital (Wilson, 1995).

However, by 1850 or so other countries, primarily the United States and Germany, had begun to catch up with the UK. Up to the outbreak of the First World War, although British industry continued to develop, parallel developments in the United States, Germany, and Japan took place at an accelerated pace. Following the Civil War, American business began to take advantage of the resources and the much larger scale of the expanding American home market, the most affluent in the world. This scale facilitated the rise of the mass production of standardised goods, which culminated in the rise of the Fordist system of assembly-line mass production in the period immediately preceding World War I. In addition to internal growth, successive waves of mergers and acquisitions led to the development of giant industrial companies like United States Steel and General Electric.

During the 1880s, concern first began to be expressed about Britain's industrial future relative to Germany and the United States, as progress in those countries accelerated during the early decades of the twentieth century, especially in connection with electricity and the petrol engine, (Sayers, 1950; Aldcroft, 1966, 1970).

Although Britain entered and successfully fought the First World War as a major power, the War had a lasting impact on the competitiveness of British industry (Greasley and Oxley, 1995). During the war itself this was not a problem: falling output stemmed from a loss of labour, reduced investment, and lower productivity and reduced supply coupled with wartime demands led to profit and price inflation. However, once normal international trading relations were resumed by 1921, British industry's lack of competitiveness led to loss of markets. The inability of British industry to share fully in subsequent world trade growth, and the persistence of high unemployment point to more enduring detrimental consequences of the First World War.

As well, during the early decades of the twentieth century British industry was hit by a sequence of shocks, including wage and price inflation during the 1919-1920 inflationary boom (Dowie, 1975; Broadberry, 1983, 1986, 1990); the consequences of the return to gold at pre-war parity in 1925 (Jones, 1985; Pollard, 1970); and US disinflation after 1929 that was powerfully transmitted to the rest of the world by the gold standard and fixed exchange rates (Eichengreen, 1992). Protection and depression elsewhere reduced British industrial exports (Dimsdale, Nickel, and Horsewood, 1989).

Following the Second World War, UK manufacturing suffered a steady decline relative to its major competitors (Chandler 1990). This was especially pronounced during the period 1950 - 1985 (Porter, 1990, p. 482), when it was characterised by low productivity and low productivity growth, low quality and a declining manufacturing base.

After 1945, Britain enjoyed a period of peace and full employment, and a higher standard of living than its neighbours such as France and Italy. The period 1950-1973 has been labelled a 'golden age' for Western Europe (and, of course, Japan), who could be seen to be catching up the leading economy, the United States. The three elements required for rapid growth by catching up—large technological gaps, enlarged social competence, and conditions favouring rapid realisation of potential—came together during this period (Crafts, 1995b).

The Second World War also inaugurated a major extension of government responsibility for economic life, most importantly a commitment to a 'high and stable' level of employment in the White Paper on *Employment Policy* of 1944: not only did governments have the techniques to secure full employment, but they presumably would be punished politically if they did not deploy those techniques effectively. In this way the government assumed a new responsibility for the economic welfare of the mass of the population, and the main issue of the 1950s then became not to 'solve' the employment problem, but to move on to expand the output produced from fully employed resources. (Tomlinson, 1996). In Britain, the concern with growth focused on domestic political concerns of the political parties and on changes in the British 'standard of living', primarily measured by the rate of consumer price inflation and the rate of money wage increases brought about by the annual wage round.

Britain joined the European Economic Community in 1973 (having spurned it in the 1950s, and been rejected in the 1960s); that step acknowledged, as few others could have done, the relative weakness rather than the international strength of Britain (Supple, 1994). For the manufacturing sector, the picture is one of rising output up to 1973.Output fell between 1973 and 1975 and subsequently recovered in the second half of the 1970s (generally taken as peaking again in 1979).

The deep recession of the early 1980s, which struck both old and new industries and was accompanied by major workforce reduction and wide-scale privatisation, was followed by a weak recovery. This recovery led straight into the Lawson boom, taking manufacturing to a new peak in 1989 before falling again in the early 1990s recession (Kitson and Michie, 1995).

The renaissance of UK manufacture can be traced to some of the radical re-structuring initiated by Margaret Thatcher during her tenure as Prime Minister. The first of these changes was to the *relative position of trade unions* in the UK. Having faced up to the miners union in 1984-5 and won, she used this as a launching platform to permanently change the face of industrial relations in the UK. Rights such as post-entry closed shop were removed and legislation introduced to regulate the procedures necessary for unions to take industrial action. This was accompanied by the introduction of penalties for failure to comply with regulations.

The net result of these actions was not just a dramatic reduction in the amount of industrial action in the UK, but also a major improvement in the industrial relations climate. Both unions and companies moved substantially away from their previous confrontational stance during this period. The Thatcher revolution and the weakening of Britain's trade union movement led to the adoption of a US-style industrial relations system that complemented its US-style financial markets (Dixon, 1995).

The second policy change was the adoption of *strong monetarist policies* aimed at reducing inflation. Increasing inflation in the mid-1980s and currency problems in the early 1990s led to sharp increases in the cost of money, which reduced manufacturing investment. On the other hand, monetarism has led to a current climate of low inflation. The world-wide recession of the early 1990s had some effect in Britain, but the government maintained a 'hard line' on inflation and government borrowing.

The third policy change was *privatisation*. This was in many ways the most innovative change, but in other ways the one with the least impact on manufacturing, as privatisation has been aimed primarily at utilities and state-run services. Where privatisation has been carried out in manufacturing, it has been highly successful. British Steel was fully privatised in the 1980s, and by the 1990s had become one of the more successful steel companies in Europe and indeed the world, establishing a reputation as a low cost producer as well as demonstrating increases in quality. In contrast, most of its other European competitors have required state subsidy or intervention to maintain their position.

There is some evidence (albeit very mixed) of some reversal of the relative decline as a result of the Thatcherite reforms of the 1980s, including:

1. *Productivity growth.* The level of productivity, after declining relative to Britain's major competitors, has been climbing since the 1980s. Though it has been climbing, there is still a considerable way to go before matching the absolute productivity levels of competitors. On the other hand, it has been argued that the growth of manufacturing productivity in the 1980s was largely due to job cuts rather than increased output, and these jobs were not being lost in a period of full employment when the labour would be taken up productively elsewhere. The benefits of this productivity growth went overwhelmingly into cutting employment and increasing dividends rather than developing new products and expanding output (Kitson and Michie, 1995).

2. *Devaluation and inflation control.* After a long period of high inflation, the government's monetarist policies coupled with the last recession led to the reduction of UK inflation rates to a level comparable to that in other European countries. The devaluation resulting from the UK's forced withdrawal from the European Currency Mechanism in 1993 did not, as expected, lead to an increase in the rate of inflation, as the increased import costs were absorbed by growth, productivity and margins. As a result, a substantial one-off boost was given to UK competitiveness by devaluation.

3. *Operational practice and performance.* There is increasing evidence that low levels of quality, an endemic problem in the past, has begun to be addressed. The UK initiatives that had led to ISO 9000 series of quality standards have resulted in high quality system certification rates in the UK. Evidence of the impact was found in a recent study of manufacturing practice and performance in Europe. Hanson et al. (1994) studied manufacturing practice and performance in a large sample of manufacturing sites in the UK, Germany, Netherlands and Finland, and found delivered quality from UK firms to be better on average than that of German firms. They also found strong evidence of leading UK firms adopting good practice and translating it into manufacturing performance. The number of sites that had reached 'world class' standards was similar to the rest of Europe.

4. *Quality.* One area of manufacturing practice in particular where the British has improved is quality. The UK initiatives that led to the ISO 9000 series of quality standards have resulted in high rates of quality system certification in the UK. The IMSS data also indicate that the UK sample had above-average levels of ISO 9000 adoption, and use of Statistical Process Control (SPC) and Value Analysis, although it still lags in the use of Total Quality Management (TQM). This is reflected in the pattern of expenditures on the costs of quality, where UK firms report higher spending on prevention and lower spending on inspection, which in turn leads to lower overall costs of failure (Table 16-2). A similar result was reported by Hanson et al. (1994), whose study found higher levels of ISO 9000 registration in the UK than in Germany, along with higher levels of delivered quality.

Table 16-2. UK Quality Management

Technique adoption (1 = low, 5= high)	UK	IMSS Sample
ISO 9000	4.08	3.14
SPC	2.92	2.89
Value analysis		
TQM	2.76	3.17
Costs of quality (sums to 100%)	UK	Sample
Prevention	27.5%	22.1%
Inspection costs	30.8%	31.0%
Internal and external costs of quality	41.7%	46.9%

Thatcher's policies were continued to a large extent by the Conservative government. led by John Major. Under his leadership, policies to increase the national competitiveness of the United Kingdom were pursued in a number of areas, including low inflation, low interest rates, low taxes, and failing unemployment. The declared goal of the British government was to transform Britain into the 'Enterprise Centre of Europe', through creating a climate in which business might flourish. Particular trends include:

1. *Privatisation* of the last remaining state-owned industries, including British Rail and the London Underground. The use of the Private Finance Initiative (PFI) to finance public spending on roads and transport and health.
2. Reduction of *social security* spending, and management of benefit spending such as incapacity benefit, job seekers' allowance, child support, and pension benefits. Social security spending, which had grown at an average of 5 percent a year after World War II, higher than GDP growth, has been reduced to 1.25 percent annual growth.
3. Improvement of the *education* system, including more access to higher education, with 1 in 3 young people pursuing higher education opportunities, up from 1 in 8 in the early 1980s; the National Vocation Qualification system (NVQA), which certified 'blue-collar' skills; investment in information technology and minimum skills levels in through the national curriculum, standardised national testing and the publication of league tables to measure school performance; and empowerment of a schools inspectorate (OFSTED).

On the other hand, the relationship between the British government and the European Union has been more problematic, especially the extension of the single market to energy, pension funds, and telecommunications; minimising the increase in regulations and regulatory power; maintaining the opt-out from the 'Social Chapter', as agreed upon at Maastricht, including the application for the working-time directive on Britain; and containing the crisis in the British beef industry due to the export ban on British beef and products due to the outbreak of BSE. It is too early yet to predict how the election of the new Labour government, led by Tony Blair, will alter the current situation.

15.2.2. CAUSES OF DECLINE IN THE MANUFACTURING SECTOR

What internal factors may have contributed to Britain's decline and subsequent renewal in the area of manufacturing? A number of arguments from the economic and managerial literature are presented in this section. First, the Confederation of British Industry (1993) has put forward a number of possible explanations for the decline in British manufacturing, of which the most prominent are:

- *State of demand.* The volatility of demand in the UK has increased business uncertainty and has led to destabilising of the capital investment cycle.
- *Inadequate return on investment.* This has been cited by many companies as a cause of under-investment.
- *Cash flow.* There is a continuing debate as to whether the high levels of divided pay-out in the UK have led to reduced cash flow, and thus less capital available for investment. Although in theory the cost of capital remains the same regardless of levels of dividend pay-out, this may not be apparent to managers.
- *Financial structure.* The financial structure of the UK, where relationships between banking institutions and manufacturing firms traditionally have been poor, has been associated with the cash flow debate. This poor relationship, alongside the influence of the stock market, has led to accusations of short-termism. The latter will be examined in more detail later in Chapter 21.

One set of arguments that has been put forward proposes that the decline of British industrial hegemony was largely due to historical factors, and there is a vast historical (e.g., Landes, 1969; Elbaum and Lazonick, 1986) that identifies factors that were common to the entire period such as fragmented industrial relations, the limited provision of technical education and training, and the lack of connection between industry and banking (Eichengreen, 1995). Presumably, these same institutional factors that held back the British economy before World War II also handicapped its performance externally: institutions are slow to change because of sunk costs and network externalities.

Supporting this line of argument, Alfred E. Chandler Jr. (1990) argues in *Scale and Scope* (1990) that in the final decades of the 19th century exploiting opportunities for rapid growth required the development of large-scale, high-speed throughput, continuous-process industries; a financial system with the capacity to meet the capital requirements of the large-scale enterprise; and a system of flexible labour relations adaptable to the technological and organisational imperatives of modern mass production. Compared with the United States and Germany, Britain acquired a very different set of institutions as a legacy of her early industrialisation, including a banking system that specialised in the provision of trade credit rather than industrial finance; an industrial system with limited economies of scale and scope; and a system of fragmented, craft-based unionism (Eichengreen, 1995). This situation persisted even after World War II—although there was a strong impetus for change in Germany and France—and the dominance of the stock market encouraged managers to think in terms of short horizons and heightened their sensitivity to financial-market conditions; the fragmentation of the union movement made it impossible to secure agreements to moderate wages and co-ordinate the changes in work roles necessary to adopt new technologies; lack of co-operation between unions, employers associations and government made it infeasible to adopt a German-style system of apprenticeship training.

Alternately, Kitson and Michie (1996) argue that causes of deindustrialisation in the United Kingdom were: (1) underinvestment in UK manufacturing gross capital stock; investment being directed toward cost-cutting rather than capacity-enhancing; (2) UK macroeconomic and industrial policy, resulting in an overvalued exchange rate (due to the Thatcher government's initial monetarist policies in 1979-1980 and membership of the Exchange Rate Mechanism at an overvalued rate) and high interest rates (high levels and volatility which discourage investment and business confidence). Further, Kitson and Miche (1995) suggest that the reasons for government policy failure lie in Britain's economic history and in the resulting distorted nature of both the economy and society. This fundamental problem, of a lack of any strong modernising force, has if anything been exacerbated since 1970. They argue that the causes of poor macroeconomic and industrial policy include the historical legacy of a continued overseas orientation not only of the financial sector but also of Britain's multinational corporations; a disproportionate burden of military spending and the distorting effect this has had on R&D; the continued inability of successive UK governments to modernise the economy; and the role of the City of London in the functioning of the economy and in the formulation of policy. More recently, with little attempt to use the public sector as a modernising force, the industrial structure has shifted to more segmented and niche product markets that require specialist capital equipment and sector-specific skills; with its effects on the productive system being the reduced bargaining power of workers and the creation of a more flexible labour market, compared with the Japanese lifetime employment system and government support for industry and German mechanisms for minimising the impact of recession through the Kurzarbeit (short-time work).

As well as institutional factors operating at the national level, a number of explanations for the erosion of UK competitiveness after 1950 have been put forward that are attributable to the management of manufacturing enterprises. These include low levels of investment in manufacturing, poor labour relations, and poor management at the company and national level. Each of these is discussed further below.

Table 16-1. IMSS Comparison of Investment in Technology and Human Resources

	Gross domestic investment in Machinery and Equipment (1990) (% of GDP)	Industry (% of GDP)	Secondary education (%)
Korea	29.20	45	
Japan	13.70	42	
Portugal	13.10	37	
Netherlands	10.70	31	
Austria	10.40	37	
Finland	10.10	36	
Italy	10.00	33	
Germany	9.80	39	
Australia	9.70	31	
Mexico	8.87	30	
United Kingdom	8.50	37	
Denmark	8.10	28	112
Spain	8.10	37	105
USA	7.80	29	
Canada	7.20	27	
Norway	6.80	34	
Argentina	--	--	
Brazil	--	39	105
Chile	--	37	109
Sweden		35	

Source: *World Competitiveness Report*, 1993, Table 1.13, p. 322; Table 1.45, p. 354

First, the level of investment in manufacturing in the UK has remained low against its competitors until fairly recently. As Table 16-1 below suggests, Britain ranks relatively low among the IMSS countries in terms of its investment in machinery and equipment. There is no consensus in the UK as to its cause, although one explanation has been the high cost of capital and associated high hurdle rates for investment.

Eltis (1995) suggests that the UK's decline was largely due to a weaker generation of new products in the United Kingdom. In world trade, markets have risen fastest in high-tech trade, where success depends on the development of industries where new products predominate. Although the world's manufacturers gained sales in these areas during the 1980s and the 1990s, many UK companies invested defensively to cut costs and therefore employment to produce a product range that changed relatively little. Hence they failed to gain new markets and UK manufacturing employment fell more and output grew less than in other leading economies. This tendency began to be reversed in the later 1980s and the 1990s.

Other authors have attributed a significant portion of the loss of British manufacturing competitiveness to weaknesses in human resources, including poor labour relations and a lack of skills in the British workforce. In particular, labour relations in the UK have been poor, with periods of heavy strike activity especially during the 1970s, when strikes became known as the 'British Disease'. The UK environment was characterised by adversarial labour relations, excessive demands from work-forces and unions, and a class-ridden management system, which led to frequent strikes, flourishing inflation, and low productivity gains.

According to Kitson and Michie (1996), the comparatively low rate of return on capital has been largely caused by the lowness of value-added per worker in relation to what labour is paid in the United Kingdom. Real wages in the United Kingdom are closer to those in France and Germany than are comparable productivity levels. Two considerations deserve special attention, weaknesses in labour skills and comparative weaknesses in management. Econometric estimates suggest that the lack of skills in the British workforce, as measured by vocational qualifications, accounted for a substantial part of the UK's manufacturing productivity gap with West German in 1987. Graduate engineers and scientists compare in quality and quantity with other leading economies, but the same cannot be said of the intermediate skill levels, with far fewer craftsmen being produced than in France in Germany, as a result graduates do more work in manufacturing industry that could be performed by intermediate workers. Most first-class graduates avoid industrial management in favour of industry and commerce, compared with France, Germany, the United States and Japan (Eltis, 1995).

Firms have also invested less in employee training, including in-house training and/or a decline in support for external provision by training agencies. This leads to a focus on a narrow range of specific skills to meet firms' immediate needs, diluting the skill content of jobs, and interacting with the deterioration of the terms and conditions of employment and the increasing pessimism about future prospects of the industry to discourage new entrants from traditional areas of recruitment. Any subsequent relaxation of hiring standards to meet the labour shortage serves to further reinforce the downgrading of the job, the dissipation of skills, the loss of competitiveness, and industrial decline (Kitson and Michie, 1995).

A third explanation has been poor management at the company and national level. It is frequently argued that the decline of UK manufacturing during this period could be attributed to poor management (e.g., Chandler, 1980). This argument has always been difficult to pin down with explicit indicators, but both poor investment and poor labour relations are outcomes of poor management. In addition, the poor labour relations climate frequently led to management focus on keeping labour under control and 'fire-fighting', rather than seeking to move towards new and better management practices.

It has also been argued that management has focused on financial control at the expense of quality, product development and customer satisfaction. At the national level, it has frequently been argued that government has paid little attention to the support of manufacturing industry or to building up an appropriate support infrastructure in training and technology development. This is particularly reflected in the relatively poor level of training, particularly skill training in manufacturing, as shown in Table 16-1.

15.3. Forces for Renewal in Manufacturing

Although some authors such as Supple (1995) suggest that Britain's decline was inevitable as the number of industrial economies grew, even if Britain's decline truly inevitable, it seems possible to reverse it. The recent development of the New Growth Economics (e.g., Romer (1990)), may provide some direction as to what future policies may be appropriate to sustain long-term improvements in manufacturing competitiveness. The New Growth Economics centres on the role of investment, human capital, the effects of policy on growth, a catch-up and convergence, and social capability. In contrast to traditional Solow style neo-classical growth economics, which assume that long-run growth of income per person requires that exogenous improvements in technology generate productivity growth, the new endogenous growth theory modifies the traditional Solow model needs to be modified explicitly to include human capital as a form of accumulation. The basic idea of endogenous growth is that there are constant returns to the reproducible factors of production, where capital includes investments of acquisition of knowledge, skill, etc., is appealing.

One strain of endogenous growth theory (e.g., Aghion and Howitt, 1992; Romer, 1990) has emphasised investments in R&D, and another the accumulation of human capital (e.g., Lucas, 1988; Stokey, 1991). Both have been found to be important in terms of empirical evidence. Redding (1996) suggests that the incentives to invest in each are interdependent. Steedman and Wagner (1989) cite differences in workforce skills or training as one major explanation for the greater innovativeness of German firms. Finegold and Soskice (1988) argue that Britain is trapped in a 'low-skills' equilibrium; that is, 'a self-reinforcing network of societal and state institutions which interact to stifle the demand for improvements in skills levels' Redding (1996) explains the empirical evidence on human capital and R&D by combining a model of human capital accumulation in the context of labour market search with a 'quality ladder' model of R&D.

Thus, by investing in the development of human capital and R&D, it may be possible to achieve a *sustainable* reversal in the relative decline of British manufacturing competitiveness, especially in comparison with close rivals such as Germany and France. Supporting this viewpoint, whilst much of the above has addressed broad trends that are apparent at the national level, there have been success stories as well, both at the individual and sectoral level. UK manufacturing has made substantial progress since the mid-1970s, particularly in the areas of quality and labour relations; nonetheless, great strides remain to be made before it matches the current levels in Japan, the United States, or Germany. Both the government, through bodies such as the Council of British Manufacturing (CBI) and the Department of Trade and Industry (DTI), and corporate management and leadership have a role to play in improvement efforts.

In the next part of this section, we address some recent trends that are contributing to the renewal of British manufacturing, specifically foreign direct investment and government policies supporting manufacturing.

15.3.1. THE ROLE OF FOREIGN DIRECT INVESTMENT

270 German and 180 Japanese companies are now manufacturing in the United Kingdom, and many of these came during the 1980s and the 1990s. Foreign owned companies are responsible for much of manufacturing employment, value-added in manufacturing, manufacturing investment, and manufactured exports. They are also demonstrating to UK workers and management that high quality and productivity levels are achievable.

Positive government policies, one of the lowest cost labour forces in Europe, a lower level of regulation than other European countries, and the adoption of the English language as the universal business language have led to a high level of foreign direct investment in the UK. The UK has the highest absolute level of FDI of any country in the IMSS study, surpassed as a percent of capital formation only by Netherlands and Belgium (See Table 16-3). The UK is also characterised by high outward foreign direct investment.

The proportion of inward foreign direct investment directed to the UK by the United States and by Japan is also high compared to other EC members (See Table 16-4).

Table 16-3. Inward and Outward Direct Investment

Country	Inward Direct Investment % of gross capital formation (1990-92)	Outward Direct Investment % of gross capital formation (1990-92)
UK	12.76%	9.32%
USA	3.06%	4.13%
Japan	0.19%	3.06%
Germany	1.99%	5.74%
Sweden	5.82%	16.23%
Argentina	9.86%	-
Brazil	1.29%	0.71%

Source: World Competitiveness Report, 1994

Table 16-4. Direct Investment in the EC

Country	US direct investment in the EC - (1951-1992)	Japanese direct investment in the EC - (1951-1992)
UK	38.8%	40.5%
Germany	17.6%	9.1%
France	11.6%	7.6%
Netherlands	9.5%	22.5%
Other	22.5%	20.3%

Source: CBI based on Japanese Ministry of Finance, US Department of Commerce

Table 16-5. Manufacturing Practices and Performance by Parent Company Origin

Deviation from the average	Japanese Companies in the UK	North American Companies in the UK	Other European Companies in the UK
Manufacturing Practice	+ 16%	+ 7.0%	+ 3%
Manufacturing Performance	+ 15%	+ 0.5%	+3%
n = 663			

Source: Voss et al., (1995)

The sustained high level of foreign direct investment in the UK can be considered one of the foundations of manufacturing recovery. Foreign direct investment has led to considerably increased investment in manufacturing, and in many cases the importing of latest manufacturing technology. In a related study (Voss et al., 1995), foreign-owned companies in the UK were found to have significantly better manufacturing practices and performance than domestically-owned companies (See Table 16-5).

There may be several links between foreign ownership and manufacturing practice and performance. First, foreign firms may bring in best practice from overseas, a practice which may explain the distinctive performance of Japanese firms. As shown in (Voss et al., 1997), Japanese firms have a distinctive set of manufacturing practices, which most Western companies have yet to fully adopt. Second, foreign firms more often invest in green-field sites, whilst domestic firms tend to expand existing sites. It is likely that green-field site investment allows a company to maximise the effectiveness of its manufacturing investment. Finally, foreign managers may find it easier to challenge many of the bad habits of domestic managers—'we have always done it this way'—particularly given the class history of the UK and the traditionally antagonistic relationship between labour and management.

The impact of foreign direct investment will be higher if there is a substantial supply chain effect. The greater sophistication of foreign firms will, over time, work to improve the quality of domestic suppliers. In a study of UK manufacturers, Hanson et al. (1993) found that the suppliers to sophisticated, mainly foreign-owned, electronics companies had significantly better practices than suppliers to other companies. This may indicate that foreign-owned companies are using their expertise to strengthen the management of domestically-owned suppliers in the UK. Such improvement does not happen overnight, and although a similar pattern is expected from Japanese investment in the UK automotive industry, it has not yet manifested itself.

Recent work by the authors confirms the suggestion that plants with foreign parents have higher levels of manufacturing practices and higher manufacturing performance than those with domestic parents. Whilst these improved practices have sometimes been associated with the ability of foreign entrants to implement best practices in green-field sites, there is evidence that such renewal is also possible in existing, or brownfield, sites where the impetus to change is strong enough.

Overall, it can thus be argued that foreign direct investment is both a source of both overall improvement in itself and a vital source of improved practice that can diffuse throughout UK industry. There are, however, a number of criticisms of excess foreign direct investment. First is the 'screwdriver factory' criticism, which argues that much new investment in the UK has been in the form of assembly plants where kits of parts imported from Japan, the USA or elsewhere are assembled. This results in very little real value added in the host country, as both design and high value-added activities such as component manufacture are still being done in the parent country. Second is that there is a net reduction in the technological capability of a country as more local manufacturing is done by companies with their R&D based overseas.

15.3.2. GOVERNMENT POLICIES TO SUPPORT MANUFACTURING COMPETITIVENESS

From the 1980s to the present, there have been increasing tensions between the social policies of the EC, reflecting the socialist and social democratic governments of most of the larger member countries, and those of the UK, reflecting the de-regulation policies of the Conservative government. In manufacturing, the contrasts have been greatest in the area of workforce and related social legislation. In a paper to the European Community, the UK Treasury and Department of Employment (1993) set out some of the UK's worries concerning EC policies. These included:

- *Labour market inflexibility.* They argue that EC policies create inflexibility in, and over-regulation of, labour markets.
- *Social security and health care.* They argue that although there is no question of abandoning social protection, in virtually all member states the costs of social protection have been rising faster than the capacity of national economies to sustain them. As a result, non-wage labour costs are much higher as a percent of labour costs than in the EC's major competitors.
- *International trade.* The UK firmly rejects calls in the EC for greater protectionism or import controls.

This paper calls on the UK to remove all internal and external barriers to trade, reduce the burden of existing regulation on business and develop firm control of subsidies and state aids to business.

As part of its separate view on the social side of the EC, the UK has secured an opt-out of some of the elements of the Maastricht treaty. UK policies have differed from other EC countries in:

- Limiting both social support and the degree to which this is passed onto the payroll costs of companies.
- Striving for an open labour market, but placing limits on control of labour force reduction in companies and on restrictive rules in employment.
- Resisting regulatory requirements for things such as works councils.

This has led to a labour market and environment sharply different from some other European countries, though not as open or deregulated as the USA. The net result has been that UK has sharply lower labour costs than many other EC countries and a more open labour market. This has generally been seen by business as a solid platform on which to build a new and more competitive manufacturing industry.

Although the UK has made a major recovery in the performance of its manufacturing industry, there are still areas of concern. These areas include:

- *The variability of UK manufacturing.* In a study of 663 manufacturing sites in Germany, Netherlands, UK and Germany, Hanson et al. (1994) found that the median of performance and practice for firms in the UK was similar to other countries in Europe. In addition, the best UK companies were on a par with the best in Germany and the Netherlands. However, manufacturing in the UK was characterised by a large tail of poor performers. This wide variability is not matched elsewhere and may be due to many reasons such as the age of plants and investment, the conservatism of UK management, or the lack of pressure applied by purchasers through the supply chain.

- *Low investment.* The low levels of investment relative to other countries continues to be a cause for concern. The Confederation of British Industry (1993) argued that ' a much greater level of investment is required to effect a substantial increase in the UK's manufacturing capability and improve efficiency'.

- *Insufficient R&D and innovation.* There is widespread support for the idea that investment in science and technology is linked to the British 'decline'. Compared to other industrial nations such as the United States and Germany, both the intensity of investment in R&D and manpower commitments to R&D have long been low in the UK (Edgerton, 1987). On the other hand, Britain science and technology is noted for its inventiveness, especially at the university and pure science level, and in particular industry segments such as chemicals and pharmaceuticals.

Further, it is not enough to have reversed the process of decline: positive effort is required to maintain the momentum of improvement.

15.4. Conclusions

After a long period of decline relative to other Western industrialised nations, there has been evidence of a reversal of the performance of UK manufacturing, as evidenced by, for example, productivity growth rates, export growth, and quality improvement. There is no single cause for this improvement, but we have identified a number of plausible ones. The IMSS study and similar research have highlighted the fact that UK practices are on a par – and not lagging – other Western nations. Foreign direct investment, the English language, and new attitudes within manufacturing have contributed to this parity. Some of this has its roots in the changing management/labour union climate originating during the Thatcher era. The level of the pound has also played an important role, although the sharp rise up to mid-1997 may have harmed competitiveness at the time of this writing.

Despite the renewal, there is little room for complacency, as demonstrated above. First, the IMSS data indicate that the US substantially lags the best countries, e.g., Japan in manufacturing practice and performance. Second, related data collected independently by the UK research team shows that the averages reported in here and in other research hide an exceptionally wide variability among manufacturers in the UK. For example, although the UK average practice and performance are similar to German averages, the UK has a substantial 'tail' of poor performers. Finally, many of the problem areas identified in the historical analysis, such as low investment, though possibly being remedied, still remain.

References

Chandler, A.D. Jr. (1990) *Scale and Scope: The Dynamics of Industrial Capitalism,* The Belknap Press of Harvard University Press, Cambridge, MA.

Crafts, N.F.R. (1995a) 'Deindustrialisation and industrial growth', *The Economic Journal,* 106, January, pp. 174-185.

Crafts, N.F.R. (1995b) 'The golden age of economic growth in Western Europe, 1950-1973', *Economic History Review,* XLVIII, 3, 429-447.

Dixon, H. (1995) 'Deindustrialisation and Britain's industrial performance since 1960', *The Economic Journal,* 106, January, 170-171.

Edgerton, D.E.H. (1987) 'Science and Technology in British Business History', *Business History,* 29(4), October.

Eichengreen, B. (1995) 'Explaining Britain's economic performance: a critical note', *The Economic Journal,* 106, January, pp. 213-218.

Eltis, W. (1995) 'How low profitability and weak innovativeness undermined UK industrial growth', *The Economic Journal,* 106, January, 184-195.

Geroski, P.A. and Walters, C.F. (1996) 'Innovative activity over the business cycle', *The Economic Journal,* 106, July, pp. 916-928.

Greasley, D. and Oxley, L. (1995) 'Discontinuities in competitiveness: the impact of the First World War on British industry', Economic History Review, XLIX, 1, 82-100.

Kitson, M. and Michie, J. (1995) 'Britain's industrial performance since 1960: underinvestment and relative decline', *The Economic Journal,* 106, January, 196-212.

Mowery, D. (1986) 'Industrial Research, 1900-1950', in B. Elbaum and W. Lazonick (eds.), *The Decline of the British Economy,* Oxford.

Porter, M.E. (1990) *The Competitive Advantage of Nations,* The Free Press, New York.

Redding, S. (1996) 'The low-skill, low-quality trap: strategic complementarities between human capital and R&D', *The Economic Journal,* 106, March, 458-470.

Supple, B. (1994) 'Fear of failing: economic history and the decline of Britain', *Economic History Review,* 3, 441-458.

Tomlinson, J. (1996) 'Inventing 'decline': the falling behind of the British economy in the postwar years', *Economic History Review*, **XLIX,** 4, 731-757.

Voss, C.A. and Blackmon, K.L. (1997) 'Differences in manufacturing strategy decisions between Japanese and Western manufacturing plants: The role of strategic time orientation' *Journal of Operations Management*, in press.

Wilson, J.F. (1995) *British Business History, 1720-1994,* Manchester University Press, Manchester, England.

PART III - PATTERNS OF CHANGE

CHAPTER 16

REGIONAL TRADING BLOCS AND MANUFACTURING STRATEGIES IN THE EUROPEAN UNION, MERCOSUR, AND NAFTA

Chris Voss, Centre for Operations Management, London Business School, London

16.1. Introduction

World-wide, nations are combining into trading blocs to gain the benefits of open markets, the harmonisation of standards and the creation of larger markets. The countries in the IMSS study represent three of the world's largest trading blocs, the European Union (EU), the new Southern Common Market (Mercosur), and the North American Free Trade Area (NAFTA). These trading blocs have many similarities, but they also have many differences in their objectives and in the mix and starting points of the nations involved.

In this chapter, we examine whether the characteristics of the different trading blocs are reflected in the manufacturing strategies of the firms in the different countries. If so, what might be the implications for companies wishing to manufacture inside any of these areas?

In Section 16.2, we briefly describe the three major trading blocs represented in the present study. In Section 16.3, we discuss more generally some factors that have led to the formation of such arrangements, and their effect on international trade and industry sectors. Section 16.4 presents some conclusions about the effects of regional trading blocs on manufacturing strategy.

16.2. Regional Trading Blocs in the IMSS: The European Union, Mercosur, and NAFTA

16.2.1. THE EUROPEAN UNION

The continuous evolution in the co-operation between European states has resulted in the formation of the European Union (EU) under the Maastricht Treaty, which was signed on 11 December 1991 and went into effect on 1 November 1993. This agreement provides for the creation of a single currency, a European Central Bank, and Community-wide citizenship.

The European Union (EU) has its roots in the original proposal for a union of European states made by Jean Monnet of France following the Second World War.

P. Lindberg et al. (eds.), International Manufacturing Strategies, 295-311.
© 1998 *Kluwer Academic Publishers. Printed in the Netherlands.*

Predecessors to the current European Union include the European Coal and Steel Community (ECSC), created by the Treaty of Paris on 18 April 1951; the European Atomic Energy Community (EURATOM) and the European Economic Community (EEC or Common Market), simultaneously created by the Treaty of Rome on 25 March, 1957; the creation of supranational institutions such as the Council of Ministers, the European Commission, the European Parliament, the Court of Justice, and the European Council by the Merger Treaty on 8 April 1965; and the establishment of the 1992 Programme by the European Single Act (ESA), signed on 26 February 1986.

Since the Treaty of Rome, the original six nations have expanded to fifteen: Britain, Ireland, and Denmark joined in 1973, Greece, Spain and Portugal in 1986, and in 1994 agreements were finalised to admit Austria, Finland, Norway and Sweden subject to national approval. Although many of the restrictions between member states have been relaxed, recent progress towards the Community-wide unified market has been retarded by the rejection of the Maastricht Treaty by Danish voters in June 1992 (later reversed), and the holdout by Britain until August 1993, compounded by the strength of the German mark versus the currencies of other countries.

The Effect of the European Union on Manufacturing Strategies

Access to larger markets has been countered by increased pressure from other countries. Spain provides an interesting example of the tension between the need to access larger markets and increased competitive pressures from other countries in the larger market (see Chapter 13). The incorporation of Spain into the European Common Market fuelled the expansion of the Spanish economy in the 1980s. It is now the third largest automotive producer in Europe, and has seen growth outstripping the EU average. However, the incorporation of Spain into the EU has also led to weaknesses that began to manifest themselves in the early 1990s. First, Spain's rapid growth was based on its position as a low-cost environment in a rapidly expanding market. The manufacturing strategies of Spanish firms, though reflecting movement towards best international practice, have not transformed as fast as the economy has. At the same time, membership of the EU has led to rapidly rising real wages, with little or no pressure to address the problems of a highly regulated labour market. In addition, Spain is heavily dependent on imported technology and foreign patents. Membership of the EU for Spain includes the objectives of pan-European convergence and international competitiveness. It has led Spanish manufacturing into a position where it must compete on high value-added, quality and technological sophistication, but both the manufacturing strategies and national technological infrastructure lag.

Denmark faces a similar set of issues, but from a different base. Danish firms are generally much smaller than the EU average. Denmark is also traditionally a high wage, high welfare economy. This is not an environment that lends itself to firms competing in large manufacturing markets. In addition, the small size of the local markets leads to the need for a high level of exports if firms are to gain sufficient scale to support R&D

expenditures. The actual pattern of manufacturing strategies found in Denmark reflects this. Our data indicate that manufacturing companies in Denmark are internationally oriented and seek to meet the needs of a limited number of specialised customers by emphasising high quality, customer service and dependable deliveries. Overall, this would seem to be an appropriate configuration of manufacturing strategies and is reflected in the high level of innovation in Danish companies and the positive export performance. There is also evidence that they are moving towards multi-focusedness, the ability to shift between conflicting goals without losing sight of other goals. This multi-focusedness (or flexible specialisation) seems to be one response from firms in smaller countries inside the EU.

At the head of the Rhine, with the world's largest port, Rotterdam, the Netherlands has become the logistics centre of Europe. This has particularly supported the development of large-scale process industries (which fall outside the scope of this study). As a result, for a small country there are a relatively large number of major multinationals such as Shell, Heineken, AKZO and Unilever, as well as other large companies such as Philips. In the sectors studied in this research, the company pattern resembles that of Denmark, with a high proportion of SMEs. The Dutch research concludes that the strategic needs of Dutch firms in the sectors studied are similar to those in Denmark; in particular, most companies wish to compete on a differentiation rather than a cost strategy, creating the need to focus on product design and engineering. However, much still needs to be done in this area, as objectives such as new product development cycle and rapid design changes are ranked as low by companies, although they are critical elements of such a strategy.

Prior to joining the EU, Sweden had a welfare system that was designed to support the unemployed in a context of low unemployment. The changes in Sweden, culminating in joining the EU in 1995, have led to massive changes. Coupled with strong devaluation, EU membership has led to an export boom, sluggish home markets, high interest rates and unemployment. Sweden seems to be following a very similar pattern to Denmark. Its companies are niche players, competing in world markets as well as the EU. In Sweden, unlike the Netherlands, the manufacturing strategies of companies are moving strongly to time-based competition, coupling quality improvement and lead time reductions. The strategies are aimed at supporting customised products in relatively small volumes and in fragmented industries.

Even in markets competing on differentiation, the high cost base of the EU may create problems for companies. The overall pattern of the impact of the EU on manufacturing strategies is mixed. The common pattern between these countries is the need to develop manufacturing strategies that will enable firms in those countries to develop competitiveness in the context of markets that are large scale and/or technically sophisticated, and demanding in terms of factors such as quality and delivery. The EU, as well as opening up markets for these countries, has created pressures through competition from large established countries such as Germany, and taken away

Table 16-1. The Manufacturing Strategies of Selected EU Countries

Market	Country			
	Denmark	Netherlands	Spain - Historic	Spain - Future
Mass/low cost markets	None in industry group studied	Netherlands strong in process industries	Investment Productivity	Quality Technology
High variety/ technology markets	Quality Service Delivery dependability Product innovation	Delivery Quality Service		
Focus	Multi-focusedness	Incremental improvement	Cost and volume	Cost and quality
Conclusions	Appropriate manufacturing strategies	Some mismatch, low priority on product development	Appropriate for the 80's, but lagging the needs of the 90's	Appropriate but actions lagging

advantages associated with lower labour costs. The manufacturing strategy responses have been varied (see Table 16-1).

Clearly, there are multiple markets within the EU, and countries compete across many of them. However, we detect some patterns. In all three countries we observe challenges arising from being within the EU. In Spain, the EU has created new markets, but has taken away in part the cost advantages that Spain had in many markets, and Spain has yet to build up the complementary advantages of quality, service, etc. In the sectors studied in Denmark and the Netherlands, the small size of firms has led to pressures to compete in well-defined niches and through innovation, and indeed to become strategically flexible through multi-focusedness. In Denmark, the research indicated that companies on the whole had appropriate manufacturing strategies, reflected by the high export performance. In the Netherlands, the change process seems less well developed and there are some potential strategic mismatches.

The impact of harmonisation. A second issue central to the development of the EU has been harmonisation. This covers standards and social policies, and may lead to a single currency, as discussed below. At the level of the manufacturing firm, the move towards harmonisation has manifested itself in contrasting ways throughout Europe. This contrast can be seen in particular by comparing the UK with other countries.

The UK has always followed a path of supporting harmonisation where it felt that it was beneficial to the national interest but rejecting it where it felt that the overall effect was negative. Indeed, this is seen by other partners in the EU as a means for the UK to gain competitive advantage within the EU. This is demonstrated in a number of areas. First is the opt-out from many of the social policies of the EU, whose objectives are to avoid the negative impact on costs of labour market inflexibility and social costs. The results are that the UK seems to be developing a cost advantage allowing business to expand, particularly up to 1995. The opt-out has allowed the restructuring of industries and companies that had previously been heading towards problems. A secondary impact of this has been the high level of foreign direct investment in the UK, partly because low costs and lack of over-regulation makes it more attractive for non-EU firms who wish to establish within the EU.

In manufacturing strategy terms the UK seems to have addressed one key order winner—costs—through economic policy. This has in turn led UK companies to focus on improvement in qualifying criteria such as quality, service and delivery improvement programmes—areas where historically it had been weak. The UK has also taken advantage of harmonisation of standards, particularly the quality standard ISO 9000 (based on the old UK BS5750). As a result, the UK has the highest level of ISO 9000 certification in the world. It is not clear, however, whether companies are fully exploiting this competitive advantage. For example, the one of the largest UK car manufacturers, Rover, is trying to position itself firmly up-market rather than exploit its low-cost base. Second, there is a tail of companies who have neither the practices or performance to compete within the EU, even with a cost advantage. It has been argued elsewhere (e.g., Hanson, Voss, Blackmon and Oak, 1994) that this tail is a particular UK problem.

A contrasting picture is found in countries where the harmonisation of social policies has led to a constraint on some of the possible manufacturing strategies. This is found strongly in both Spain and Portugal. Both countries came from a low labour cost environment, particularly Portugal, where historically manufacturing strategies were associated with competing on cost, with less attention to quality, service and delivery. Harmonisation of EU social policies has played a part in taking this advantage away. This has meant that the manufacturing strategies of companies has had to change to stress non-price aspects of competition. In Portugal, for example, there is still a long way to go, particularly in delivery reliability.

Another potential constraint on manufacturing strategies is in the area of rapid new product development. The EU social policies put severe restrictions on working hours and workforce flexibility. In the context of striving to meet a new product introduction deadline, in the short term this may severely constrain a company's ability to work long hours and transfer people. A recent study of new product development in Germany found that German mangers saw these as two of the top three inhibitors to effective innovation (the third being costs). As a result, it may not be surprising that many EU companies do not see competing through rapid product development as a viable manufacturing strategy.

Monetary union. Another change that will affect manufacturing firms is the prospect of economic and monetary union under the Maastricht Treaty. Whilst the European Community (EC) was strictly a free trade area and a customs union, the Maastricht Treaty sets forth the prospect of monetary union between member states. This economic and currency union (ECU) will be based on a single currency under an international agreement whereby sovereign states delegate monetary policy to a supranational agency, the European Central Bank (ECB). The primary goal of the ECB's monetary policy is to be price stability.

The Maastricht Treaty set stringent deadlines for countries to meet, especially in terms of budget deficit reduction, which has affected both those core group countries that expect to participate in the monetary union (Germany, the Netherlands, France) and those who do not (Italy, Spain, Sweden), particularly in the tightening of fiscal policy to reduce debt/GDP ratios, primarily through long-term measures such as cutting social expenditures and the wage components of government consumption, and short-term measures such as labour-tax increases and capital spending cuts (Perotti, 1996). Such adjustments have had considerable effects on private consumption, growth and investment, unemployment, long-term interest rates, and unit labour costs across the EU.

One of the primary areas targeted for cuts is social expenditures. In Europe, high labour costs and social benefits are often blamed for loss of competitiveness and profitability, and high unemployment rates. On the other hand, there is widespread agreement among politicians that social security and welfare benefits should be as universal and unconditional as possible to minimise distortions.

Second, the EU is forcing choices about labour and employment policy, especially the choice between working and not working rather than hours of work. In countries such as the United States, Canada, and Switzerland, labour markets are highly decentralised and negotiations take place at the enterprise level. In the Scandinavian countries, labour markets are highly centralised, and unions are economy-wide. In most of the rest of Europe, labour markets are in between and unions are sectoral. EU-wide harmonisation will undoubtedly affect the structure of labour markets and thus have a major effect on manufacturing.

There are a number of unresolved issues associated with the EMU, including:

1. *Membership*. Not all member states would be involved: some that would like to join will not be accepted, some that would be accepted will not join. Denmark and the United Kingdom have opted out from the EMU. Another is co-ordination between those EU member states that have adopted the single currency and those that have not.

2. *Agreements on monetary policy and instruments of monetary policy* (Portes, 1996). e.g., the new common currency replacing national monetary units; exchange rates; financial markets; and the role of central banks.

3. *Overcoming exchange-rate volatility*. Devaluation of the currency and pegging the exchange rate to bring down interest rates and increase competitiveness and profitability.

Summary

The EU has widened internal markets, exposing domestic firms to both opportunity and competition. In addition, harmonisation of social policies and convergence of wage levels have created constraints and pressures. We have seen three major impacts of larger internal markets and harmonisation on the manufacturing strategies of firms within different EU countries. First, in low-wage countries such as Spain and Portugal, the simultaneous moves away from labour cost-based advantage and into markets where quality, service and delivery are far more important have left firms within those countries the task of adapting their manufacturing strategies. In general they have some way to go in both the changes required and the resulting performance. Second, smaller firms in smaller countries must change their manufacturing strategies as they adjust to the need to compete effectively in niche markets both inside and outside the EU. The effectiveness of theses changes varies from firm to firm and country to country. Finally, we see how UK firms are trying to exploit a contrary position within the EU to create manufacturing advantage. Monetary union has yet to be implemented, but manufacturing firms are already feeling the impact of economic policies for fiscal tightening leading up to it.

16.2.2. MERCOSUR

The Southern Common Market (Mercosur) comprises Argentina, Paraguay, Uruguay, and Brazil, and represents a population of 190 million people. Although it is only one of several trading blocks in Latin and South America, it is the largest, and the most similar to the EU and NAFTA. Mercosur allows for the free movement of goods and services and factors of production among its member states, and establishes a common external tariff for third countries. The objective is to create a large-scale domestic market, with longer-term economic harmonisation, but without the strong social harmonisation found in the EU. One desired output of the market is to force the rapid development of manufacturing capabilities of the firms in all countries and possibly lead to a shake-out with the survival of the fittest.

International trade agreements in Latin America date back to the Latin American Free Trade Association (ALALC) treaty, signed in 1960, which provided a free-trade zone between member states. This was succeeded in 1980 by the Latin American Integration Association (ALADI), which created an economic preference zone. These previous Latin American free-trade associations increased trade mainly through the granting of selective bilateral preferences. The establishment of a true common market did not begin, however, until Brazil and Argentina began taking concrete steps towards integration by signing commercial protocols in 1986 and a Treaty for Integration, Co-operation, and Development in 1988. Paraguay and Uruguay signed a new treaty on March 26, 1991, in Asuncion, Paraguay, officially establishing the Southern Common Market (Mercosur). The Treaty of Asuncion was primarily an instrument to achieve free trade among members and to seek the efficiencies needed for competition in the world economy. In 1994, this was extended to moving towards a customs union.

Although the aims of Mercosur are similar to the early aims of the EU and of NAFTA, its starting point is rather different, being completely composed of developing countries. Unlike most of Europe, South America has had to overcome a political legacy of military dictatorship and ruinous economic conditions, including protectionism, inflation, instability and debt. As such, the economy has had to overcome a legacy of state industry, state intervention, and isolated markets. With the shift to democratic governments, new economic policies have brought about the end of inflation, followed by economic growth, and consumer booms. Hyper-inflation ended in Argentina in 1991 and Brazil in 1994. Other differences include:

1. Brazil is much more dominant than Germany in the EU, similar to the US in NAFTA. Brazil has a much higher proportion of trade going to external partners than Mercosur partners; unlike Argentina. Brazil tends to be largely inward looking, due to its industrial structure and its large and growing domestic market. Uruguay and Paraguay joined because they cannot afford to be excluded.
2. Overcoming a legacy of import substitution, with companies not competitive on the world market.
3. Differences of opinion over whether it should be a union of nation states: Brazil is intent on not ceding political sovereignty, whilst Argentina wants more of a EU-model.

Already, Mercosur has lifted artificial barriers between Argentina and Brazil, who are natural trading partners and together account for more than half of Latin American's GDP. Until the mid-1980s, Brazil and Argentina remained strictly separate. In 1985, presidents Raul Alfonsin of Argentina and Jose Sarney of Brazil began co-operating on strengthening democracy and reducing military tensions, using bilateral trade preferences in terms of sectoral agreements. In the early 1990s, their successors Carlos Menem and Fernando Collor began moving towards trade opening and economic reform, scrapping import bans and quotas, managed trade, and industrial policy. Under Mercosur, the two countries will have a common foreign trade policy, a common external tariff for the majority of products, and market-friendly macroeconomic policies. The co-operation between Argentina and Brazil also has the effect of increasing political stability, democracy, defence, and diplomatic power.

By the turn of the century, Mercosur will cover the world's fourth-largest integrated market, following NAFTA, the EU, and Japan, with at least 240 million people and an output of over $1 trillion. Much of the rest of South America will be linked to Mercosur by free-trade agreements. During 1996–2000, Mercosur plans to integrate Chile; similar agreements with Bolivia and the Andean Group (Columbia, Ecuador, Peru, and Venezuela) will follow. Mercosur also has a framework agreement with the EU, its single largest source of external trade and investment, and is exploring potential agreements with NAFTA and Japan. Talks with the North American Free Trade Association of the United States, Canada, and Mexico will potentially lead to a free-trade area embracing the entire Western hemisphere after 2005.

As of yet, the South American 'common market' is a goal, rather than an achievement, but it has a framework and a timetable (see Figure 16-1), with political commitment to market opening and export-based growth. Thus far, the countries have made no commitment on free movement of labour, a major difference from common market.

> 1. 1991–1994. Transition phase. Cut tariffs on trade between Mercosur members, except for cars, sugar, and sensitive products.
> 2. 1994–2001. Customs union. Applying a common external tariff (EDT) to imports from third countries. Some exemptions were allowed by country, but some will converge by January 2001, with convergence on imported capital goods such as computers and telecommunications by 2006.
> 3. 1996–2000. Common market. Standardisation within the free-trade area and customs unions, and also on economic policies.

Figure 16-1. The Timetable for the Implementation of Mercosur Reforms

Effects on Manufacturing Industry

The advantages of Mercosur to its members come from the potential for growth, economies of scale and scope, and specialisation. Inter-Mercosur trade has soared from $4 billion in 1990 to $14.5 billion in 1995. Combined GDPs have grown an average of 3.5% annually since 1990. The Mercosur region exports mainly primary products, but has sizeable local markets for manufactured goods, in industry sectors including cars and car parts, chemicals, and machinery. Primary sectors will be agribusiness and mining, with a diversified and modernised manufacturing industry fuelled by cheap and abundant energy.

Mercosur's effects are already becoming apparent in much of manufacturing industry. Trade between member states has increased, mainly in the form of intra-industry trade, as regional trade liberalisation allows increasing specialisation within a larger market The larger and more open market is also attracting direct foreign investment, from multinationals upgrading existing plants, building new ones, or entering the region for the first time. The motor industry in particular has benefited from direct foreign investment.[1] For example, Renault began work in August 1996 on a $1 billion auto plant in Brazil to make the Mégane range. Although Japanese firms have yet to produce cars in Brazil, Asian firms such as Honda, Toyota, and Kia have announced investment plans. Existing plants are upgrading to the same production standards as Europe or US, requiring modernisation; for example, VW's new truck and

---

[1]   The *Economist*, October 12, 1996.

bus plant in Resende, Brazil, uses lean production. The change in regulations on local content, dropping the minimum local content from 100% to 60% has led to a shakeout of automobile component producers.

The trade agreement is also leading to the growth of regional brands. This allows multinational companies to organise production more efficiently and enter new markets. In response, domestic firms are shifting away from unrelated conglomerates to focus on core businesses. (The rise in Mercosur-wide distribution, however, will be limited by the development of an integrated logistics system.)

The impact of Mercosur on manufacturing strategies is twofold. First, it has greatly enlarged the market, and as a result the strategies of small domestic producers who served high-variety, low-volume markets are having to change to higher-volume, lower-variety. This clearly requires considerable process change. Second, the larger market will attract more sophisticated competitors and will develop more sophisticated customers. In both countries studied, until recently firms operated in closed economies. The isolation of most South American firms has meant that the level of changes required to reach 1990s levels of quality, reliability, product development, etc. is very great. An agenda in Chapter 4 characterised Brazil as moving from the factory of the past to the factory of the future. As shown in Chapters 18 and 21, the strategies adopted by companies in different countries are dependent on their starting point and national context.

The Lessons of Mercosur for Other Developing Economies
The development path of the Mercosur nations may provide a model for other developing economies. This also reflects a shift in thinking about trade policy (Krueger, 1997). Trade policy, especially the liberalisation of trade and payments has become critical for both industrialisation and economic development. During the 1950s, it was assumed that the optimum trade policy for economic development in developing countries was 'import substitution', the domestic production of import-competing goods and protection again imports through quotas or prohibitions, which would lead to industrialisation. Development economists advocated balanced growth for the developing countries based on exports of primary products, and import substitution strategies rather than 'outward-looking' strategies. This policy was followed by a number of countries, especially those that had been part of former colonial systems, e.g., Latin America and China, and also others such as Thailand and Turkey. During the 1990s, this has shifted to a focus on outer-directed trade regime and fairly uniform incentives across exporting and import-competing goods, exchange rates, and liberalisation of trade and payments.

Mercosur may provide an alternate path to that followed by the East Asian economies, which are often held up as an example. These countries—including the four 'Tigers', Hong Kong, Singapore, Korea, and Taiwan—have successfully followed the route of beginning by exporting labour-intensive manufactured goods, and then manufacturing goods with an increasing degree of capital-intensity and technical sophistication, based on the absorptive of increasingly sophisticated technology rather than invention and innovation. To an extent, the presence of the Japanese in this region has effected a large-scale transfer of technology and working practices that has not taken place in other developing regions; therefore, the success of Mercosur in stimulating growth and industrial development through regionalisation and trade liberalisation will be used as a model for other LDCs, including Eastern Europe and Africa.

16.2.3. NAFTA

The final trade area studied in our research is the North American Free Trade Area (NAFTA), consisting of the United States, Canada, and Mexico. As Chapter 18 is devoted to NAFTA, rather than repeat the full chapter we will contrast NAFTA and the EU.

NAFTA differs significantly from the EU. Although Germany is the dominant country within the EU, it is not that much larger than its partners. By way of contrast, NAFTA consists of one very large country, the United States, with two smaller partners, the two highly developed nation of Canada and the developing nation of Mexico. Second, the strong convergence and harmonisation drive behind the EU is not shared by NAFTA. It is not expected that Mexico should try to converge with the US, only that through free trade there will be benefits in all three countries, with a particular benefit to Mexico that may reduce political pressures in both the US and Mexico.

NAFTA also reinforces existing trade patterns, such as intra-industry trade in the automobile industry between Canada and the US, and 'off-shore' production for US firms in Mexico. As a result, what we see in the manufacturing strategies of the three countries is, rather than a convergence of manufacturing strategies, the retention and reinforcement of the distinctive manufacturing strategies of firms in each country, in particular Mexico. If we compare with the issues facing companies in Spain, we see strong contrasts. In Spain, companies are trying to adjust their manufacturing strategies to be able to compete as mainstream competitors within Europe once their cost advantage is seriously eroded. In Mexico, the strategies are consistent with seeking to retain their low labour cost advantage, but rather to acquire the additional 'order qualifying' capabilities needed to compete effectively against sophisticated low-cost producers from around the globe.

16.3. Evidence from the IMSS

The three regional trading arrangements discussed above represent not only a sizable chunk of the world's industrial production, but also markedly different approaches to managing international trade. As Figure 16-2 below shows, with the exception of the United States and Japan, companies in the IMSS sample were highly oriented towards international trade, reporting a high degree of both imported inputs to the manufacturing process (X-axis) and exports of manufacturing goods (Y-axis).

The rise of regional trading agreements is reflected in the sourcing patterns of the companies in the IMSS sample, as shown in Figure 16-3. Over the entire sample, the companies reported that on average 67% of their imports of manufacturing inputs came from within their economic area, and 69% of their manufactured goods were exported to their economic area.

Why have regional trading blocs become important? Trade between countries arises for reasons of both comparative advantage and imperfect substitution (Frankel et al., 1996, p. 54). Comparative advantage is determined by differences in factor endowments—i.e., the well-known Heckscher-Olin model, which states that countries will produce goods depending whether their factor endowments are labour-intensive or capital-intensive. Additionally, the imperfect-substitutes model of trade suggests that consumers treat different varieties within industries as imperfect substitutes, e.g., Japanese and American automobiles.

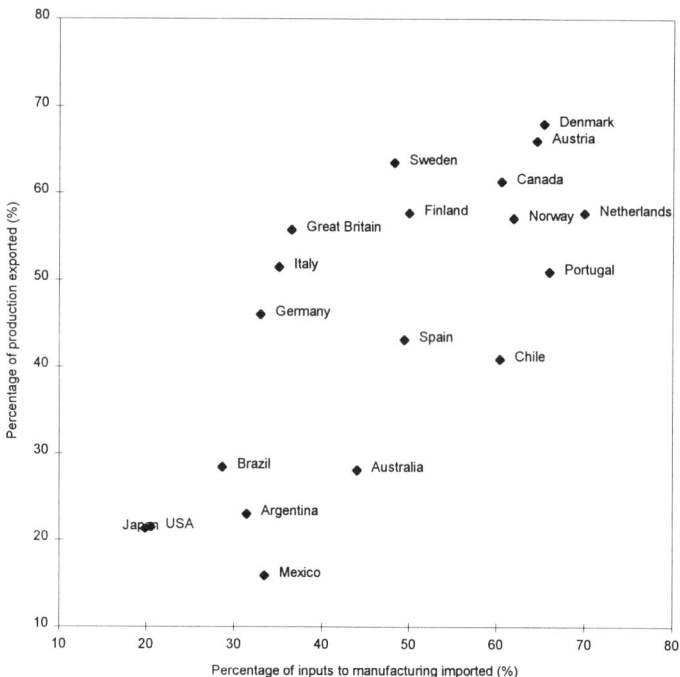

Figure 16-2. Imports and Exports by IMSS Firms

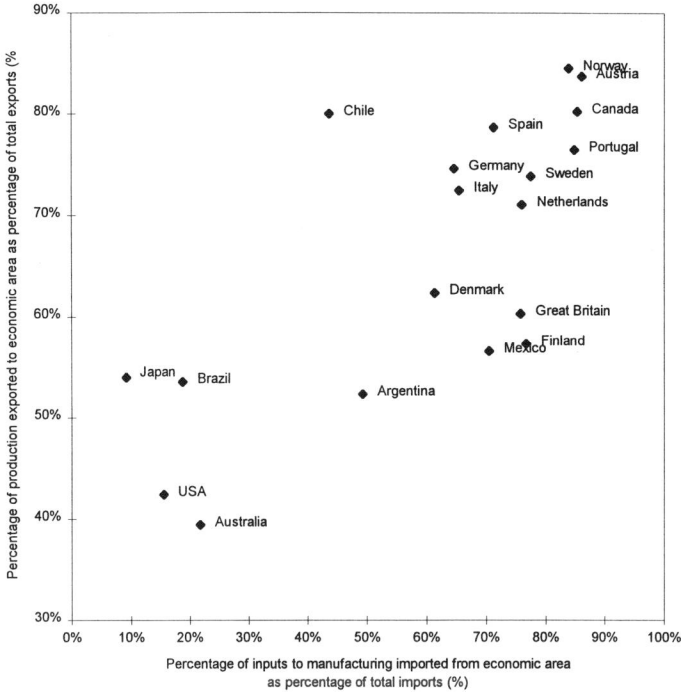

Figure 16-3. Imports and Exports - From/To Economic Area vs. Total

Intra-regional trade naturally arises because of the geographic proximity of two countries, or a shared common border or common language. Thus, 'natural' regional trading arrangement tend to develop along continental lines (Krugman, 1991b). The three major dimensions of such arrangements are (1) free-trade areas (FTA); (2) customs union, and (3) monetary union.

Positive effects of regional trading arrangements include the elimination of distortions in the relative price between domestic goods and the products of other members of the customs union; *negative* effects arise from the introduction of distortions in the relative price between the goods of members and non-members. Benefits to industry include:

- Increased effects of increased competition, specialisation, and trade;
- Economies of scale will reduce costs;
- May reduce technical inefficiencies and put pressures on profit margins;
- The elimination of internal tariffs on inputs make exports cheaper;
- Increased investment at a lower cost of capital due to more attractive locations;
- Import-competing industries may reduce their size and political influence;
- Minimising economic shocks to individual countries.

The effects of regional trading blocs differ when examined at the country, trading bloc, or global level. Overall, the rise in regionalism may have a number of benefits, especially to peripheral and poorer nations who will benefit from trade creation. Smaller countries join regional blocs to insure access to vital foreign markets; in joining, they are often forced to make concessions to larger countries in their negotiations, for example, the experience of Paraguay and Uruguay in Mercosur. Spilimbergo and Stein (1997) found that poor countries are likely to become better off from a move to intercontinental blocs, even though the rich are worse off.

Whether a RTA has positive or negative effects relative to the multilateral trading system depends on the trade-off between trade creation and trade diversion (Viner, 1950). *Trade creation* results from an increase in intra-group trade and/or trade with third countries. *Trade diversion*, on the other hand, consists of switching import purchases from outside country C to higher-cost partner B, thereby resulting in a terms of trade loss. There is a high level of trade outside many RTAs; for example, the largest trading partners of the European Union (EU) are the United States and Japan.

However, regional trading blocs are a form of discrimination compared with the multilateral trading system under the General Agreement on Tariffs and Trade (GATT)[2] and World Trading Organisation (WTO). Regional trading arrangements are designed to restrict imports and protect domestic producers, with protection mainly taking the form of tariffs rather than other types of barriers, such as import quotas. Under GATT, free-trade areas and customs unions may be used to facilitate trade between members and not to raise barriers to non-members. Although the original purpose of GATT was to encourage the multilateral liberalisation of world trade, it did allow some non-discriminatory trade practices along the principles of Most Favoured Nations (MFNs),[3] as world-wide free trade is not attainable for political reasons. However, recent preferential trading arrangements such as NAFTA and Mercosur depart from this principle, moving away from most favoured nation (MFN) tariffs to preferential trading arrangements (PTAs).[4]

Some amount of regionalisation is probably inevitable. A number of economists have suggested that the world is heading towards a system of three trading blocs: Europe, the Americas, and Asia. There is a trend towards the development of networks of regional arrangements between the major powers and groups of neighbouring countries, e.g., the incorporation of the EFTA countries into the EU, or the US and Canada. These groups then enter new arrangements with other regional partners, such the Eastern European countries and the EU or NAFTA and Mercosur.

---

[2]   GATT was set up in 1947 to provide a set of rules and a code of conduct governing trading relationships between signatory nations, and to provide a framework under which trade can be progressively liberalised.

[3]   Non-discriminatory most-favoured nation (MFN) tariffs - if any tariff is imposed, it will be imposed equally on all MFNs; tariffs on non-MFNs do not apply - i.e. on internal trade but not on trade with non-members.

[4]   The major exception is generalised system of preferences(GSP), which allows preferences on imports from developing countries.

Such 'hub and spoke' arrangements occur when an FTA is created that overlaps part of an existing one without a change to a multilateral agreement. This benefits the hub but creates spoke-spoke losses, because trade liberalisation is only taking place between each spoke and the hub, rather than between spokes, which would not be solved by the creation of an FTA consisting of just the spokes. In such a hub and spoke system, as Perroni and Whalley (1996) point out, increased regionalism may lead to increased friction among trade blocs, full trade hostilities, inter-bloc non-co-operative trade war, and a breakdown of the multilateral trading system. However, from recent experience, large powers such as the US and EU are capable of resolving their disputes without creating major conflicts. than tariffs are being used, such as anti-dumping provisions, subsidies, intellectual property.

Krugman (1991a) shows that with no transport costs each of the three blocs can exploit its monopoly power by raising tariffs to a greater extent than it would if acting as smaller blocs or individual countries. The negative effects of trade diversion will outweigh the positive effects of trade diversion. On the other hand, Krugman (1991b) shows that with high intercontinental transport costs, continental blocs are the optimal outcome. Frankel, Stein and Wei (1996) found that the desirability of RTAs depends on whether the extent of regionalisation exceeds an optimal level that is determined by the magnitude of transportation costs.

16.4. Conclusions

World trade has become more regionalised as the result of preferential trade agreements such as the European Union, NAFTA, and Mercosur. Whilst the degree of integration varies between blocs, and trade between blocs such as the EU and the USA remains significant, this movement towards greater regionalisation has important implications for manufacturing firms. These include:

1. Increased economies of scale and scope within a bloc.
2. Increasing specialisation, either in products or stage of the supply chain, less conglomerates; increasing intra-industry trade.
3. Increased potential for supply chain expansion, depending on
4. Increasing the attractiveness of direct foreign investment in developing countries, leading to expansion of multinational activity, technology transfer, increased capital goods, modernisation of industry. Direct foreign investment by multinational companies brings both capital and, if the multinational is in an advanced countries, can serve as a vehicle for technology transfer; thus, foreign direct investment can raise the rate of growth in the host country, depending also on the level of human capital there (Borensztein et al., 1995).
5. Upgrade in the standards of quality, cost, etc.

6. Increased emphasis on the quality of human resources, both for the absorption of technology, dealing with upgrades in the standards of equipment, and more complex / more demanding products.
7. For developing countries, supporting the transition from import-substitution to outer-directed trade, and from primary exports to intermediate, final, and consumer goods.

Several conclusions can be drawn from the analysis presented above, supplemented by the individual chapters in this book. First, companies within the EU will experience strong forces of change due to standardisation and harmonisation that will not be felt by companies in NAFTA and Mercosur. This will force firms that were previously competitive due to being at least partially buffered from pan-European competition to become much more competitive or fail. In turn, this limits the range of manufacturing strategies that are available to them, especially when the path-dependency effects of national history and geography are considered.

Similarly, companies within the Mercosur region will be subjected to greater competitive pressures as protective tariffs are lifted; however, for many companies this will be more than off-set by the greater scale of the common market available to them. Thus, compared with the EU, the effect of Mercosur may be to actually increase the range of manufacturing strategy options.

Finally, as the NAFTA agreement in North America is mainly an expansion of existing international trade agreements, the main effect on manufacturing strategies may be to provide for greater diffusion of world-class manufacturing from leading-edge companies in the US and Canada to Mexico, and thence to other Latin American industries.

Whether trading blocs are beneficial depends, ultimately, on the perspective that is taken. However, in view of the inevitable shift towards greater regionalisation, it is evident that more research needs to be done on their effect on manufacturing strategy, at the firm, industry, national, regional, and global levels.

References

Barro, R. and Sala-i-Martin, X. (1991) 'Convergence across states and regions', *Brookings Papers on Economic Activities*, 1, 107-158.
Borensztein, E., DeGregorio, J., and Lee, J-W. (1995) 'How does foreign investment affect economic growth?', *NBER Working Paper No. 5057*, Cambridge, MA, March.
Boulton, P. and Roland, G. (1996) 'Distributional conflicts, factor mobility, and political integration', *AEA Papers and Proceedings*, 86, 2, May, 99-104.
Campa, J., and Goldberg, L.S. (1997) 'The evolving external orientation of manufacturing industries: evidence from four countries', *NBER Working Paper No. 5919*, February.
Doughtery, C. and Jorgenson, D.W. (1996) 'International comparisons of the sources of economic growth', *AEA Papers and Proceedings*, 86, 2, May, 25-29.
Finlay, R. (1996) 'Modelling global interdependence: centers, peripheries, and frontiers', *AEA Papers and Proceedings*, 86, 2, May 1996, 47-51.
Frankel, J., Stein, E., and Shang-Jin, W. (1995) 'Trading blocs and the Americas: The natural, the unnatural, and the supernatural', *Journal of Development Economics*, 47, 61-95.

Frankel, J.A., Stein, E., and Shang-Jin, W. (1996) 'Regional trading arrangements: natural or supernatural', *AEA Papers and Proceedings,* 86, 2, May, 52-56.

Grimwade, N. (1989) *International Trade: New Patterns of Trade, Production and Investment,* Routledge, New York.

Irwin, D.A. 'The United States in a new global economy? A century's perspective', *AEA Papers and Proceedings,* 86, 2, May, 41-45.

Jones, R.W. and Engerman, S.L. (1996) 'Trade, technology and wages: a tale of two countries', *AEA Papers and Proceedings,* 86, 2, May, 35-40.

Kruger, A.O. (1997) 'Trade policy and economy development: how we learn', *NBER Working Paper No. 5896,* January.

Perotti, R. 'Fiscal consolidation in Europe: composition matters', *AEA Papers and Proceedings,* 86, 2, May, *AEA Papers and Proceedings,* 86, 2, May, 105-110.

Perroni, C., and Whalley, J. (1996) 'How severe is global retaliation risk under increasing regionalism?', *AEA Papers and Proceedings,* 86, 2, May, 57-61.

Wonnacott, R.J., 'Free-trade arrangements: for better or worse?' *AEA Papers and Proceedings,* 86, 2, May, 62-65.

CHAPTER 17

NAFTA: AN EMPIRICAL OPERATIONS PERSPECTIVE

*F. Johnson, J. Kamauff, N. Schein, A.R. Wood, Richard Ivey School of Business,
University of Western Ontario, London, Ontario, Canada*

17.1. Introduction

The liberalisation of trade barriers is an important element of the new global economy. The most recent manifestation of this trend has been the North American Free Trade Agreement (NAFTA) between Canada, Mexico, and the United States, which was implemented January 1, 1994. While many management scholars have addressed the trade liberalisation issue, few have specifically addressed its implications for operations managers.

Manufacturing strategy is part of an ongoing research effort in the Richard Ivey School of Business. During 1993, over 100 Canadian companies were contacted to participate in an international manufacturing strategy survey, part of which was devoted to study the effects of NAFTA on the operations strategies of the participating organisations. The objectives of this research stream were to advance the understanding of operations strategy, to assess the operations manager's role in strategy formulation and implementation, and to further understand the issues operations managers feel are important for their organisations and themselves.

This article focuses on the anticipated impact of NAFTA on the operations strategies of Canadian, US, and Mexican firms. The implication of this trade agreement for operations is critical: the very existence of many organisations will depend on their strategic response to NAFTA.

This research indicates that several differences exist between the NAFTA partners with respect to the operations strategies employed by their domestic industries: Canadian firms emphasise dependable deliveries and customer service; US firms focus on customer service and manufacturing quality; and, Mexican firms are moving to widen their product range and develop capabilities that will provide faster deliveries. Each of these factors has an impact on how Canadian, US and Mexican firms are positioned to compete under NAFTA.

17.1.1. NAFTA

Canada's trade strategy for NAFTA was aimed at further developing the existing Free Trade Agreement (FTA) with the United States. Bilateral trade between Canada and the

P. Lindberg et al. (eds.), International Manufacturing Strategies, 313-322.
© 1998 *Kluwer Academic Publishers. Printed in the Netherlands.*

US in 1993 was the largest between any two countries, representing $268 billion. Canadian exports to the US in 1993 were $145 billion, representing 80% of all Canadian exports. The US supplied $123 billion of Canadian imports, representing over 73% of total imports (The Economist Intelligence Unit, 1994).

NAFTA protects the existing FTA, further enhances access to the US, and opens the Mexican marketplace with its over 80 million potential consumers. Bilateral trade between Canada and Mexico was a meagre $2 billion in 1993, leaving opportunities for Canadian exporters to expand their markets. However, Canadian manufacturers must adjust to this new competitive balance. As demonstrated previously, to successfully compete under NAFTA, Canadian organisations must shift from a low-cost strategy to other strategies based on customer service, quick deliveries, flexible response to customer needs, and the development of new products or design that better meet the needs of target markets (Dooley et al., 1991).

17.1.2. OPERATIONS STRATEGY

There are typically three levels of strategy within an organisation: corporate, business unit, and functional. Corporate strategy defines the markets and industries in which the organisation should participate and accordingly allocates resources among the business units. It is at this level that the distinctive competencies of the organisation are specified and pursued. Business unit strategies define the way in which the business unit will compete within its industry segment, in support of the corporate-level strategy (Hayes and Wheelwright, 1984).

Business units formulate strategies for each of the core activities, which include marketing, human resources, finance, technology, information, and manufacturing/operations. These functional-level strategies are designed to be interactive and to support the firm's higher-level business and corporate strategies. An effective operations strategy is characterised by a *consistent* pattern of decisions, which are supportive of the higher-level strategies of the firm and *congruent* with the other functional strategies (Hayes and Wheelwright, 1984). Figure 17-1 presents the typical strategic hierarchy within organisations.

Skinner's (1969, 1974) seminal articles identified the strategic importance of the operations function and provided the foundation for the guiding principles in operations strategy: operations strategy is a top-down process that should be design to support the firm's corporate strategies; a focused factory will outperform a plant that attempts a broader mission; operations can compete on dimensions other than low cost; simplicity and repetition breed competence; and, a factory cannot perform well on every yardstick. Skinner maintained that proper implementation of a firm's operations strategy should be co-ordinated with its higher level strategies in order to develop and to maintain predetermined competitive advantages.

The challenge to operations is to effectively manage the dynamic relationship between the firm's corporate-level strategy and its operations strategy. Too often, operations strategy is at best inconsistent with the corporate strategy within the firm, and management can be criticised for the failure to use operations in developing and achieving business- and corporate-level objectives. Within most organisations, a gap frequently exists between operations strategy and the operations tasks that are essential to corporate success.

Operations strategy can, therefore, be defined as the manner in which a firm allocates the resources of the operations function in the attainment of the goals and objectives established by the higher-level strategies. Operations strategy is a functional-level strategy that must not only support corporate and business strategies, but must also be consistent with the strategies of the other functional areas such as technology, human resources, finance, and marketing.

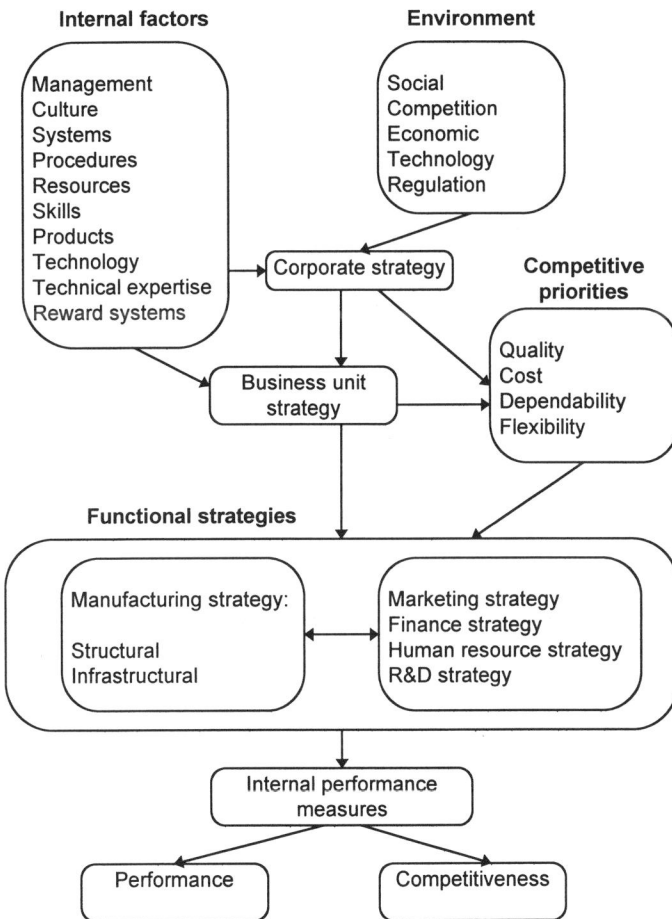

Figure 17-1.

17.2. International Manufacturing Strategy Survey (IMSS)

The investigation of operations strategy at the Richard Ivey School of Business is part of the more widely encompassing International Manufacturing Strategy Survey (IMSS) research effort world-wide, involving 20 countries and 600 organisations. The IMSS study was initiated in 1992, and the Canadian data were collected in 1993. The objectives of the study were to explore the elements of the manufacturing strategies used by firms around the world and to analyse differences between countries.

The survey instrument consisted of five major parts: the strategies, goals, and finances of the organisation; the current manufacturing practices of the organisation; the organisation's past and future activities in manufacturing; manufacturing performance; and the impact of NAFTA (Canadian respondents only). Within the survey instrument, 258 questions were arranged in 56 categories of the manufacturing strategy domain. These categories dealt with the current goals of the organisation, changes in market and product strategies of the organisation, current cost structures, facilities-related decisions, planning and control systems, quality initiatives, two-year goals and activities of the organisation, current performance, and strategic NAFTA impact. An average of four departments or functional areas were involved in providing information to complete the survey.

A total of 36 Canadian manufacturers responded to the survey, a response rate of 30%. The industries surveyed represented a subset of the ISIC Major Division 3 (Manufacturing), Division 38 (Manufacturers of Fabricated Metals, Machinery, and Equipment), Subdivisions 1 through 5 (Metal Products, Machinery, Electrical Machinery Apparatus, Transport Equipment, and Professional and Scientific Measuring Equipment). A breakdown of the responding organisations is presented in Table 17-1.

The breakdown between industry types indicates that the responding Canadian organisations spent, on average, 3.7% of sales on research and development, 12.3% of sales on new process equipment, and 2.8% of sales on training or related human resource development. The organisations served a balanced number of markets and customers and had a mixture of both national and international activities. Market share ranged from 1% to 90% of the available market, with only 38.5% of the respondents indicating that they were experiencing either rapid or steady growth in their markets. The average return on investment was 9.6%. The respondents indicated that their product volumes had grown by 46.8% on average since 1991, and that 25.1% of future

Table 17-1 Industry Profile (n = 36)

ISIC Code	Description	Percentage of Respondents
381	Metal Products	37
382	Machinery	25
383	Electrical	12
384	Transportation	14
385	Professional	12

revenue would be generated through new product introduction. The respondents indicated that their average supplier base was 534 vendors and that, on average, they intended to reduce that number by 22.3% within the next five years. The average cost structure of the respondents was 56.1% for materials, 17.4% for labour, and 26.5% for overhead.

17.2.1. COMPETITIVE PRIORITIES

The surveyed organisations indicated their current competitive emphasis through their choices in the following areas: capacity, technology, human resources, processes, products, and geography. They also indicated their competitive priorities through the selection and ranking of the degree of importance of cost, speed of delivery, customer service, quality, dependable deliveries, and breadth of product line. The particular set of strategic choices that any individual firm chose depended on that firm's interpretation of the competitive nature of its business environment. A five-point Likert scale was used to measure the respondents' choices with respect to these variables, with a score of "1" indicating "not important" and a score of "5" indicating "very important". The ranking of the Canadian firms' competitive priority selections appears in Table 17-2. As the table indicates, the respondents in the pre-NAFTA economy were positioning themselves to compete in markets requiring high levels of manufacturing quality, customer service, and dependable deliveries.

Table 17-3 indicates the relative capacity, process, product, and geographic choices of the responding firms. Firms were asked to describe their policy in terms of overall manufacturing capacity. "Higher than market" indicates a policy of maintaining excess capacity, whereas firms unable to meet market demand have "lower than market" capacity. Respondents were asked to identify their process type by selecting one of three categories: one-off, batch flow, or line flow. The "one-off" category is characteristic of firms that produce small volumes of a large range of different products, such as machine shops and printing companies. "Batch flow" is characteristic of a standardised job shop, where the firm has developed a stable line of products, such as heavy equipment or speciality chemicals. "Line flow" manufacturing operations are design to support the production of highly similar products such as automobiles. Geographic focus refers to the markets served by the firm, with a score of "1" indicating a national focus and a score of "4" indicating an international focus.

Table 17-2 Competitive Priorities

Competitive Priority	Mean Score	Rank
Lower manufacturing costs	4.22	5
Customer service	4.75	1
Dependable deliveries	4.67	2 (tied)
Faster deliveries	4.28	4
Manufacturing quality	4.67	2 (tied)
Wider product range	3.19	6

Table 17-3 Capacity, Process, Product and Geographic Emphasis

Strategic Focus	Percentage of Respondents		
	Canada (n = 36)	US (n = 41)	Mexico (n = 62)
Capacity			
Higher than market	38.9	46.3	77.4
Equal to market	47.2	41.5	12.9
Lower than market	11.1	7.3	9.7
Process Type			
One-off	19.4	17.1	1.6
Batch	61.1	36.6	51.6
Line flow	19.4	19.5	46.7
Product focus (average number of different products)	450	1248	53
Geographic focus (national vs. international)[1]	3.94	3.77	2.79[2]

Notes:
1. Five-point scale: 1 = national, 5 = international
2. Significant at 0.05 level for Canada and US

The Mexican respondents demonstrated characteristics of a marketplace closed to foreign competition: a national, as opposed to an international, focus; high capacity compared with the total market; and a limited product focus. The US respondents offered a significantly higher level of product choice and maintained an international market focus, as would be expected in a competitive marketplace open to foreign competition. Although the Canadian scores were similar to those of the US respondents in the area of capacity, process type, and geographic focus, a lack of product line breadth was identified. Such findings are consistent with Canada's relatively smaller and closed economy prior to adopting the FTA.

17.2.2. THE IMPACT OF NAFTA

The Canadian respondents were asked if the trade agreement would have an impact on their operations. As demonstrated in Figure 17-2, a total of 46% of the companies indicated that their manufacturing strategy would change under the NAFTA agreement. Of the remaining 54%, 74% indicated that the agreement would not have an impact on their manufacturing strategies, while 26% indicated that they did not know what impact NAFTA would have on their manufacturing strategy.

We found that firms within certain industries were more inclined to forecast that NAFTA would have an impact on their manufacturing strategy. The most prevalent reason for NAFTA not having an impact on the firms surveyed was that the product and market in which these firms competed had been operating in a free-trade environment. A high percentage of firms in ISIC Code 381 (Metal Products) and ISIC Code 383 (Electrical) indicated that NAFTA would have an impact on their manufacturing strategy. Whereas manufacturers in these product area would lose

46

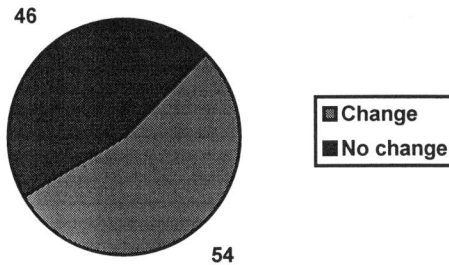

Change
No change

54

Figure 17-2. NAFTA Impact on Manufacturing Strategy

protection under NAFTA, firms in the transportation sector (ISIC Code 384), such as automotive product manufacturers, anticipated a limited impact because of the free trade environment previously created by the Autopact.

Of the firms who felt that NAFTA would have an impact on their strategy, 25% indicated that their manufacturing strategy would shift from its current emphasis. The remaining 75% did not know exactly how NAFTA would have an impact on their manufacturing strategy. This finding demonstrates a high degree of uncertainty among Canadian manufacturers with respect to the degree of preparedness for the impact of NAFTA on their operations: even if Canadian manufacturers feel that NAFTA would have an impact on their operations, they were uncertain as to how their operations strategies should change.

17.2.3. COMPETITIVE EMPHASIS

The survey asked the Canadian respondents to indicate their strategic emphasis in a NAFTA economic environment. As a group, the respondents ranked the strategic variables exactly the same as they did in the pre-NAFTA economic environment, but the mean score of each of the variables was lower than that under the pre-NAFTA environment. Although these differences are not statistically significant, the findings indicate a shift in the commitment of the respondents to their existing operations strategies and a degree of uncertainty with respect to their post-NAFTA strategy. Table 17-4 indicates the differences between the pre-NAFTA and NAFTA scores, along with the US, Mexican, and world scores.

Figure 17-3 illustrates the relative priorities for each NAFTA partner, and Figure 17-4 compares Canada with the world sample. Figure 17-3 indicates that lower

Table 17-4 Pre-NAFTA, NAFTA, USA, Mexican, and World Strategic Variable Mean Scores

Strategic Variable	Canada, Pre-NAFTA	USA	Mexico	Canada, NAFTA	World
Lower manufacturing costs	4.22	4.32	4.27	4.14	4.33
Customer service	4.75	4.78	4.19[3]	4.57	4.16
Dependable deliveries	4.67	4.25[2]	4.37	4.47	4.44
Faster deliveries	4.28	4.08	4.26	4.14	4.60
Manufacturing quality	4.67	4.83	4.55[4]	4.43	4.26
Wider product range	3.19[1]	3.50	3.79	3.17	3.39
N	36	41	62	36	602

Notes:
1. Canada different at 0.05 level from Mexico
2. US different at 0.05 level from Canada and Mexico
3. Mexico different at 0.05 level from Canada and US
4. Mexico different at 0.05 level from US

manufacturing costs ranked equally important among the NAFTA partners; however, differences exist between the other strategic variables. Differences among the other competitive priority variables highlight the distinctions between the NAFTA partners' operations strategies.

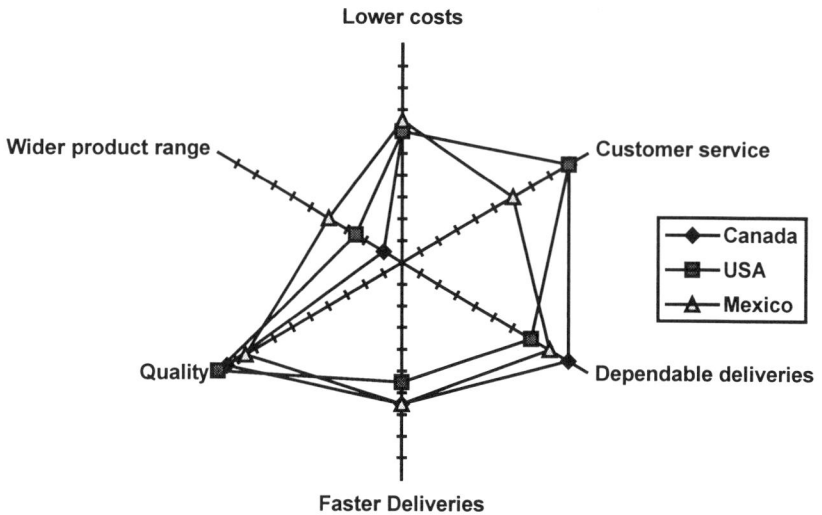

Figure 17-3. Strategic Variables - NAFTA Partners

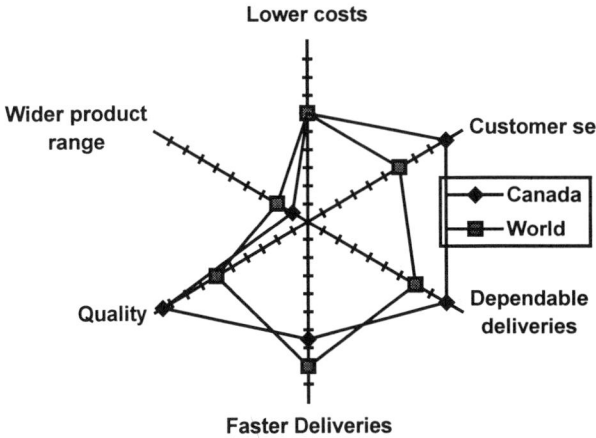

Figure 17-4. Canada and World Scores on Strategic Variables

Based on these results, it is evident that Canadian firms expect to compete in markets where dependable deliveries and customer service will be critical factors. The Mexican firms place a lower emphasis on customer service and manufacturing quality than either their Canadian or US counterparts, choosing instead to strengthen their product range. As illustrated in Table 17-3, we found that Mexican firms offered significantly fewer products than the Canadian or US firms, and, therefore, seem poised to correct this competitive disadvantage. The US respondents indicated that customer service and manufacturing quality would be their main strategy, while de-emphasising dependable deliveries and wider product range. US firms, therefore, seem to be positioning themselves to compete in markets where attention to customer delivery schedules and defect rates is important. Such capabilities are consistent with firms competing in a Just-in-Time environment.

As demonstrated in Figure 17-4, Canadian firms expect to meet world competition in terms of manufacturing costs, choosing instead to compete in other areas. Canadians are positioning their operations function for leverage in markets where customer service, dependable deliveries, and quality are important. Meanwhile, the Canadians should avoid markets where fast deliveries and product proliferation are critical. The competitive position of Canadian firms relative to the world market is, therefore, much the same as that of the NAFTA market.

17.2.4. OPPORTUNITIES AND THREATS

The Canadian firms were asked to identify where the greatest opportunities were under NAFTA and which NAFTA partner represented the greatest competitive threat. The US market was viewed as representing the greatest opportunity by 70% of the respondents, while Mexico was identified by only 3% of the respondents. Similarly, 89% of the respondents indicated that their US competitors represented the greatest threat, while only 11% of the firms considered Mexico a competitive threat. Therefore, although the Canadian firms expect the US marketplace to provide the greatest opportunities under NAFTA, the Americans are also expected to provide the greatest competition. This attitude is strongly influenced by geographic proximity between Canada and the US, and the size of the US marketplace.

17.3. Conclusions

Operations should play a critical role in the strategic planning of an organisation. This is particularly true when the competitive environment shifts, such as is the case under NAFTA. The future success of the manufacturing sector in Canada, the US, and Mexico will be a function of the operations strategies selected by each organisation as it adjusts to the realities of the new economic trading block created by NAFTA.

Our research indicates that Canadian operations managers are moving from cost-based competition to a focus on other strategic dimensions: customer service, dependable deliveries, and manufacturing quality. This strategic thrust suggests that Canadian and American organisations should move towards non-price competition to meet the competitive threat of the NAFTA environment: future success in manufacturing will depend on management's ability to achieve excellence in all aspects of operations (Dooley et al., 1991).

Canadian manufacturers appear to be prepared, at least for the time being, to ignore the competitive opportunities in Mexico, choosing instead to concentrate on the American marketplace because of its geographic proximity and size. In doing so, the Canadian firms must avoid volume-based competition, an area where US competitors' capabilities will provide a competitive advantage. Somewhat disconcerting, however, is the sense that operations managers do not anticipate significant changes in competitive priorities because of the global environmental changes exacerbated by NAFTA. Although more longitudinal research is necessary to examine the true impact of NAFTA on operations strategy, this investigation lays the foundation for comparative analysis and subsequent empirically-oriented inquiry into the implications of globalisation.

References
Dooley, Arch R., Albert R. Wood, and Miguel Leon, "A North American Free Trade Agreement", Business Quarterly, Autumn 1991.
The Economist Intelligence Unit, *Country Report: Mexico,* Second Quarter, 1991.
The Economist Intelligence Unit, *Country Report: Canada,* 1994-1995.
Schein, N. H., Johnson, P.F., Kamauff, J.W., Britney, R.R., and Wood, A.R., "NAFTA: An Empirical Study of Operations Managers' Competitive Priorities", *Administrative Sciences Association of Canada Conference Proceedings,* Operations Management Division, Halifax, Canada, 1994.
Skinner, Wickham, "Manufacturing - Missing Link in Corporate Strategy", *Harvard Business Review,* May-June 1969, pp. 136-145.
Skinner, Wickham, "The Focused Factory", *Harvard Business Review,* May-June 1974, pp. 113-121.

CHAPTER 18

TRAJECTORIES OF CHANGE

18.1. Trajectories of Change - An Introduction

Any company that is striving to improve its manufacturing operations must over the longer term make improvements and investments in both the organisation for manufacturing and the systems and technology that are required. However, for any firm there is a limit to the amount of change that an organisation can manage or cope with at any one time. Given this, an organisation must make choices as to what improvement and change activities to invest in, and what to delay. Over time, the issue becomes one of sequence, and the sequence chosen can be represented as a trajectory.

It has been proposed by Ferdows and de Meyer (1990) that there is a best sequence, beginning with organisation. This they call their 'sand cone' model. In our international study, one of the interesting questions that we set out to address was whether there were common or different sequences—*trajectories of change*—amongst the sample countries.

Trajectories of change can be viewed in terms of both the individual country and patterns across countries. As the study collected data on the state of both organisational and technological investment, we were able to position a company or a country in terms of progress in these two dimensions. From the data in the study, we have identified many items that are associated with technological—'hard'—or organisational—'soft'— approaches. Some of these are summarised in Figure 18-1.

Trajectory characterised by above average or earlier investment or higher performance in:	
Soft trajectory	*Hard Trajectory*
TQM programmes	% invested in process equipment
Teamwork	% invested in R&D
Multiskilling	Use of FMS/FMC
Intensity of team approaches	Use of NC machining centres
Level of training	Use of robotics
Number of employee suggestions	Level of CIM investment
Staff turnover	Employees per machine
Absenteeism	Capacity higher than demand
	Throughput efficiency

Figure 18-1. Hard vs. Soft Trajectories

P. Lindberg et al. (eds.), International Manufacturing Strategies, 323-330.
© 1998 *Kluwer Academic Publishers. Printed in the Netherlands.*

Table 18-1. Organisational and Technological Investments in Argentina and Brazil

% of turnover	Argentina	Brazil
% invested in process equipment	13.3%	9.6%
% invested in R&D	4.5%	3.9%

18.2. Trajectories at the national level

18.2.1. HARD TRAJECTORIES

Brazil and Argentina

A detailed comparison of these two countries is made in Chapter 19. This provides a particularly interesting contrast, as both are developing countries and members of Mercosur, the South American Common Market. In reviewing the manufacturing strategies of the two countries, there is much evidence that companies in the two countries have been pursuing different strategies. The data indicate that Argentina has been investing more heavily in both equipment and R&D.

This, in turn, is reflected in many areas. First, the impact can be seen in the production processes. Here, Argentina has less one-off production, higher numbers of sophisticated machines, and more companies setting capacity levels higher than demand. This is reflected in higher productivity (e.g., fewer people per machine), higher inventory turnover, greater throughput efficiency, and shorter lead times. Table 18-2 below presents comparative figures for the two countries:

If we look at the organisational—the 'soft'—variables a different picture emerges. Here, Brazilian firms seem to have made a more consistent investment in these areas, though the differences are not as great as with investment, as shown below in Table 18-3.

Table 18-2. Argentine and Brazilian Investment in 'Hard' Variables

Area	Argentina	Brazil
Production processes		
Line	30%	30%
Batch	60%	35%
One off	9%	35%
Use of M/C centres		0.6%
Employees per machine	2.8	8.1
Capacity > demand	41%	21%
Inventory turn	3.7	7.4
Throughput efficiency	68%	49%
% rev. from new prods	20%	11%

Table 18-3. Organisational or' Soft' Variable Measures

Area	Argentina	Brazil
% companies using:		
TQM programme	78.1%	94.7%
Team work	31%	34%
Multiskilled	41%	43%
Intensity of use of teams		
Team approaches (1-5 scale)	2.6	3.9

The impact of these investments shows up in people-related measures such as staff turnover and absenteeism (see Table 18-4):

Table 18-4. Measures Related to People

	Argentina	Brazil
Staff turnover	7.3	3.9
Absenteeism	2.4	4.4

Clearly, over the past few years, firms in Argentina and Brazil have taken different approaches to improving their manufacturing practices, one characterised by hard, the other by soft, trajectories. These different approaches are reflected in the operating performance of firms in the respective countries.

Other countries
Similar patterns are to be found in other countries. A country with a distinctive pattern of hard investment is Finland. As shown in Chapter 6, the level of use of sophisticated manufacturing equipment is high, confirmed in Table 18-5 below.

This high level of investment was based around fast economic growth in the 1980s, where firms typically chose to invest in high-technology solutions to meet customer needs (though some firms also started focusing on soft approaches during this period). In the early 1990s, Finland saw a sharp economic downturn, and simultaneously there was over-capacity, limited capital for further investment, and the need to rationalise existing manufacturing. As a result, there has been a move towards lean production and related approaches such as cellular manufacturing.

Table 18-5. Investment in Machinery in Finland

	Use per 1000 employees	
Area	Finland	Rest of world
Flexible Manufacturing Systems/Centres	4.0	1.8
Machining Centres	8.9	4.5
NC Machines	21	17
Manufacturing Robots	7.8	5.0

There are a number of lessons from the Finnish study. First, the 'hard' trajectory followed by firms in Finland in the 1980s came clearly from the industry structure and economic market background of a country growing very rapidly and operating in diverse markets requiring flexibility and responsiveness. As these conditions are changing, so are the manufacturing strategies of Finnish companies. Having made this high level of investment, firms are increasingly looking to 'soft' approaches to support rationalisation and development.

18.2.2. SOFT TRAJECTORIES

Examples of countries following the soft route are the UK and Netherlands. In the industries studied in this survey, the Netherlands does not have a history of mass production. Although having about average investment levels, it appears to emphasise soft approaches to support logistics and quality management. However, despite this, there are some inconsistencies such as low intended use of self-managed teams.

The survey data can be reviewed in the historical context of the Netherlands. As described in Chapter 7, Dutch companies were engaged at a very early stage in work experiments with semi-autonomous work groups. TQM took off in the early 1980s and is still seen as one of the cornerstones in the competitive battle. The Netherlands seem to be lagging behind in terms of Advanced Manufacturing Technology (AMT) adoption; however, this may partly be due to the lessons learnt by the early adopters that technological innovation through automation needs to integrated with social and organisational innovation.

The UK has followed a similar pattern to the Netherlands. In the UK's case, a possible explanation may be language and foreign direct investment. As much of the Japanese 'Lean Production' soft approach has diffused to the West through English language publications and through Japanese foreign direct investment, the UK has been well placed to be at the forefront of these activities. In addition, the changing nature of industrial relations and work organisation has led to major increases in performance during the last 15 years. On the other hand, Britain has always been seen to have lagged in capital investment. It can be argued that lack of investment will place a limit on the benefits to be gained by soft approaches. The causes of lack of investment are uncertain and the degree to which it is due to management, short-termism or a different financial environment are much debated (see Chapters 15 and 21).

The study data on the UK are supported by the work of Oliver et al. (1994). In a study of Japan, France, Germany, and Italy, the authors found that the UK was well ahead of the others in adopting organisational aspects of lean production such as teamwork and operator responsibility for quality. However, they found UK weak on process discipline, the 'harder' aspect of lean production, and this generally had a negative impact on performance.

18.2.3. THE MIDDLE PATH

All of the above examples have been of countries that are following a predominately 'hard' or 'soft' trajectory. It may also be possible to follow a mixed one. A prime example of this is Japan. The data from the study indicate that Japanese companies have are probably further down the line in 'soft' approaches that those in any other country. In addition, Japan is a leader in process technology.

In Chapter 9, Yamashina describes the various periods that companies have been through in Japan in order to get to the current situation of excellence in both dimensions. First, as can be seen from Figure 9-3, the investment in technology has followed a continuous trajectory from the 1960s to the present day, and is no doubt still continuing. In the first period, described as 'product out', focus was on meeting demand and rapid growth. In the second period—'market-in', the focus was on diversification—increasing variety and responsiveness. Especially during this period, there was simultaneous development of flexible manufacturing technologies and soft approaches, in particular JIT and Kaizen. By 1988, the emphasis had changed to the 'constantly launching new products' period. In this period the emphasis was on product development processes. Investment moved more to R&D and new product development. During this period mass customisation was developed. *Mass customisation* is an excellent example of an approach that requires an equal investment in hard and soft approaches. Finally came today's period, which is searching for profit. The response has included continued development of soft approaches such as Total Productive Maintenance and learning processes in manufacturing. There is also a continued emphasis on product and production engineering.

The overall pattern from Japan has been a *simultaneous* investment in hard and soft approaches over the last 30 years. There was probably a slight bias to hard in the early days, and a slight bias to soft in more recent times.

A similar but less strong pattern can also be found in Sweden. Sweden is best known for its long term commitment to new and often radical approaches to work organisation. Some of these, such as those in Volvo's plants at Kalmar and Udevalla, have received a high international profile. As Chapter 14 indicates, this investment in soft technology has continued, with Swedish companies continuing to innovate and to invest in simplification of process structures and organisational process and to move towards decentralisation. However, in addition to this many Swedish firms have invested heavily in new manufacturing technology. Thus, in many ways the pattern is similar to Japan, though in Sweden firms tend to be small and niche players with little line production.

18.3. Change trajectories at the company level

Frick et al. have examined this in detail in Chapter 20. They have constructed indices of organisational and technological indices and have positioned 29 companies on a matrix (Figure 20-3). This shows, first, that the organisations are positioned near or below the diagonal. Most companies seem to be following a trajectory up the middle, balancing both organisational and technological change. A few seem to be positioned well below the diagonal, indicating a bias towards technological change. The overall pattern seems to be as in Figure 18-2.

An important finding is that, as may be expected, companies in the high/high position had the highest performance, and those in the low/low position the lowest. For those in the middle, who had probably invested more in technology than organisation, their performance depended on the complexity of the market. If the market complexity was low, their performance was good, but if the complexity was high, performance suffered. This may indicate that in complex markets requiring effective use of technology, the benefits of these technologies could not be obtained without further investment in organisational change.

An unanswered question from the Norwegian study was whether companies with low technology change, but high organisational change would fare better in complex markets. Unfortunately, the sample did not contain any companies who were following this trajectory.

Two further conclusions can be drawn from the Norwegian data. First, even if we can conclude that the overall trajectory is biased towards technology—a hard trajectory—there is wide variability amongst firms. Some may be following a different trajectory from the rest. Second, there are firms positioned along the whole trajectory, from the starting point to a point of high and successful implementation of both hard and soft approaches. A firm's set of activities sat any point in time will reflect its position on the trajectory. If, for example, it has started with technological investment,, in the later stages of the trajectory it should begin to concentrate on organisational aspects.

18.4. Discussion

A review of the above analysis indicates that there are a variety of different patterns found amongst the countries studies, as shown in Figure 18-2 below. The Argentina-Brazil comparisons highlight the finding that the resulting performance is consistent with the trajectory followed. A hard trajectory is likely to impact on factors such as productivity, a soft trajectory will favourably impact employee turnover and absenteeism. The best performance seems to come from following both, as in the case of Japan, and at the individual firm level the best firms in Norway. The work of Oliver et al. (1996) indicates that soft changes without accompanying process discipline are unlikely to reach their full potential.

In Japan, firms seem to be able to follow both 'hard' and 'soft' approaches simultaneously. In other countries, firms seem to follow one or the other. The

'Hard' trajectory - emphasising technological change	Argentina
	Finland
	Norway
'Soft' trajectory - emphasising organisational change	Brazil
	Netherlands
	UK
Combined trajectory - balancing both organisational and technological change	Japan Sweden

Figure 18-2. Patterns of Change at the National Level

trajectories followed by countries are not always unidirectional. For example in Norway, although it is has a technological bias, it is not that far from the diagonal, and most firms are also making organisational changes as well. Indeed, the best firms seem to have followed a trajectory that has led them eventually to have adopted both hard and soft approaches.

In others such as Finland, the pattern again seems sequential, with one being followed by another (in the case of Finland 'hard' being followed by 'soft'). Elsewhere it has been argued that firms can only cope with a limited number of change activities at any one time. As a result, when they try to do too much they often fail in all of these activities. Because of this, it might be argued that it is the late adopters of new approaches, who try to implement many things at once, who have these problems; rather than the early adopters, who continually improve (*Kaizen*) and are not pressured to do too many things at once.

Each of the country studies has indicated that the trajectory used is in part based on the socio-economic background and history of that country. As indicated, in a number of countries such as Finland and Japan, over time, as the context changes so does the particular set of choices made.

It is interesting to return to the work of Ferdows and de Meyer (1990). Is their assertion that there is an optimum route supported by our study? The first point is that the conclusions that they drew from their data analysis should be questioned. Our analysis indicates that at the level of the firm and country, the trajectory being followed will be a function of its history—it is highly path dependent, in other words. If they have invested highly in technology, then it is likely that they will also be investing in organisation and quality. It is dangerous to assume a sequence based on cross-sectional data without looking at the relevant company and national history. Our data show that both hard and soft investment are needed to attain superior manufacturing performance, particularly when the firm is operating in complex markets. There is a strong argument that investment in technology mandates organisational change. It has also been argued that organisational change without the underlying hard discipline may not be effective (Oliver, 1996).

Even if Ferdows and De Meyer's 'sand cone' model is sound, the actual choices made by a firm, or the firms within an industry sector and country, will depend very much on their previous path and the current socioeconomic context.

18.5. Conclusions

Best performance at both the firm and national level comes from having followed a trajectory of improvement that embraces both technological investment and change, and organisational investment and change. Our data indicate that firms and countries have followed trajectories that emphasise organisational change prior to technical change— soft trajectories; the reverse—hard trajectories, or have followed a middle route— simultaneously hard and soft. These differences in trajectories at the national level can be in part explained by the socioeconomic background of the various countries. As these change, so do the strategies of firms within those countries.

We do not find evidence that supports one trajectory over another. There is considerable variability between countries, and between firms within a country. In particular, firms may be following different trajectories and have different positions on those trajectories.

At any point in time, a firm will already be on a trajectory, and its future choices will be dependent on whence it has come, what investments it has already made in technology and organisation. At the level of the firm, its position may be defined in terms of three dimensions:

- the relative emphasis that it has made so far in terms of organisation and technology
- its position - how far along the trajectory it has progressed
- its strategy - the current policies, programmes and emphasis, which implies its future trajectory

There are a number of policy implications. First, given a clean slate, the optimum improvement trajectory would seem to be a combination of both soft and hard approaches, rather than one that emphasises one at the expense of the other. It may be made up of a series of steps of smaller changes. However, for the individual country, or firm within it, the future trajectory will be based on the starting point. If a country has invested heavily in technology, then the future improvement trajectory is likely to be different from one that has a reverse pattern.

Second, the choice may be more difficult for firms and countries who lag rather than are ahead. Catching up with the best will require a faster rate of change. This in turn puts pressure on firms' ability to implement successfully. This in turn implies the need to make explicit choices as to the priority and sequence of activities within a trajectory.

Finally, we must remember that countries are made up of individual firms and sites. This was demonstrated by the Norwegian data. There is considerable variation between individual firms in the positions and trajectories chosen. The future trajectory of a firm will depend as much on the starting position, firm level skills and resources as the national context. In a country where overall, the best trajectory is for example a 'soft' one, there will be firms for whom the reverse may be true.

References

Ferdows, K. and De Meyer, A. (1990) 'Lasting improvements in manufacturing: in search of a new theory', *Journal of Operations Management*, **9**, 2, 168-184.

Oliver, N., Delbridge, R., and Lowe, J. (1996) 'Lean Production Practices: International Comparisons in the Auto Components Industry', *British Journal of Management*, **7**, March, S29-S44.

Voss, C.A., Blackmon, K., Hanson, P. and Oak, B. (1995) 'Competitiveness of European Manufacturing', *Business Strategy Review*, **6**, 1, 1-25.

CHAPTER 19

ENVIRONMENT AND MANUFACTURING STRATEGY: COMPARING
MODERNISATION PATHS AND PERFORMANCE IN BRAZIL AND
ARGENTINA

*Paulo Fernando Fleury and Rebecca Arkader, COPPEAD/UFRJ, Rio de Janeiro,
Brazil; Marcelo Paladino, Roberto Luchi, and Eduardo Remoulins, IAE, Universidad
Austral, Buenos Aires, Argentina*

19.1. Introduction

Brazil and Argentina rank, respectively, as the first and third largest economies in Latin
America. They are part of the Mercosur Common Market and have both experienced
considerable environmental change in recent years following the adoption of
stabilisation programmes to fight soaring levels of inflation and regain control of vital
economic variables. A feature of both programmes has been a competitive shock
through trade liberalisation, deregulation, and privatisation programmes. However,
there has been a time lag in the adoption of such policies in each country.

Since 1990, the Brazilian economy has undergone a series of measures intended to
relieve the high level of protectionism and reduce the presence of the State in the
economy, through deregulation and privatisation. However, the most serious attack on
inflation and economic turmoil came only in 1994, with the Real Plan and the gradual
introduction of a new currency. In Argentina, on the other hand, the Convertibility
Stabilisation plan was put into action three years earlier. Despite differences in the plans
adopted, firms in both countries have had to adapt to the new competitive rules. They
have had to face intensified competition from foreign products and firms, which has led
to the adoption of new strategies for coping with the new environment and surviving in
it.

Despite the difficulties faced by Brazilian firms since the beginning of the 1980s,
the effects of instability and prolonged recession in Argentina on its manufacturing
sector were far more devastating. Though both countries faced adverse conditions and
prospects in their economies, Argentina, at the outset of its upturn, could be deemed to
be behind Brazil in terms of both structural and infrastructural aspects of the country's
business sector.

Because IMSS data were collected in 1993, they reflect different moments in the
transition from protectionist to liberal policies in the Argentinean and Brazilian
economies. Brazilian firms, though severely hurt by the crisis, seem to have been better
equipped, based on their past experiences, to cope with the new environment.

331

P. Lindberg et al. (eds.), International Manufacturing Strategies, 331-343.
© *1998 Kluwer Academic Publishers. Printed in the Netherlands.*

Argentinean firms, on the other hand, were enjoying a much more promising business environment, with a rising investment climate and growing demand.

Faced in the beginning of the 1990s with the need to stand up to enhanced competition, the alternative for the promotion of change before firms in both countries was the adoption of more up-to-date hardware or software—or both—as a way to improve chances in the new market conditions. In terms of hardware, this would mean revamping old facilities or starting up newer ones with modern automation equipment and technologies; on the software side, it would mean adopting the 'lean manufacturing' paradigm in managerial practices.[1]

The purpose of this paper is to search for clues about the relationship between environment and strategy, based on a comparison of data for Argentinean and Brazilian firms in the IMSS sample. More specifically, it intends to identify similarities and differences in modernisation paths adopted by firms in each country and to discuss possible explanations for such results. It is hoped this can bring about a better understanding of firm behaviour under severe environmental change.

After a brief highlight in Section 19-2 of some specific economic indicators to help position each country's business environment in 1993, Section 19-3 introduces some comparative data on firms, including their operational structure and performance, their stated plans for improvement, the programs adopted for modernisation, and the results obtained so far. Section 19-4 discusses the similarities and differences uncovered, as well as possible reasons for the findings. The paper concludes with a characterisation of the paths to modernisation in each country and comments on possible implications for future competitiveness of Argentinean and Brazilian firms in view of the international business environment in the mid-1990s.

19.2. The Economies of Argentina and Brazil in 1993 numbers[2]

The economy of Brazil, the eighth largest in the world, was 60% larger than that of Argentina in 1993 (with GNPs of $409 and $260 billion respectively). On a per capita basis, however, Argentina fared much better than Brazil—$7,760 compared with $2,618. The effects of the stabilisation plan could indeed be felt in Argentina, where GNP growth rate for the period 1989–1993 was 5.85% compared with negative 0.08% in Brazil. Both countries showed a decrease in real growth in industrial production (in 1992): the index number for Brazil was 95, whereas Argentina's was 99 (1980=100).

Inflation rates were even more adverse for the Brazilian economy: an amazing 2,830% in Brazil versus only 10.6% in Argentina after the stabilisation plan. However,

---

[1]   See, for instance, Womack et al., 1990, for a comparison between the old- mass production - and the new - lean production - manufacturing paradigms and resulting strategies and practices.

[2]   Data in this section, unless otherwise indicated, were taken from the IMD *World Competitiveness Report 1993*, which contains statistical data as well as the opinion of executives world-wide on several structural and environmental aspects of countries.

the seemingly enduring strength of the Brazilian economy is supported by its comparatively more favourable foreign trade figures: exports were almost three times higher ($39.9 versus $14.4 billion), accounting for a larger proportion of the GNP. Even more telling is export growth in the period 1990–1992: 6.55% compared with negative 0.75%; the balance of trade was $12.85 billion in Brazil, compared with a deficit of $3.71 billion in Argentina.[3]

Direct investment in Argentina and Brazil shows the effect of the stabilisation/convertibility plan in the former on the country's investment levels: FDI was $4.18 billion versus only $1.45 billion in Brazil. The explanation may lie in the opinions of executives polled for the *World Competitiveness Report* (IMD, 1994) on subjects such as price control, state control of enterprises, anti-trust laws, and intellectual property protection: in all of them Argentina obtained higher ratings than Brazil. This was true for the cost of capital and investment in infrastructure. Brazil's distribution system, however, was regarded as somewhat superior to that of Argentina.

Brazil still ranked low concerning illiteracy figures in 1993: 18.3% compared with 4.7% in Argentina. In fact, executives' opinion on qualified manpower in Argentina slightly favoured this country over Brazil. The opposite, however, was expressed in the case of senior management and competence and qualification of engineers, as well as in employee turnover.

Productivity was higher in Argentina - $10.23 versus $3.78 for GDP per employee per hour and $11,477 versus $8,118 for GDP per person employed[4]. The rate of change in overall productivity in the 1985–1993 period was negative for both nations, but Argentina, with a rate of -2.9%, had a more adverse rate than Brazil, with -0.61%. Total hourly compensation for manufacturing workers in Argentina was somewhat higher than in Brazil: $3.30 compared with $2.55.

19.3. Benchmarking Manufacturing Strategy in Argentina and Brazil

19.3.1. FIRM PROFILES

The Argentinean and Brazilian sub-samples in the IMSS database comprised 41 and 28 companies in the metal products, machinery, and equipment industry respectively.

Firms in both countries tended to be *market leaders* in terms of their main product line - 47% versus 30% for nearest competitor in Argentina and 40% versus 21% in

---

[3]   In fact, Brazil was the only one of the four major Latin American countries (the others being Argentina, Mexico, and Chile) to post a positive trade balance. This is perhaps the result of the more advanced stage of liberalising trade policies in those nations, as similar results in trade balance later experienced in the Brazilian case seem to indicate.

[4]   Numbers for 1992.

Table 19-1. Firm profiles in Brazil and Argentina (averages)

		IMSS	ARG	BRA
Market share				
Dominant product line	(%)	33.92	47.27	40.06
Main competitor	(%)	23.60	30.15	21.12
Market development	(1-5)	2.88	2.37	2.93
Market aims				
Market coverage	(1-5)	3.35	3.00	3.43
Customer focus	(1-5)	3.49	3.37	3.25
Geographical focus	(1-5)	3.68	2.83	3.14
Percentage of				
Materials imported	(%)	43.20	31.42	28.68
Production exported	(%)	42.55	23.02	28.41

Brazil (see Table 19-1). These figures seem to indicate that, at least for the industrial sub-section under consideration, the market in Argentina was more concentrated than it was in Brazil. However, markets for the main product line were experiencing growth in Argentina, whereas in Brazil they were reported to be stable.

There were also strong similarities in *geographic and customer focus*. Both countries ranked in the intermediate range in a 1 to 5 scale on geographic focus; the percentages of production exported were also similar - 23% for Argentina and 28% for Brazil - as well as those for imported needs - respectively 31% and 29%. The score for customer focus, in a scale of 1 (few customers) to 5 (many customers) was practically the same - 3.4 and 3.3, respectively, for Argentinean and Brazilian companies.

Another point of similarity was that most companies in Argentina and Brazil favoured a "customer order" *production planning system* over an "order forecasting" system. There was however a higher incidence of such a practice among Brazilian companies (77%) compared with Argentinean companies (64%).

Several differences did appear in a comparison of data for companies in the two countries. *Size* stood out as one of the most significant, Brazilian companies being much larger than their Argentinean counterparts. In average, they had almost five times the number of employees (1,476 versus 300), 80% more machines (182 versus 107), and a larger number of plants (46% of Brazilian companies had more than one plant compared with 26% of companies in Argentina).

The issue of *integration* also distinguished both sub-samples. Argentinean companies tended to be more vertically integrated, as 69% of their total value added came from manufacturing and 31% from assembly operations, compared with 54% and 46% respectively in Brazilian companies.

Capacity policy seemed to be more conservative in Argentina than in Brazil - 41% of Argentinean firms adopted a positive capacity cushion and 44% planned their capacity to equal demand, compared with respectively 21% and 71%.

Argentinean and Brazilian firms also differed in the way they *organised manufacturing*. Sixty-one percent of the former adopted a batch production system, and 30% and 8%, respectively, used line and one-off production systems. The adoption of the three types of processes was more balanced in Brazilian companies, with batch and one-off systems being each used by 35% of companies and the remaining 30% corresponding to the users of a line system.

Cost structures were somewhat dissimilar in Argentinean and Brazilian companies. Direct salaries held a higher proportion in total manufacturing costs in the former— 26%—compared with the latter—20%; the contribution of direct materials to cost was on the contrary higher in Brazilian than in Argentinean companies—57% versus 48%.

19.3.2. OPERATIONAL PERFORMANCE OF COMPANIES

A comparison of Argentinean and Brazilian firms in terms of operational performance indicates that the latter tended to fare better in terms of variables related to efficiency in the use of resources (both human and material), whereas the former showed superiority regarding variables related to differentiation (speed of delivery and innovativeness). Nevertheless, firms in both countries were in general far behind performance levels corresponding to best international manufacturing practice.

Higher *efficiency in the use of resources* in Brazilian companies was indicated by results concerning four variables: inventory turnover, twice that of Argentinean firms (7.4 versus 3.7); personnel turnover, nearly half the average level for Argentinean firms (3.9 versus 7.3); short term absenteeism, also found to be nearly half that of Argentinean firms (2.4% versus 4.4%); and throughput efficiency, almost 40% higher in Brazilian firms (68% versus 49%).

Higher strength in *differentiation* in Argentinean firms was pointed out by three indicators: total lead time (37 days compared with 99 days in Brazilian companies); investment in research and development (4.5% of revenues versus 3.9% in Brazilian companies); and the percentage of total revenue coming from new products (20% versus 11%).

Companies in both countries reported a considerable weakness concerning one *social aspect* in manufacturing: the number of suggestions per employee. Where the average for the IMSS sample as a whole was 7.6, the corresponding Argentinean and Brazilian figures were 1.9 and 1.6 suggestions per employee per year. This clearly signalled to poor participation levels in manufacturing in either case.

19.3.3. OBJECTIVES FOR IMPROVEMENT

Answers for both Argentinean and Brazilian companies, when asked to indicate the existence of *quantified goals* on 16 objectives listed in the IMSS survey questionnaire and their *relative importance*, pointed to a clear and direct relationship between the existence of such goals and the degree of importance attached to them: the higher this was, the higher the percentage of firms with quantified goals.

Both similarities and differences were identified in the answers for Argentinean and Brazilian companies. Of the five most important objectives for improvement, three were shared by firms in both countries: *improving direct labour productivity; reducing unit costs*; and *reducing manufacturing lead time*. But the most important objective for Brazilian companies, the *reduction of inventory levels*, was not even among Argentinean firms' five most important goals. On the other hand, the objective of *improving conformance quality*, the most often quantified and third most important objective for Argentinean firms, did not hold the same place among Brazilian priorities, in either of the senses considered. Argentinean firms also stated a higher concern for *reducing unit cost* than their Brazilian counterparts. Finally, the objectives of *reducing the number of suppliers, reducing procurement lead time*, and *improving supplier quality*—which are all related to the supply chain, one of Brazil's weakest points in manufacturing—were quite more important to Brazilian than to Argentinean firms.

In both countries, the main concern was toward variables related to the *Cost* and *Time* dimensions. However, the *Quality* dimension was also considered to be important. In addition, in either case, the *Innovativeness* and *Flexibility* dimensions were among the least important.

19.3.4. CURRENT MANUFACTURING PRACTICES

Hardware variables
The levels of *investment in process equipment* in Argentinean firms were higher than those reported by Brazilian firms (See Table 19-2). Such investments seemed to be more directed to the acquisition of conventional machines, of which Argentinean firms had a higher proportion than Brazilian firms. The latter made more intense use of numerical control machines, in comparative proportions, but Argentinean firms had a larger number of machining centres. In general terms, however, levels of automation were equivalent in both countries, and quite behind levels for more advanced countries such as Germany, Japan, or the USA.

Argentinean companies in the sample were more capital intensive than Brazilian companies: the number of machines per employee was respectively 8.1 and 2.8. Such results, however, should not be taken uncritically, as they may be due, at least in part, to differences in process types.

Table 19-2. Hardware variables

	Brazil	Argentina
Investment in process equipment (% of turnover)	13.3%	9.6%
Conventional machines (% of total number of machines)	88.3%	91.7%
Numerically controlled machines (% of total number of machines)	10.6%	4.1%
Machining centres (% of total number of machines)	0.6%	3.8%
Number of machines per employee	2.8	8.1

Table 19-3. Software variables

	Brazil	Argentina
Mixing small and larger orders in the same equipment	84.6%	71.1%
Use of work teams	34%	31%
Use of JIT methods for raw materials and components	19.6%	22%
Preventive maintenance (% of total expenditures)	32.3%	33.8%
Corrective maintenance (% of total expenditures)	67.7%	66.2%
Salaried workers (% of workforce)	88%	63%
Use of incentive payments to workers	7%	34%
Hours of training to regular workers (per year)	43	33
Hours of training to new workers	31	112

Software variables

Some organisational aspects could be highlighted in comparative terms (see Table 19-3):

- Though Brazilian firms positioned themselves as being closer to their *suppliers* than those in Argentina, their performance regarding variables related to the supply chain was in fact poor.

- The organisation in Brazilian firms was less *focused*, as they tended to mix more frequently smaller and larger orders in the same.

- *Span of control* in Brazilian firms' manufacturing operations was significantly wider than in Argentina: while a foreman, in the former, had 40 workers under his supervision, this number came down to 20 in the latter. However, both presented an equivalent number of organisational levels in manufacturing. Regarding the number of job classifications, Brazilian firms were adopting 41, whereas Argentinean firms worked with only 8.

- No serious differences could be identified concerning the level of use of *work teams* or the use of *JIT* methods for the reception of raw materials and components. The same held true for the proportion of preventive to corrective *maintenance*. Again, the breakdown of *quality expenditure* categories was similar in either case, although Brazilian firms tended to show slightly higher external costs of quality.

Considered by some as the main characteristic feature of the new manufacturing paradigm, some social aspects in manufacturing can also be brought up comparatively:

- Brazilian firms tended to have less flexibility due to the prevailing type of *workforce employment practices* – 88% were salaried workers compared with 63% in Argentinean firms.

- Argentinean firms were much more likely to use *incentive payments* to workers than Brazilian firms.

- Whereas Brazilian firms tended to give more hours of *training* to regular workers, Argentinean firms invested nearly twice as many hours in the training of their new workers.

- *Job rotation* was considered as higher by Brazilian firms (with a score of 3.6 versus 3.0 in a scale of 1 to 5).

- *Production scheduling* was equally centralised in the planning department in either country.

Adoption of modernisation programs

Among the 27 *improvement programs* considered in the IMSS survey questionnaire, Health and Safety programs were, by far, those with the highest adoption rates both in Argentina and Brazil, respectively by 82% and 100% of firms. This program was also the most intensely used as well as that with the highest payoff.

The two other programs among the five most adopted in both Argentina and Brazil were Definition of a Manufacturing Strategy (respectively 95% and 69%) and Statistical Process Control (respectively 91% and 76%).

It is worth noting that three among the five most adopted programs in Argentina related to *quality issues*—TQM, SPC, and QPD—versus only one—SPC—in Brazil. The other two most adopted programs in Brazil were Value Analysis and Team Approach.

Brazilian firms showed, in general, a higher performance in terms of adoption rates, intensity of use and reported payoffs in the 27 programs. The highest gaps favouring Brazilian firms over Argentinean firms in the rate of program adoption were in the case of *Kanban* (58.9%); DFM/DFA (58.7%); CAM (58.3%); Simultaneous Engineering (53.4%); and ISO 9000 (46,0%). In terms of intensity of use, these were in *Kanban*; DFM/DFA; CAM; ISO 9000; and Team Approach.

Some programs, despite low adoption rates and intensity of use both in Brazilian and in Argentinean firms, were reported as having high payoffs, indicating perhaps they were drawing the attention of management. These were SMED, DFM/DFA, and CAM both in Brazil and in Argentina; ABC and QFD in Brazil; and Kanban and Simultaneous Engineering in Argentina. The attention received by programs such as SMED and CAM might be indicating that firms perceived the importance of introducing flexibility into their operations.

Performance improvements

There was a coincidence in four out of five operational indicators reported as presenting the highest gains in *performance* in Argentinean and Brazilian companies: conformance to specifications, inventory turnover, on-time deliveries, and customer service. The fifth indicator was speed of product development, in the case of Argentina, and delivery lead time, in the case of Brazil.

The *improvement rates* of Argentinean firms were above those of Brazilian firms in 10 out of 13 indicators in the survey. The highest gaps in performance improvement rates favouring Argentinean firms were in market share (15.9%); equipment changeover (14.1%); product variety (12.6%); speed of product development (10.8%); and manufacturing unit cost (9.3%).

The three indicators where Brazil showed higher improvement than Argentina were inventory turnover (16.6%); conformance to specification (11.3%); and delivery lead time (2.5%).

19.4. Comments on Similarities and Differences in Manufacturing

The data on strategies, practices, and performance in the IMSS survey sample represent a wealth of information concerning the recent status of manufacturing in the metal products, machinery, and equipment industries of several countries, including Brazil and Argentina. However, considerable care should be taken in the observation and analysis of the data, especially while comparing data relative to developed and developing countries and assessing the meaning of results in such diverse contexts.

It is important to have the above in mind in any attempt to perform a comparative analysis of the Argentinean and Brazilian data. For example, the higher proportion of external costs of quality found in the case of Brazilian companies (see Section 19.3.4) might be indicating, instead of poor field performance, a counterbalancing reduction in internal costs, as a result of stronger activity in quality improvement programs. Another example comes from the surprisingly high percentage of multiskilled workers both in the Argentinean and Brazilian work-forces. One possible explanation is to consider that employees, paradoxically due to low qualification, were able to serve in several different tasks—in fact, the numbers should be indicating multitask workers instead of multiskilled workers. Therefore, care should be taken to put results in context and search for possible explanations for anomalous or surprising indicators.

An examination of the similarities and differences in the information relative to the Argentinean and Brazilian firms in the IMSS survey allows us to identify at least two main sets of intervening reasons, one related to the structures and histories of the industrial sector in each economy and the other to the economic environment in each country at the time of data collection.

In the set of structural issues, the first aspect to be considered is the significantly larger size of Brazilian firms. This may be mainly due to the overall size of the economy, but can also be credited to a greater proportion of firms with a one-off, unique product type of process. The strong industrial drive in the Brazilian economy in the 1960s and 1970s seems to have created a more diversified and perhaps stronger industrial base, which is reflected in the industries under analysis. This difference between firms in the two countries seems to persist, despite higher levels of concentration and vertical integration verified in the case of Argentinean companies, as discussed below.

There is in fact an indication of a larger number of significant local competitors in the metal and engineering industries in Brazil, as the lower concentration implied by smaller comparative market shares seems to suggest. Argentinean firms, on the other hand, tend to be more vertically integrated than their Brazilian counterparts, probably a result of the absorption, by stronger companies, of small suppliers unable to survive in the severe recession of the 1970s and 1980s. Brazilian firms tend to be more export-oriented, especially in terms of sales outside the economic area—this is consistent with the traditional orientation of economic policy in each country, although this difference is more typically found in other, more export-oriented industries in Brazil.

However, despite the seemingly better developed manufacturing structure in Brazilian companies, firms in both countries seem to be far behind those in developed countries in terms of advances towards automation and flexible equipment (with only a slight advantage on the side of Brazilian firms).

There are some key aspects to be considered in terms of economic environmental conditions in Argentina and Brazil at the time of the research. Both countries had been through a long period of recession and high inflation. The extended recession has perhaps had more adverse structural effects on the Argentinean industry, where general competitive conditions were comparatively weaker at the start of the new economic policy in 1991 and more firms had succumbed due to previous difficulties. However, the Argentinean process in the direction of redesigning its economic policy started first and the country was far ahead in terms of stabilisation, consequently benefiting from reduced uncertainty and a generally more favourable economic climate. Despite the liberalising policies introduced since 1990 in Brazil, it was only in 1994 that environmental uncertainty decreased in response to a stronger stabilisation plan, similar in some respects to that previously adopted in Argentina.

It is therefore possible to identify two distinct moments in terms of the economic environment affecting sample firms in the two countries—whereas Argentinean firms were working under much more confidence in the future prospects of their economy, Brazilian firms were facing enhanced competition from within and abroad, while having, at the same time, to cope with the hazards of high inflation and poor future prospects.

It is considered here that the similarities and differences in characteristics, positioning, and performance discussed in the previous sections can be mostly traced to such issues. In many cases, both are at play.

The combined effects of faster market growth and less instability in Argentina have led to higher volume growth in the country's firms; the prospects for future changes in volume, however, were considered to be the same in both cases.

Higher concentration in Argentina might be a mixed result of past and present conditions—the survival of the strongest and the search for scale, by means of mergers and acquisitions, especially on the part of foreign capital (a feature of the Brazilian industry in the early seventies).

The somewhat stronger trend to working with forecast orders against customer orders in Argentina might be traced to a stronger integration of manufacturing and marketing as a consequence of greater economic stability. But structural factors, mainly the higher incidence of 'project-type' products in the Brazilian sample, might also account for this difference.

The observation of results in terms of operational performance, goals, and improvement efforts and gains clearly points out to each country's main concerns. Porter (1990) has already stressed the fact that firms tend to act toward minimising the effect of negative aspects rather than enhancing their strengths—the fear of failure being stronger than the hope of success. The results in this study seem to confirm such an approach.

The weakest point in Brazilian firms is found in the area of supply chain management. Company records are poor in terms of the number of firms in the supplier base and of lead times. While part of the difference in lead times might be again imputed to the higher incidence of the one-off type of production, the disadvantage is still present even if only batch and line process firms are considered. Brazilian firms have less just-in-time deliveries from suppliers and blame material shortages for half the delays in product deliveries. They also suffer from less flexibility, as they are less inclined to maintain capacity buffers.

On the other hand, Brazilian firms tended to be more efficient in their use of resources. Firms competed mostly on prices and the environmental pressures have led to giving priority to cost reduction practices, mainly in terms of quality and waste reduction efforts. The adverse conditions for investment have further stressed this trend, implied by a stronger use of existing capacity. Brazilian firms have shown higher inventory turnover and throughput efficiency. They also make better use of human resources, as indicated by lower turnover and absenteeism rates.

Argentinean firms were also in the basics as far as competitive priorities were concerned—price and quality. But economic conditions have favoured a lot more investment in hardware and in human resources. However, investments to date have not been able to change the equipment base of the industry in favour of more automation— maybe due to the still present excess in product variety and lack of adequate scale.

Argentina seems to be a little ahead of Brazil in the increasing the flexibility of employment relationships, indicated by the larger proportion of part time employees and the smaller number of job classifications. In terms of social aspects in manufacturing, both countries are fairly equivalent, even though Argentinean firms tend to favour more incentive payments and to give more hours of training to new employees.

Quality issues began to be dealt with first in Brazilian companies, which seem to have been earlier in the adoption of 'Japanese style programs' and to have obtained more promising returns in this respect. It is therefore less of an issue to these firms, in comparison with supply problems. Topmost in importance for immediate future action among Brazilian firms was inventory reduction—besides confirming the stress on supply problems, this result agrees with concerns of firms in a period of high inflation. Argentinean firms, on the other hand, seemed to stand behind in quality efforts and to suffer less from supply problems, in this last case also due to their higher degree of vertical integration.

On the other hand, the issue of product variety and development is still at stake in both countries (though with slightly better prospects in Argentina), due to its close relation to the level of endogenous technological capabilities.

These priorities are indicated in the stated goals and in the adoption of modernisation programs of firms in the two countries. Brazilian firms' most important goals dealt primarily with cost and time; Argentinean firms, besides cost and time, were mostly concerned with quality. In terms of adoption of modernisation programs, only one program in Brazil concerned quality among the top five, against three in Argentina.

It is worth noting that in both countries 'health and safety' programs ranked first in adoption, intensity of use, and payoff—this is perhaps a typical result in a developing country, where policies toward enhancing the well being of workers have been shown to yield remarkable rewards in terms of workforce motivation and productivity.

Argentinean firms indicated higher improvement rates than their Brazilian counterparts. While probably the economic environment in Argentina has led to better overall performance, it might also be assumed that Argentinean firms were somewhat behind in terms of modern manufacturing strategies and practice and as a consequence enjoyed the possibility of faster gains in performance. Whatever the reason, Argentinean firms have undoubtedly advanced faster in the last years than Brazilian firms.

19.5. Concluding Remarks

There in no one single path to overall modernisation in manufacturing, implying a total change in technology, operational systems, managerial and production processes, and labour organisation, as firms may opt for a technological path (changes in hardware) or for a functional modernisation path (changes primarily in software), as a preferential road to eventually achieving both ends (Abranches et al., 1994).

The results of the IMSS survey and the comparison of data for Argentinean and Brazilian firms in the sample lead to some conclusions regarding the preferential paths to modernisation that firms in each country seem to be adopting. Brazilian firms were apparently moving along a 'software' path to modernisation, with emphasis on practices identified with the 'lean production' paradigm—mainly attention to waste reduction and quality issues. Argentinean companies, on the other hand, were apparently moving in a similar direction, but in addition were heading toward hardware modernisation, as indicated by more ambitious plans for investment in new equipment.

These general orientations might be, indeed, traced to the structural and environmental issues discussed in the previous section. Their different starting points, conditioned by the historical development of their industrial sectors, accounts for the somewhat different priorities in terms of managerial action toward modernisation. The economic climate in Argentina allowed its companies to pursue modernisation strategies involving both the software issues needing managerial attention—with quality deserving special emphasis—and the hardware needs in terms of addition of newer and more up-to-date equipment. In Brazil, on the other hand, economic conditions still did not favour capital investments; firms fought instead for greater competitiveness streamlining operations in search of higher efficiency and response.

Therefore, it might have been expected that Brazilian firms would complement their modernisation efforts in the direction of investment in hardware once the environment showed signs of economic recovery and stability. The first impressions after the recent stabilisation efforts, as conveyed by news in the press, seem to confirm this expectation; further research might indicate if, indeed, their previous option for a software path was conditioned by environmental constraints or was merely a reflection of their technological and managerial capabilities. This is also the case concerning firm reaction to the recent economic difficulties experienced by Argentina.

In the case of both countries, however, there is a critical matter deserving official and managerial attention, this being the quality and level of skill of the work force, as well as the issue of worker participation. The modest advances in the social aspects of manufacturing represent an obstacle to manufacturing improvement in current conditions. If competing on flexibility and innovativeness is indeed a requirement for success in present day markets, this may perhaps be the most serious challenge that firms in Argentina and Brazil will have to face in their quest for enhanced international competitiveness.

References

Abranches, S., Fleury, P.F., Amadeo, E, and Caldas (1994) C. 'O novo contexto da competição internacional e o posicionamento do Brasil', Working Paper No.1. Rio de Janeiro, Brazil: Finep, Janeiro. (In Portuguese.)

IMD/WEF (1994) *World Competitive Report 1993*. Lausanne, Suisse.

Porter, M.E. (1990) *The Competitiveness Advantage of Nations*. New York: Free Press.

Womack, J.P., Jones D.T. and Roos D. (1990) *The Machine that Changed the World*, MIT Press.

CHAPTER 20

A STUDY OF TECHNOLOGICAL-ORGANISATIONAL DEVELOPMENT AND MARKET DYNAMICS IN DENMARK AND NORWAY

Jan Frick,[1] Frank Gertsen,[2] Poul H. K. Hansen,[2] Jens O. Riis,[2] & Hongyi Sun,[1] [1]Dept. of Business Administration, Høyskolen i Stavanger, Stavanger, Norway and [2]Dept. of Production, University of Aalborg, Aalborg, Denmark

20.1. Introduction

In recent years, experience in introducing elements of Computer-Integrated-Manufacturing (CIM) has brought about a recognition that implementation of CIM should be balanced in regard to both technological and organisational dimensions to obtain true competitive benefits. Empirical studies indicate that many industrial enterprises have been able to apply advanced manufacturing technology successfully, at least technically; however, few companies have been capable of improving their competitive strength (*cf.* Voss, 1988). As a consequence, attention has been directed towards ways of establishing a balance between technological and organisational means, and towards establishing a link between changes in manufacturing and corporate improvements, for example, in terms of increased competitiveness.

Increasingly, CIM implementation has been viewed as a process of continuous change, instead of an all-out effort to install CIM in a single shot. This suggests an evolutionary approach in which the introduction of advanced manufacturing and information technology is planned and implemented in a series of steps in parallel with organisational change, and with due respect to the overall company situation including market requirements.

Any industrial enterprise will continuously face the question of what to emphasise when taking concrete steps towards CIM. Should the organisation or the technology alone be changed; or, more likely, should the company choose an evolutionary path combining both to improve its performance? Which moves towards CIM may be justified by market requirements, and how important is the role of corporate strategy?

We have carried out an empirical study to observe how industrial enterprises actually deal with these issues and to understand their behavioural patterns.

An early systematic perspective on company development was suggested by Leavitt (1964). Many others have subsequently studied the development of companies by pairs of factors, e.g., organisation-technology or environment-technology or environment-organisation. Organisational-technological relations have been treated by Daft (1989), Twigg & Voss (1991), and Bessant (1990), just to mention a few.

P. Lindberg et al. (eds.), International Manufacturing Strategies, 345-362.
© *1998 Kluwer Academic Publishers. Printed in the Netherlands.*

Environmental-organisational issues have been the subject for contingency theorists, e.g., Lawrence (1969), Mintzberg (1983). The subsequent research has only included partial perspectives and has been mostly qualitative. We should like to contribute to this issue with both qualitative analysis and quantitative methods.

Other authors have suggested common phases of the development of an enterprise and identified crises during the company life cycle (e.g., Greiner, 1972; Churchill, 1983). To find out from exploitative empirical studies how companies develop, inspiration may be drawn from their general characterisation of each development phase. These development stages may be measured empirically in terms of organisation, technology, market status, and strategy formulation as well as performance. Some research questions are: How do companies move for organisational development and CIM-technology utilisation? Are they moving step by step or stride by stride? Do organisation and technology development take place simultaneously or sequentially? Why are they moving in a specific direction? Which factors determine their moves? What will or should their next moves be?

The following sections present our empirical study. Analyses and discussions will put forward explanations and answers to the research questions proposed previously. Implications will be drawn from the study to help a small- or medium-sized industrial enterprises plan and implement appropriate moves towards CIM. A final purpose of this research is to explore the empirical pattern of organisational changes and technological innovations in terms of dynamic path, static status, stages and paradigms.

20.1.1. AN ANALYTICAL TOOL: THE O-T MAP

In order to look at the previous questions and fulfil the research aim, the Organisational-Technological map (O-T map) was developed. The O-T map is a simple two-dimensional co-ordinate diagram with Organisation (O) and Technology (T) as vertical and horizontal axes respectively. The O-T map is a useful tool to describe organisation and technology both dynamically and statically. Ettlie (1988), Twigg and Voss (1991) and Bessant (1992) have used similar diagrams to subjectively illustrate the path or trajectories of technological innovations and organisational changes. Although their definitions and conclusions differ, their research has contributed to the description of organisational changes and technological innovation. Based on the above research and our own experiences in Denmark and Norway, we define three evolutionary paths. They are organisationally-oriented (*O*-path), balanced (*B*-path) and technologically-oriented (*T*-path) as illustrated in Figure 20-1a. We also define an individual company's location at a particular point in time as its status (Figure 20-1b).

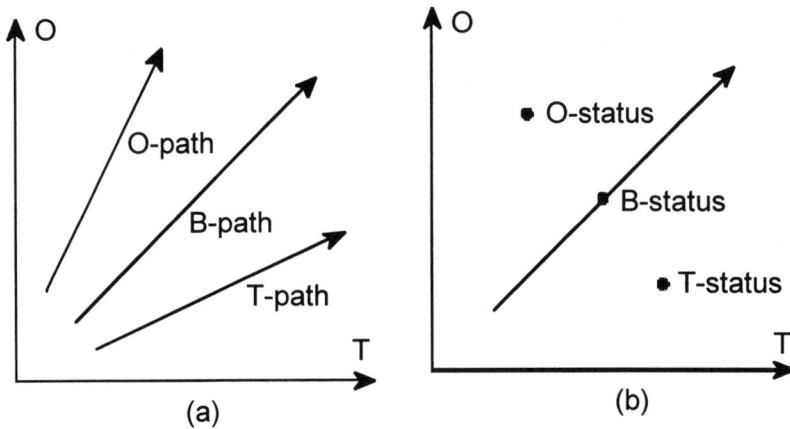

Figure 20-1. The dynamic path and static status on the O-T map

20.2. The Empirical Study

To find out where companies are located, and what kind of path they are following when adopting new technologies, empirical data are needed.

Sixteen industrial companies were initially included in the survey, which was extended to a total of 28. They constitute a wide spectrum of industrial enterprises covering different industries and products (communication, medico-technique, robots, farming equipment, engineered facilities, furniture, suppliers of mechanical parts, etc.). They also varied in size from 50 to 800 employees. The companies were mainly selected on the basis of prior contacts.

The interviews were conducted in 1991–92. For each company, 160 questions were asked in a guided interview, either on-site or by telephone, using a questionnaire specifically developed for this study. The questions covered technology utilisation, organisational status, the market situation, strategy and management, investments, and performance.

The Technology Dimension. The questions about technology asked about the utilisation of CIM technologies in engineering, manufacturing and information processing as well as the level of integration. The measurement in our questionnaire indicates the extent of automation and integration with CIM technologies. In our qualitative analysis, CIM technologies have been divided into five stages according to the degree of automation and integration: (1) conventional/manual, (2) stand-alone machining tools and small systems, (3) islands of automation, (4) partially-integrated and automated, and (5) fully-integrated and highly automated. All the SMEs in our survey fall into the three middle stages.

The Organisational Dimension. The questions regarding organisational issues were formulated in the areas of human resources, structure, management systems and culture. The intention was to measure the organisational characteristics such as specialisation, formalisation, standardisation, decentralisation, configuration and flexibility.

Based on their responses, the organisations were divided into three categories. The first is a one-dimensional organisation characterised by a hierarchically and functionally-oriented organisational structure with clear division of jobs and narrow qualifications for employees. The second is a two-dimensional organisation that combines a functionally and product-oriented organisational structure in a matrix-type structure. The third is a multi-dimensional organisation that is somewhat organic and capable of combining productivity and adaptiveness. The organisational and operating structure tends to be a non-hierarchical set of networks.

Market Dynamics. This index seeks to describe the dynamics of the market place as well as the extent of internationalisation. The specific measuring points include uniformity of customers, geographical market distribution, frequency of new product introduction, level of uncertainty, and influence of national and international laws.

Strategy and Management Attitude. The strategy index measures whether the company has a formally formulated strategy and whether the strategy is spread widely through the whole organisation so that every department and employee are aware of it.

Company Performance. The performance index focuses on strategic performance including increase in market share, increase in revenue, profit increase, and a measure of adaptability.

Investment Index. The investment index is a qualitative measure of the kind of technology investments that the enterprise has made in the past three years.

20.3. Statistical analysis and classification

The 160 questions from the questionnaire were clustered into groups and the indices mentioned above. The maximum score from each group was set to 5, with each index formed by summarising several groups. Finally, a minimum and a maximum score of each index were defined to which the individual indexes of each enterprise were related on a 0 to 50 scale. An X-Y plot of the organisational indexes versus the technological indexes shows a strong and statistically significant correlation. The best-fit regression with 16 companies seems to be a polynomial regression of the fifth order with R=0.93 and 80% confidence intervals (see Figure 20-2).

The slight S-form of the regression line supports our intuition that there are two groups of enterprises, each with different characteristics. From this analysis we have a high-positioned and a low-positioned group in the O-T diagram. (See Figure 20-2). A separate analysis of the two groups strengthens this intuition. The correlation of the organisational and technological index in the high group is statistically significant while the test shows no significance for the low group due to the relatively small sample. However the average value of the pre-defined indexes in general shows differences between the two groups.

The market dynamics index suggested a further classification of the companies into four groups of companies, low and high O-T against low and high market dynamics.

The groupings based on the market dynamics index are illustrated in Figure 20-3. Supplemented with our detailed knowledge of the companies in our sample, we have tried to capture the characteristics of each of the four types of companies, representing a tentative grouping of industrial enterprises.

Group 1 (Novices): Low O-T, low market dynamics group
Companies in this group have a lower score on market dynamics, performance and strategy indices. They have adopted CIM technologies to some extent, but they are not yet effectively utilising them for competition and performance improvement. Even though some of the companies have more than ten years of investment in CI,. they still haven't progressed from their original location, thus the description as Novices. A serious threat to this group will appear if the market becomes more dynamic; but they do not appear to be prepared for it.

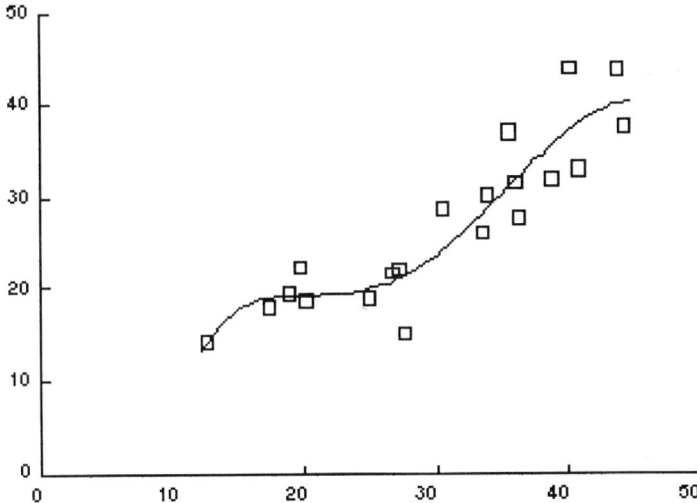

Figure 20-2. The O-T diagram. A plot of the technological index versus the organisational index.

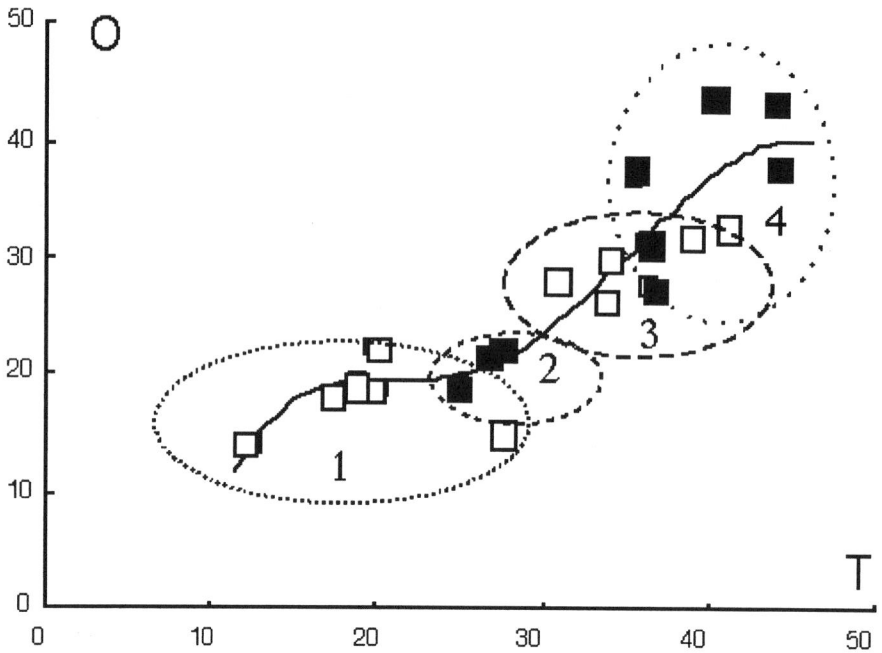

Figure 20-3. The position of the four groups of companies in the Organisational-Technological diagram.

Group 2 (Market-oriented): Low O-T, high market dynamics group

Companies in this group have relatively high performance and are operating in a highly dynamic market. They have adopted new technology to some degree, which apparently has been utilised to support their market position. Companies in this group seemingly have many options available; the high performance index allows them to take their time. However, the high score on the investment index suggests plans to move in the technology direction, which may bring the O-T balance to a critical point. The high market orientation may help these companies to make full use of the new technology.

Group 3 (Technology-led): The high O-T positioned, low market dynamics group

Companies in this group have a relatively high score on both the technology and the organisation indices yet they are operating on a market with low dynamics and with a lower performance than companies in Group 2. The strategy index is rather high. This indicates that companies in this group deliberately have focused on technology and plan to continue in this direction, as shown by their heavy investment in CIM technologies. In addition, the organisation index is significantly below the technology index. We find that companies in this group are in a critical situation with limited options open. If they increase the technology index without increasing the organisational index, they will reach a critical point.

Group 4 (World Class Manufacturing): High O-T, high market dynamics group
Companies in this group have the highest average score on all indexes. We call them candidates for World Class Manufacturer, because they typically (1) are operating on the world market and exporting more 90 per cent of their sales; (2) have high performance; (3) have adopted the WCM means, i.e., both high technology and organisation scores; and (4) have a formal and widely spread strategy to match means and ends.

All the companies in our sample have invested in CIM technologies over the past decade, yet they show different application of technology and organisational means, they operate in different degrees of market dynamics, and they have different performance index scores. Market dynamics has been identified as an important factor with a rather complex connection to the adoption of technological and organisational means. A critical zone centred around companies in the Technology Push group has been identified, with an apparent momentum towards technology but with few options available because of the relatively low performance level. The positions of the companies in the O-T diagram follow an S-shaped curve, which has led us to focus on the development of the companies over time. Is an industrial enterprise likely to follow an S-shaped development path in its application of technological and organisational means? To answer this question, we have selected three companies from the sample. These have been followed over a period of time, and the observations are reported in the following section.

20.4. The Dynamic Path

The data analysis started with organisation and technology indexes. The T-index measures the extent of technology utilisation, while the O-index measures the status of education and skills, matrix and project team, employee involvement and communication, etc. The O and T index data were plotted on the O-T map and a polynomial regression was conducted. The fourth-order regression (best fit with 28 companies, $R^2 = 0.81$) is a concave curve (the bold curve in Figure 20-4). Although the curve is a snapshot with slight trace, it can be proposed to be an indicator of the general path along which the sample companies are moving. The general empirical path and the three hypothetical paths are shown in Figure 20-4.

The path as a whole does not fit any of the hypothetical paths because the static status (reflected by the distance to the balanced path, dB) and the dynamic trajectories (reflected by the slope dO/dT) vary along the path. This suggests that the path should be divided into smaller stages for further analysis. The path was divided according to the

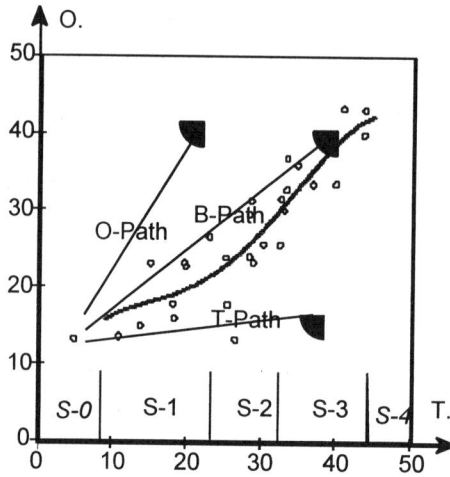

Figure 20-4. Paths, status, stages and paradigms on the O-T map

dB and dO/dT measures. Five stages were identified, as Figure 20-4 and Table 20-1 illustrate (S-0 to S-4). These stages *are not* identical to the groups in the previous section. The analysis focus on the three middle stages, where most of the samples are found. The main findings are highlighted in Table 20-1, and will be elaborated below.

According to the data distribution, almost all of the 28 sample companies are in the three middle stages, 23%, 45% and 32% respectively. No sample company is located at the organisational or technological-oriented extremes.

Table 20-1. A summary of the findings

	Stages	Stage 0	Stage 1	Stage 2	Stage 3	Stage 4
1	% of samples	0.0%	23%	45%	32%	0.0%
2	Organisational character	Traditional organisations		(Shifting)	Emerging Organisations	
3	Technological character	Conventional	Stand-alone	Island of automation	Partially integrated	Fully integrated
4	Performance (Feature)	/	Middle Decrease, Failure	Lowest Increase, Recovery	Highest Increase, Success	/
5	Status (Tendency)	/	T or B-Status Off B-status	T-status To B-status	B-status In B-status	
6	Path (Feature)	/	T-path T-push	T-,B-,O-path Shifting	O, B, T-path, Balanced	/
7	Groups		Novice	Market orientation	Technology Push	World Class Manufacturing candidate

The average performance of companies in Stage 1 is in the middle, but tends to decrease, which we define as 'failure'. The average performance in Stage 2 is the lowest but tends to increase slowly, which we characterised as 'recovery'. The average in Stage 3 is the highest and tends to increase, which we characterised as 'success'. This relates to the groups in Section 20-4, where higher initial investment in technology seems to correlate with low performance until further investment in organisation produces a recovery in performance.

The organisational character. The emergence of a new form of organisation has been observed in the above analysis. The new organisation is characterised by (1) high education and training, (2) co-operation with suppliers, (3) use of matrix and teams, (4) employee participation, communication of goals etc. The statistical correlation (R) between these four dimensions and performance is 0.56, 0.82, 0.59, and 0.71, respectively ($p < 0.02$).

The technological character. The development stages roughly correspond to the five technological stages, i.e., conventional, stand-alone, islands of automation, partially-integrated and fully-integrated. No sample company has reached the fully-integrated stage. It looks as if there is a certain sequence in the application of technology in functional areas. According to the priorities that companies gave in different areas, the adoption of technology follows the functional sequence of (1) finance and administration, (2) marketing and production management, (3) production automation and integration, (4) design automation, (5) partial integration, and (6) full integration.

Static. In Stage 1, companies are between the B- and T-status, but *moving away from it.* In Stage 2, companies are in the T-status, but moving towards the B-status. In Stage 3, companies are approaching the B-status. The status influences the performance in certain way. It can be proposed that, other things being equal, companies in the B-status (i.e., Stage 3) tend to have higher performance than those in the T-status (i.e., Stages 1 & 2). This can be further demonstrated by the statistical analysis. The statistical correlation between status, represented by the distance to the balanced path (dB, see Section 20-2), and performance is -0.34.

Dynamic path. In Stage 1, companies followed a T-path. Following Stage 2, companies start to implement more organisational changes. At Stage 3, companies are following a B- and then O-path. Looking back, the path that a company followed determines its current status. For example, those companies that consistently followed the T-path would reach a T-status.

20.5. Case Examples

The initial purpose of the O-T-diagram was to describe the current status of an enterprise. However, an additional application might be to plot the moves of specific enterprises on the diagram in order to suggest appropriate paths according to given conditions. As indicated in the statistical analysis, there seems to be a critical area in the O-T diagram where enterprises need to redirect their focus from technological to organisational change.

Three of the enterprises in our survey have been observed for up to eight years. Their individual moves have been plotted in the O-T diagram (Figure 20-5). Enterprise A is for the moment positioned in the critical area and has not made significant changes during the last two years; Enterprise B has made a significant move from a low-level position to being almost a WCM-candidate; and Enterprise C can be regarded as a WCM-candidate.

20.5.1. CASE A

Enterprise A is order producing and operating on a world-wide market. The enterprise has specialised in development and production of industrial processing equipment within a niche in the food industry.

- *Technological change and investments.* Company A's technological investments may be characterised as islands of automation, including an MRP system, CNC machines, and a CAD system. They have all been technical successes.
- *Organisational change and development.* The major change in the past six years has been the establishment of a new logistics department and the reduction of production capacity. 60% of the parts are now produced by sub-suppliers. Attempts have been made to improve the integration between the technical and the commercial sides of the enterprise, including project control concepts, order specifications concepts, definition of new project leader roles, and so on.
- *Market development.* Many of the customers are large international corporations, a trend that has increased during the period. Given these characteristics, the enterprise has only a small possibility of influencing the market. As a consequence, the strategy is to be flexible and to be able to deliver whole systems.

Development features of Case A. Its current position in the O-T-diagram was reached in late 1989. Since that time, only slight changes have taken place. During this period the enterprise has experienced a yearly decreasing profit rate due to an increase in overhead costs.

A number of issues were critical in relation to the past move and especially in relation to further development. The organisational and the technological changes are

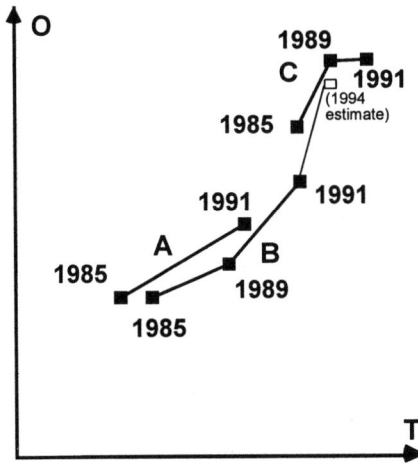

Figure 20-5. The moves of enterprise A, B, and C in the years 1985 to 1991 (with an estimate for 1994 for company B)

not related, e.g., CAD has increased drawing capabilities, but has had no impact on cross-functional integration. There are only two or three change champions in the enterprise, and practitioners are favoured compared with more systems-oriented people, so changes are often specified by external consultants. However, the implementation carried out by the enterprise has partly failed—there is no formal technology strategy, and strategic planning is generally regarded as a sales issue. The technical director is fully occupied with making contacts with customers and with co-ordinating the daily work on orders. In the early phases the enterprise successfully recruited new talents within technology areas (MRP, CAD, CNC).

20.5.2. CASE B

Enterprise B produces equipment in steel for a world-wide market. During the period of observation it has moved from the novice class through the market-oriented class to the class of WCM-candidates. (This company is also represented in the IMSS analysis.) It is given a similar high score in the IMSS benchmark for 1992.

Technological change and investments. The enterprise has invested heavily in technology during this period. Until 1988, these investments may be characterised as islands of automation, both in production equipment, administration, planning and control systems, and equipment for development and testing. Most of these decisions have been made based on a foundation of payback and strategy. Afterwards, and especially in 1990 whilst moving to the WCM-candidacy, the strategic focus has shifted towards functional integration and utilisation of manufacturing technologies to serve market demands regarding flexibility and short lead times.

- *Organisational change and development*. The organisation has been educationally updated continuously along with the technology investments. Significant changes in the organisational structure have been planned and executed along with continuous changes in procedures. These were often triggered by major changes in the market. Additionally, employees have developed skills for multiple tasks during the period, and several tasks that needed consultants before are now carried out by employees. The enterprise has been strategically focused during the whole period, but strategic statements have increasingly addressed the relation between market considerations and internal affairs.

- *Market development*. In the period the market has spread from mainly the Scandinavian market with some world-wide activities to a mainly European and world-wide market. This means that the increase in volume and variety primarily has been outside the original home market and therefore involves a new competitors. With the exception of 1986 and 1990–91 there has been a steady market increase, but the market in 1992 was at a very low level.

Development features of case B. In the critical transition area, some steps have been important for the moves taken:

- the development of a visionary concept prior to the major changes in organisation and production technology.
- the use of internal employees in the development process (approximately one hundred persons are directly involved, of whom only five are external).
- the strategy has added more focus on functional integration to serve market dynamics.
- both organisational and technological movements are characterised by rather large leaps.

20.5.3. CASE C

Enterprise C has moved within the WCM-candidate class from 1985 to 1991. It operates on a dynamic world-wide market, and it has a clear identity as a high-technology and high know-how enterprise. This has made it easier for the enterprise to engage highly educated people in the development and engineering department. Also, production personnel are better qualified than in other local enterprises.

- *Technological change and investment*. Because the enterprise has been in the forefront of technology, it has made significant, but step-by-step, investments in advanced technology during the period of our observation. Some CIM technologies, adopted in the early 1980s, were islands of automation. The enterprise started earlier than most of the other enterprises in this survey to consider integration. Enterprise C is one of the first in its area to use and achieve payback for several new technologies.

- *Organisational change and development.* The enterprise has adjusted its organisation on several occasions and it may be characterised as almost continuous adaptable to market conditions and possibilities. This includes changes in both procedures and organisational structure, which is probably based on the high educational level of the employees and a significant effort in further educational activities. Strategy statements have been formulated and executed during the entire period. This includes also the ability of the employees to participate in making or commenting on the strategy.
- *Market development.* By various methods and channels the company has increased its volume and market share on a world market during this period. The basic technology demanded by the marked has changed significantly in this period, but the enterprise has managed to increase its competitiveness in the period.

Development features of Case C. Some important characteristics may be identified from this case. Because of a culture for high technology and skills, the enterprise has been able to cope with

- a dynamic world-wide market
- a high degree of involvement in strategic thinking;
- step-by-step moves in both organisational and technological directions, though more emphasis was put on technology between 1989 and 1991.

20.6. Implications

The analysis of the empirical study and the case descriptions have provided us with a new insight into the route that small- and medium-sized industrial enterprises have taken and are taking towards CIM. On this basis we will describe a set of implications based on our general understanding that may be useful for an industrial enterprise wanting to determine its next move, and for national policy-makers. In view of the rather small sample of the survey, the implications have a tentative, postulating nature.

Quasi-balanced technology-organisation development
The empirical data show a statistically significant correlation between technological and organisational development; i.e., the more advanced manufacturing technology is employed, the more likely it is that the enterprise changes from a one-dimensional to a two-dimensional organisation, and further to a multi-dimensional organisation. Thus, it appears that, by and large, SMEs do manage to maintain a balanced development of their organisation when introducing new technology. However, the data support an S-

shaped relationship, which indicates that companies tend to move in the technology direction first, or to a larger extent, before adopting organisational means. The data have helped us to identify a critical zone centred around the 'technology push' group of companies discussed in Section 20-3. They seem to move continually in the technology direction, with no improvement in performance and market dynamic index. This has led us to formulate the following proposition:

> *Proposition 1.* A critical zone has been identified for companies in the middle zone. With only a few options open, they need to shift focus from technology to organisation.

Case B is an example of a major 'technology leap', whereas Case C has chosen an approach of small steps in both directions. For the critical zone, this suggests the following proposition:

> *Proposition 2.* Enterprises should be cautious about adopting a Big Technology Leap implementation approach. Instead, they should focus on mixing small steps of organisational and technological development.

The extent to which this is possible depends very much on the divisibility of a given technology. It is our general impression that enterprises have tended to look at new technologies as a total system that has to be decided at once and for all, e.g., CAD, MRP or FMS. Industrial experience suggests that most systems may be divided so as to allow for a gradual implementation.

> *Proposition 3.* Enterprises should focus on the divisibility of a given technological means to allow for a gradual implementation.

Market dynamics is an important factor
There was a significant statistical correlation between the O-T dimensions and market dynamics; i.e., on the average, the more advanced technological and organisational means are adopted, the more frequent the enterprise is operating in a dynamic market with strong competition, or vice versa. However, the data do not tell us what came first. The exception to this is the 'technology push' group of companies identified in Section 20-3.

Some of the enterprises studied had experienced instances of a drastic change in the market place, e.g., a drop in sales or increased competition. This was followed by a change in the organisation, and not in the technology. Several questions occur: Did the companies have enough time to introduce new technology? Does technology only provide opportunities for market improvements, requiring the simultaneous development of organisation and market?

No clear picture emerges. However, the close connection between market dynamics, technology and organisation has been reinforced, and suggests further research directions and the need for careful attention by industrial managers.

Proposition 4. The Market-Technology-Organisation triad holds a key to successful moves towards World Class Manufacturing.

A diagnostic tool for planning the next move for an industrial enterprise

As a by-product, this questionnaire (or similarly the IMSS questionnaire) may be used as a useful diagnostic tool for an industrial enterprise considering making a next move in the technology-organisation space. Although not complete and not fully reliable, this questionnaire has been able to locate the present position of a company with a rather limited effort. Furthermore, experience shows that it raises many questions about the potential benefit of a given move compared with alternative moves, especially the connection to the market.

Hence, the use of the questionnaire for analysis and diagnosis may provide a good basis for planning the next move, and even outline a path for the company to follow for the coming years, in essence providing an implementation strategy.

Proposition 5. Identification of the present position in the Market-Technology-Organisation space may serve as a diagnostic tool to provide a good basis for deciding the next moves.

In addition, Propositions 1, 2 and 3 may also be applied to the individual company by drawing attention to critical transitions.

According to the above findings, the general pattern of organisational changes and technological innovations can be summarised as follows:

- *Stage 1* - Companies are staying between B and T status and their average performance is in the middle. They are moving away from B-status to T-status by following a T-path. As a result, their performance tends to decrease.
- *Stage 2* - Companies are in a T-status and their average performance is the lowest. However, they start to turn towards the B-status by shifting from T- to B- and then O-paths, and, as a result, their performance tends to increase.
- *Stage 3* - Companies at manage to keep the B-status by follow the B- and O- paths and have highest and increasing performance.

There is a causal chain behind the pattern, as illustrated in Figure 20-6. The performance is partially influenced by the static status. The status is determined by the path that the company followed. The path is guided by the vision or paradigm of manufacturing that a company or rather its management has.

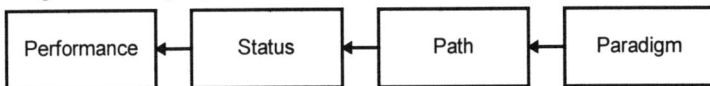

Figure 20-6. The mechanism behind the organisational changes and technological innovations

National policy-making

During the past decade, government agencies in many Western European countries have sought to stimulate the development of their industries so that they become competitive world-wide. Most of these programs have been technology-oriented, based on the philosophy that an increase in the proficiency of technology development would automatically increase competitiveness. Our study suggests the following two propositions:

> *Proposition 6.* The present context of an industrial enterprise should guide the direction of governmental support. That is, a contextual approach should be adopted to allow for the selection of supporting means appropriate to the present position of the individual enterprise.

> *Proposition 7.* The government program should reflect the intricate and complex nature of the Market-Technology- Organisation relationship.

20.7. Conclusions

This chapter has explored some issues based on our data and developed a few initial hypotheses. The data have supported the balanced development of technology and organisation, although an S-shape curve was detected. The frequently seen two-variable approach has been expanded to include organisation, technology and market dynamics, for which measurable indicators have been defined. The survey holds a clear evidence of a connection between these three factors, although it appears rather complex and not yet fully explored. The data have suggested two ways of classifying companies: (1) into groups, respectively denoted Novice, Market-Oriented, Technology Push, and Candidate for World Class Manufacturer, and (2) into 5 stages. This helped us identify a critical transition zone centred around the technology push group.

The results of this analysis encourage further research.

References

Bessant, J. (1990) 'Organization adaptation and manufacturing technology', in Haywood, (ed.), *CIM: Revolution in Progress, Proceedings of the final IIASA conference*, Austria, 351-360.

Bessant, J. (1992) *Managing Advanced Manufacturing Technology—Five Waves*, NCC Blackwell, Oxford U.

Churchill, N.C. & Lewis, V.L. (1983) 'The five stages of small business growth', *Harvard Business Review*, May-June.

Daft, R.L. (1989) *Organization Theory and Design*, Third Edition, West Publishing Company.

Ettlie, J.E. (1988) *Taking Charge of Manufacturing: How Companies Are Combining Technological and Organizational Innovations to Compete Successfully*, Jossey-Bass Publishers, San Francisco.

Frick, J. (1991) 'CIM strategy development and classifications', *Central Topics in Developing Manufacturing Companies Towards CIM According to Experience*, IPS, Fuglsøe, Danmark, AUC/DTH.

Frick, J., Gertsen, F., Hansen, P.H.K., Riis, J.O., & Sun, H. (1992) 'Evolutionary CIM implementation - an empirical study of technological-organisational development and market dynamics', in: *The Proceedings of CIM-Europe, 8th Annual Conference*, Birmingham, May.

Frick, J. & Irgens, C. (1995) 'The 'learning organisation' as an approach to increasing manufacturing competitiveness', *Nordic Conference on Business Studies*, København, **12**.

Frick, J. & Irgens, C. (1996) 'Increasing technology benefits (achieving benefits from technology) through organizational learning', *The Fifth International Conference on Management of Technology*, Miami, **18**.

Frick, J., & Riis, J.O. (1991) 'Organizational learning as a means for achieving both integrated and decentralized production systems', *Computer Application In Production Engineering*, Bordeaux, North-Holland, IFIP WG 5,7.

Frick, J.; & Sun, H. (1995) 'Paradigm Shifts in Manufacturing', *FAIM95 - Flexible Automation & Intelligent Manufacturing*, Stuttgart, **12**.

Gjerding, A.N. *et al.* (1992) 'The productivity mystery: industrial development in Denmark in eighties', DJØF Publishing, Copenhagen, Denmark.

Greiner, L.E. (1972), 'Evolution and revolution as organizations grow', *Harvard Business Review*, July-August.

Hansen, P.H.K. (1993) *Managing Integration in Manufacturing System: A model of objects and mechanism*, Ph.D. thesis, University of Aalborg, Aalborg, Denmark.

Haywood, B. (1990) 'National differences in the approach to integrated manufacturing: a case study of FMS in UK and Sweden', in B. Haywood, (ed.), *CIM: Revolution in Progress, Proceedings of the Final IIASA Conference*, Austria, 435-451.

Hörte, S.A. (1991) 'On the choice between human oriented and technology oriented manufacturing system', *Proceedings of the 3rd International Production Management Conference on Management and New Production Systems*, Gothenburg, Sweden, May 26-29, 354-360.

Jaikumar, R. (1986) 'Post-industrial manufacturing', *Harvard Business Review*, Nov.-Dec., 69-76.

Jones, D.T. 'Beyond the Toyota Production System: the era of lean production', in the *Proceedings of The 5th International Operations Management Association Conference on Manufacturing Strategy*, Warwick, June 26-27.

Lawrence, P.R. & Lorsch, J.W. (1969) *Organization and Environment*. Illinois, Irwin.

Leavitt, H.J. (1965) 'Applied organisational change in industry: structural, technological and humanistic approaches', in J.G. March (ed.), *Handbook of Organisations*, Chicago, Rand McNally.

Mintzberg, H. (1983) *Structure in Fives: Designing Effective Organisation*, Prentice-Hall.

Riis, J O. (1991) 'Research methodology', Working Paper, *The 4th IPS Research Seminar*, Fuglsø, Denmark.

Riis, J.O.; & Frick, J. (1990) 'Organizational learning: a neglected dimension of production management systems design', *Advances In Production Management Systems*, Helsinki, North-Holland, 141-150.

Schonberger, R.J. (1987) *World Class Manufacturing Casebook: Implementing JIT and TQC*. The Free Press, NY.

Sun, H. (1993) *Patterns of Organisational and Technological Development with Strategic Considerations: Managerial Implications for Advanced Manufacturing Technologies*, Ph.D. Dissertation, Department of Production, University of Aalborg, Aalborg, Denmark.

Sun, H. and Riis, J.O. (1994) 'Organisational, technological, strategic and managerial issues along the implementation process of advanced manufacturing technology: a general framework of implementation guide', *International Journal of Human Factors in Manufacturing*.

Sun, H. and Hansen, P.H.K. (1992) 'Research methodology in an empirical study of technological-organisational development and market dynamics', in: *The Proceedings of 6th IPS Research Seminar*, 23-25 March, Fuglsø, Denmark.

Sun, H.; and Frick, J. (1995) *Paradigm Shifts in Manufacturing, Implications for Technology Management*, Ashton, Birmingham, UK.

Twigg, D. and Voss, C. (1991) 'Integrating technology and organisation in the implementation of CAD/CAM', in *Proceedings of The 3rd International Production Management Conference on Management and New Production Systems*, Gothenburg, Sweden.

Voss, C.A. (1988) 'Success and failure in advanced manufacturing technology', *International Journal of Technology Management*, **3**, 3, 285-297.
Yin, R.K. (1989) *Case Study Research - Design and Methods*, SAGE Publications.

PART IV - NEW IDEAS IN MANUFACTURING STRATEGY

CHAPTER 21

NEW IDEAS IN MANUFACTURING STRATEGY

21.1. Introduction

Through examining data from a wide range of countries we are able to develop new ideas on manufacturing strategy. In the remaining chapters, various contributors use the IMSS data and perspectives from different countries to explore manufacturing strategy. The diversity of the sample has thrown up many new ideas.

21.2. Multi-focusedness

The current thinking in manufacturing strategy has been developed very much in the context of large industrialised economies, often with limited levels of import and export relative to the total economy. As shown in Chapter 16, economies such as the US rely predominately on large, often homogenous, domestic markets, where order winners are well-defined. However, this is not the situation in many smaller economies around the world. Countries such as Italy and Denmark have both small and fragmented markets, that also rely heavily on exports.

One result seems to be the development of multi-focusedness, a concept that is proposed in Chapter 26. Multi-focusedness is *not* the strategic flexibility described by Roth as where firms develop the capability to switch between one focus and set of order winners to another. But, rather, it is the ability to respond simultaneously to the demands of different countries and markets. As the authors define it, multi-focusedness drives companies to pursue a number of different objectives, traditionally regarded as antithetical, simultaneously, rather than focusing on specific objectives considered mutually exclusive.

This challenges some of the established thinking of manufacturing strategists such as Skinner, Hill, and others. This is an area that deserves considerable future examination.

First, we need to explore further to compare focused and multi-focused companies to determine whether multi-focusedness is a viable strategy. Second, the factors that lead to its successful use need exploring. We have already implied that the nature of the domestic and international market in some countries could be a possible driver of this phenomenon. A second driver could be the presence of a healthy small firm sector

365

P. Lindberg et al. (eds.), International Manufacturing Strategies, 365-368.
© 1998 *Kluwer Academic Publishers. Printed in the Netherlands.*

found in some countries such as Italy. Firms in such sectors may be multi-focused in their own right; or may play an important role through the supply chain in enabling large firms to become multi-focused.

Finally, if this is indeed a different form of manufacturing strategy, what are its implications for firms in large economies such as the US and Germany?

21.3. Beyond the cell - networking, integration and process ownership

The concept and benefits of cellular manufacture are well established, and all countries participating in the IMSS were using cellular concepts to a greater or lesser extent. However the Finnish team have sought to place cellular manufacture in a broader context, drawing on a separate study of manufacturing firms in Finland. Taking as their base the traditional engineering view of the cell, they examine how over time it has evolved into what they call the network cell and the network factory. The context of the smaller country is apparent again in Finland. The authors argue in Chapter 22 that the predominantly one-off nature of manufacture in engineering companies in Finland has led to a context in which cellular manufacture has been expanded in context and in scope.

The study of Finnish factories showed that process management was part of the route to the network factory. The findings from Finland are remarkably consistent with findings from the Italian and Dutch team in Chapter 26. The latter have developed a 'new manufacturing paradigm' which brings together process-ownership and integration with multi-focusedness.

This 'new manufacturing paradigm' and the network factory would seem to be the way in which new practices originating in global manufacturing have been developed and adapted to meet the internal and external contexts of firms in smaller countries who compete in world markets.

21.4. Quality and Technology

There has been much debate about the effectiveness of Total Quality Management and its impact on the bottom line. Some of the difficulty has probably arisen because TQM has been seen to be universally good, and the interaction with other key manufacturing strategy variables has not been considered. Although both quality and technology appear in most models of manufacturing technology, to date little thought has gone into possible connections between the two. In Chapter 23 John Ettlie, using the Abernathy-Utterback model of evolution of technology, explores this issue through analysis of IMSS data.

Having divided the firms into two groups—high and low technology—he tests hypotheses concerning performance, technology and TQM. The results indicate that

High-tech firms had significantly higher market share when they invested in R&D, Low-tech firms had better market share when they invested in R&D.

These findings are both theoretically sound and interesting. They indicate how we might use studies such as the IMSS to explore in greater depth the relationships between the many content variables of manufacturing strategy and performance.

21.5. Improvement programmes, Strategy and Performance

Voss has highlighted the relationship between the traditional contingency view of manufacturing strategy and the use of improvement programmes. The Korean team have used the IMSS data, not just to review the practices in Korea, but to examine this relationship. In Chapter 25, they examine both which practices and which contextual factors discriminate between high and low performing groups. They find that thirteen out of the twenty seven productivity-improving practices appear to significantly contribute to achieving one or more of six strategic goals. Of these, zero defects seemed to be the most important. In terms of context, diversification in terms of markets, customers and process technology seemed to be a dominant feature of the high performance group. Whilst this analysis was confined to the Korean sample, it indicates the questions that should be asked concerning the relationships between strategy and improvement programmes.

21.6. Emerging theories

The development of new approaches and ideas in manufacturing continues unabated. In Chapter 24, John Ettlie reviews the new ideas that are appearing both in academia and in industry and their relevance to manufacturing strategy. He examines two in detail—alliances and global networks, and resource based theory. In addition he reviews business process re-engineering and benchmarking.

The resource based view which has developed in strategy is a particularly high potential for application in manufacturing strategy, and Ettlie reviews it in this context as well as in the context of another new area: Mass-Customization.

21.7. Summary

The International Manufacturing Strategy Survey was conducted in a wide variety of countries from large to small, from sophisticated to those newly exposed to international competition. Although most of our understanding and the underlying theories of manufacturing strategy come from larger nations, the variety of the sample has enabled the various teams in the study to propose and develop new paradigms of manufacturing and strategy. Some of these such as multi-focusedness, are contrary to the established wisdom of manufacturing strategy. Others, such as the network factory, may more likely be adaptations of existing strategies and practices to fit specific local needs.

The new ideas put forward here present challenges both for companies and academics to develop and test further and hold out the possibility of adapting manufacturing strategy practice and theory to match different conditions in different countries.

CHAPTER 22

THE NETWORK CELL AS A STEP TO THE NETWORK FACTORY

Raimo Hyötyläinen and Magnus Simons, VTT Automation, Industrial Automation, Espoo, Finland

22.1. Introduction

Historically, the organisational structure of the company has been based on a functional hierarchy based on the principles of mass production. Companies are formed of a number of hierarchical levels, and the organisation is clearly divided into separate departments such as sales and marketing, product design, process planning, purchasing, production, and after-sales service. These in turn are divided into specialised sub-functions. This structure is based on a strict division of labour and on a simple and restricted set of information flows (Rummler and Brache, 1990).

Due to the pressures to shorten lead times, to customise products, and from the increasing technical complexity of products and production, there is an increasing need for communications, co-operation, and co-ordination between functions. To cope with these pressures, new kinds of relations between functions have emerged in companies (see Figure 22-1). In this new situation, companies have tried to create an environment for direct communications between functions (bold arrows), but it is mainly *ad hoc,* informal, and lacks systematic development (dashed arrows).

Traditionally, functional managers have attended to matters and problems occurring between organisational functions. As the need for co-operation between functions has begun to exceed the capacity of managers, the only way to work is to involve workers across functions. Co-operation between functions is no longer restricted to mere exchange of data or information; it also includes changes in the division of labour. Tasks typically performed in one function can now be *delegated* to another function when needed. In this way, capacity bottlenecks in one function can be relieved through help from skilled workers in another. This kind of co-operation has been found in Finnish companies between areas such as product design and production planning, production planning and production, product design and production, etc. (Simons and Hyötyläinen, 1995a, 1995b).

However, functional divisions still persist in the way companies operate and in their ways of thinking. The change to closer connections within the organisation has not fundamentally changed the functional division of tasks nor the way of thinking in the company, but has mainly affected management's role. The frequency of

P. Lindberg et al. (eds.), International Manufacturing Strategies, 369-384.
© *1998 Kluwer Academic Publishers. Printed in the Netherlands.*

Figure 22-1. The New Communications Paradigm

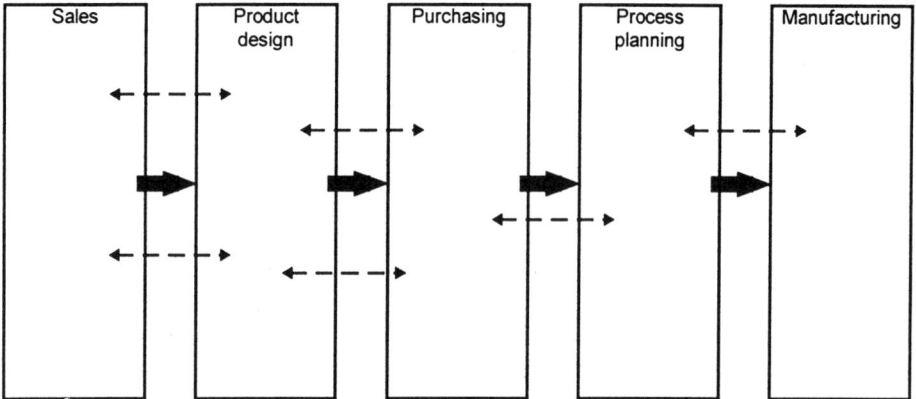

communications between functions, however, has been drastically increased and is now performed by the workers. It has become two-way communication, with feedback between functions faster and more direct. The problem with this new way of working is the lack of planned action. To a large extent, the communication is *ad hoc*, informal, and depends on the voluntary involvement of people. It lacks the necessary tools and methods for systematic and efficient co-operation.

The changes in the relationships between different functional areas have not touched only manufacturing tasks and production organisation. Inside the manufacturing function, a traditional hierarchical structure still prevails, especially in the machine shop type of organisation, where an organisational model that we call *craft-rationalised* has formed (Alasoini et al., 1994, pp. 5-6). It is 'craft' because it has been built largely on the manual skills of production workers, and 'rationalised' because it has been an effort towards the clear specialisation of tasks among and between blue- and white-collar workers. This kind of production model is best suited to low-volume production where the company operates in a relatively stable production environment.

However, the production environment faced by most companies is no longer stable, but changes rapidly all the time. Changes in product markets and the rapid development of technology have forced companies to cast about in haste for new concepts of production through which to improve their competitiveness and flexibility. Two cornerstones of striving for new production forms are cellular systems and teamwork. Debates about and experiments with production cells have already been going on for decades. However, in the last few years the need to reorganise production has gained a new momentum. At the same time, there is an urgent need to find new solutions and models. This has emphasised more clearly than before the necessity of new kinds of co-operative and *networking modes of operation*. One reason for this is the world-wide debate on *lean production* (Womack et al., 1990).

It is evident that industry is at a cross-roads with regards to this new mode of operation. This raises three questions. First, what kind of production cell is appropriate in these new conditions? One has to take into account the need for network structure in the organisation and *continuous development activity*. We will present the concept of the *network cell*, which we have created jointly with several Finnish companies (Alasoini et al., 1995). Second, to what extent and in which forms are companies adopting new forms of cell structures? There are only a few studies on this matter. We have analysed 26 company cases and present the main results of our study. Third, what are the possibilities of renovating the factory organisation at large? New network structures for production will have twofold effects. On the one hand, a network cell cannot operate without corresponding changes to the rest of the organisation. On the other hand, new production forms create new opportunities to change the hierarchical structures and division of work at the factory level. We will present the preliminary results of an analysis from an ongoing study of Finnish companies.

22.2. Towards the Network Cell

The production cell forms the basis for building organisational structures in this new mode of operation. Traditional cell concepts have emphasised autonomous cells, in which multi-skilled workers operate relatively independently in their own work area, and stand apart from the activities of the rest of the organisation (Sandberg, 1982). This kind of cell is built according to the principles of socio-technical systems design, which is aimed at improving the quality of work and at defining 'whole tasks'. This is also meant to enrich the jobs of workers by including different planning tasks such as quality and materials management. The socio-technical model was born in reaction to the hierarchical principles of the Tayloristic tradition (van Eijnatten, 1993).

This kind of a cell model does not, however, meet the demands for efficient operation and continuous development activity posed by the new mode of operation well. The socio-technical cell as such does not require changes in the divisional structure of an organisation, and it does not guarantee success in today's changing business environment. The first criterion for high-functioning cell organisation is that it integrate well with the tasks of production control and targets. The second criterion is that cells, by acting in co-operation with other cells and supporting functional areas, are able to develop the product and operations. It is needed to support building a 'learning organisation' (Leonard-Barton, 1992; Garvin, 1993).

In contrast to the socio-technical cell model, we introduce a new kind of cell concept that definitely alters the traditional, craft-rationalised way of working and organising. We call this new model a *network cell* (Alasoini et al., 1995). The name comes from the fact that the cell can be seen not only as a control unit, but also as an organisational unit, which besides its control task functions as a part of the collaboration network in the factory. From the perspective of the network cell, its nearest co-operative partners are the supervisors, manufacturing chief, maintenance staff, product designers, materials management personnel, customer service, and other production cells. A model of the network cell is presented in Figure 22-2 (Alasoini et al., 1995).

The network cell has three dimensions:

1. *Production activities.* In the network cell, workers change tasks within a job rotation system. Workers are multi-skilled or are able to attend to several machines and work stations. The work is by its nature group work.

2. *Support activities.* The cell acts partly by itself but mainly in co-operation with its support functions for support activities in its area. The central support activity is manufacturing control. It is responsible for the schedule in its areas within an overall schedule. The cell takes care of many maintenance activities, including pre-maintenance, minor repairs, and general work arrangements. The cell is responsible for quality and quality control in its area. In the domain of product design, the cell takes

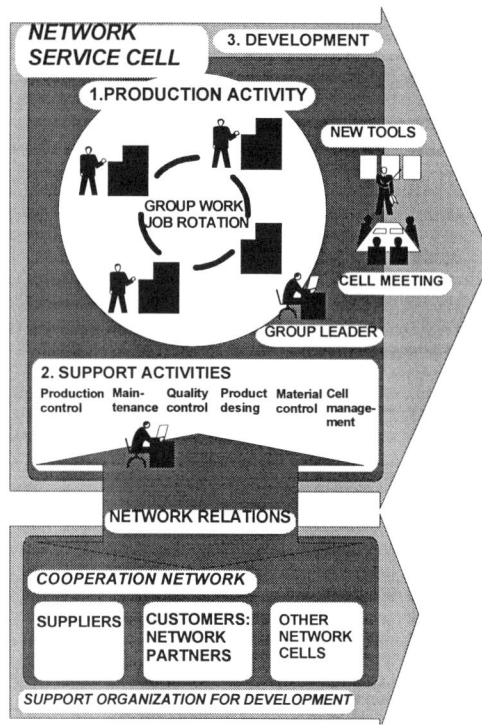

Figure 22-2. The Network Cell Model

part in developing the manufacturability of product and product concept together with product designers. The cell also attends to materials management and follow-up within its working area.

3. *Improvement activity.* The cell systematically follows up problems occurring in its products and processes. This is the basis for continuous improvement work within the cell.

The cell has its own internal rules for cell management and has agreed-upon procedures for the rest of the organisation. Cell meetings are held at regular intervals, and the group leader is responsible for, besides operational tasks, many support activities in manufacturing control, work arrangement, and chairing cell meetings.

With regards to the new kinds of activity within the cell, *new tools* are of the highest importance. These new tools are needed, on the one hand, for normal activities within the cell and in conducting support activities and, on the other hand, in improvement activities. The first set of tools includes a multi-skills table, a training plan, schedule plans, tools for quality and productivity follow-up, action models, and a cell board. The second set of tools includes a follow-up book, a table for development measures, and systematic models and methods for problem solving.

A network cell does not act separately from the other activities of the organisation. The network cell model does not emphasise the autonomy of the cell in the same sense that the socio-technical cell model has done. Form the point of view of the activity of the network cell, *organisational network relations* are essential. The importance of network relations becomes especially evident in carrying out support activities because a network cell has to continually co-operate with support staff in order to perform its support activities,. The need for co-operation increases further as production becomes faster and more customer focused. A network cell also needs network relations in its improvement activities. First, a network cell often needs the help of support staff in its problem-solving activities. Second, many production problems have their roots elsewhere in the organisation. For example, a problem occurring in production can necessitate changes in the product concept and in product drawings or materials. Materials problems can result from the need to change a supplier's production process.

22.3. The Adoption of the Cell Model in Finnish Enterprises

The central elements in the development of production in Finnish engineering companies were Just-in-Time (JIT) and the implementation of production automation in the 1980s. At the same time, companies organised their factories with the help of group technology into product shops and further into larger product factories. As a result, companies were able to streamline production and to shorten throughput time. Cellular production has been well-suited to the streamlined manufacturing layout (Buchanan, 1994). In this structure, cells have mainly been traditional control units in which workers have limited autonomy in responding to work arrangements in their work station.

During the 1990s, manufacturing companies have paid a great deal of attention to changes in their activities to attain manufacturing improvements. Companies have been seeking new strategies and models of operation to improvement their competitiveness (Kiviniitty et al., 1994), as they have been faced with increasing demands from customers and pressures from the rapidly changing business environment. The ongoing debate about new production concepts has also spurred enterprises to increase their rejuvenation of activities. Here, the debate on lean production has reached a special position. The traditional cell model and development work based on it have turned out to be insufficient for breaking hierarchical and functional models of organisation.

It is interesting to see how Finnish companies have started to change their mode of operation. What role does the cell concept have in this activity? And which cell models are Finnish companies adopting? Here, the model of the network cell can be used as a reference system for assessing company practices. The problem is, however, that there are only a few systematic studies of companies that have moved to cellular production and their change processes.

We have collected descriptions of 26 companies that have adopted cell principles in their production processes during the past few years (Alasoini et al., 1995; Hyötyläinen, 1994). Some of these cases have been reported in published studies. Part of the descriptions are based on reported cases in books, seminar paper, company papers, and articles in professional journals.

A summary of the collected cases is presented in Table 22-1. These cases cover nearly 200 cells, which of course is only a part of the cells implemented in Finnish engineering companies during the 1990s. Although the collected material is incomplete, there is good reason to assume that most of advanced Finnish firms are represented in the sample, and that adding new cases would not significantly extend the spectrum of cases presented in this table. In fact, many of the cases presented here are generally acknowledged as 'best practice' examples of the ongoing improvement in Finnish industry.

In nearly all of these cases, the cells have been implemented as part of a larger production change. A typical situation is the move from a functional layout to streamlined production, in the form of product shops and product factories. Another feature has been the renewal of the principles of production control in alignment with JIT principles. Other typical features include the building of a new factory, the implementation of automated systems, and the production of a new product. In many cases, this kind of development activity had already begun in the mid-1980s or soon after. However, the cases in Table 22-1 represent the adoption of cellular manufacturing in the 1990s.

We have characterised the case examples in Table 22-1 along five dimensions of activity. These dimensions are as follows:

Table 22-1. Finnish examples of cellular manufacturing during the 1990s

Case		Dimensions of activity				
		1	2	3	4	5
1	Steel foundry	O	O			
2	Machine manufacturer	O	X			
3	Tool manufacturer	X	X			
4	Tap factory	X	X			
5	Electronics factory	X	X			
6	Big hammer factory	X	X			
7	Machine job shop	X	X			
8	Electrical machine manufacturer	X	X			
9	Cable machine manufacturer	X	X	O		
10	Valve manufacturer	X	X	X		O
11	Electronics factory	X	X	X		O
12	Diesel engine manufacturer	X	X	X	X	X
13	Elevator factory	X	X	O	X	
14	Instrument manufacturer	X	X		O	
15	Hydraulic machine manufacturer	X	X		X	
16	Drilling machine manufacturer	X	X	X	X	X
17	Car manufacturer	X	X	X	X	X
18	Machinery factory	X	X	O	X	X
19	Metal sheet factory	X	X	O	X	X
20	Automotive supplier	X	X		X	X
21	Lock factory	O	X	X		O
22	Exercise equipment manufacturer	X	X		X	O
22	Electronics manufacturer	X	X		X	X
24	Pump factory	X	X		X	X
25	Electrical transformer manufacturer	X	X		O	X
26	Power transmission factory				X	X

Notes:
1. Job enlargement; 2. Support activities; 3. Group work; 4. Network Activity; 5. Development Activity
X - is included in the tasks of the cell and is prevailing practice
O - is planned to be included in the tasks of the cell or is set as a goal

1. *Job enlargement.* Workers in a cell must master several operations. At the extreme, each worker can perform every operation, which makes possible extensive job rotation within the cell. This presents a question of horizontal job enlargement. A *traditional control cell* based on a streamlined production layout is confined to this type of job enlargement.

2. *Support activities.* Different kinds of support activities can be included in the responsibilities of a cell. These include responsibilities for quality, local scheduling, disturbance handling, maintenance, and work arrangements. These support activities by their nature are planning and control tasks. In the socio-technical tradition, including these tasks in the duty of workers is classified as job enrichment. In this context, it is also common to speak about socio-technical and semi-autonomous cells (van Eijnatten, 1993).

3. *Group work.* Job rotation and job enrichment *per se* do not require teamwork. Group work means that workers within the cell are working as a tight group towards common goals. The socio-technical tradition also emphasises team work as a means to increase workers' motivation and the quality of work life (van Eijnatten, 1993).

4. *Network activity.* The network cell is part of the total collaboration network in a product shop or product factory. This requires tools and methods common to the cell and the support functions. These tools can be for example computer displays, control tools, and different follow-up tools.

5. *Continuous improvement activities.* The central dimension of activity in a network cell is continuous improvement of work methods, processes, and products. This requires the enlargement of the skills of cell workers in the improvement dimension. Network relations and common tools and methods are also necessary for the success of improvement activities.

To be considered a network cell, the cell must have group work, network activity, and development activity as well as job enlargement and support activities. The dimensions of network activity and continuous improvement activity distinguish the network cell from the socio-technical cell because the socio-technical cell does not pay any attention to these matters.

When indications of one or more dimensions of activity were found in the case examples, we have also tried to assess whether that it is a question of prevailing practice or only of a plan or target. The results are presented in Table 22-1.

The 26 case examples can be classified into two groups:

1. *Traditional cell* (control cell or socio-technical cell). Ten cases (cases 1-9 and 14) belong to this group. Among these cases, group work is, however, planned or set as a target in one case (case 9) and network activities (case 14). As the cases are analysed more closely, it is noted that in fifteen cases a traditional cell has been the starting point, but elements of the network cell have been added later. These fifteen cases will be considered below.

    Traditional control cells based on the principles of streamlined production layout have been the starting point in eight cases (cases 1-4, 14, 18, 22, and 25). In other words, nearly one-third of all cases had restricted their cell to only the principles of job enlargement at the beginning of the 1990s. The practices in some of these cases have changed later on, in particular through sharing the responsibility for quality to cells. This is a clear indication that Finnish companies have paid a great deal of attention to quality and especially on attaining ISO 9000 quality system certification during the 1990s.

It can be seen in seven cases (cases 5-11) that elements of the socio-technical cell have been used as a starting point. In these cases, the autonomy of the cell and job enrichment with planning and control tasks have been emphasised. In many cases, the cells themselves, rather than the traditional supervisor, are responsible for scheduling. At the same time, the cell has been allocated many other support activities such as the responsibility for quality, disturbance handling, maintenance, and management.

2. *Network cell.* There are elements of the network cell in sixteen case examples (cases 10-13 and 15-29). On the other hand, these elements are incomplete in three cases (cases 10-11 and 21), consisting of only group work and continuous improvement activity planned or set as a target.

There are fourteen mentions of *network activity*, of which two mentions represent a planned use or goal. All fourteen mentions can be divided into three groups. First, a cell is directly in contact with other functional areas, suppliers in carrying out its manufacturing tasks and support activities.

There are also fourteen mentions in these cases in the dimension of development activity. Of these cases, four are still at the level of planned use or a goal. All these cases can also be divided into three groups. The first group has project-like development groups in the area of a product job or a product factory. These groups solve production problems. There are also some production workers within these groups. Second, these groups have set as a target continuous improvement as a part of cell activity without mentioning concrete methods and tools. Third, development work is connected into the quality system and quality improvement. Another goal is to get cells and workers involved in continuous improvement. Some methods for this are training in quality and problem solving, development groups, quality monitoring, follow-up reports, and meetings. The cases in which continuous improvement activity was assigned to a cell's tasks are closest to the network cell model. There are five cases of this kind (cases 12, 15, 17, 19, and 23) in Table 22-1. Otherwise, the factories have used multi-functional groups set up on the basis of the product line.

In summary, we emphasise three points. First, different kinds of cells have been set up during the past few years in Finnish engineering companies. There are remarkable differences in the scope of the activities that have been assigned to these cells. In many cases, the cell's tasks only cover job enlargement and some elements of job enrichment. Second, we can state that many case examples mention network activity and continuous improvement, which represents a new phase in the development of cellular manufacturing. These new dimensions have been implemented without systematic methods and tools. Third, in those eight cases (cases 10, 13, 14, 18, 22, 23, 24, and 25) in which temporally different phases in the implementation and development of the cells within a company have been described, there has been a clear movement from the traditional control cell to the network cell. For example, in case 22 the first cells were implemented as traditional control cells. Next, cells were implemented as autonomous socio-technical cells. In the most recent cells, network activities have been adopted, and there are plans to also implement continuous improvement activities. This trajectory can be interpreted as part of a learning process occurring in organisations. It is obvious that the limitations of the traditional cell models have become obstacles to the further development of production systems. The ongoing debate on new production models has also influenced the adoption of new solutions. In some cases, consultants and researchers have served as conduits for bringing new ideas into firms.

22.4. Process Management: The Route to the Network Factory

As we have seen, there are different production concepts. Finnish companies are seeking new production solutions. The mass production paradigm, characterised by hierarchy and departmentalised structures, is well known. The socio-technical approach has been tried through implementation in different companies, and, as we see it, has not had good results (cf. van Eijnatten, 1993). Many companies are moving from isolated and 'autonomous' structures towards network cells and network relationships, as seen above. An emerging paradigm of lean production offers good theoretical guidance and practical tools for this renewal process (Womack et al., 1990).

Notwithstanding, the organisational structure of the factory has remained largely unaltered. Functional hierarchy in alignment with mass production still prevails in a little modified form, as seen above. The production organisation has been based on this departmentalised and hierarchical structure. Within this structure, it has been natural to organise manufacturing according to the craft-rationalised model (Alasoini et al., 1994; see also above). When manufacturing operations are being organised according to the lines of the network cell, the functional and hierarchical structure at the plant level becomes a barrier to the further improvement of operations.

Functionally-divided organisations have been criticised for 'throwing tasks over the transom'—that is, communication between functional areas is only one-way and thin (Rummler and Brache, 1990; Davenport, 1993). As we have seen above, companies have tried to address this problem through inter-functional feedback and feed-forward between parallel tasks. The problem that remains is that company activity as a whole is being sub-optimised because there is no systematic development of the complete system of activities producing customer value.

New models are needed to overcome the problems being experienced in companies moving to the practices characterised by the network cell. A promising new approach is business process development and management (Rummler and Brache, 1990). We have developed and used a business process approach for integrating company functions, and have applied it in Finnish companies (Simons & Hyötyläinen, 1995). The business process approach is based on analysing the whole company and developing solutions based on the business needs of the organisation.

To avoid the problem of sub-optimisation, the whole production system and all of its activities must be examined. To accomplish this, we have chosen a three-level business process analysis as the basis of the method (Simons & Hyötyläinen, 1995a, 1995b). Business processes are studied at the company level, as multi-functional business processes, and as activities performed within the function. These business processes are then compared to the company's business objectives.

In the literature, a great variety of approaches to improving business processes can be found (e.g., Rummler and Brache, 1990; Davenport, 1993). The critical business process to study is the customer order delivery process, which is also closely connected to product development, especially in the one-off production typical of Finnish engineering companies (Simons & Hyötyläinen, 1995a, 1995b). The customer order delivery process involves all functional areas, and affects all of the business objectives. The delivery process consists of the following sub-processes (see Figure 22-3):

- defining customer needs
- product design
- product development
- process and production planning
- materials management
- production
- accounting

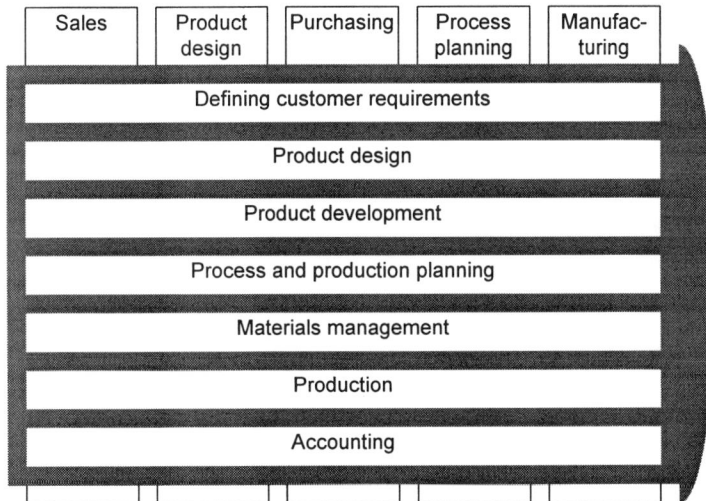

Figure 22-3. The Customer Order Delivery Process and Sub-Processes

These in turn can be further divided into sub-sub-processes and tasks. The sub-processes listed above are typical of most manufacturing companies, especially one-off production. In general, however, they are not considered to be cross-functional. Adopting the business process approach means a total change in the view of the organisation (Hammer and Campy, 1993). One can even say that in a sense the organisation is turned upside-down.

In the complicated new and continuously changing business environment, a major competitive advantage for the company is the ability to perform continuous improvement of products and production processes. The lean production concept also strongly emphasises this (Womack et al., 1990). However, it is easier to say than to put into process. There is a lot to do before companies can properly implement continuous improvement methods into their normal operations. Why is it so difficult? We have successfully developed new kinds of problem-solving methods and organisational practices in companies (Alasoini et al., 1994). In principle, these methods are very simple and they can be used by everyone in the organisation, and become common tools for the problem-solving process based on co-operation between different functional areas.

It can be said that the identification of business processes and organising for the cellular way of working both create new opportunities for organising ongoing improvement activities in companies. Another central goal of business process renewal is continuous improvement (Simons & Hyötyläinen, 1995a, 1995b). To do this, sub-processes have to be analysed more closely (Figure 22-4).

It is also important to involve people carrying out the tasks and operation in the sub-process in the re-engineering of the sub-process. In this way, it is possible to create

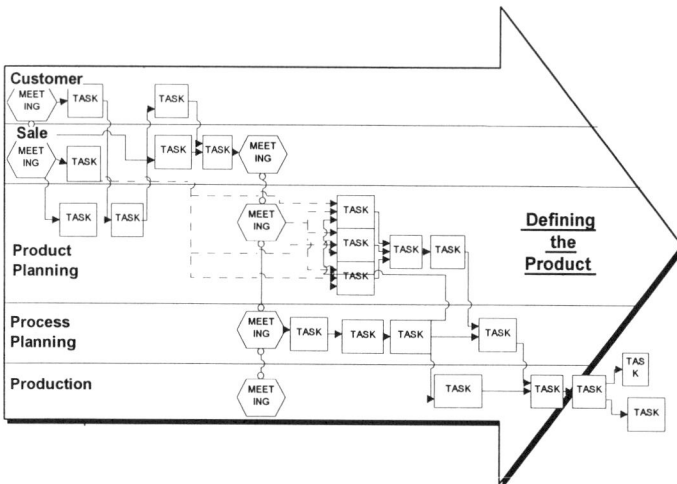

Figure 22-4. The Process Model of the Business Process for Defining the Product (Simons & Hyötyläinen, 1995a, 1995b)

both team and network structures at the same time for future operations. This requires cross-functional team-work (Simons & Hyötyläinen, 1995a). Cross-functional teamwork was chosen as the method for organising the re-engineering for three reasons. First, a cross-functional team approach ensures that all functional aspects are considered. Second, the team approach brings the people involved in the development process closer together, forcing them to confront each other and to reach consensus in difficult matters. The business process analysis helps the team focus on process performance and business objectives rather than people and mistakes. Third, the team approach involves the people performing the process tasks in the development of new means and methods, i.e. the goal is to use the total organisational potential to analyse, plan, and implement new solutions to problems in company activities.

Our aim is to make business processes, including the development process, problems and business objectives explicit for all the organisation. As it is, a central problem in the case companies is that people in different functional areas view company activities and problems only from their own perspective, and the problems they experience in their work as being caused by other functional areas or departments. Trying to solve a problem from this starting point only produces confusion and frustration. For instance, one company studied had a set of so-called 'eternal' problems, which were familiar to most employees and to which many had their own solutions, but still these problems were never properly solved, because no attempt to solve the problem was ever sufficient to cover all the necessary angles of the problem (Toikka et al., 1995).

22.5. Concluding Remarks

In the one-off environment typical of Finnish engineering companies, it has been hard to apply the principles of mass production fully. On the one hand, due to the need to produce high volumes and wide variety to customer demands, workers need to have high skill levels. Production under these circumstances is based on a functional layout where skilled workers have mastered several operations and many tasks. On the other hand, engineering work is emphasised in customised production due to the need to make individual changes to products (Simons & Hyötyläinen, 1995a, 1995b). Traditionally, engineering and design work are done in a craftsman-like style. In these circumstances, the combination of the work and the organisational style we call the 'craft-rationalised' model is pervasive. In principle, this model is based on the two production paradigms of mass production and craftsman-like production. In one-off production, the two paradigms have, however, merged into a single model with its own characteristics. This model changes all the time due to pressures from the market. The basic question becomes: is it possible to add new characteristics to the prevailing model and to improve its function? Or, is there something wrong with this model? If so, where should we seek for a new model?

We assume that the lean production model can offer a lot of help, but it will not be straight-forward. A new way of thinking will be needed. What must be taken into account is that both mass production and lean production were originally developed in an environment of high-volume production, in fact in the automobile industry. Firms formerly organised around mass production are now adopting new concepts, becoming more flexible and manufacturing new products at increasing speeds. This evolution from mass production to flexible production has been termed 'mass customisation' (Pine et al., 1993). Firms are becoming much leaner and more flexible.

The special needs of one-off production and the producers of customised products have long been neglected by management science. This is urgently felt in Finnish manufacturing industry, and especially in the engineering sector, where production is to a large extent based on low volumes and specialised products. Common features of Finnish engineering companies are a business-to-business environment with large-scale and complex products such as papermaking machinery.

What kind of production model could be an alternative to the existing practice of one-off production? We have developed the concept of the network cell and the network factory. We assume that the concepts might offer a good starting point for future production practice. They might also provide solutions for companies producing at higher volumes than most Finnish engineering companies.

The network factory concept is aimed towards solving the problems inherent in the 'craft-rationalised' model, which can only be done through solving them in co-operation. The network factory model implies the formation of a new kind of co-operative relationship between functional areas and forming network connections between teams within different support functions and network cells at the operational level. The idea of the network cell is to emphasise the co-operation between the cell and different functional areas and support staff within the company. The network factory model consists of three parts, which are closely intertwined with each other. These parts are a team and cellular way of working; the formation and management of business processes, and the organisation of continuous development activity. We have seen from the case examples that many Finnish companies have chosen a development path along these lines in the 1990s.

From the point of view of future needs, the network factory concept alone is not sufficient. Companies must strive for closer contacts with their suppliers and customers in order to be competitive in the rapidly changing and complex environment. New concepts for one-off production are badly needed, so that companies can solve their complicated problems and take full advantage of their production model. We are outlining a new concept, the *network enterprise*. This is larger than the network factory concept, which concentrates mainly inside the factory walls (*cf.* Womack and Jones, 1994).

Here we stress only two points. The first one is the concept of the boundaryless enterprise. We can no longer consider enterprises as strictly bounded entities. Enterprises must co-operate with each other much more closely than in the past. This is a necessity for efficient operation in rapidly changing technological and competitive environment. At this level of co-operation, a basic tactic may be to seek win-win situations, rather than the old confrontative style.

A second point is to understand that organisations as such do not co-operate with each other. People as representatives of their own organisations make agreements on various matters with people from other organisations. This kind of thinking is no longer adequate in the networking world! It may be better to conceptualise the situation as the co-operation of different teams with each other to carry out common production goals. The relationships between the teams from different organisations can become the ground for 'knowledge-creating' structures in the future (cf. Nonaka, 1991). This kind of situation places hard demands on network structures, and methods and tools used in accomplishing common processes and tasks. We will surely need a new kind of strategy and methods development for future inter-company networks.

References

Alasoini, T., Hyötyläinen, R., Kasvio, A., Kiviniitty, J., Klemola, S., Ruuhilehto, K., Seppälä, P., Toikka, K., and Tuominen, E. (1994) 'Manufacturing change: interdisciplinary research on new modes of operation in Finnish industry', Working Paper No. 48, University of Tampere, Work Research Centre.

Alasoini, T., Hyötyläinen, R., Klemola, S., Seppälä, P., Toikka, K., and Kiviniitty, J. (1995) *Building the Network Cell,* The Union of Finnish Metal Industry, Helsinki. (to be published).

Buchanan, D. (1994) 'Cellular Manufacturing and the Role of Teams', in J. Storey, (ed.) *New Wave Manufacturing Strategies: Organisational and Human Resource Management Dimensions,* Paul Chapman Publishing, London, 204-225.

Davenport, T.H. (1993) *Process Innovation: Re-engineering Work Through Information Technology,* Harvard Business School Press, Boston, MA.

van Eijnatten, F.M. (1993) *The Paradigm that Changed the Workplace,* Assen - Swedish Center for Working Life - van Gorcum, Stockholm.

Garvin, D.A., (1993) 'Building the learning organisation', *Harvard Business Review,* July-August, 78-91.

Hammer, M. and Campy, J. (1993) *Re-engineering the Corporation: A Manifesto for Business Revolution,* HarperBusiness, New York, NY.

Hyötyläinen, R. (1994) 'The models of cell and the implementation of cells in the 1990s', Unpublished report, VTT Automation, 12/12/ 1994, Espoo, Finland, 41 pp.

Kiviniitty, J., Hyötyläinen, R., and Alasoini, T. (1994) 'Shift to adaptable production as a social and cultural process: the Finnish research programme "Work, culture, and technology' in an international comparison', In T. Kauppinen, and M. Lahtonen, (eds.), *Action Research in Finland,* Labour Policy Studies, **82**, Ministry of Labour, Helsinki, 275-302.

Leonard-Barton, D. (1992) 'The factory as a learning laboratory', *Sloan Management Review,* Fall, 23-38.

Nonaka, I. (1991) 'The knowledge-creating company', *Harvard Business Review,* November-December, 96-104.

Pine, J., Victor, B., and Boynton, A. (1993) 'Making mass customisation work', *Harvard Business Review,* September-October.

Rummler, B.A. and Brache, A.P. (1990) *Improving Performance: How to Manage the White Space on the Organisation Chart,* Jossey-Bass Publishers, San Francisco, CA.

Sandberg, T. (1992) *Work Organisation and Autonomous Groups.* LiberFörlag, Uppsala, Sweden.

Simons, M. and Hyötyläinen, R. (1995a) 'Adaptable manufacturing: a model for business process development', In: VTT Symposium: Product Models in Design and Production Planning, VTT, Technical Research Centre of Finland, Espoo, Finland. To be published.

Simons, M. and Hyötyläinen, R. (1995b) 'Adaptable manufacturing: the development of delivery process in one-of-a-kind manufacturing', Research Report 1636, VTT - Technical Research Centre of Finland, Espoo, Finland, in Finnish.

Toikka, K., Kiviniitty, J., Simons, M., Hyötyläinen, R., and Alasoini, T. (1995) 'Building continuous development activity', The Union of Finnish Metal Industry, Helsinki, to be published.

Womack, J.P., Jones, D.T., and Roos, D. (1990) *The Machine that Changed the World,* Rawson, New York.

CHAPTER 23

QUALITY, TECHNOLOGY, AND GLOBAL MANUFACTURING

John E. Ettlie, School of Business Administration, University of Michigan

23.1. Introduction

In spite of early contributions (Juran, 1951; Feigenbaum, 1956; Deming, 1950; Dodge, 1969), the quality movement has only been ablaze for slightly more than ten years in the United States (e.g., Crosby, 1979; Deming, 1986). Results have been mixed. Total Quality Management (TQM) has been practised since the 1980s in the US (Dean and Evans, 1994). Although quality levels have improved in selected industries such as automobiles, customer satisfaction is still higher with Japanese and European cars (Rechtin, 1994).

Several recent surveys summarised by Buran (1994) indicate widespread dissatisfaction with the results of US quality initiatives. Over 50% of companies report that quality programmes have not led to better business performance. Less than one-third of US Fortune 500 firms believe quality programmes have had a significant impact on competitiveness. Over 85% of ISO 9000 registrants think that it will take eight years or more to recover their costs.

Quality programmes appear to have failed to meet expectations in two-third of US firms primarily because they have not been related to customer outcomes (Buran, 1994). Only a small number of companies qualify for the Malcolm Baldrige National Quality Award, but those that do tend to be able to forge the bond between customer orientation and operational performance. Areas of persistent weakness in Baldrige applications include an unclear linkage between quality and strategy; lack of data and analysis; and partial systems that do not integrate information technology and quality (Reiman, 1993).

Most R&D has historically been focused on the development of new products, and this trend continues. In a survey of IRI members, Wolff (1994) reports that the proportion of R&D spending for new products increased from 39% to 44% during the four years from 1988 to 1992. There is also evidence that more resources are being expended towards new product development in Japan (Hamilton, 1993). R&D investments have been shown to pay off in a majority of industries, and especially in pharmaceuticals, consumer products, chemicals, and services (Waddock and Graves, 1994). Therefore, it is not surprising that there has been considerable attention paid to

This chapter is also published as an article in *Production and Operations Management*, vol. 6, 1997, and appears here with permission.

P. Lindberg et al. (eds.), International Manufacturing Strategies, 385-400.
© 1998 *Kluwer Academic Publishers. Printed in the Netherlands.*

strategies for successful new product development and introduction (Souder and Sherman, 1994). However, the increase in service R&D (Wolff, 1994) also suggests that companies are introducing products faster, but at the expense of quality, which has to be improved after a new product is introduced.

In this study global manufacturing investment patterns in R&D and in new plant and equipment were evaluated. Results from 600 durable goods firms in 20 countries indicated that industry significantly moderated the association of R&D intensity and TQM with market share. In high-technology firms, R&D intensity was significantly associated with market share; in low-technology firms, TQM was significantly associated with market share. R&D intensity was significantly and inversely related to TQM. However, TQM and CAM (Computer-Aided Manufacturing) were significantly and positively correlated, as predicted. Regional differences indicated that European and Scandinavian firms tended to have lower market share than North American, Asian or South American companies.

23.1.1. THEORIES OF QUALITY

In spite of the great movement towards quality programmes around the work, beginning with the quality circles in Japan and in the United States 20 years ago or more, a widely accepted theory of quality and total quality management has not emerged (Dean and Bowen, 1994). There are any number of possible reasons for this state of affairs. There are at least two general approaches to quality issues in organisations, a situation that promotes confusion immediately.

Total quality and total quality management are often claimed to have their well-springs as far back as the scientific management movement started by Frederick Taylor, a mechanical engineer and the father of industrial engineering (Romm, 1994). His work in the early 1990s separated planning and execution of tasks, a separation that is essential to total quality philosophies (Goetsch and David, 1995). Typical works in this category are Dean and Evans (1994), Imai (1986), Flood (1993), and Garvin (1988).

A second widely accepted approach to quality is based on statistical principles, and is often called statistical quality control. Recent examples of this approach are Gitlow, Oppenheim and Oppenheim (1995), Bergman and Kleifsio (1994), and Farnum (1994). To add to the confusion, many authors, including Deming in his later work, merge the two traditions (e.g., Vroman and Lushsinger, 1994; Besterfield et al., 1995; Evans and Lindsay, 1993). This is probably because statisticians often begin with a strictly quantitatively-bounded discipline approach to quality and then discover that, at a minimum, the assumptions of these statistical models are not often satisfied (e.g., Taguchi's design of experiments).

The legacy of this historical development pattern of the quality movement has been a reign of confusion, made worse in some cultures, such as that of the US where the priority placed on problem solving has emphasised a "quick fix" to quality issues with little time for careful analysis. The methods of measuring the results of any intervention in organisations have also added to the problem: accounting has its own rules for measuring performance; manufacturing other rules; and so on. The clash between continuous, incremental improvement in operations as opposed to radical interventions such as business process re-engineering has also contributed. Romanelli and Tushman (1994) show that most fundamental change in organisations comes as part of a radical, punctuated equilibrium shift rather than through the accumulation of incremental changes.

One recent attempt to advance theory in the area was published by Sitkin, Sutcliffe, and Schroeder (1994). This is a contingency model that suggests that TQM and associated practices should be matched appropriately to situational requirements. The authors contrast a TQM approach, which emphasises control, with a TQL (Total Quality Learning) approach. High uncertainty conditions would favour a TQL approach. Both principles and practices differ under the two approaches. For example, in contrasting management practices for capability enhancement, the authors make the following distinctions (Sitkin, 1994, p. 548, excerpted from Table 2: Practices Associated with Total Quality Control and Total Quality Learning):

Management Practices	Total Quality Control	Total Quality Learning
Capability Enhancement	Enhanced exploitation of existing skills	Enhanced exploration of new skills
Increased efficiency in use of existing resources	Increased availability of slack resources	
Increased effectiveness in control over processes, products, and services	Increased effectiveness in learning and capacity enhancement	
Increase performance reliability	Increased resilience in the face of new and/or unexpected changes or requirements	
Doing things right the first time	Doing things that are likely to provide insight, but only have a moderate probability of succeeding	

This separation between TQM (control) and TQL (learning) approaches to quality has the potential for incorporating technology issues in a quality-performance model. New technology is generally required in uncertain environments, and a life cycle model, such as that discussed in the next section, would be consistent with this approach.

23.1.2. THE UTTERBACK-ABERNATHY MODEL

Abernathy and Townsend (1975) reviewed studies of product, process, and technological change, concluding that there were similarities in the patterns of development of productive segments. Although Abernathy and Townsend (1974) concentrated on the evolution of the productive segment and its relationship with innovative capability and productivity, in Utterback and Abernathy (1975) the firm's strategy for competition and growth is introduced more specifically, and the firm's propensity to host product or process innovations is discussed in the context of the evolutionary staging.

Abernathy and Townsend originally hypothesised that the productive segment of a firm "tends to evolve and change over time", according to a "predictable profile", and "that the state of development which a productive segment has reached along this profile will determine its propensity to host particular types of innovation" (1975, p. 381). The profile was hypothesised as being common for different industries and is derived in part from the premise that "the factors which critically enable innovation are best described as patterns of conditions rather than in terms of single important variables" (p. 381).

Abernathy and Townsend state that their unit of analysis is the technology user or the productive segment of the firm. They define the productive segment as the "overall production process which is employed to create a product, whether the product is goods or a service" and it includes the "physical product, the characteristics of input materials, and the characteristics of the product demand that are incident on the process". This definition was subsequently modified by Utterback and Abernathy (1975, p. 641) to include process equipment, work force, task specifications, and work and information flows.

Three states of development were identified as being common to the productive segments of firms regardless of industry: unconnected, segmental, and systemic. The definition and description of these stages are not included here because this model was modified and refined in the later article by Utterback and Abernathy.

In this later article, Utterback and Abernathy (1975) developed and tested an evolutionary model of true production process, which was defined as "the system of process equipment, workforce, task specifications, material inputs, work and information flows, etc. that are employed to produce a product or service" (1975, p. 641). "Production process" replaced the term "productive segment" originally used by Abernathy and Townsend (1974). Another term is also used, "process segment" (1975, p. 642), which appears to be most like the original definition of productive segment. These differences are summarised in Ettlie (1979).

Utterback and Abernathy (1975) originally proposed that successful firms tend to invest heavily in product R&D early in the life cycle of an industry or product group. As the dominant design of a new product emerges, investments switch to process technology and strategies switch to cost minimisation as opposed to product feature variety. The basis of competition varies with the stage of maturity of the product-process core of a firm and an industry.

Each stage now includes not only a description of the state of evolution of the production process, but also the dominant competitive strategy. These stages are summarised below, from Utterback and Abernathy (1975).

Stage 1. Uncoordinated production process and *product performance-maximising* strategy. The process is "composed largely of unstandardised and manual operations, or ... general purpose equipment" (1975, p. 641), and is relatively organic and flexible, responding easily to changes in an environment in which there is "great product diversity among competitors" (1975, p. 641). Although the process is initially inefficient, processes and products evolve rapidly toward improvement, with corresponding market expansion and redefinition. The competitive strategy is characterised by rapid product change emphasising product performance, and both product and process innovation respond to market need.

Stage II. Segmental production process and *sales-maximising strategy*. The process becomes more efficient, tasks more specialised and more integrated through automation, although some segments of the process remain essentially manual. The process is more rigid and further development is subject to maturing of a product group with increased sales. The competitive strategy is one of increasing visibility to the consumer; products become more varied and improved with new components at first, and them more standardised as market uncertainty is reduced. Most innovations are stimulated by technological opportunities.

Stage III. Systemic production process and *cost-minimising* strategy. The system production process is well-integrated and the most resistant to change of the three. It constitutes a major investment, and even minor changes have costly consequence. Therefore, process changes come only slowly. The primary competitive strategy focuses on reducing product price in the face of reduced margins on standardised products. Because specification of the production process is now easier, the process segment is likely to host innovations that will make the process more efficient, and therefore production and cost-related factors are likely to be the major stimuli for innovation.

In addition to refining the model with particular attention to predictions concerning the innovation process, Utterback and Abernathy (1975) used data from the study of successful industrial innovations reported by Myers and Marquis (1969) to support specific hypotheses derived from the model. Although limited to nominal data, the firm's stages of product and process development were compared using categories based on the nature of the innovation and other variables.

In this paper, a merging of quality and technology theory is proposed. Instead of studying quality as a separate issue in an organisation, which rarely obtains in practice, a technology life-cycle approach is used to examine the quality, technology, performance relationship consistent with the Sitkin et al. (1994) model. This is taken up next.

23.1.3. EXTENDING THE UTTERBACK-ABERNATHY MODEL

Although there are problems with the Utterback-Abernathy (U-A) model, e.g., contingencies required for successful performance can be explained independently of an evolutionary process (Ettlie, 1979), it does serve as a framework to compare the results of investments in manufacturing innovation. Therefore, the U-A model is explored as a way of reconciling the confusion about theories of quality.

In general, it is hypothesised that the productive segment moves at an unspecified, slow rate from a flexible, unstandardised, environmentally sensitive condition to a more rigid, integrated state that enjoys the benefits of high productivity but has a lower "innovative capability". The desirability of the conditions imposed by a particular stage of development depends on the environment of the productive segment. Development to the system stage appears to be appropriate in a stable environment, but if the environment then begins to change at a more rapid rate or becomes unstable (e.g., competition from innovative products), management has two options: either move the productive segment to a foreign country or "backtrack along the traditional course of evolutionary process development to a more flexible state" (Abernathy and Townsend, 1975, p. 892). Thus, it was illustrated that there is a trade-off between the productive segment's capability for innovation and productivity improvement. In addition, Abernathy (1976) presented an in-depth historical study of the Ford Motor Company that tends to support this model.

The de-maturation of durable goods manufacturing, and the emergence of economies of scope afforded by flexible manufacturing technologies (Ettlie and Penner-Hahn, 1994), offer a significant alternative to scale economies, requiring a rethink of earlier theories. Distinguishing between radical and incremental innovation (Ettlie et al., 1984) and punctuated equilibrium models (Anderson and Tushman, 1990) does not sufficiently account for this trend. This de-maturation was originally addressed by Abernathy and Townsend (1975, p. 392) with the inclusion of the atavistic tendency of the productive segment to "backtrack" from the system or last stage of development to earlier stages when the environment becomes less stable. The authors go on to say that at times the best choice may be to "slow or reverse evolutionary process or to remain in that particular stage which offers the best trade-off between conflicting objectives (of adaptability and innovativeness vs. higher productivity rates)" (Abernathy and Townsend, 1975, p. 395).

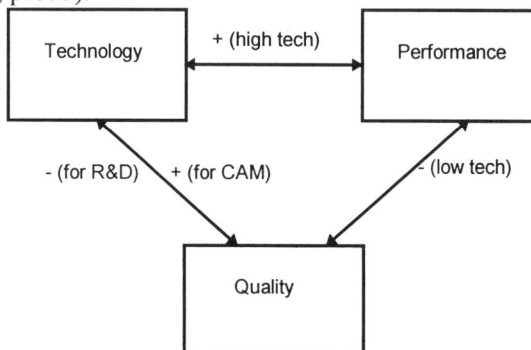

Figure 23-1. Quality, Technology and Global Manufacturing

In building on the systems-oriented view of successful companies (Liker, Ettlie and Ward, 1995), several avenues of hypothesis generation are possible. One parsimonious approach, which combines organisational learning perspectives and both resource-based theory (Peteraf, 1993; Wernerfelt, 1984) and loose coupling, is summarised by Cole (1994). In Cole's model, the combined emphasis of individual and organisational learning is predicted to be most successful. US companies have, up until recently, emphasised individual learning and seem to have mastered break-through innovation—especially in some high-technology industries—while Japanese manufacturing firms have mastered organisational learning and incremental innovation. Cole (1994) argues that successful global firms master both. Perhaps industry matters more than various theories (e.g., Imai, 1986; Scott, 1987) have taken into account. An alternative is summarised in Figure 23-1.

Using the original Utterback-Abernathy model as a springboard to reconcile the quality-technology explanations of performance variations in global manufacturing, three constructs are included in this model: technology, quality, and performance. Although the choice of performance measure will likely influence the outcomes of empirical testing (Cameron, 1986), propositions are developed below that take into account industry difference in making predictions between these three constructs. In general, the model predicts that in high-technology industries, the technology-performance connection is strongest, whilst in low-technology, mature industries the TQM-performance association is strongest. In general, it is predicted that firms will attempt to make their technology and quality strategies consistent. Therefore, R&D intensity will be inversely related to TQM whilst computer-aided manufacturing (CAM), which is introduced at more mature stages in the U-A model, will be positively related to TQM.

23.1.4. PROPOSITIONS

Several propositions can be derived from this model (Figure 23-1) of the circumstances under which technology, quality and performance are related. The Utterback-Abernathy model predicts that product and then process innovation to be greatest during the early stages of growth of an industry. Therefore, it would be expected that in high-technology industries, performance would depend more on R&D than on quality. In mature industries, the opposite would be true. Firms generally evolve from containment to preventative quality investments (Crosby, 1979; Deming, 1986; Imai, 1986). This is summarised by the first two hypotheses:

Hypothesis 1. In high-technology industries, R&D intensity and market share are significantly associated.

Hypothesis 2. In mature, low-technology industries, TQM (Total Quality Management) programmes are significantly correlated with market share.

The Utterback-Abernathy model does not inform directly on the industry-free relationship between quality and technology, but the general notion that consistency between corporate and functional strategies is associated with survival and growth in manufacturing is instructive in making predictions from Figure 23-1 (Hayes and Wheelwright, 1984). However, the measure of technology matters here; therefore, the relationship will vary by whether technology is measured by product R&D (R&D intensity) typical of stage I or process R&D (CAM) for stage II.

Hypothesis 3. TQM is significantly and inversely associated with R&D intensity.

Hypothesis 4. TQM is significantly and directly associated with process R&D (CAM).

These four hypotheses summarise the operationalised model presented in Figure 23-1 earlier, with market share taken as the dependent variable, and with two alternative models of technology: R&D intensity and CAM (Computer-Aided Manufacturing). Regional differences are also explored.

23.2. Methodology

The IMSS data were used to test the propositions in this research. The sample is summarised in the Appendix. Data collection from the US is typical of the 20 countries, where one principal investigator in each country was charged with data collection and follow-up for the study. The US participants were all durable goods manufacturers, with high added value shipments and with strong market positions in their respective industries. The response rate was 32%. Response bias by SIC code was checked for frequency of returns, resulting in an observed chi-square of 7.89 (d.f. = 9), which was not significant. This indicates that industry and propensity to return a questionnaire were not related in the US sample.

The US response rate of 32% was about the same as other surveys of this time (Tomaskovic-Devey Leiter and Thompson, 1994). The response rate for the total sample of 600 firms in 20 countries was 44.7%, and ranged from a high of 100% in Denmark to a low of 17% in Norway.

23.2.1. MEASURES

The mailed questionnaire included sections dealing with TQM, investments in automation (general levels and highest level), and investments in R&D (R&D intensity), and maintenance (validation purposes), and firm size (number of employees). Cost of quality was represented in four categories: inspection costs, internal costs, preventive costs, and external costs. Actual cost of quality was not sought—only relative proportions. One performance measure was compiled: market share in the main product line, which was used in the regressions as the dependent variable.

23.2.2. TQM MEASURE AND VALIDATION

In order to validate scales, a procedure similar to that used most recently by Flynn, Schroeder and Sakakibara (1994) to validate quality measures was used. That is, perceptual measures of adopted quality practices were correlated with criterion (performance) measures.

In the case of this study, a scale was constructed from items in the manufacturing strategy section of the questionnaire on the degree of use ("no use" scored 1 to "significant use the last two years" scored 5) of various practices including quality initiatives. An SPSSx item analysis produced a five-item scale including Total Quality Management (TQM) programme, zero defects and Kaizen (continuous improvement) programme, quality function deployment (QFD) and quality policy deployment. The Cronbach's alpha for the scale was 0.80 and the average inter-item correlation was 0.45 (n = 317). The same scale was also computed for just the US data with similar results (Cronbach's alpha = 0.76).

This TQM scale was significantly correlated with the proportion of money spent on preventive maintenance, with a correlation coefficient of 0.22 (p < 0.01, n = 304). The validity of this TQM measure thus appears to be quite good.

One-way analysis of variance was used to validate the industry context grouping assumption. It was found that only one industry grouping (SIC 36, electrical equipment including computers) was significantly higher in R&D intensity (f = 4.95, p = 0.0007) than the others, with mean R&D intensity = 7%.

Table 23-1. Correlation Matrix and Descriptive Statistics

		1	2	3	4	5
1.	R&D Intensity (%)	1.00				
2.	CAM	0.16 **	1.00			
3.	TQM	-0.19 **	0.19 **	1.00		
4.	Size (number of employees)	-0.03	0.22 **	0.11	1.00	
5.	Market share (%)	0.13 **	-0.11	0.12 *	-0.07	1.00
	Mean	4.85	8.2	13.8	867	33.9
	S.D.	6.84	4.2	5.5	1843	23.6
	Sample size (N)	500	325	327	586	545

** p < 0.01; * p < 0.05

23.3. Results

In Table 23-1, the correlation matrix of Pearson product-moment coefficients is presented for the variables of the study (computed using SPSSx). Descriptive statistics are also included.

In Table 23-2, the results of the moderated regression analysis are presented. The mean level of R&D intensity (4.85%) was used as the group cut off, so two moderated regressions were evaluated against the OLS (ordinary least-squares) model taking market share as dependent and R&D intensity, firm size (number of employees), and TQM as independent. The OLS model results are presented with the dependent variable in Table 23-2 in each case. Mean substitution was used for missing data.

The results in Table 23-2 strongly support the first two hypotheses. In the high-technology industries, only R&D intensity is significantly associated with market share (beta = 0.24, p < 0.01), controlling for size of firm (beta = -0.03, n.s.). In low-technology industries, TQM is the only significant predictor of market share (beta = 0.13, p < 0.05), again controlling for size of firm. R&D intensity is positively associated with market share in low-tech firms, but the relationship is non-significant (beta = 0.07, n.s.). TQM is positively related to market share in high-tech industries, but again it is non-significant under those circumstances (beta = 0.08, n.s.).

Examination of Table 23-1 also indicates support for hypotheses 3 and 4. Industry and region notwithstanding, R&D intensity is significantly and inversely related to TQM, r = -0.19 (p < 0.01). However, TQM and CAM are significantly and positively correlated (r = 0.19, p < 0.01).

Table 23-2. Moderated Regression (Market Share is Dependent)

	Independent variables	All cases		Group 1 High-Tech (R&D >=4.85%)		Group 2 Low-Tech (R&D <4.95%)	
1	R&D intensity	0.12	* *	0.24	* *	0.07	
2	TQM	0.10	*	0.08		0.13	*
3	Size of firm (no. of employees)	-0.07		-0.03		-0.08	
	F-ratio (p)	5.7	** (0.0007)	4.12	** (0.0007)	2.7	0.043)
	R² (R²adj)	0.03	(0.02)	0.07	(0.05)	0.03	(0.02)
	d.f.	3,	596	3,	177	3,315	

** p < 0.01; * p < 0.05

Notes: Mean substitution for missing data. Entries are standardised regression coefficients. When just complete data cases were used, F = 2.01, p = 0.11; TQM (beta = 0.15, p = 0.0193), firm size (beta = -005, p = 0.37) and R&D intensity (beta = 0.02, p = 0.77) resulted, with d.f. = 3, 263.

23.3.1. REGIONAL DIFFERENCES

In order to explore regional difference among the 20 country groupings, analysis of variance (ANOVA) was conducted using SPSSx. First, one-way analysis of variance tests were done, primarily due to missing data concerns. These results are presented in Table 23-3 below. Five regions were used: South America, South East Asia, Europe, North America, and Scandinavia.

These one-way ANOVA results indicate that TQM and CAM exhibit the greatest regional differences. R&D intensity and firm size (number of employees) exhibit no regional differences, and market share showed significant differences between North American (mean = 40%) and Europe (mean = 28%). Overall analysis of variance was limited by missing data, but still indicated that Europe and Scandinavia were lower in market share (not shown), so a dummy variable was created coding these two regions as "0" and the other three regions (North America, South East Asia, and South America) as "1". The results of these regressions are presented in Table 23-4.

Table 23-3. One-Way Analysis of Variation (ANOVA) Results by Region

Variable	F	(p)	(d.f.)	Significant multiple comparisons
Market share	5.45	**	(.0003)	(4,503)	North America (x = 4045) > Europe (x = 28%)
R&D intensity	2.17		(0.07)	(4,464)	
TQM	18.19	**	(.0001)	(4,298)	North American (x = 15.7) > Europe (x = 12.1), Scandinavia (x = 10.6); South East Asia (x = 17.3) > Europe, Scandinavia
CAM	9.27	**	(0.001)	(4,319)	South East Asia (x = 9.6), Europe (x = 9.4), Scandinavia (x = 8.0) > South America (x = 4.8)
Firm size	1.68		n.s.	(4.541)	

$** p < 0.01; * p < 0.05$

Regression results in Table 23-4 show that region had a significant association with market share in all three equations: for the total sample, and for the high-tech and low-tech groups. In all cases, the Europe/Scandinavia group had significantly lower market share. The general model is still intact, however, with the exception that TQM fails to have a significant beta (0.08, $t = 1.4$, $p = 1.6$) in the low-tech regression. The rest of the coefficients remain essentially identical to those in Table 23-2. It is interesting to note that firm size (number of employees) does exhibit significant effects in the total sample regression (beta = -0.09, $p < 0.05$). This is discussed later.

In general, the Utterback-Abernathy model appears to be easily extended in order to begin to reconcile the quality-technology-performance relationships in global manufacturing, using market share as the performance measure. These results are consistent with a contingency model of the quality process in organisations (Sitkin et al., 1994).

Table 23-4. Regressions including Region (Market Share is Dependent)

	Independent Variables	All Cases		Group 1 High Tech		Group 2 Low Tech	
1	R&D intensity	0.13**		0.26**		0.06	
2	TQM	0.06		0.02		0.08	
3	Firm size (number of employees)	-0.09*		-0.04		-0.09	
4	Region	0.18**		0.22**		0.17**	
	F (p)	9.2**	0.0001	5.32**	0.0005	4.39**	0.0018
	R^2 (R^2adj)	0.06	0.05	0.11	0.09	0.05	0.04
	d.f.	4,595		4,176		4,314	

23.4. Discussion

The Utterback-Abernathy model of the evolution of the productive segment of the firm was used to help reconcile the quality-technology issues of performance in global manufacturing. It was predicted that R&D intensity or TQM would be alternatively good predictors of market share depending on the industry context of the firm (approximated by high and low technology groupings). The model was strongly supported. High-tech firms had significantly higher market share when they invested in R&D. Low-technology firm had better market share when they invested in TQM programmes. Not surprisingly, R&D intensity was inversely correlated with TQM efforts, and CAM was directly associated with TQM, as this evolutionary stage model of productive segment would predict.

Given the significant support for this model, it would be interesting to speculate on other quality-technology-performance relationships. Market share and ROI were significantly related in this sample ($r = 0.17$, $p < 0.01$, not shown). Perhaps some of these results can be generalised to other performance measures, although this would be a rare finding (Cameron, 1986).

Perhaps one of the most provocative issues to emerge in the midst of the quality movement has been whether quality can be "driven" to the bottom line. That is, what is the relationship between quality indicators and organisations effectiveness, variously measured? The "paradox" issues notwithstanding, any number of attempts have been made to try to explain the circumstances under which quality and various measures of performance are related (Garvin, 1988). Most recently, DeLean (1994) has presented a model by which AT&T Easylink service unit has been able to predict market share directly from customer satisfaction data. In general, the way this is done is that a strategy is tailored to a product line for impacting market share with service-quality-driven value. Great care is taken to define, measure, and then to improve the specific quality attributes require from human effort and operational systems. Then a Customer

Value Added (CVA) measure is computed. DeLean (1994) defines the CVA in the following:

What is CVA?

Customer Value Added (CVA) is a measure of the satisfaction of AT&T's customers relative to that of our competitors' customers. CVA, as a metric, is linked to Economic Value Added (EVA) because it directly measures the drivers of purchasing behavior.

$$CVA = \text{AT\&T WWPF * Score x 100}$$

Competitors WWPF Mean Score

If score is:

= 100 AT&T customers' perception of value equals to the mean perceived value by all competitors' customers.

> 100 AT&T customers' perception of value is greater than that of competitors' customers.

< 100 AT&T customers' perception of value is less than that of competitors' customers.

* WWPF - Worth What Paid for or Overall Value

Drivers of CVA have been found to be customers' overall satisfaction with quality and price, improved relative perceived quality, price or a combination of the two, a function of both product and service quality, and price satisfaction comprising initial price and life cycle cost.

Although aggregations at levels above the product line have not shown any consistent relationship (Fornell, 1995), these results from AT&T do indicate a relationship when products are isolated. Other have even suggested (e.g., Ducker, 1994) that employee satisfaction and customer satisfaction, and therefore, market share, are related. Results from this study would predict that all of these relationships would depend on life-cycle issues concerning the productive segment of the firm.

There are fine-tuning issues that could be introduced in the model as well, independently of the various performance outcomes. How do the various types of processing innovation interact with quality programmes? For example, CAM is broad enough to include both flexible assembly and flexible manufacturing. Earlier results reported by others (Chen and Adams, 1991) indicated either a negative or no relationship between quality goals and flexibility in manufacturing.

Earlier literature (Imai, 1986; Cole, 1990, 1994; Currie, 1991; Kono, 1992) suggested strong cultural differences in quality and technology emphasis between the

US and Japan. Missing data problems prevented any extensive country or regional comparisons, but this could be a logical extension. The CAM and TQM scales, in particular, had missing data problems.

It is interesting to not that the variable measures that exhibit the greatest regional difference (Table 23-3, TQM and CAM) are those developed from scaled-item responses, even though this is considered an "expert" report on "objective" conditions. Number of employees, for example, does not show this pattern, even though it is generally true that firms in Europe and Scandinavia tend to be smaller than those in the US (part of the North American group), which may explain why size does depart from the established patterns of Table 23-2 and Table 23-3. When region is introduced in Table 23-4, the size-market share relationship was significant and inverse for the "All Cases" regression (beta = -0.09, $p < 0.05$).

The general issue of culture bias in cross-national survey research of this type deserves attention in future research of this type (Ellis, 1992). In the follow-up calls to some non-respondents in the US, managers often said that the data requested were too sensitive (e.g., production capacity), which raises another, related issue for future research.

In general, there appears to be sufficient preliminary evidence, albeit subject to differences in performance measures and cultural groupings, to suggest further exploration of the model of quality-technology integration is warranted. Cost of quality proportions and TQM programme reports cannot actually substitute completely for actual cost of quality levels, which may be product- and culturally-dependent, but the moderating effects of core technology and industry differences in these firms appears to be a clear empirical trend that supports the U-A model. In future research, this is one methodological refinement that deserves to be included.

Finally, there is emergent case evidence to suggest a total quality movement for natural (environmental) management (TQEM) movement beginning in the US, Japan, Sweden, and Germany, to name but a few countries (e.g., Romm, 1994). Will proactive concern for the natural environment provide the necessary link between quality, technology, and performance measures not adequately described or predicted by the U-A model? The cases (e.g., 3M, AT&T, Compaq Computer, Dow Chemical, DuPont, Xerox, and Boeing) of pollution prevented through waste elimination or prevention or proactive and not adequately explained by earlier theory.

References

Abernathy, W.J. (1976) 'Production process structure and technological change', Decision Sciences, 7, 4, 607-619.

Abernathy, W.J. and Townsend, P.L. (1975) 'Technology, productivity, and process change', Technological Forecasting and Social Change, 7, 379-396.

Bergman, B., and Klefsjo, B. (1994) Quality: From Customer Needs to Customer Satisfaction, McGraw-Hill, London.

Besterfield, D.H., Besterfield-Michna, C., Besterfield, G.H., and Besterfield-Sacre, M. (1995) Total Quality Management, NJ, Prentice-Hall, Englewood Cliffs.

Buran, W. (1994) 'The state of the art on process re-engineering and quality', presented at the Quality and Process Re-Engineering Conference, Michigan Business School, Ann Arbor, Michigan, USA, March 25, 1994.

Cameron, K.S. (1986) 'Effectiveness as paradox: consensus and conflict in conceptions of organizational effectiveness', Management Science, 32, 5, May, 539-553.

Chen, F.F., and Adam, E.E. (1991) 'The impact of Flexible Manufacturing Systems on productivity and quality', IEEE Transactions on Engineering Management, 38, 1, February, 33-45.

Cole, R.E. (1994) 'Reflections on organizational learning in US and Japanese industry', in J. Liker, J. Ettlie and J. Campbell (eds.), Engineered in Japan, Oxford University Press, New York (forthcoming).

Cole, R. (1990) 'US quality improvement in the auto industry: close but no cigar', California Management Review, Summer, 71-85.

Crosby, P.B. (1979) Quality is Free, New American Library, New York.

Currie, W. (1991) 'Managing production technology in Japanese industry -- an investigation of eight companies, Part 2', Management Accounting - London, 69, 7, July/August, 36-38.

Dean, J.W. and Bowen, D.E. (1994) 'Management theory and total quality: Improving research and practice through theory development', Academy of Management Review, 19, 3, July, 1-27.

DeLean, M.L. (1994) 'AT&T's quality journey: an insider's view on bottom line improvement', presented at the University of Michigan, School of Business Administration, December 14, 1994.

Deming, W.E. (1986) Out of the Crisis, Massachusetts Institute of Technology Center for Advanced Engineering Study, Cambridge, MA.

Deming, W.E. (1950) Some Theory of Sampling, John Wiley and Sons, New York.

Dodge, H.F. (1969) 'Notes on the evolution of acceptance sampling plans, Part II', Journal of Quality Technology, July, 155-156.

Ducker, W. (1994) Personal communication, 1994.

Ellis. (1992) 'Identification of unique cultural response patterns by means of item response theory', Journal of Applied Psychology, 77, 2, 177-184.

Evans, J.R. and Lindsay, W.M. (1993) The Management and Control of Quality, 2nd edition, West Publishing Company, Minneapolis/St. Paul MN.

Faltermayer, E. (1994) 'Competitiveness: How US companies stack up now', Fortune, April 18, 52-64.

Farnum, N.R. (1994) Modern Statistical Quality Control and Improvement, Duxbury Press, Belmont, CA.

Feigenbaum, A.V. (1956) 'Total quality control', Harvard Business Review, November-December, 94-98.

Flood, R.L. (1993) Beyond TQM, John Wiley and Sons, New York.

Flynn, B.B., Schroeder, R.G., and Sakakibara, S. (1994) 'A framework for quality management research and an associated measurement instrument', Journal of Operations Management, 11, 339-366.

Fornell, C. (1995) Personal communication, 1995.

Garvin, D.A. (1988) Managing Quality, The Free Press, New York.

Gitlow, H., Oppenheim, A. and Oppenheim, R. (1995) Quality Management: Tools and Methods for Improvement, 2nd edition, Irwin, Burr Ridge, IL.

Goetsch, D.L. and Davis, S. (1995) Total Quality, Prentice Hall, Englewood Cliffs, NJ.

Imai, K. (1986) Kaizen: The Key to Japan's Competitive Success, McGraw-Hill Publishing Company, New York.

Juran, J.M. (ed.) (1951) Quality Control Handbook, McGraw-Hill, New York.

Kono, T. (1992) 'Japanese management philosophy: can it be exported?', in T. Kono (ed.), Strategic Management in Japanese Companies, Oxford, Pergamon Press, 11-23.

Liker, J.K., Ettlie, J.E., and Ward, A.C. (1995) 'Managing technology systematically: common themes', in Liker, J.K., Ettlie, J.E., and J. Campbell (eds.), Engineered in Japan, Oxford University Press, New York, forthcoming.

Myers, S. and Marquis, D.G. (1969) Successful Industrial Innovations, National Science Foundation, NSF 69-17.

Peteraf, M.A. (1993) 'The cornerstones of competitive advantage: a resource-based view', Strategic Management Journal, 14, 3, March, 179-192.

Pfeffer, J. and Salancik, G.R. (1978) The External Control of Organizations: A Resource Dependence Perspective, Harper & Row, New York.

Rechtin, M. (1994) 'Europeans roar into no. 2 in CSI', Automotive News, July 11, 1994, 2, 59.

Reiman, C. (1993) 'The Baldrige Award criteria: a dynamic model for integrating business management requirements -- focus on quality and productivity', presented at the Production and Operations Management Society, October 5, 1993, Boston, MA.

Romanelli, E. and Tushman, M.L. (1994) 'Organizational performance as punctuated equilibrium: an empirical test', Academy of Management Journal, 37, 5, 1141-1166.

Romm, J.J. (1994) Lean and Clean, Kadansha International, New York.

Scott, W.R. (1987) 'The adolescence of institutional theory', Administrative Science Quarterly, 32, 4, December, 493-511.

Sitkin, S.B. Sutcliffe, K.M. and Schroeder, R.G. (1994) 'Distinguishing control from learning in total quality management: a contingency perspective', Academy of Management Review, 19, 3, 537-564.

Vroman, H.W. and Luchsinger, V.P. (1994) Managing Organization Quality, Irwin, Burr Ridge, IL.

Wernerfelt, B. (1984) 'A resource-based view of the firm', Strategic Management Journal, 5, 171-180.

CHAPTER 24

US MANUFACTURING IN THE 1990S: THE CHASE AND THE CHALLENGE

John Ettlie, School of Business Administration, University of Michigan, Ann Arbor, Michigan, USA and Peter T. Ward, Fisher College of Business, Ohio State University, Columbus, OH

24.1. Introduction

Mercedes is benchmarking Ford Motor Company in building its Tuscaloosa, Alabama plant. 'Now Detroit is the benchmark', and Ford is the best in labour productivity. But Ford still benchmarks Toyota. (New York Times, March 4, 1995, p. 17).

Taking the auto industry as a microcosm of global durable goods manufacturing for the sake of argument, this statement from Mercedes-Benz management is quite instructive. Ford is currently seen as the benchmark in North American manufacturing, but both Ford and industry experts say that only location and labour productivity puts them in this venerable position. Globally, Ford still trails Honda, Nissan, and Toyota.

These industry productivity data may sum up better than any other single example the state of the art of manufacturing in the US. Progress during the last decade has been real and impressive by any standard. But, beyond this type of observation, it is hard to generalise. There is good news and bad news.

24.1.1. THE GOOD NEWS: US LEADERS ARE GLOBAL LEADERS

Perhaps the best example of a 'comeback' industry in the US is the electronic chip-making industry. The course of this industry and shows a remarkable turnaround for US competitors when compared with the former leaders (by a large margin), the Japanese.

A case history in a related industry, computer printers, is instructive in regard to this trend of US companies displacing Japanese firms in market and technology leadership. This is the case of Hewlett Packard's ink-jet printers, which have become one of the most successful US products of the decade. (Yoder, 1994 - See Case Study 1.) This case illustrates the intensity of US-Japanese competition in manufacturing and shows how both imitation and creativity combine in HP's research strategy, which has been very successful.

Seiko Epson has since countered with its own strategy on the high-end printer market with unique demonstration marketing and sales, which have been very successful. So far, the last chapter in this story has not yet been written.

P. Lindberg et al. (eds.), International Manufacturing Strategies, 401-415.
© *1998 Kluwer Academic Publishers. Printed in the Netherlands.*

Seiko Epson has since countered with its own strategy on the high-end printer market with unique demonstration marketing and sales, which have been very successful. So far, the last chapter in this story has not yet been written.

24.1.2. THE BAD NEWS: DIFFERENCES IN US LEADERS AND FOLLOWERS

Although quality levels have improved in selected industries such as automobiles, customer satisfaction is higher still with Japanese and European cars (Rechtin, 1994). Since the late 1980s, the devaluation of the dollar has contributed as much as half the gain in competitiveness of American industry, and there continue to be problems with the balance of payments (Faltermayer, 1994).

Perhaps an even more surprising example of the loss of US leadership is the aerospace industry. Although one company, Boeing, continues to lead the airframe manufacturers, Europe's Arianespace has built on a decade-old lead in the space race to dominate the launch business. The Challenger disaster in 1986 put the US launch industry on hold for nearly a year, and it has never really recovered. There is a similar pattern in the US machine tool industry, which shows the US a distant fourth in world production of machine tools (Critical Technologies Institute, RAND, August 1994).

Industry-by-industry comparisons provide one basis for evaluating the relative comparative advantage of a country in a global market (Porter, 1990). Perhaps more disturbing is that several recent surveys indicate widespread dissatisfaction with the results of US quality improvement initiatives. Over 50% of companies report that quality programmes have not led to better business performance (American Quality and Productivity Council, 1992). Fewer than one-third of US Fortune 500 firms believe quality programmes significantly impacted competitiveness (Arthur D. Little, 1993). Over 85% of ISO9000 registrants think it will take eight years or more to recover their costs (Deloitte and Touche, 1993).

Quality programmes have failed to meet expectations in two-thirds of US firms, primarily because they have not been related to customer outcomes (Buran, 1994). Only a small number of companies qualify for the Malcolm Baldrige National Quality Award, but those who do tend to be able to forge the bond between customer orientation and operational performance. Areas of persistent weakness in Baldrige applications include an unclear linkage between quality and strategy, lack of data and analysis, and partial systems that do not integrate information technology and quality (Reimann, 1993).

At least four points are worth noting. First, one of the reasons that so many surveys report problems with quality programmes is that only about 10% of all companies are truly outstanding. Therefore, many surveys report 'average' results or percentages, and are accurate to their population of companies: the results just hide the real lessons worth noting. A good illustration of this fact is that there are only a small number and only a *very* small percentage of companies that qualify for Baldrige National Quality Award site visits.

Second, the key to understanding the circumstances under which quality programmes work is understanding how they link to technology programmes in the same companies. Note in the case example that for the electronics supplier, both *Kaizen* programmes and investment in new processing equipment were made. What is typical in US companies is that technology decisions such as those made for new products and processes are made by people and functions who have no quality responsibility, and vice versa. When these two groups or people or units do combine, it typically is in isolation from the rest of the company. This is one of the problems with the 'platform' approach to design.

Third, companies that have high conformance to specification quality and high customer satisfaction are the firms that drive quality to the bottom line (as *Business Week*, August 8, 1994 puts it). The voice of the customer is not lost in the translation from customer needs to manufacturing specification for product and process. All three key functions in design -- R&D, marketing, and production -- are involved at the appropriate stages of projects. Cross-functional design does *not* mean that everyone has a big say all the time during the life of a programme -- sometimes marketing should have more to say in a product specification, and sometimes it should be engineering or production.

When a top 10% company learns a quality lesson it does not fade away, resulting in the all-too-common situation that these same quality problems reappear during the next product cycle or project. The whole company learns the lesson, because top managers make it their business to see that this happens. Learning and quality are intertwined and are never separated.

A final lesson in this that is worth mentioning is to avoid an obvious leap to the conclusion that it stands to reason that all one needs to do is benchmark the best (that top 10% company in your industry) and then leapfrog over them. Stop right there. Quality is not something you put in the oven overnight and take out as a complete baked cake the next day. It doesn't work that way. Even if you could understand how the best become the best, following, let alone skipping over the steps that got them there, is extremely difficult. Furthermore, quality alone should not dominate all thinking in manufacturing. US labour productivity has grown steadily for the last five years, which is no small accomplishment. 'The current economic expansion's productivity increase outdoes those of the three previous recoveries. Only the booming 1960s look better' (*Business Week*, September 25, 1995, p. 8).

Quality and productivity are not necessarily incompatible. The case history of Lincoln Continental is an example of this tendency in successful US new product launches). Here quality, performance, and aesthetics have a relationship, too. Deming, Senge, and customers all had a say in the new design, which came in on time and under budget. When team work did bog down, Deming asked team members to address the question: 'Who depends on me?' (Simon, 1995).

A total of 87% of the parts were ready for the first prototype, and 93% were available at the second prototype stage (Simon, 1995). One thing that was discovered during the Continental programme was that delays in meeting deadlines were often created because people were afraid to announce delays as they occurred because of fear of reprisal. That is, people delay telling others that there will be a delay, and there is no way to compensate (Simon, 1995).

24.2. Emerging Theory

Two emerging theoretical areas are presented that have implications for research on manufacturing: (1) alliances and global networks, and (2) resource-based theory. Two important practice trends are also reviewed: (1) business process re-engineering and (2) benchmarking.

24.2.1. ATTITUDES AND GLOBAL NETWORKS

In US manufacturing, the 1980s was the decade of down-sizing, and this continues into the 1990s (Bryne, 1994). As many as 500,000 jobs may have been lost during this period, although it is difficult to estimate this number. If non-manufacturing jobs are included, the number is even higher, as many of these jobs depend on the manufacturing sector.

The decade of the 1990s has emerged with continued competitive pressure from relentless technological change and globalisation. The US now exports a higher percentage of its domestic output than Japan (Carson, 1994). The response has been to build alliance (Teece, 1992; Gerybadze, 1995) and networks (Limerick and Cunnington, 1993). According to a Coopers & Lybrand (New York) survey, 55% of the fast-growing companies in the US are involved in an average of three alliances each (*Wall Street Journal*, April 20, 1995, p. A1). A related trend is an increase in mergers and acquisitions, which were up 35% over the previous year during the first quarter of 1995 (*Wall Street Journal*, April 4, 1995, p. A3). Japanese direct investment in the US peaked in 1988, but US-Japanese joint ventures in manufacturing continue to be important (Ettlie and Swan, 1995).

Globalisation and the pace of new product and service introduction, especially, as well as competitors' choice to form alliances, appear to be the primary drivers of this trend to partnering. In some cases, these alliances and joint ventures are among competitors, which may be the ultimate in management challenges (Browning, Beyer, and Shetler, 1995). This seems especially true in light of the continuing stream of findings such as those reported by Birnbaum-More, Weiss, and Wright (1994). They found in 13 US product-market segments that successful firms used a focused approach towards a market with increased product and process innovation regardless of concentration and stage of industry maturity. On the other hand, certain types of R&D projects may be favoured among consortium members after this multiple alliance form becomes established (Bolton, 1993). But in Europe, at least, and for members of the

Eureka technological consortium, projects led by small firms tend to be problematic (Peterson, 1993). Further, Bleeke and Ernst (1995) report that the median life for alliances is 7 years and that 80% of joint ventures ultimately end in sale by one of the partners.

A number of different theoretical perspectives have been used to characterise alliances and joint ventures. Among the most widespread is the application of *transactions cost theory,* which can be used to predict or explain multinationalisation of a firm (Teece, 1986). For example, Garrette and Queslin (1994) found evidence of hybrid forms of governance structures predicted by transactions cost theory, split evenly between market and hierarchy dominance among strategic partnerships in the telecommunications equipment manufacturers. Gulati (1995) reports findings on multi-industry alliances that challenge singular emphasis on transaction costs. He finds that repeated alliances are *less* likely to be organised using equity after initial shared R&D equity partnerships.

Network theory has been applied in a variety of ways. Networks of innovators appear to be universal, and are more than the sum of bilateral relationships (De Bresson and Fernand, 1991). Networks can take different forms and have a variety of governance structures, e.g., ring-core combinations (Stroper and Bennett, 1991). For technology alliances, Hagedoorn and Schakenraad (1994) used nearly 10,000 co-operative agreements from the CATI (Co-operative Agreements and Technology Indicators) data system and found that the more patent-intensive the agreement, the more the firms are involved in strategic partnering. Information technology firms co-operate more, process industries less. US firms tend to be larger and are not more inclined than are Japanese or European firms. Finally, as far as the impact on performance, the results are mixed. For European and American process industries, the authors found a positive association between R&D-driven co-operation and profitability. Kogut, Walker and Kim (1995) found strong support for the 'network centrality' hypothesis in the semiconductor industry: network centrality, technology dominance, and the successful entry of start-up firms.

Manufacturing networks of decentralised plants are among the scenarios for future production configuration (MacCormack, Newmann, and Rosenfield, 1994). One way of understanding these networks of suppliers, producers, and distributors is through *contract* theory. For example, Tsay and Lovejoy's model (1994) demonstrates a framework of supply chain control using quantity flexibility contracts. The contract between customer and supplier stipulates a percentage increase that each schedule element is allowed in subsequent planning iterations, and places bounds on reductions. This system encourages customers to be more accurate in their forecasts to suppliers.

In a related development, Kogut and Kulatilaka model a multinational network with a stochastic dynamic programming model that shows how the management of across-border co-ordination has led to changes in the heuristic rules for performance evaluation and transfer pricing. On the implementation side, AT&T has installed a global information system for 60 sites world-wide to co-ordinate activities (Sykes, 1995).

24.2.2.   RESOURCE-BASED VIEW

Theoretical developments in the resource-based view (RBV) of strategy have important implications for the way we view manufacturing, and particularly, competitive advantage based on manufacturing capabilities. The emergence of the RBV in the competitive strategy community is evidenced by the increasing number of papers that take this view, and recent special issues on the topic in *Strategic Management Journal* (Barney and Zajac, 1994) and the *Journal of Management* (Barney, 1991). Prior to the inception of the RBV, theory underlying virtually all competitive strategy was based on the assumption that resources (or, more aptly for this discussion, *capabilities)* are homogenous and mobile in their environment (i.e., industry or competitive group) (Barney, 1991). Therefore, any advantage built on a firm's relative superiority in manufacturing would be short-lived because competitors would acquire similar resources (achievable because of resource mobility) and thus maintain resource homogeneity. These assumptions, which fly in the face of manufacturing-based advantage, have limited applicability of competitive strategy theories to manufacturing strategy.

In its essence, RBV turns the two assumptions on their head, resources are assumed to be heterogeneous among competing firms and imperfectly mobile in their environment. This view allows for firms to build and maintain different (and/or greater) packages of capabilities than their competitors and to use this difference as the basis of sustainable advantage.

According to the RBV, a resource possessed by a firm must have four attributes to be the source of sustainable competitive advantage. As depicted in Figure 24-2, the resource (or capability) must be valuable (i.e., exploits opportunities or neutralises threats), it must be rare among competitors, it must be imperfectly imitable, and there cannot be strategically equivalent substitutes available to competitors (Barney, 1991). In terms of manufacturing, capabilities that provide differential advantages with respect to cost, customerization, time, etc. often fulfil the requirements for sustainability because they are valued by customers and hard to duplicate by competitors. Manufacturing

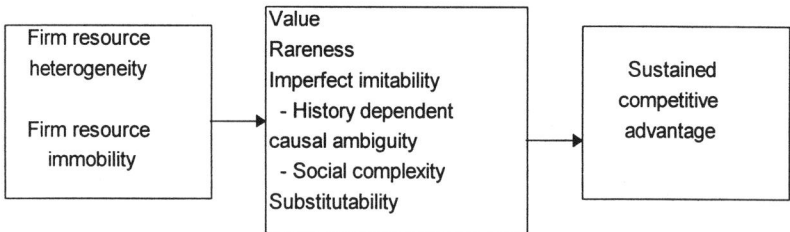

Figure 24-2. Relationship among resource heterogeneity and immobility, value, rareness, imperfect imitability, and sustained competitive advantage (Source: Barney, 1991b)

capabilities are often hard to duplicate because they result from complex sets of investments in technologies and people that are built over time and acquired from numerous sources.

The RBV fits well with the concepts of manufacturing strategy articulated by such authors as Hill (1994), Hayes, Wheelwright and Clark (1988), and many others. In this way, RBV may provide a robust theoretical context for the concepts and empirical work in manufacturing strategy that has been lacking. Perhaps more importantly, because of the influence of RBV the research generated in competitive strategy is more likely to be of consequence to those interested in manufacturing strategy issues. Certainly the notions of capabilities (Dierickx and Cool, 1989; Stalk, Evans, and Shulman, 1992) and core competence (Prahalad and Hamel, 1990) are consistent with RBV and salient to manufacturing strategy.

Finally, RBV provides a model for better understanding competitive advantage based on manufacturing capabilities. For example, Stalk and Weber (1994) describe a situation where time-based competition has become ubiquitous in some Japanese industries. Because all competitors provide a large variety of products in rapid-fire succession, the enormous skill yields no distinctive advantage. As a result, Japanese firms in these industries bear the expense of time-based competition without the rewards of higher margins. Stalk and Weber go on to point out that the true advantage of time-based competition lies in the knowledge that it can provide about customers. Extracting and applying this knowledge can yield a differentiation advantage that is sustainable because it is rare as well as valuable and hard to duplicate, even in Japan.

24.2.3. MASS CUSTOMISATION - A RESOURCE-BASED VIEW OF NBIC

Mass customisation, providing products that are created to the customers' specification at a price close to that of similar products that are mass produced, provides a current and illuminating subject for a resource-based view. The very notion of mass customisation is challenging to a traditional view of manufacturing economics because it requires that manufacturers be able to achieve relatively low costs and customisation simultaneously. Mass customisation also presents other challenges to the producing company in areas such as engineering, developing marketing channels, information processing, and logistics. Despite the inherent difficulties of mass customisation, Pine (1993) provides evidence that many companies are trying mass customisation.

We focus on the particular case of the National Bicycle Industrial Company (Kotha and Fried, 1993; Kotha, 1993), a Japanese-based manufacturer of a broad line of bicycles with the majority of its sales in the domestic market. The company was facing an environment in which sales were relatively flat because of a mature market and overseas rival with a cost advantage, particularly in the lower end of the market. The idea for mass customisation at NBIC originated with the firm's president, who noticed that a department store offered women the opportunity to custom order dresses that would be delivered to their home within two weeks. He asked if a similar opportunity could be offered to bicycle buyers.

This idea sparked the development of a mass customisation system at NBIC for high-end bicycles, which was developed by a team of product designers, engineers, and production workers in only four months. The system guaranteed Japanese consumers that a custom-made bicycle would be delivered in precisely two weeks, and these bicycles were priced at a 20 to 30 per cent premium relative to similar mass-produced bicycles. Although the mass customisation segment makes a relatively small portion of NBIC's total bicycle sales, it resulted in increased sales in the most profitable portion of the market, the high end. High-end bicycle buyers were generally delight with NBIC's mass customisation offering, but NBIC's major competitors, Bridgestone and Miyata, were unable to duplicate NBIC's success when they tried similar mass customisation efforts.

The benefits of mass customisation to the customer are clear: a custom bicycle for only a small price premium. Some of the benefits to NBIC are similarly clear: increased revenues and profits resulting from a competitive edge in the high end of the market and enhanced reputation, which helped sales in other segments as well. According to Kotha (1995), NBIC also benefited from learning attributable to mass customisation. The information system provided detailed information about what 'lead users' wanted with respect to product features and innovation. This information is useful intelligence that is applicable to customers in NBIC's much larger mass production segment, thus providing a company-wide marketing edge. In addition, NBIC's best production workers were rotated through the mass customisation facility, thus learning from the application of advanced technologies and bringing the product of this learning back to the mass production part of the business (Kotha, 1995).

Taking a resource-based view of NBIC's mass customisation strategy, we see that the resources or capabilities developed by NBIC meet the required criteria; i.e., they are valuable, rare among competitors, cannot be imitated perfectly, and are not easily substituted. The value of mass customisation capability is evidenced by the marketplace success of the offering. The fact that these capabilities are not easily imitated is demonstrated by the lack of success of the imitations by very capable competitors. NBIC's effort could not be imitated successfully because of its first-mover advantage and also because the system they put in place was complex and its elements could not be inspected easily by competitors. Similarly, competitors have been unable to offer acceptable substitutes for custom bicycles.

Prior to the advent of the resource-based view, a strategy scholar would have attributed NBIC's apparent advantage to information lag, rather than NBIC's ability to develop and sustain an advantage based on specific capabilities. The resource-based view accepts that resources such as these necessary to achieve mass customisation are sustainable because they cannot be easily 'bought'. Thus, capability building through manufacturing strategy makes theoretical as well as practical sense.

24.2.4. RE-ENGINEERING

Breaking functional silos to improve business processes has been a critical issue in some companies for a number of years. For example, Ford's well-documented transformation during the 1980s involved major changes to core processes that required breaking through previously inviolate functional boundaries (Pascale, 1990). The publication of Michael Hammer's *Harvard Business Review* article on re-engineering (Hammer, 1990) and his book, co-authored with James Champy, on the same subject (Hammer and Champy, 1993) lent clarity to the topic by providing a set of principles and examples.

Advocates of re-engineering such as Hammer and Champy or Davenport (1993) describe re-engineering as a radical or revolutionary change in the way we think about key processes and organise those processes, which result in order of magnitude improvements in business outcomes. Key processes are identified as both manufacturing / operations and non-manufacturing processes. Examples of key processes for three companies are provided by Davenport (1993) and shown in Table 24-1.

The reality of what passes for re-engineering in actual practice is often quite different from the major revolutions described by re-engineering experts. In fact, the term re-engineering has become a 1990s buzzword, and worse, the term is used in many companies as a convenient, evidence-free justification for virtually any incremental changes that managers believe need to be made.

This does not mean that re-engineering is not an important means for accomplishing major change in organisations. Dixon and his colleagues (1994) provide some perspective on re-engineering based on a study of 23 cases involving major re-

Table 24-1. Key business processes of leading organisations (Source: Davenport, 1993)

IBM	Xerox	British Telecom
Market information	Customer engagement	Direct business
Market selection	Inventory management and logistics	Plan business
Requirements	Product design and engineering	Develop processes
Development of hardware	Product maintenance	Manage process operation
Development of services	Production and operations management	Market products and services
Production	Market management	Provide customer service
Customer fulfilment	Supplier management	Manage products and services
Customer relationship	Information management	Provide consultancy services
Service	Business management	Plan the network
Customer feedback	Human resource management	Operate the network
Marketing	Leased and capital asset management	Provide support services
Solution integration	Legal	Manage information resource
Plan integration		Manage finance
Accounting		Provide technical R&D
Human resources		
IT infrastructure		

engineering projects. Based on this evidence, they distinguish between re-engineering and more incremental, continuous improvement approaches on three dimensions:

1. Improvement trajectories change with re-engineering, e.g., flexibility replaces cost reduction as a goal or time-to-market supersedes production performance.

2. Re-engineering is spurred by sea changes in the competitive environment. Just improving where we are already capable doesn't work any more.

3. Top management is more involved in leading re-engineering. Continuous improvement is more of a bottom-up process.

These same authors provide insights about the nature of major re-engineering projects in the US:

- In most cases, crisis is not the impetus for re-engineering. More often, top management perceives current capabilities to be out of synch with future competitive requirements.

- IT does not drive re-engineering, in most cases, although it does enable aspects of projects. The same argument is made for performance measurement systems.

- Re-engineering projects are characterised by ambitious but fuzzy objectives (e.g., 30 time improvement).

- Benchmarking is often helpful in setting re-engineering project performance measures.

- High-level project champions are identifiable. top management is heavily involved.

- Consultants are commonly used in designing the project; infrequently, in its management.

- Training in process analysis and teamwork are often used. TQM training often cited as valuable background.

- Communication between re-engineering team and those not directly involved is important.

Widespread application of re-engineering has major implications for manufacturing. Staff positions and supervisory positions will both become fewer in number. Concomitantly, direct workers will become more involved in accomplishing tasks that were previously considered overhead. As a result, the emphasis and key to success in many organisations will have to shift to *doing* work as opposed to *supervising* work.

24.2.5. BENCHMARKING

Benchmarking was most probably popularised for the first time by Jacobson and Hillkirk (1986) in their book about the revival of Xerox Corporation, 'the first American company to regain market share against the Japanese', (p. 3). Benchmarking is listed as the first of 10 key factors in Xerox's dramatic turnaround:

'1. Competitive Benchmarking. Xerox swallowed its pride, admitting that others (especially the Japanese) might have a better way of doing things, and sent investigative teams on world-wide scouting missions to find out. Benchmarking can help determine not only the cheapest sources for quality parts, but the best manufacturing and service methods. Xerox even examined L.L. Bean's distribution methods to see what it could adapt'. (Jacobson and Hillkirk, 1986, p. 9).

In spite of the popularity of benchmarking (about three-quarters of firms introducing new products and services report that they do benchmark other companies, Ettlie and Johnson, 1994, p. 111), it was not until the appearance of two other books that the importance of benchmarking in a larger context was realised.

The first of these important books is quite well known. This is Peter Senge's *Fifth Discipline* (1990). The fifth discipline is systems thinking, but it leads to an orientation toward organisational learning as a fundamental founding principle and more clear in a less well-known, but equally important, work by McGill and Slocum (1994) called *The Smarter Organization*. 'Smarter organizations are learning organizations, able to process their experiences -- with customers, competitors, focus. The relationship between organisational learning and benchmarking is made partners, and suppliers -- in ways that allow them to create environments in which they can be successful. Learning is their sustainable competitive advantage.' As can be seen in McGill and Slocum's framework (reproduced in Figure 24-3, their Table 3.1, p. 86).

Strategic Dimensions	Different approaches to strategy	
	Conventional	Learning
Underlying premise	Fit the firm to the environment.	Change the environment to fit the firm.
Competitive advantage	Mass, size, vertical integration.	Intent to learn: anytime, anything, anywhere.
Resources / means	Invest in fixed assets.	Invest in evolving/emerging opportunities.
Problem-solving logic	Formal planning; quantitative analysis.	Intuitive, sense-making; learn from failures.
Basis of thinking	Linear, incremental.	Continuous experimentation.
Role of alliances	Cost reduction.	Learning new insights from partners.
Role of customers	Conceived as marketing tools.	Conceived as groups, individuals to learn from.

Figure 24-3. Crafting a Strategy for the Tough New Customer (Source: McGill and Slocum, 1994, p. 86)

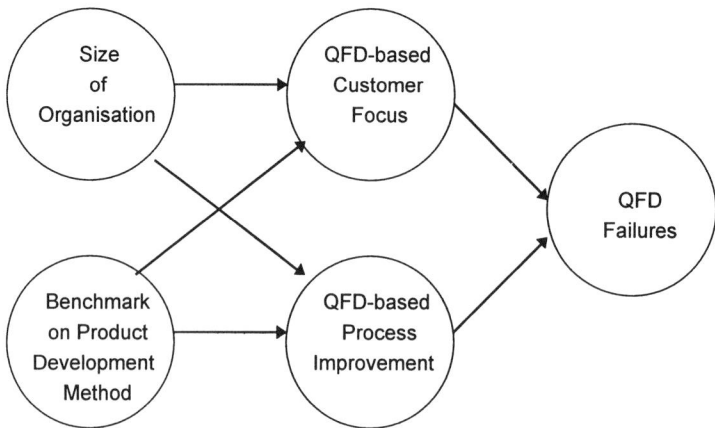

Figure 24-4. QFD Case Study Results (Source: Ettlie and Johnson, 1994)

Note that in the McGill and Slocum (1994, p. 86) summary of different approaches to strategy, learning from failures, experimentation, and learning from partners all fall under the learning organisation. They later go on to define benchmarking as 'a method of discovering how to improve performance by learning from others dealing with similar issues'. However, they define 'similar issues' quite broadly. For those interested in any type of service issue, for example, Southwest Airlines is recommended for study by the authors, and so on.

There are lots of practical guides around for benchmarking, but few big-picture summaries or systematic empirical evidence on the process of benchmarking and its impact on performance. One exception to this rule is the study by Ettlie and Johnson (1994) on QFD (Quality Function Deployment) projects and the role of benchmarking in new product/service development. The results of the Ettlie and Johnson (1994) partial least-squares model are replicated in Figure 24-4.

These results show how both QFD-focused customer voice and process improvement decrease the incidence of QFD failures. However, benchmarking tends to interfere with the voice of customer deployment during the process: yes, it does promote process improvement (0.371), but it also conflicts with customer focus (-0.527) in the model. Only a handful of companies have been able to resolve this conflict effectively, but most experience it.

Ultimately, it is the distinction between Total Quality Control (TQC) and Total Quality Learning (TQL) that may resolve some of these conflicts, conceptually, in this field. This distinction has been made in an article by Sitkin, Sutcliffe, and Schroeder (1994), and part of their summary is reproduced here in Figure 24-5.

These perspectives share three TQM (Total Quality management) precepts: customer satisfaction, continuous improvement, and the total systems perspective, but the principles derived under the two views are very different (e.g., monitor customers vs. scan for new customers).

Principles derived from common precepts		
Shared TQM Precepts	Control-Oriented Principles (TQC)	Learning-Oriented Principles (TQL)
Customer satisfaction	Monitor and asses know customer needs.	Scan for new customers, needs, or issues.
	Benchmark to better understand existing customer needs.	Test customer need definition.
	Respond to customer needs.	Stimulate new customer need definitions and levels.
Continuous improvement	Exploit existing skills and resources.	Explore new skills and resources.
	Increase control and reliability.	Increase learning and resilience.
Treating the organisation as a total system	First-order learning (cybernetic feedback).	Second-order learning.
	Participation enhancement focus.	Diversity enhancement focus.

Figure 24-5. Linking the Distinctive Principles Associated with TQC and TQL to Common Underlying TQM Precepts (Source: Sitkin et al., 1994, p. 546.)

24.3. Case Study 1: The Hewlett Packard Inkjet Printer1

The introduction of the HP inkjet printer and the follow-on, improved colour version, is a dramatic example of a US company recapturing a market from the Japanese. Combining the best of US creativity with Japanese tactics, HP went from no printer business in 1984 to nearly $8 billion in sales ten years later.

How did HP do it? First, HP managers, and especially Richard Hackborn, departed from company tradition to overcome a typical Japanese firm limitation -- layers of bureaucracy. Printer business teams were located in 'outposts' such as Boise, Idaho, not at HP headquarters in Palo Alto, CA. Other HP products had traditionally been high-profit, high-cost, niche market entries. This strategy is what lost HP the calculator market. Mr. Hackborn decided to learn from this experience and go for a mass market -- that's were the global businesses were going, he argued.

But it takes more than 'American' creativity to be successful. The Japanese often have criticised US companies for being complacent and 'local' in orientation HP clearly avoided this mistake by not resting with an early win with its inkjet - the company continued to reinvest profits in new product innovations and improvements. Another advantage that HP exploited was its proximity and understanding of the rapidly changing US market, often considered to be a bellwether.

> HP managers realised they could not dethrone market leaders Seiko Epson Corporation without new technology. Dot matrix printers were the mass market standard at the time. HP experimented with 'thermal' inkjet technology, which had great cost advantages over laser printing, being cheaper and more easily converted to colour. HP partnered with Canon and blanked the resulting technologies involved with many patents, as well as adopting a continuous improvement philosophy. These strategy are much more typical of Japanese firms. From Epson, HP learned that a simple product is best -- this becomes a platform for improvements. Now HP 'owns' the market for inkjet printers.

References

Barney, J.B. (1991a) 'The resource-based model of the firm: Origins, implications, and prospects'. *Journal of Management*, **17**, 1, 97-98.

Barney, J.B. (1991b) 'Firm resources and sustained competitive advantage'. *Journal of Management*, **17**, 1, 99-120.

Bleeke, J. and Ernst, D. (1995) 'Is your strategic alliance really a sale?' *Strategic Planning*, **73**, 1, January-February 97-105.

Bolton, M.K. (1993), 'Organizational innovation and substandard performance: When is necessity the mother of invention?' *Organization Science*, **4**, 1, February, 57-75.

Browning, L.D. Beyer, J.M. and Shetler, J.C. (1995) 'Building cooperation in a competitive industry: Sematech and the semiconductor industry'. *Academy of Management Journal*, **38**, 1, 113-151.

Bryne, J.A. (1994) 'The pain of downsizing'. *Business Week*, May 9, 60-69.

Carson, J.G. (1994) 'The big switch'. *Barron's*, **74**, 7, February 14, 16.

Davenport, T.H. (1993) *Process Improvement*, Harvard Business School Press, Boston, MA.

DeBresson, C. and Amesse, F. (1991) 'Networks of innovators: A review and introduction to the issue'. *Research Policy*, **20**, 5, October, 363-379.

Dierickx, L. and Cool, K. (1989) 'Asset stock accumulation and sustainability of competitive advantage'. *Management Science*, **35**, 12, 1504-1510.

Dixon, J., Arnold, P. Heineke, J., Kim, J. and Mulligan, P. (1994) 'Business Process Engineering: New strategic directions'. *California Management Review*, **36**, 4, 93-108.

Ettlie, J.E. and Swan, P. (1995) 'US-Japan manufacturing joint ventures and transfer of practice'. Chapter 10 in J. Liker, J.E. Ettlie, and J.C. Campbell (eds.), *Engineered in Japan*, Oxford University Press, New York.

Ettlie, J.E., and Johnson, M.D. (1994) 'Product development benchmarking versus customer focus in applications of Quality Function Deployment'. *Marketing Letters*, **5**, 2, 107-116.

Gerybadze, A. (1995) *Strategic Alliances and Process Redesign*, Berlin, New York.

Gulati, R. (1995) 'Does familiarity breed trust? The implication of repeated ties for contractual choice of alliances'. *Academy of Management Journal*, **38**, 1, 85-112.

Hammer, M. (1990) 'Reengineering work: Don't automate, obliterate'. *Harvard Business Review*, **68**, 4, 104-113.

Hammer, M. and Champy, J. (1993) *Reengineering the Corporation*. Harper Collins, New York.

Hayes, R.H., Wheelwright S.C. and Clark K. (1988) *Dynamic Manufacturing*. The Free Press, New York.

Jacobson, G. and Hillkirk, J. (1986) *Xerox: American Samurai*, Macmillan, New York.

Kogut, B. Walker, G., and Kim, D-J. (1995) Cooperation and entry induction as an extension of technological rivalry, *Research Policy*, **24**, 1, January, 77-95.

---

[1]   This brief case history is based, in part, upon the following article: Yoder, S.K., 'Shoving Back: Hogging Patents and Slashing Prices, H-P Grabbed the Inject-Printer Market from Japan', *Wall Street Journal*, September 8, 1994, p. A6 ff.

Kogut, B. and Kulatilaka, N. (1994) 'Operating flexibility, Global manufacturing, and the option value of a multinational network'. *Management Science*, **40**, 1, January, 123-139.

Kotha, S. and Fried, A. (1993) 'National Bicycle Company: Implementing a Strategy of Mass Customization'. International University of Japan/New York University Case Series.

Kotha, S. (1995). 'Mass customization: A strategy for knowledge creation and organization learning'. *International Journal of Technology Management*, forthcoming.

Limerick, D. and Cunnington, B. (1993) *Managing the New Organization*, Josey-Bass, San Francisco, CA.

MacCormack, A.D., Newmann, L.J., and Rosenfield, D.B. (1994) 'The new dynamics of global manufacturing site location'. *Sloan Management Review*, **35**, 4, Summer, 69-80.

McGill, M.E. and Slocum, J.W. Jr. (1994) *The Smarter Organization*, Wiley, New York.

Pascale, R.T. (1990) *Managing on the Edge*, Simon & Schuster, New York.

Peterson, J. (1993) 'Assessing the performance of European collaborative R&D policy: The case of Eureka'. *Research Policy*, **22**, 3, June, 243-264.

Pine, B.J. II (1993) *Mass Customization: The New Frontier in Business Competition*, Harvard Business School Press, Boston, MA.

Prahalad, C.K. and Hamel, G. (1990) 'The core competence of the corporation'. *Harvard Business Review*, **68**, 3, 79-81.

Senge, P.M. (1990) *The Fifth Discipline*, Doubleday Currency, New York.

Simon, F. (1995) 'Empowering Teams: The Lincoln Continental Program'. Presented at the Manufacturing Seminar, University of Michigan, Ann Arbor, Michigan, USA, February 27.

Sitkin, S.B., Sutcliffe, K.M., and Schroeder, R.G. (1994) 'Distinguishing control from learning in Total Quality Management', a contingency perspective. *Academy of Management Review*. **19**, 3, 537-564.

Stalk, G., Evans P., and Shulman L.E. (1992) 'Competing on capabilities: The new roles of corporate strategy'. *Harvard Business Review*, **70**, 2, 57-69.

Stalk G and Wever, A.M. (1993). 'Japan's dark side of time'. *Harvard Business Review*, **71**, 4, 93-102.

Storper, M., and Bennett H. (1991) 'Flexibility, hierarchy and regional development: The changing structure of industrial production systems and their forms of governance in the 1990s'. *Research Policy*, **20**, 5, October, 407-422.

Sykes, C. (1994) 'AT&T adopts a global manufacturing architecture'. *Manufacturing Systems*, **12**, 1, January, 34-39.

Teece, D.J. (1992) 'Competition, cooperation, and innovation'. *Journal of Economic Behavior and Organization*, **18**, 1-25.

Teece, D.J. 'Transactions customs economics and the multinational enterprise: An assessment'. *Journal of Economic Behavior*, **7**, 1, March, 21-45.

Tsay, A.A. and Loveloy, W.S. (1994) 'Supply chain control with quantity flexibility'. Working Paper, University of Michigan Business School, December, 50 pp.

Yoder, S.K. (1994) 'Shoving back hogging patents and slashing prices, H-P grabbed the ink-jet-printer market from Japan'. *Wall Street Journal*, September 8, 1994, p. A6.

CHAPTER 25

MANUFACTURING STRATEGIES, PRACTICES AND PERFORMANCE IN KOREA

B-H Rho, Sogang University, Y-M Yu, Dankook University, D-S Chang, Kyungki University, and S-H Chung, Hansung University, Korea

25.1. Introduction

Korean manufacturers have been very successful in overseas markets for the last 30 years, supported by the Korean government's export-driven policy. They have contributed tremendously to today's Korean economic growth, and some of them are now even world-leading companies in some areas such as memory chips, household electrical appliances, and shipbuilding.

However, a considerable number of Korean manufacturers are still struggling to reshape their competitiveness by shifting their strategic focus away from advantage based on lower labour costs to becoming more innovative, dependable, and flexible. Average wages have increased too sharply since 1988 to give them cost leadership and many developing countries in Southeast Asia and China have already caught up with them with cheaper labour costs in labour-intensive industries such as textiles and shoes. Thus, they are now challenged by a rapid change in the external competitive environment and eager to find a new way of competing in more globalised markets.

25.1.1. THE IMPORTANCE OF MANUFACTURING STRATEGY

Manufacturing strategy reflects corporate strategy and enables the manufacturing function to contribute to the long-term competitive advantages of the strategic business unit. Wheelwright (1978) and Buffa (1984) asserted that manufacturing should co-ordinate its decisions in a coherent manner so as to be consistent with, and in support of, the overall corporate strategy. According to them, manufacturing strategy's role is to support corporate strategy. Buffa (1984) asserted that manufacturing's strategic role is to select competitive priorities and pursue them as part of the basic framework of long-term business strategy. Thus, manufacturing strategy is a functional-level strategy that supports corporate strategy in achieving the organisation's strategic goals.

Since the initial call for a strategic concept of manufacturing by Skinner (1969), much has been written on this subject. Schroeder et al. (1986) reported that, in terms of the progression of thinking, the entire literature may be briefly summarised as follows:

P. Lindberg et al. (eds.), International Manufacturing Strategies, 417-434.
© 1998 *Kluwer Academic Publishers. Printed in the Netherlands.*

1. An initial awakening to the observation that manufacturing is the missing link in corporate strategy;

2. The prescriptive call for manufacturing strategy to be supportive of, and consistent with, the overall corporate strategy; and

3. The current state-of-the-art thinking that manufacturing can be more proactive in leading other functional areas in the contribution to the development of the corporate strategy.

They pointed out that manufacturing has now become a necessary function, changing from a support function to developing corporate strategy. Hayes et al. (1984) postulated that the manufacturing function should play a more proactive role in leading other functional areas in the development of the overall corporate strategy. Manufacturing is the most important strategic resource (Hill, 1988). According to Berry et al. (1992), CEOs are looking for improvements in manufacturing as an integral part of a total corporate response. Hum et al. (1992) asserted that the adoption of a strategic view of manufacturing may be the first and most important step in responding to the new set of competitive challenges, and they reported in their empirical survey that most manufacturing and operations managers perceived the role of manufacturing strategy as being the most critical to building competitive advantages.

Increased global competition, rapid science and technology changes, shortened product life cycles, and increased product variety force manufacturers to produce higher quality products at lower costs faster than before. Thus, manufacturing strategy is being given a more proactive strategic role as it is most important to increase flexibility, productivity, and quality simultaneously to create a competitive weapon for attaining business and corporate success. As Skinner (1969) argued, a close link should be established between corporate (or business) strategy and manufacturing practices, and a consistent link between manufacturing practices and manufacturing strategy is a prerequisite for meeting competitive challenges.

25.1.2. MANUFACTURING CHARACTERISTICS THAT LEAD TO COMPETITIVE ADVANTAGE

Co-operation between manufacturing and marketing functions
Hill (1989) and Terry (1992) asserted that it is critically important for manufacturing improvements to link markets and manufacturing strategy. According to Berry et al. (1992), it is better to improve master production scheduling rather than to invest in computer-based shop-floor control systems without an adequate understanding of the market, in order to better reflect actual sales mix in plant schedules, and so avoid imbalances in finished goods inventory, characterised by current inventory shortages and excess stock. In addition, they asserted that investment in the shop-floor control system was not only unnecessary, but also that the increase in paperwork raised overheads and took supervisory and management attention away from the key area of

master production scheduling. Thus, in order to achieve competitive advantage, master production scheduling should be linked with the marketing function to fulfil customers' rapidly changing orders.

Ronen et al. (1992) presented a strategic master production schedule that is derived from updated manufacturing strategy and provides co-operation of all relevant functions in the manufacturing priority determination process to assign a specific priority to each customer order. Chan (1993) asserted that manufacturing and marketing should be customer focused, and that they must be integrated in order to be a world-class competitor who combines both the strategies of least cost (world-class manufacturing) and differentiation (world-class marketing). Rho et al. (1994) reported in their empirical study that higher interface congruence between manufacturing and marketing can bring better manufacturing performance: better design of new products, lower manufacturing cost, and faster delivery.

Egalitarian operations management of bottom-up communications

According to Chan (1993), it is necessary to link manufacturing strategy with marketing strategy to plan customer-focused and time-based intelligent corporate strategy (ICS). This provides a paradigm for world-class manufacturing companies in coping with change and uncertainty in their search for market share and profits. It unifies workers, suppliers, distributors and even competitors in the pursuit of excellence in customer service for profit and market share.

In addition, he asserted that the philosophical core of ICS is teamwork zealotry based on principle-centred relationships of sincerity, trust, and integrity. It promotes open communication among functions within the organisation and across organisational boundaries for quality decision making. It is a holistic, people-oriented approach that is influenced by and managed with high spirits, emotions, values, and drive by the people. This, this people-oriented open communications enables the two functions of marketing and manufacturing to be integrated effectively.

Human resources management with education and training

White et al. (1982) found in their survey that most major problems of MRP implementation arose. because of education and training, lack of management, and the method of change adopted. According to Kinnie et al. (1992), major changes in manufacturing strategy failed to achieve their objectives because of problems of a non-technical nature more often than because of problems of a technical nature. They reported that four non-technical aspects—extent of planning, ownership of the changes, managers' perception of the changes, and the standard used for evaluation—are critically important to implement changes in manufacturing strategy. Thus, manufacturers may fail to achieve their objectives of new manufacturing strategy without effective education and training of human resources.

Technical innovation for quality
Young et al. (1992) found in their empirical survey that the principles of quality and technological innovation in automobile and electronics led to Japanese success in the machine tool and textile industries as well. According to them, Japanese managers of machine tools and textiles have achieved much higher performance than South Korean or Chinese managers through more effectively implementing MRP(II) or JIT in production planning and scheduling and materials management. Further, Japanese manufacturers have implemented technological innovations such as MRP(II) and JIT based on total quality control.

Integration of manufacturing strategy and manufacturing resource planning
Fawcett (1992) argued that it is necessary to implement co-ordinated global manufacturing strategies, which are operationalised through the establishment of global manufacturing networks and strategic logistics management for building a competitive edge. He stressed that the result of building logistics to strategic link global manufacturing resources is an enhancement of the firm's competitive position in both home and global markets.

Manufacturers in countries where national resources are scarce should especially enhance their ability to integrate manufacturing resources through global manufacturing networks. Leavy (1988) asserted that a co-operative relationship between business strategy and production and inventory planning has to be established by linking industry analysis to production planning through demand management and resource planning. Wheelwright et al. (1981) asserted that integration of manufacturing planning and manufacturing control is one of the key component of manufacturing strategy. Ronen et al. (1992) stated that strategic master production scheduling can be performed through an integrated module in MRP(II) and JIT systems, and this then forms the base function for an integration of manufacturing strategy with manufacturing resource planning.

Manufacturing competence and business performance
Kim et al. (1993) developed a conceptual framework from their empirical analysis that explains the relationship between competitive priorities and manufacturing competence. According to this framework, the strategic goal of business is represented by competitive priorities and manufacturing competence is represented by the degree of consistency between the importance given to a capability and the firm's strength with regard to that particular capability. They selected price, flexibility, quality, delivery, inventory, and services as competitive priorities.

They determined that manufacturing competence depends on the degree of manufacturing capability to support these competitive priorities. Their survey results show that a firm will gain more by focusing its resources on improving a key competitive variables than by attempting to improve across all dimensions simultaneously. When an effective business strategy is developed, they concluded, manufacturing competence will be critical in achieving its objectives. They asserted that the concept of competitive priorities and manufacturing competence are central components of manufacturing strategy, and that competent manufacturing contributes to the realisation of business strategy objectives.

25.1.3. LIMITATIONS OF MANUFACTURING STRATEGY THEORY

Since Skinner's (1969) introduction of the manufacturing strategy concept, the study of manufacturing strategy has evolved considerably. However, although manufacturing strategy is as important as previous studies have stated, the basic framework of manufacturing study is still incomplete (Leong et al., 1990; Kim et al., 1993). Fewer studies have been conducted on manufacturing strategy than on any other area of operations management (Young, 1992). In particular, few empirical analyses of manufacturing strategy and practices have been carried out (Hum et al., 1992), Because of the shortage of survey-based empirical work and the lack of effort placed in integrating manufacturing strategy ideas with concepts and theories, the research paradigm in manufacturing strategy is not solidly established (Kim et al., 1993).

25.2. Research Methodology

This study was conducted as a part of the IMSS project initiated by Chalmers University of Technology in Sweden and London Business School in England. The same questionnaire and sampling procedures were used for gathering data as in the original IMSS project. The research sample concentrated on Korean manufacturers of fabricated metal products, machinery and equipment (SIC 381-385) that employed 100 or more full-time workers. Smaller firms were excluded, since they have less opportunity to apply strategic concepts. Copies of the questionnaire were sent to 1000 randomly selected firms, and 61 usable questionnaires were returned, giving a response rate of 6.1%.

The process of the statistical data analysis was conducted as follows:

1. Descriptive statistical analysis was used to identify the overall characteristics of manufacturing strategy, practices, and performance.
2. Correlation analysis was used to evaluate the relationship between the perceived importance of competitive priorities and their respective manufacturing performances.
3. Multiple regression analysis was used to examine the relative contributions of various productivity improvement activities to each of the competitive priorities.
4. Cluster analysis and t-tests were conducted to find out differences in manufacturing strategies and practices between the high and low performance groups.

5. A stepwise discriminant analysis was conducted to identify critical manufacturing practices that differentiate high performance manufacturers from low performance ones.

25.3. Results

25.3.1. DESCRIPTIVE STATISTICAL ANALYSIS

This part presents the results of the statistical analyses that were used to investigate different sets of questions from the returned questionnaires. Analyses were performed to answer several different research questions. The major research questions to be addressed in this section are as follows:

1. Sample distribution
2. Strategic character of samples
3. The relationship between competitive strategy and manufacturing performance
4. Differences between high and low performance groups.

25.3.2. SAMPLE DESCRIPTION

Table 25-1 presents the data organised by business unit. Of the 61 companies, most reported falling into the "company" category (59%), followed by "plant" with 27.9% and "division" with 11.5%. This result indicates that more than half of our sample consists of single companies that have not diversified yet.

Table 25-2 organises the data by the organisation size. According to Table 25-2, 25 companies (41%) can be classified as large, with 300 or more employees. On the other hand, 59% (36 firms) of the sample have 300 or fewer employees. This result implies that our sample represents small-to-medium-sized firms in Korea rather than large.

Table 25-1. Sample distribution by business unit

Classification	Frequency	Percentage	Cumulative Percentage
Missing	1	1.6%	1.6%
Company	36	59.0%	60.7%
Division	7	11.5%	72.1%
Plant	17	27.9%	100%

Table 25-2. Sample distribution by company size

Number of employees	Frequency	Percentage
100 to 300	36	59.0%
300 to 500	9	14.8%
Above 500	16	26.2%

Table 25-3. Sample distribution by industry

SIC Code	Industry classification	Frequency	Percent
missing		1	1%
381	Metal products	14	23%
382	Machinery	7	11%
383	Electrical	6	9%
384	Transport equipment	17	27%
385	Measuring & controlling	16	26%

According to Table 25-3, the manufacture of transport equipment (27.9%), professional and scientific measuring and controlling equipment (26.2%), and metal products (23.0%) accounted for most of the sample, followed by mechanical and electrical equipment (21.3%). This indicates that the sample was more or less equally distributed in the metal processing and assembly industry.

25.3.3. STRATEGIC CHARACTERISTICS OF THE SAMPLE

Table 25-4 describes investment in the strategic factors of R&D, process equipment, and training and education. The results illustrate how Korean industry is investing relatively actively in process equipment (8.7% of total revenues), but is relatively weak in investing in R&D and training and education compared with advanced countries.

Table 25-5 indicates that Korean manufacturers emphasise quality and design more than manufacturing cost as strategic objectives. The implies that Korean industry is in a transition period, refocusing its strategic objectives towards customer orientation and emphasising faster and more dependable deliveries. However, customer service (4.44) and wider product range (3.59) received relatively lower importance as strategic objectives. This means that Korean firms are emphasising manufacturing efficiency-oriented objectives more than customer-oriented objectives.

Table 25-4. Investment in major strategic factors (percent of total revenue)

Strategic factor	Investment (percent of total revenue)
Research and development	2.54%
Process equipment	8.67%
Training and education	0.62%

Table 25-5. Average degree of importance of the manufacturing objectives

Manufacturing objectives (1= not important; 5 = very important)	Average
Having lower manufacturing costs	4.66
Offering faster delivery	4.48
Having superior customer service	4.44
Offering superior product design and quality	4.77
Offering more dependable deliveries	4.75
Offering a wider product range	3.59

Table 25-6. The characteristics of market scope

Market aims	Scale	Average
Market coverage	1 = few; 5 = many	2.90
Customer focus	1 = few, 5 = many	2.92
Geographic focus	1 = national; 5 = international	3.08

Table 25-7. Geographic orientation of sourcing and sales strategy

Sourcing and sales area	% of purchasing	% of sales
Other part of this country	59.9%	64.9%
Outside this country but within economic area	17.8%	12.4%
Outside this economic area	4.4%	5.9%

The characteristics of *market scope* pursued by Korean manufacturers are presented in Table 25-6 and Table 25-7. The results show that Korean manufacturers score about average in terms of market coverage, customer focus, and geographic focus,

However, sales and purchasing volumes, as indicated in Table 25-7, present a different perspective on *geographic focus*. The actual business transactions of the companies in the sample tend to be primarily locally-oriented (local sales 64.9% and local purchasing 59.9%), which implies that Korean manufacturers have yet fully to expand internationally.

25.3.4. LINKS BETWEEN MANUFACTURING AND STRATEGIC PLANNING

Table 25-8 illustrates the links between manufacturing and other functional parts of the organisation. First of all, translating corporate and marketing goals into a manufacturing strategy seems to be performed through a somewhat formal process (average = 3.51). In addition, the average score of 3.93 for the degree of manufacturing influence on the development of corporate strategy and goals indicates that the role of the manufacturing function is emphasised when corporate strategy is established. However, manufacturing objectives for Korean firms seem to be controlled still by short-term goals such as financial planning.

As illustrated in Table 25-9, the degree of automation employed by business units in the Korean sample shows that most are still at the conventional or repetitive automation level, while the use of flexible automation or computer-integrated manufacturing systems is still beginning.

Table 25-8. Strategic links between manufacturing and corporate strategy

Classification	Average
Translating process other goals into manufacturing strategy (1 = none; 3 = informal process; 5 = formal process)	3.51
Manufacturing influence on the development of corporate strategy (1 = not at all; 5 = a lot)	3.93

Table 25-9. Use of automation

Classification	Automation equipment	Number of units
Manufacturing	FMS/FMC	2.46
	NC machines (not in FMS/FMC)	15.67
	Conventional machines	63.4
	Machining centres	7.3
	Robots	1.8
Assembly	Robots	6.0
	Flexible Assembly System	1.2

Table 25-10 indicates the degree of implementation over the past two years for the manufacturing improvement activities and their relative payoffs. In general, the results show below average for both the degree of implementation and relative payoff except for CAD. Among the 27 surveyed major improvement programmes, CAD, TQC, SPC, MRP, and JIT deliveries to customers show relatively high degree of implementation and payoff. The results suggest that Korean industry emphasises efficiency-oriented improvements rather than effectiveness-oriented ones. For example, in manufacturing

Table 25-10. Manufacturing improvement programmes: Level of implementation and relative payoffs

Programme	Degree of use (1 = no use; 5 = high use)	Relative payoff (1 = low; 5 = high)
Total quality management programme	3.11	2.89
Statistical process control	2.23	2.07
ISO 9000 certification	1.48	1.57
MRP	2.97	2.61
MRP II	2.23	1.97
JIT manufacturing (lean production)	2.43	2.16
JIT deliveries to customers	2.62	2.43
SMED (single minute exchange of dies)	1.38	1.30
Pull scheduling	1.70	1.48
Zero defect programmes	1.70	1.75
Computer-aided manufacturing (CAM)	2.33	2.13
Computer-aided design (CAD)	3.18	3.05
Design for Assembly/Mfg (DFA/DFM)	1.64	1.59
Quality function deployment (QFD)	1.27	1.48
Value analysis	1.62	1.56
Quality policy deployment (QPD)	1.72	1.66
Plant within a plant	1.46	1.41
Defining a manufacturing strategy	2.21	2.16
Simultaneous engineering	1.26	1.36
Activity-based costing	1.69	1.70
Implementing a team-based approach	1.80	1.64
Benchmarking	1.48	1.52
Kaizen (continuous improvement)	1.61	1.57
Total productive maintenance (TPM)	2.31	2.03
Energy conservation programmes	2.02	2.10
Environmental protection programmes	1.72	1.89
Health and safety programmes	2.41	

design, CAD shows a high degree of implementation, while DFM, value engineering, and concurrent engineering have low levels of implementation. Similar results were found for quality-related improvement programmes: innovative activities such as ISO 9000 certification and Quality Policy Deployment (QPD) had low levels of implementation and payoff.

25.3.5. MANUFACTURING IMPROVEMENT PROGRAMS AND PERFORMANCES

Only thirteen out of twenty-seven improvement programs appeared to contribute to at least one or more manufacturing performances as shown in Table 25-11. The ZD program shows an outstanding effect on the manufacturing performances. It was found to be effective for improving all the addressed manufacturing performances except manufacturing costs. JIT deliveries has a positive relationship with two manufacturing performances: quality and costs. Each of the remaining eleven programs shows a significant contribution toward only one its respective manufacturing performance. CAM, plant within a plant, team approach, benchmarking and energy conservation programs have shown an inverse relationship with their respective manufacturing performances. It may be a result of temporary unstableness of the system at the beginning stage of adopting these new programs. Or it may be an unavoidable side effect which can occur when not having well prepared program implementation process.

Table 25-11. The effect of improvement programs on manufacturing performances

Manufacturing Improvement programme	Quality	Cost	Depend able Delivery	Custom er Service	Fast Delivery	Wider product range
SPC			.759			
ISO 9000		-.375				
MRP	.369					
MRP II		.523				
JIT deliveries	.286	.263				
ZD programs	.258		.310	.334	.454	.550
CAM						-.248
DFA/DFM				.299		
Value analysis						.293
Plant within a plant			-.328			
Implementing team approach	-.432					
Benchmarking		-.339				
Energy conservation programmes						-.367
Adjusted R-square	.284	.351	.346	.230	.192	.323
Significance	.000	.000	.000	.000	.000	.000

Note: Each coefficient is a Beta-coefficient and statistically significant at the level of alpha=.05 or less by t-test.

25.3.6. DIFFERENCES BETWEEN HIGH- AND LOW-PERFORMING GROUPS

Comparing managerial practices of high and low manufacturing performance groups can help us begin to grasp how manufacturers can achieve higher performance. We have analysed the contribution of the various performance improvement programmes to each of six manufacturing performance criteria. However, manufacturing competence is a complicated concept: it hardly seems achievable simply through adopting operational activities aimed for higher manufacturing performance. It seems more likely that high manufacturing performance will be achieved through co-ordinated and integrated managerial efforts in the areas of strategy, tactics, and operations. Therefore, knowing what high-performing companies are doing differently than low-performing companies in strategic and operational differences is but a starting point in understanding manufacturing competence.

Six manufacturing performance measures, more specifically the percentage change in six areas of manufacturing performance, were used as clustering variables. A pre-clustering procedure was performed before clustering to eliminate outliers that might lead to an unreliable clustering result. Respondents were classified as outliers if one or more of their performance levels exceeded their relative mean value plus or minus three standard deviations. Since the lower limit of the mean minus three standard deviations had a negative value for all six performance measures, only the upper limit was used for performance measures in classifying outliers, as shown in Table 25-12.

Two clustering procedures were used to classify the whole sample into two distinct performance groups: hierarchical and non-hierarchical clustering. The SPSS modules CLUSTER and QUCIK CLUSTER were used for hierarchical and non-hierarchical cluster analyses, respectively. First, Ward's method using squared Euclidean distances was used to estimate the clustering pattern and get initial seed values of the six clustering variables for the high and low performance groups. Next, the six pairs of performance centroids that were obtained through the hierarchical clustering procedure were used as initial seeds for the non-hierarchical clustering procedure.

Table 25-13 shows the final clustering result. Except for delivery lead-time, five of the six manufacturing performance measures showed a significant difference between the high and low performance groups. The same clustering result was obtained when delivery lead-time was excluded. Thirty-four respondents were classified into the high-

Table 25-12. Performance Improvement - Means, Standard Deviations, and Upper Limit

Cluster variable	Mean (%)	S.D	Upper Limit
Conformance to specification	13.74	17.45	66
Average unit manufacturing cost	8.95	9.26	36
On-time deliveries	18.44	22.97	87
Customer service	12.18	15.41	58
Delivery lead time	12.30	16.82	62
Product variety	11.10	19.58	68

Table 25-13. Comparison of High and Low Performance Groups

Manufacturing Performance (% Change)	Good	Poor	Prob.
Conformance to specification	16.79	3.04	0.00
Average unit manufacturing cost	10.17	4.50	0.00
On-time deliveries	23.23	1.00	0.00
Customer service	16.61	2.18	0.00
Delivery lead time	12.67	6.81	0.08
Product variety	11.91	1.50	0.00

performance group and the other twenty-two respondents were classified into the low-performance group.

The high manufacturing performance group also showed significantly higher levels of several business performance measures (Table 25-14), including significantly higher increases in inventory turnover and market share.

Table 25-14. Business Performance Comparison of High and Low Performance Groups

Business Performance (% Change)	Good	Poor	Prob.
Inventory turns	12.20	5.22	0.00
Market share	13.76	2.40	0.00
Profitability	11.91	7.59	0.46

Table 25-15 compares the classification results by hierarchical and non-hierarchical clustering to assess the clustering validity. It compares the cluster memberships assigned by the two clustering procedures. As shown in this table, 96.42% of the respondents were classified into the same performance group and thus this classification shows a high level of clustering validity.

Table 25-15. Comparison of Hierarchical and Non-Hierarchical Classification Results

Hierarchical clustering	Non-hierarchical clustering		
	High	Low	Total
High	34	2	34
Low	0	20	22
Total	34	22	

Classification consistency: $(34+20)/54 = 96.42\%$

25.3.7. COMPETITIVE PRIORITIES AND MANUFACTURING PERFORMANCE

For successful strategy, manufacturing strategy should have clearly-defined strategic goals and a well-thought-out implementation scheme. A good match between strategic focus and operational activities will enable long-term competitive advantage through increased effectiveness; in other words, a more goal-congruent allocation of limited manufacturing resources.

Table 25-16. Pearson Product-Moment Correlations

Goal importance (5 - very important)	Mfg. quality	Mfg. cost/unit	On-time deliveries	Customer service	Delivery lead time
Manufacturing quality	-00.24 0.283				
Manufacturing cost		0.208 0.229			
Dependable deliveries			0.017 0.371		
Customer service				0.271 -0.082	
Faster deliveries					-0.052 -0.228
Wider product range					

Top row = correlation coefficient for high-performance group
Bottom row = correlation coefficient for low-performance group

Table 25-16 shows the Pearson product-moment correlation coefficients for the degree of importance of manufacturing goals and their respective performance. The magnitude of the correlation coefficients for both the high and low performance groups were too low to be statistically significant. That is, the analysis did not show a clear relationship between the importance of strategic goals and manufacturing performance in that area, even for the high-performing group This implies that Korean manufacturers may lack a clear linkage between strategic goals and operational activities.

25.3.8. MARKETS, CUSTOMERS AND PRODUCTS

The high-performing manufacturers had wider market coverage and a more diversified customer focus, as shown in Table 25-17. They also had greater differences between orders and/or customers in order types, technologies, and quality requirements. That is, differentiation rather than focusedness can be a more effective competitive strategy in Korea. Competing through differentiation in Korea means competing through product customisation rather than through differentiating products for different customers and/or market segments. One empirical study has also reported that Korean manufacturers are now seeking more differentiation than before, and emphasising customisation in their competitive strategy more than Japanese manufacturers (Rho et al., 1995).

The high performance group seems to have more pressures to fulfil short-term financial, budget, and output requirements. This may be a result of the orientation toward differentiation and customisation, since they should be able to satisfy less stable and more diverse market segments and/or customer demands. The high performance manufacturers have also experienced a higher increase in production volume over the past five years. They are also expecting more rapid growth of production volume in the future. Thus, they have a more optimistic view of the change in market and/or customer demands. They expect that a higher proportion of revenues will come from new

Table 25-17. Differences Between Markets, Customers, and Products

Strategic characteristic	High	Low	t-test significance
Market coverage (5 = many markets)	3.32	2.40	*
Customer focus (5 = many customers)	3.32	2.45	*
Differences between customers/markets	2.64	3.31	**
Manufacturing's need to meet short-term requirements	3.82	3.18	**
Production volume change over the past 5 years	38.55	12.77	*
Estimated production volume change over the next 5 years	56.23	14.09	*
Estimated percentage change in revenue from new products over next five years (%)	27.29	6.36	*
Proportion of direct salaries/wages (%)	18.70	14.18	**
Proportion of manufacturing overhead (%)	24.47	16.00	*

$** p < 0.01; * p < 0.05$

products, which seems to be related to their differentiation orientation, which indicates that they will need a wider product range to yield more revenues in the future.

25.3.9. MANUFACTURING PRACTICES

An analysis of manufacturing practices related to organisation and management did not show a significant difference between the high and low performance groups except for the number of suggestions for process and product improvements by employees (Table 25-18). Contrary to expectations, the low performance group had a significantly higher number of suggestions than the high performance group. However, this result needs to be interpreted cautiously since the contribution of employee suggestions to product and process improvement in Korea has not been quantified. An empirical study of Korean manufacturing has reported that the percentage adopting quality circles has decreased to 86.7% in 1991 from 96.3% in 1987, and perceived benefits of quality circles have dropped markedly (Rho and Yu, 1994). Since most worker suggestions in Korea come through quality circles, this devaluation may reflect their limited usefulness.

Manufacturing practices related to the process and technology also failed to show much difference between the two performance groups. The high performance manufacturers are using more NC machines than their counterparts. Since they have to

Table 25-18. Differences in Manufacturing Practices

Manufacturing Practice	High	Low	t-test significance
Number of suggestions per employee per year	3.67	8.22	*
Number of NC machines installed	22.35	5.45	**
Throughput efficiency (%)	65.47	52.04	**
Proportion of external quality costs (%)	20.15	9.50	*

$** p < 0.01; * p < 0.05$

satisfy many different market segments and/or customers, they surely need more flexible machining capabilities to produce more customised products. More use of NC machines seems to contribute in part to higher throughput efficiency. About 65.5% of the throughput time was devoted to work on products by high-performing manufacturers, whilst only 52.0% by low-performing manufacturers.

Only the proportion of external costs to the entire cost of quality shows a significant difference between the high and low performance groups. The high-performing manufacturers are spending more money on post-sales activities such as handling warranties, repairs and returns. However, this higher proportion of external costs may be a result of more extensive involvement in customer service rather than incomplete or lower product quality.

Table 25-19 shows which productivity improvement activities have a significant difference in either degree of use of relative payoff between the two performance groups. The high performance manufacturers have implemented a higher level of process quality improvement and manufacturing planning activities such as SPC, MRP, and MRP II. They have also achieved higher payoffs from these activities. Their emphasis on differentiation may have them paying more attention to dependable planning and scheduling capabilities to satisfy unstable short-term customer orders. They use more CAD and TPM even though their relative perception of payoffs is not significantly different than the low performance group's.

25.3.10. PRACTICES DISCRIMINATING BETWEEN HIGH AND LOW PERFORMANCE GROUPS

Thus far, twenty-one strategic and manufacturing practices that have shown significant differences between the two manufacturing performance groups have been discussed. However, some of these significantly different practices may address very similar aspects of practice or performance, or may be less useful than other in discriminating between manufacturers with high and low performance.

A stepwise discriminant analysis using each respondent's group membership as the dependent variable and the 21 significantly different manufacturing practices as independent variables was conducted in order to differentiate each practice's

Table 25-19. Productivity Improvement Activities

Productivity improvement activity	Degree of use (5 = high)			Relative payoff (5 = high)		
	High	Low	t-test	High	Low	t-test
SPC	2.50	1.59	*	2.32	1.36	*
MRP	3.20	2.45	**	2.85	2.00	*
MRP II	2.47	1.68	**	2.20	1.36	*
CAD	3.41	2.68	**	--	--	--
TPM	2.52	1.81	**	--	--	--

$** p < 0.01; * p < 0.05$

Table 25-20. Results of Step-Wise Discriminant Analysis by Performance Group

Manufacturing Practice	Standardised Discriminant Weight	Discriminant Loading
Market coverage (5 = many markets)	0.391	0.268
Differences between customers (5 = little difference)	-0.397	-0.183
Manufacturing's need to fulfil short-term requirements (5 = a lot)	0.224	0.214
Production volume change over past 5 years (%)	0.452	0.204
Proportion of direct salaries and wages (%)	0.313	0.196
Proportion of manufacturing overhead	0.455	0.241
Number of suggestions per employee per year	-0.584	-0.310
Number of NC machines installed	0.365	0.163
Throughput efficiency (%)	0.465	0.196
Proportion of external quality costs (%)	0.634	0.272

Canonical correlation: 0.794
Group centroids: High = 1.031, Low = -1.594
Wilk's Lambda: 0.369 (p = 0.000)
Classification accuracy: 92.86%

contribution to group membership. Because of the small sample size, a holdout sample could not be used to validate the procedure. Ten out of the twenty-one manufacturing practices were selected as significant variables (Table 25-20). The discriminant analysis has a very high statistical significance (p = 0.000) and classification accuracy (92.86%).

Diversification in terms of markets, customers, and process technologies seems to be a dominant feature of the high performance group in contrast to the low performance group. This orientation towards differentiation calls for more flexible manufacturing capabilities and short-term customer or market demand satisfaction More use of NC machines means making more investment in flexibility and results in higher manufacturing overheads. This hardware advance brings higher throughput efficiency. Higher external quality costs seem to reflect a higher emphasis on customer service. Fewer worker suggestions might hardly be considered as a feature of high-performing companies; however, the quality of employee suggestions needs to be taken into account since some empirical studies in Korea have already addressed this issue.

25.4. Summary and Conclusions

Korean manufacturers appear to pay little attention to the importance of strategic focus. Their current manufacturing practices are not thought to be in accordance with their conceived important competitive priorities, since even the high-performing manufacturers have failed to demonstrate a significant direct relationship between priorities and practices. Thus, the first thing they have to do to improve their competitiveness will likely be to think seriously about what competitive priorities are really important to them and what manufacturing practices they should focus their limited resources on in order to achieve their strategic goals.

Thirteen out of twenty-seven productivity-improving activities appear to significantly contribute to achieving at least one or more of six strategic goals. Above all, the zero defects (ZD) programme proved to be the most effective tool for improving competitiveness: it did show a significant contribution to all strategic goals except for lower manufacturing costs.

The comparison of manufacturing practices between the good and the poor manufacturing performance groups gave us some meaningful clues to improving manufacturing competitiveness: more diversification, hardware advance for better flexibility and better customer service following sales. The good performance manufacturers have more diversified markets and customer orders. They have more flexibility to meet their short-term and diversified customer and/or market demands from adopting more flexible process technology. They seem to be more extensively engaged in post-selling customer service.

References

Akkermans, H. (1993) 'Participative business modeling to support strategic decision making in operations - a case study', *International Journal of Operations and Production Management*, **13**, 10, 34-48.

Archer, G. (1990) 'MRP: A review of failure and a proposal for recovery using CBS', *BPICS Control*, December 1990/January 1991.

Berry, L.W. and Hill, T. (1992) 'Linking systems to strategy', *International Journal of Operations and Production Management*, **12**, 10, 77-91.

Buffa, E.S. (1984) *Meeting the Competitive Challenge*, Dow Jones-Irwin, Homewood, IL.

Chan, K.C. (1993) 'Intelligent corporate strategy: Beyond world-class status', *International Journal of Operations and Production Management*, **13**, 9, 18-28.

Fawcett, S.E. (1992) 'Strategic logistics in coordinated global manufacturing success', *International Journal of Production Research*, **30**, 4, 1081-1099.

Hayes, R.H. (1982) 'Why Japanese factories work', *Harvard Business Review*, July-August, 72.

Hayes, R.H. and Wheelwright, S.C. (1984) *Restoring Our Competitive Edge*, John Wiley and Sons, New York.

Hill, T. (1989) *Manufacturing Strategy*, Irwin, Boston, MA.

Kim, J.S. and Arnold, P. (1993) 'Manufacturing competence and business performance: a framework and empirical analysis', *International Journal of Operations and Production Management*, **13**, 10, 4-25.

Lee, S.M. and Schniederjans, M. (1994) *Operations Management*, Houghton and Mifflin, Boston MA.

Leong, G.K., Snyder, D.L., and Ward, P.T. (1990) 'Research in the process and content of manufacturing strategy', *OMEGA*, **18**, 2, 109-122.

Kinnie, N.J., Staunton, R.V.W., and Davis, E.H. (1992) Changing manufacturing strategy: some approaches and experiences, *International Journal of Operations and Production Management*, **12**, 7/8, 92-102.

Leavy, B. (1988) 'The production and inventory management and strategy fields—a case for more dialogue', *Production and Inventory Management*, First Quarter, 61-64.

Rho, Boo-Ho, Hahm Y-S, and Yu, Y-M, (1994) 'Improving interface congruence between manufacturing and marketing in industrial product manufacturers', *International Journal of Production Economics*, **37**, December, 27-40.

Rho, Boo-Ho, and Yu, Y.-M. (1994) *Changes in Manufacturing Practices for Improving Productivity in Korea*, Jipmoondang, Seoul, Korea.

Rho, Boo-Ho, Murakoshi, T. and Yu, Y-M. (1995) 'Comparing manufacturing practice and productivity between Japan and Korea', *Proceedings of the 12th Pan-pacific Conference*, New Zealand, 1995.

Ronen, B., and Rozen, E. 'The missing link between manufacturing strategy and production planning', *International Journal of Production Research,* **30,** 11, 2659-2681.

Schroeder, R.G., Anderson, J.C. and Cleveland, G. (1986) 'The content of manufacturing strategy: an empirical study', *Journal of Operations Management,* **6,** 4, 405-415.

Hum, Sin-Hoon and Leow, Lay-Hong, (1992) 'The perception of the strategic role of manufacturing amongst operations managers: An empirical study based on a newly industrialized economy', *International Journal of Operations and Production Management,* **12,** 11, 15-23.

Skinner, W. (1969) 'Manufacturing—missing link in corporate strategy?', *Harvard Business Review,* May-June, 136-145.

Westbrook, R. (1988) 'Time to forget just-in-time? Observations on a visit to Japan', *International Journal of Operations and Production Management,* April, 5-21.

Wheelwright, S.C. and Hayes, R.H. (1985) 'Competing through manufacturing', *Harvard Business Review,* Jan-Feb., 99-109.

White, E.M., Anderson, J.C., Schroeder, R.G., and Tupy, S.E. (1982) 'A study of the MRP implementation process', *Journal of Operations Management,* **2,** 3.

Wood, R., Hull, F., and Azumi, K. (1983) 'Evaluating quality circles: the American application', *California Management Review,* **XXVI,** 1, Fall 1983, pp. 37-53.

Young, S.T., Kwong, K. K., Cheng, L. and Fok, W. (1993) 'Global manufacturing strategies and practices: a study of two industries', *International Journal of Operations and Production Management,* **12,** 9, 5-17.

CHAPTER 26

THE MULTI-FOCUSED MANUFACTURING PARADIGM: ADOPTION AND PERFORMANCE IMPROVEMENTS WITHIN THE ASSEMBLY INDUSTRY[i]

Gianluca Spina, Emilio Bartezzaghi, Andrea Bert, Raffaella Cagliano, Politecnico di Milano, Milan, Italy, Domien Draaijer, Philips Semiconductors, Nijmegen, The Netherlands, and Harry Boer, University of Twente, Enschede, The Netherland

26.1. Introduction

This contribution defines a new manufacturing paradigm and investigates its adoption and performance within the assembly industry on a global basis. The paradigm is based on a set of basic principles for designing and managing production systems that discard the traditional way of organising manufacturing activities and apply across companies that pursue different manufacturing strategies and implement different innovative techniques.

The paradigm is characterised by the simultaneous implementation of three principles: strategic *multi-focusedness, integration* of business processes across functions, and *process ownership.* Using fuzzy logic, the hypotheses about the adoption of and the performance improvements due to this multi-focused manufacturing paradigm are explored using a sample of 443 companies out of the 600 companies from the IMSS database.

The analysis suggests that the paradigm has been widely adopted across the countries and industries in the IMSS survey. It further shows that process ownership has been poorly implemented by most of the companies, compared with higher levels of multi-focusedness and integration. Finally, it appears that companies that have adopted the paradigm have improved their performance significantly more than non-adopters.

The main managerial implication of the multi-focused manufacturing paradigm is that companies can simultaneously achieve and continuously improve and change performances across competitive priorities that have traditionally been regarded as antithetical provided that 1) the workforce is educated in such a way that they are not only capable but also allowed to proactively manage their processes, and 2) tasks are grouped and resources are allocated to departments so that integration and process ownership are facilitated.

---

[i] This has been published as: 'Strategically Flexible Production: The Multi-Focused Manufacturing Paradigm', *International Journal of Operations and Production Management*, **16**, 11, 1996, pp. 20-41.

P. Lindberg et al. (eds.), International Manufacturing Strategies, 435-461.
© 1998 *Kluwer Academic Publishers. Printed in the Netherlands.*

26.2. Research Background

Over the past twenty years, many forces have driven industrial companies to rediscover the strategic role of manufacturing, including the globalisation of competition, the growing saturation of markets, the increasing quality and service expectations of customers, the increasing pace of technological innovation and the shortening of product life cycle, as well as many others. As a consequence, manufacturing and assembly activities have undergone many changes, not only technological but also organisational and managerial.

A vast body of innovations such as Just-in-Time, Total Quality Management, and Concurrent Engineering have been widely implemented across countries and industries. As a result, both the internal organisation of the factory and its external environment— including market demand, technology development, workforce education and expectations, labour and capital market—appear to be very different today from the general features that dominated the industrial development in the past, generally referred to as the Fordist paradigm.

The shift away from mass production to a new form of industrial organisation has followed different paths, some of which initially drew enthusiastic attention but were abandoned or reshaped later on—e.g., the experiences of Volvo in Kalmar and Uddevalla and the so-called "neo-craftsmen" models. Other alternative paradigms have thrived and could attract much interest—see Kennedy and Florida (1989) for a review. For example, the model of "flexible specialisation" (Piore and Sabel, 1984) draws upon the case of the textile district in Northern Italy and the textile machinery district of Baden-Wurtemberg. Though fascinating, "flexible specialisation" appears to be inapplicable to the most important capital-intensive sectors.

The Japanese way has been regarded as the most appropriate replacement for the Fordist paradigm. The development of the Just-in-Time concept at Toyota since the 1950s and further refinements have seemed to discard the basic principles of Fordism (see, for example Womack, Jones, and Roos, 1990). However, this view has been questioned, and Toyotism—with its extreme exploitation of workers' capabilities—has also been depicted as "hyper-Fordism" (Doshe, Jurgens, and Malsch, 1985). In addition, Western manufacturers have experienced many difficulties in adopting or adapting the Japanese style of management and way of organising the production system. Many country-specific factors have been put forward to explain these difficulties, including ethnic and sociocultural aspects, the role of educational systems, industrial structure, the role of the government in the planning of industrial activities, the long range attitude of financial managers in Japan, the steady growth of the Japanese economy since the fifties, the quality of public infrastructures, and the cost of money (see Odaka (1984) and Liepitz (1987) for instance).

In the 1990s, the shift away from Fordism is still taking place throughout the industrialised countries, partly embodying some features of the previous post-Fordist experiences and partly introducing new ones. Companies seem to be re-designing their processes, but only by discarding some traditional features of the industrial organisation—strong labour specialisation, heavy control hierarchies, functional organisation, trade-off management, co-ordinating mechanisms based on formal procedures, and so on. Despite the range of different strategic choices that industrial companies must make and the different environmental conditions they have to meet, is it possible today to identify a new paradigm underpinning the different models that companies are implementing? Also, is this paradigm a clear-cut break with Fordism? We think so, and in this chapter we provide the outlines of what we call the *multi-focused manufacturing paradigm*. In fact, it is possible to discover a limited set of shared principles for designing and managing the production systems that combines different models and paths of innovations (see Section 26-4). Those principles appeared to have discarded the traditional Fordist assumptions. We also explore the adoption of the new paradigm and the consequent performance improvement on a global basis using the IMSS data.

The emergence of the multi-focused manufacturing paradigm is not expected to provide a new *"one best way"*. Recently, those who sought to vindicate the strategic approach to manufacturing have contended that the slavish imitation of successful managerial and organisational innovations, even when feasible, drives companies to become similar to each other, thus narrowing their strategic space (Hayes and Pisano, 1994). However, the adoption of a paradigm does not mandate the implementation of an integrated set of techniques and a shared configuration, which could prevent companies from building their own competitive advantage. In fact, the whole research project on the multi-focused manufacturing paradigm is motivated by the idea that a clear distinction is needed between three levels:

- *innovative techniques* for improving production systems, that is, the wide set of individual innovations in process, technology or organisation (for example, Just-in-Time deliveries, Statistical Process Control, MRP II, Quality Function Deployment, Kanban and many others);
- *manufacturing models*, that is, the systematic implementation of combinations of techniques that companies select and adapt according to their internal and external environments. In fact, the manufacturing model of a single company will be affected by technological and organisational constraints and opportunities, internal capabilities and goals, and finally competitive conditions;
- the emerging *manufacturing paradigm*, that is, a limited set of new principles that underpins the new techniques and integrates the manufacturing models. This new paradigm is supposed to replace the prevailing *modus operandi* within different countries and assembly industries, which is generally referred to as the Fordist paradigm.

A vast body of literature has already investigated the adoption and diffusion of the individual techniques (see for example Voss and Robinson, 1987; Gilbert, 1990;

Bartezzaghi, Turco and Spina, 1992). Noticeable "applications" or "adaptations" have been detected all over the industrialised world—for example in Italy (Bonazzi, 1993)—not only within the automotive industry, but elsewhere, for example in the electronic industry in the USA (Abo, 1990). In addition the transferability of some successful models has been studied—e.g., the Toyota model—showing that there are limits to the extent to which they can be imitated, due to a number of country-specific factors.

The basic assumption of such research is that the individual innovative techniques are actually universally applicable and thus relatively easy to imitate, but they must be combined in different manufacturing models—the system-level implementation—contingent on some contextual factors relating, for example, to exogenous factors such as country, industry, company size, and process technology, and endogenous factors relating to the individual strategy and goals that the company pursues. Therefore, no new one best way is expected, but there is instead 1) considerable design space for a company to select a manufacturing model, i.e., a configuration in terms of technologies, organisational forms and managerial techniques that suits its situation best; and 2) a new set of principles underlying the different models. Just as different models were inspired by the Fordist paradigm in the past, the multi-focused manufacturing paradigm will integrate different models and innovation paths for shifting manufacturing away from Fordism.

26.3. The New Manufacturing Paradigm - Consistency and Performance Improvement

The rise of the multi-focused manufacturing paradigm results from the environmental changes that have taken and are still taking place (see Section 26-4.1). These require companies and their production systems to adapt in terms of new forms and levels of effectiveness. A wide literature has theoretically defined the *effectiveness* of manufacturing systems, which generally refers to the idea of consistency, that is the fit between the component elements of the organisation and its environment (Draaijer and Boer (1995) and Draaijer (1993) extensively report on this subject). Some of the previously-mentioned authors also indicate how manufacturing should be organised in order to meet present market needs, resource availability, workforce expectations and so on. This is the area of the literature where capabilities and design characteristics of the manufacturing system are brought together.

Design characteristics refers to the choices being made in designing the manufacturing system. Authors such as Skinner (1985), Hill (1985), Hayes and Wheelwright (1984) have proposed several dimensions or decision categories to facilitate the description of the manufacturing system. Although there are some differences in the dimensions distinguished by the authors, the key point is that a co-alignment of dimensions is required in order to achieve the requisite capabilities. Consistency in these decisions is seen as crucial for creating an effective manufacturing system. Additionally, Hayes and Wheelwright (1984) distinguish between external and internal consistency. External consistency refers to the match between the manufacturing strategy and the business environment of the company such as available resources, competitive behaviour and market demands. Internal consistency refers to the match within the manufacturing function and across functions within the business unit.

Environmental changes call for new internal and external consistency. If the environmental changes are great enough, they may not only require changes at the technique or model level, but also at the paradigm level. This is exactly what we see happening: many manufacturers are adopting new principles, explicitly or implicitly, in order to adapt themselves to present requirements (see Section 26-4 for the implications of the new principles).

However, effectiveness is a relative dimension. In fact, in order to be attractive to customers, a production system should perform better than those of its competitors. If the company is to remain effective, the following questions demand a positive answer: does the current production system perform at least as well as its competitors, and is it able to improve at least as fast as its competitors? Thus, in order to measure the ability of a new paradigm to improve the effectiveness of a production system, we must test the level of performance improvement related to the orientation to the paradigm at both manufacturing and business level.

Porter (1980) speaks of the "degree of rivalry among firms" and the "threat of new entrants" when describing the relationship between a company and its competitors. He observed that firms may employ three different strategies to defend their competitive position: overall cost leadership, differentiation or focus. Porter concluded that it is not possible for a company to successfully maintain two or more strategies at the same time. Consequently, the capabilities of the manufacturing systems should be geared towards one of these three different strategies. However, Miller and Friessen (1986) concluded after a thorough empirical study that a combination of cost leadership and differentiation is possible, especially in the consumer market (*cf.* Bolwijn *et al.,* 1986).

Whether the performance of a manufacturing system compared with competitors is adequate if survival is at stake depends on the goals of the company. Therefore, to assess the current and future strengths of a company it is advisable to describe the position of a company relative to its rivals. This is in line with Pfeffer (1977), who maintains that effectiveness can only be assessed comparatively. We can measure the position of a company relative to competitors via two dimensions: first, its relative position regarding performance in the market place; second, the relative speed of organisational change aimed at improving performance (see Figure 26-1).

Performance improvement is one of the main goals a company should pursue in order to survive in a competitive environment. Cell (1,1) indicates that a even better performing company may lose its competitive advantage because it responds less adequately to opportunities and threats than its competitors. Cell (3,3) represents a company that performs worse than its competitors but whose future relative

		Speed of change relative to competitors		
		lower	equal	higher
Performance	Better	1,1	1,2	1,3
level relative to	Equal	2,1	2,2	2,3
competitors	Worse	3,1	3,2	3,3

Figure 26-1. Performance versus speed of improvement (based on: Draaijer and Boer (1995) and Draaijer (1993)).

performance will improve. In order for a company to survive it should at least improve with a speed equal to that of the nearest competitor. For that reason, we examine whether the multi-focused manufacturing paradigm enables a better degree of performance improvement than companies that have a lower degree of adoption of the paradigm.

26.4. The Multi-Focused Manufacturing Paradigm

26.4.1. BASIC PRINCIPLES

Recently, much has been written about the general changes occurring in manufacturing systems. For example, Drucker (1990), Hayes, Wheelwright and Clark (1988), and Bartezzaghi (1992) have all proposed conceptual frameworks that each suggest a limited number of principles underpinning different manufacturing models. Although there are different emphases in those proposals, they can be regarded as coherent identifications of a unique paradigm based on external and internal consistency. Today's external consistency seems to require multiple performances simultaneously; rapid priority changes; time effectiveness and quick response; increased quality of working life and, in general, more involving and motivating tasks for an increasingly educated workforce. To match these requirements, internal consistency is also needed, involving global optimisation; process focus in the organisational design to keep quality and time consistent with customer needs; development of internal capabilities and local problem solving; alignment of manufacturing and the new product development processes. According to Bartezzaghi (1992), and integrating his framework in the light of other contributions, the multi-focused manufacturing paradigm can be articulated in terms of three basic principles:

- *Multi-focusedness* and *strategic flexibility*. This first element relates to the firm's manufacturing strategy. The multi-focused manufacturing paradigm drives companies to pursue a number of different objectives, traditionally regarded as antithetical, simultaneously, rather than focusing on specific objectives considered mutually exclusive. In addition, the paradigm implies strategic flexibility; that is, the ability to rapidly shift competitive and manufacturing priorities from one set of goals to another within the same manufacturing system. This principle revises the traditional assumption about rigid trade-offs involving manufacturing performances.

- *Integration.* This second principle relates to production organisation from a macro-structural perspective. It entails a resolute process focus, especially concerning those processes directly involved in the value-added chain. Process integration is pursued across internal functions and with both customers and suppliers. The previous emphasis on functional optimisation should be abandoned in favour of a re-design of the company based on the concepts of operating continuity, and process integrity across functional barriers.
- *Process ownership* (Schonberger, 1990). This third element also relates to production organisation, here from a micro-structural perspective. It aims at involving all employees at any hierarchical level in decision making and problem solving. Delegation, involvement and knowledge of the process are embodied in this principle. The ultimate purpose is to develop at least some degree of local problem-solving capabilities, in order to detect and resolve process anomalies as soon as possible and to avoid time-consuming hierarchical referral.

Both integration and process ownership are closely related to multi-focusedness. In fact, integration fosters the globalisation of the goals and strategic flexibility, making the organisation more capable to rapidly react to market turbulence and to seize volatile opportunities. Process ownership contributes to enhancing the quality of the outputs and to reducing the lead-time of the business processes, which in turn is the primary mechanism to reduce or, even better, avoid the trade-offs among performance areas regarded as antithetical. Thus, the above three principles should be seen as a whole, as entailed by the current external and internal consistency required within most of the industries and the marketplaces.

26.4.2. THE EMPIRICAL INVESTIGATION

The empirical description of the adoption and the resultant performance improvement of the multi-focused manufacturing paradigm is based on a set of state variables that show, at a given time, to what extent a manufacturing unit is simultaneously oriented to multi-focusedness, integration and process ownership. "As the complexity of a system increases, our ability to make precise and yet significant statements about its behaviour diminishes, until a threshold is reached beyond which precision and significance (or relevance) become almost mutually exclusive characteristics" (Zadeh, 1973). The multi-focused manufacturing paradigm is a complex and multidimensional concept and relates to a complex system—i.e., the whole of the operation. "In retreating from precision in the face of overpowering complexity, it is natural to explore the use of what might be called *linguistic* variables that is, variables that are not numbers but words or sentences in a natural language" (Zadeh, 1973).

Thus, the assessment of the orientation to the paradigm is necessarily based on a wide set of variables, only partially numerical, that provide the basis to evaluate the degree of belonging of a unit to the paradigm, at a given time. In fact, the paradigm is not a "yes or no" matter, but a concept that can range from a very weak belonging into a very strong one. In addition, the process of adoption is supposed to be progressive over time, so that an intermediate degree of belonging can be associated to a certain unit at a certain point in time.

For all the above reasons we use a fuzzy-logic approach (see, for instance, Zimmermann (1993)). First, the set of state variables connected to the paradigm was identified (see the leaves in the tree of Figure 26-2). Then, membership functions were built up to relate the single state variable to the degree of belonging, ranging from 0 (not belonging) to 1 (complete belonging). The tuning of the membership functions is based mostly upon the literature on current best practices all over the world within the assembly industry (see Appendix 26-1 for some example and Cagliano and Spina (1996) for a description of all the membership functions). Starting from the basic set of variables, a hierarchical methodology was followed that aggregates the leaves into the intermediate concepts, to the three basic principles—multi-focusedness, integration and process ownership—and to the paradigm as a whole. Figure 26-2 shows the whole filter and, in particular, the operators we used for the aggregation of the leaves to the final degree of belonging to the paradigm (see also Appendix 26-2). These are mainly "*fuzzy-and*" and "*and*" operators, given the necessity of the co-presence of the three principles and their sub-principles. "*Or*" and "*fuzzy-or*" operators were used when single items can be regarded as alternatives with respect to the paradigm adoption.

Of course, the computed degree of belonging to the paradigm embodies a noticeable subjectivity relating to the selection of the state variables, the definition of the membership functions and the logic of the aggregation. Belonging to the multi-focused manufacturing paradigm is defined as an absolute concept, but the degree of belonging to it can be regarded as a relative figure, useful to benchmark manufacturers from different countries and industries. In addition, the tuning of both the membership functions and the parameters of the *fuzzy operators* influences the absolute figure of the degree of belonging, but not significantly the rank of the units within the sample, preserving the opportunity for cross-sectional comparisons. To get a reliable rank, it is important to properly select the "*or / fuzzy-or*" and the "*and / fuzzy-and*" operators, and the shape of the membership functions (increasing, decreasing, S-curve, step function, etc.).

26.4.3. RESEARCH HYPOTHESES AND METHODOLOGY

The purpose of this chapter is to investigate the multi-focused manufacturing paradigm, which requires addressing two issues:

- the adoption of the multi-focused manufacturing paradigm across industries and countries;

- the effectiveness of the multi-focused manufacturing paradigm, i.e., its ability to provide the adopters with superior improvement capabilities (see Section 26-3).

Two sets of specific hypotheses have been formulated for the two issues, respectively.

The adoption of the multi-focused manufacturing paradigm

We expected that some contextual factors may influence the adoption of the multi-focused manufacturing paradigm across industries and countries, although the paradigm has the potential to change the way any manufacturing system is managed. Namely, we expect that the paradigm will be adopted:

1. Mainly in the industrialised countries (Japan, North America and the most advanced European), and to a lesser extent in the NICs and the less developed European countries. In fact, some unfavourable conditions are expected to hamper the paradigm adoption, such as the poverty of the public infrastructure, the lack of a well-educated workforce, and low labour costs, which are more likely to attract mass production;

2. Within the assembly industry, not only in the automotive industry, which has attracted much of the attention since it was the cradle of both Fordism and post-Fordist experiences; but also within other assembly sectors, especially the electronic and electro-mechanical industries that have markedly experienced mass production in the past;

3. By large and medium-sized companies, since they are expected to have a more robust managerial culture and to be more sensitive to managerial and organisational innovations;

4. By those companies that operate in stable markets, with low growth rates, since competition in such cases tends to become keen and multi-dimensional, with various performances simultaneously required (cost, quality, service level, frequent product renovation, etc..);

5. By companies that mainly operate in a *make-to-stock* environment, since the pioneering applications of the new manufacturing techniques actually took place in such a planning environment.

Based on the procedure discussed in Section 26-4 we evaluate below the degree of belonging to the paradigm of the single units we explored, and thus compare the average adoption of the sub-samples extracted to test the above hypotheses.

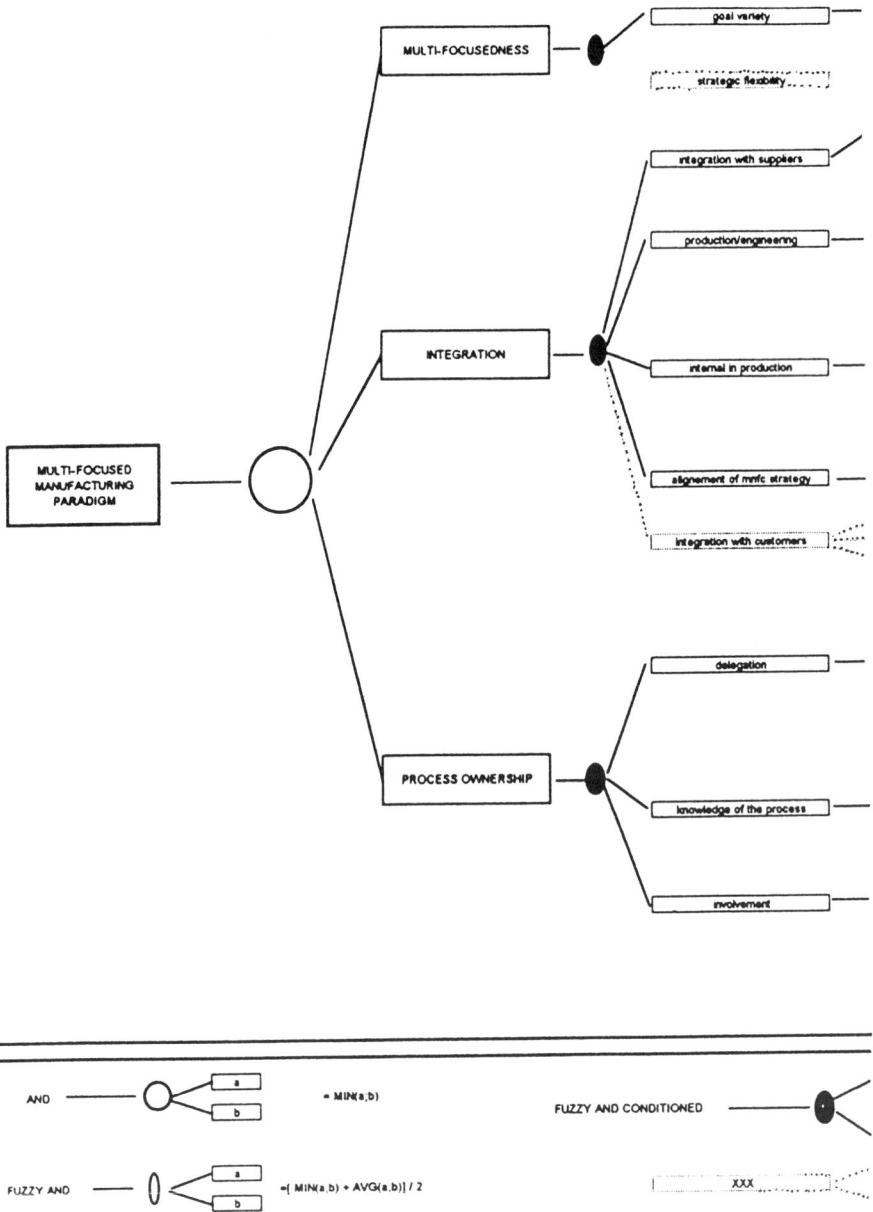

Figure 26-2. The Operationalisation of the Multi-Focused Manufacturing Paradigm

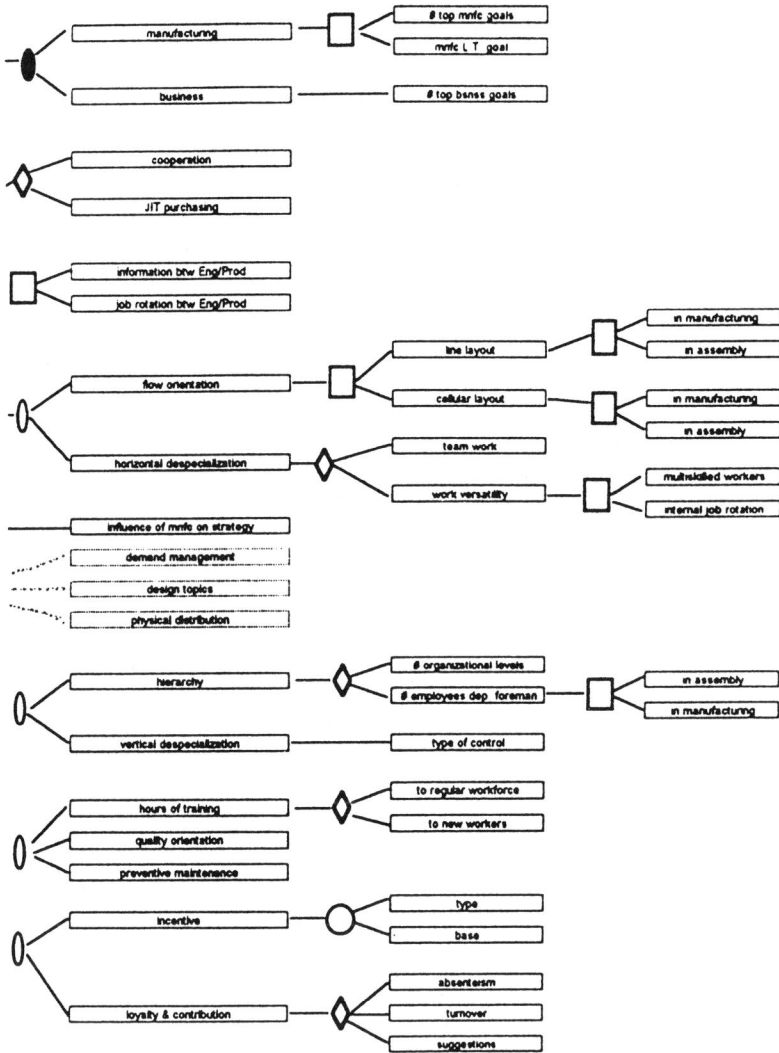

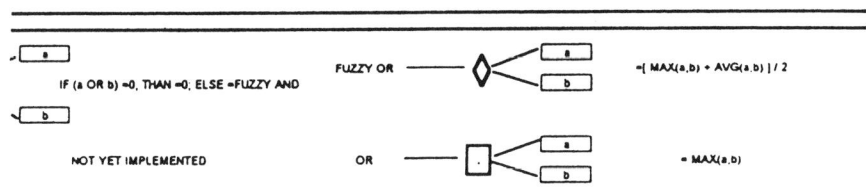

Figure 26-2. The Operationalisation of the Multi-Focused Manufacturing Paradigm (continued)

The effectiveness of the multi-focused manufacturing paradigm

The second aim of this chapter was to investigate if the adoption of the *multi-focused manufacturing paradigm* results in a higher degree of performance improvement compared with non-adopting companies.

To address this issue, we tested whether:

1. Companies that have adopted the principles of the paradigm are better capable of improving their performance compared with non-adopters in order to show the relevance of the multi-focused manufacturing paradigm;
2. Partial adoption of the paradigm also results in performance improvement;
3. The three principles of the paradigm reinforce each other.

Again, we used the previously described method to determine the degree of belonging to each of the three basic principles and consequently the degree of belonging to the paradigm as a whole. Further, we have developed a specific method for exploring different degrees of adoption of the multi-focused manufacturing paradigm and the related performance improvements. The "star model" (Figure 26-3) distinguishes between companies with partial degrees of belonging to the paradigm. This is also useful to study all kinds of innovation paths for full adoption of the paradigm. It is called the star model because the intersection of two out of three principles gives the shape of a three-pointed star.

We anticipated three classes of belonging to the paradigm: 1) complete adoption (core), 2) partial adoption (stars) and 3) non adoption (see Figure 26-3). Complete adoption refers to the companies who have adopted all three principles. Partial adoption means that the company has adopted two out of the three basic principles. Non-adoption refers to companies that have adopted either one principle out of three or none at all.

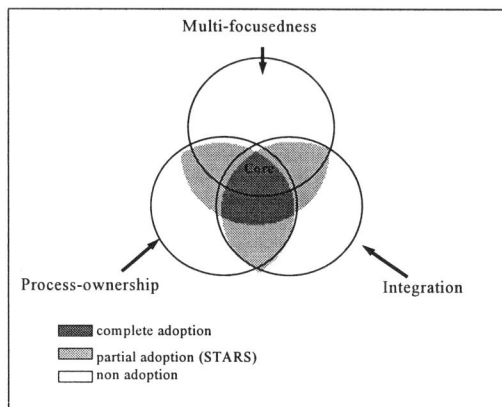

Figure 26-3. The star model: intersection of principles

Table 26-1. Distribution of the 443 processable companies by industry (the percentage in brackets for total sample)

ISIC	Description	Respondents	
381	Metal products (except machinery)	142	32.1%
382	Machinery (except electrical)	66	14.9%
383	Electrical machinery apparatus, appliances and supplies	92	20.8%
384	Transport equipment	55	2.4%
385	Measuring and controlling equipment, optical goods	40	9.0%
-	Not specified or other	48	10.8%

We consider a principle to be adopted if the company has a score higher—for the single principle at hand—than the mean within the sample. In order to be a core adopter a company must have a score higher than the mean within the sample for all three principles.

In summary, our major line of argument is that 1) a new manufacturing paradigm is emerging as a replacement for *Fordism*; 2) it involves strategic multi-focusedness, integration of business processes across functions, and process ownership; 2) it can be measured using a fuzzy-logic approach (illustrated in the discussion of operationalisation in Section 26-4); 4) it is widely adopted across countries and industries but there are some factors that influence its adoption (see Section 26-4 for hypotheses); and 5) it results in a higher improvement capability compared with non-adopters (see section 26-4 for hypotheses).

26.5. The Research Sample

In order to explore the emerging paradigm on a global basis, we analysed the IMSS database but had to restrict ourselves to 443 companies, due to missing answers for 157 out of 600 companies, using the methodology presented in Section 26.4.2. Empirically we followed the procedure illustrated in Appendix 26-3, aimed at selecting only those companies that can be scored correctly as to the adoption of the multi-focused manufacturing paradigm. The dropping of non-processable cases has not modified significantly the distribution of the original sample of the IMSS database (600 companies). Table 26-1 and Table 26-2 show the distribution of the sample by industry and country.

26.6. The Adoption of the Multi-Focused Manufacturing Paradigm within the Assembly Industry

The multi-focused manufacturing paradigm seems to have been adopted on a global basis. Looking at the sample in Table 26-3, it appears that strategic multi-focusedness has been adopted by most companies, and process ownership the least, with process integration in between. Given the necessity of all being present, the degree of belonging to the paradigm is mainly determined by process ownership. Consequently, countries with a higher orientation to process ownership show a higher score in belonging to the paradigm.

How can we explain the widespread poor adoption of process ownership? Our primary concern was to verify the appropriateness of the membership functions we used to score the companies. All of them were realistic, since we could find companies in the sample that belonged completely to the paradigm for each item in the filter, even for those related to process ownership. However, whilst a number of companies can achieve the complete orientation to the three sub-principles of process ownership—delegation, knowledge of the process and involvement—separately, no company fully achieved them jointly. Indeed, many companies were implementing multi-focusedness at the business and manufacturing level and integrating different business processes at the same time, while delegation, knowledge of the process and involvement appeared, to some extent, to be mutually exclusive.

This is difficult to justify from a theoretical perspective, since the three sub-principles should reinforce one another, and no definite process ownership should be possible without the concurrency of the three sub-principles. This led us to imagine that the adoption of the multi-focused manufacturing paradigm is a step-by-step process, in which multi-focusedness is market-driven and thus first adopted; integration is the organisational answer at a macro level to meet the challenge of multi-focusedness; and process ownership should eventually provide the local mechanism to support integration at a micro level. However, process ownership is not yet fully recognised as the key enabling factor. In addition, its implementation is expected to meet more organisational inertia and cultural barriers. Such a phased adoption of the multi-focused manufacturing paradigm might account for the low orientation to process ownership. However, it can not be skipped. The investigation of the effectiveness of the new paradigm clearly provides an empirical proof, since the core adopters of the paradigm—i.e., companies that implement the three principles at the same time—achieved higher performance improvement than partial adopters, especially than adopters of multi-focusedness and integration but not process ownership.

Table 26-2. Geographical distribution of the 443 processable companies

Country	Number	Country	Number
Sweden	42	Portugal	24
Norway	11	Spain	24
Finland	16	USA	33
Denmark	13	Canada	14
Great Britain	27	Mexico	51
Germany	18	Argentina	28
Austria	21	Brazil	21
The Netherlands	20	Chile	4
Belgium	2	Japan	16
Italy	34	Australia	24

Table 26-3. The Multi-Focused Manufacturing Paradigm around the World

	N	A1 Multi-focusedness				A2 Integration				A3 Process Ownership				Belonging to paradigm (final score)
		Business Level	Mfg. Level	Total	Suppliers	Prod. engr.	Internal prod.	Mfg. strategy	Total	Delegation	Process knowdge	Involvement	Total	
Total	443	0.82	0.88	0.78	0.71	0.66	0.69	0.59	0.49	0.31	0.36	0.27	0.19	0.15
Japan	16	0.90	0.97	0.90	0.77	0.73	0.73	0.81	0.61	0.16	0.47	0.49	0.17	0.16
Denmark	13	0.79	0.78	0.69	0.74	0.79	0.68	0.54	0.50	0.57	0.49	0.39	0.35	0.25
Sweden	42	0.87	0.89	0.83	0.73	0.80	0.80	0.61	0.57	0.50	0.40	0.39	0.30	0.25
Finland	16	0.82	0.88	0.79	0.67	0.75	0.60	0.48	0.49	0.44	0.35	0.25	0.21	0.20
Norway	11	0.87	0.85	0.83	0.87	0.71	0.60	0.50	0.52	0.36	0.24	0.40	0.16	0.14
Scandinavia	82	0.85	0.87	0.80	0.74	0.78	0.72	0.56	0.54	0.47	0.38	0.36	0.27	0.23
USA	33	0.89	0.92	0.85	0.76	0.86	0.69	0.61	0.58	0.30	0.31	0.40	0.23	0.19
Canada	14	0.99	1.00	0.99	0.64	0.64	0.74	0.46	0.44	0.27	0.23	0.18	0.13	0.11
N.America	47	0.92	0.94	0.89	0.73	0.79	0.70	0.57	0.54	0.30	0.28	0.33	0.20	0.17
Australia	24	0.83	0.71	0.67	0.75	0.87	0.76	0.66	0.60	0.36	0.30	0.30	0.19	0.15
Brazil	21	1.00	0.98	0.99	0.78	0.68	0.72	0.67	0.55	0.28	0.43	0.27	0.23	0.18
Mexico	51	0.73	0.96	0.77	0.72	0.36	0.72	0.67	0.43	0.22	0.30	0.09	0.10	0.09
Argentina	28	0.28	0.93	0.80	0.62	0.55	0.59	0.70	0.34	0.21	0.35	0.27	0.12	0.09
Chile	4	0.85	0.88	0.79	0.84	0.75	0.75	0.88	0.41	0.22	0.21	0.26	0.03	0.00
S.America	104	0.81	0.95	0.83	0.71	0.49	0.69	0.67	0.43	0.23	0.34	0.18	0.13	0.11
Portugal	24	0.86	0.87	0.83	0.60	0.58	0.54	0.58	0.40	0.36	0.51	0.31	0.25	0.20
Spain	24	0.79	0.85	0.75	0.70	0.57	0.71	0.57	0.47	0.33	0.38	0.25	0.21	0.17
Pen. Iberica	48	0.82	0.86	0.79	0.65	0.57	0.63	0.58	0.43	0.34	0.45	0.28	0.23	0.18
Italy	34	0.82	0.77	0.69	0.73	0.67	0.65	0.54	0.50	0.27	0.33	0.28	0.19	0.14
Great Britain	27	0.82	0.81	0.74	0.78	0.69	0.71	0.46	0.45	0.25	0.40	0.25	0.17	0.11
Austria	21	0.72	0.77	0.61	0.53	0.58	0.61	0.45	0.36	0.31	0.38	0.22	0.20	0.10
Belgium	2	0.90	1.00	0.92	0.25	0.50	0.50	0.37	0.28	0.19	0.27	0.17	0.24	0.12
Germany	18	0.77	0.86	0.74	0.54	0.73	0.59	0.50	0.45	0.22	0.36	0.05	0.10	0.07
Netherlands	20	0.66	0.88	0.69	0.74	0.63	0.78	0.50	0.51	0.30	0.43	0.10	0.16	0.12
D.M. Area	61	0.72	0.84	0.68	0.59	0.64	0.65	0.48	0.43	0.28	0.39	0.13	0.15	0.10

Our findings with respect to the five specific hypotheses are presented in the following sections.

Geo-economic context. Our basic hypothesis about the diffusion of the paradigm in the most advanced countries appears to be confirmed. The orientation to the multi-focused manufacturing paradigm seems to be present in different economic areas, although not uniformly. The country factor is strongly related to the degree of belonging to the paradigm and also to the three principles and all their sub-principles. In fact, the one-way ANOVA test of the probability that the differences in the mean scores of the national samples are random is less than 0.01 for all the sub-principles in Table 26-4.

In particular, the Scandinavian area appears to be most oriented to the paradigm, with far higher levels than the mean of the sample for all three aspects—delegation, knowledge of the process and involvement—of process ownership. Also, the average score of integration exceeds the mean of the sample and particularly as to the integration of production-engineering.

Japanese firms confirm to be strongly oriented to the paradigm for most of the sub-principles. Integration is more pursued than elsewhere and in particular the link between manufacturing and business strategy seems to make the difference. Mainly because of hierarchy—i.e., many organisational levels—delegation scores are very low, which negatively affects the score of the Japanese companies. In turn the knowledge of the process and the involvement score is very high. Actually they seem to dominate the rest of the sample as to the orientation to the multi-focused manufacturing paradigm except for the delegation.

Companies from the Deutsche Mark area show the lowest degree of belonging to the paradigm, due to the poor orientation to process ownership. In particular, the German companies score very low as to involvement and delegation. The German companies in the sample tend not to use group incentives, suffer higher short-term absenteeism and enjoy less suggestions that determines, on the whole, an average score of the involvement far below the mean of the sample. In addition, those companies maintain a highly centralised production control, which causes the low level of delegation.

US companies stand out for their effort to integrate production and engineering, and they are also markedly oriented to the involvement of the workers.

The multi-focused manufacturing paradigm also seems to be adopted in the NICs. For example, the Brazilian companies proved to be extremely multi-focused, to be pursuing different kinds of process integration, and to commit themselves to develop the knowledge of the process in the workers. Indeed, the Brazilian sample is biased towards the best-practice companies, often controlled or participated by foreign corporation, while most of the national samples do not show such a bias. Notwithstanding the philosophy of the multi-focused manufacturing paradigm seems to overcome some unfavourable national conditions—e.g., the shortage of well educated manpower, the poverty of the infrastructure and the low labour cost that is expected to attract mass productions rather than lean ones—at least when in the track of a global corporate culture.

Industrial context. We also looks at the relationship between the adoption of the multi-focused manufacturing paradigm and industry sector, not only in ISIC 384, which in the database is mainly formed by car assemblers or car component producers, but also in the electrical and machinery industry, which show the highest orientation to the paradigm on the whole (see Table 26-4). This suggests that it is no longer just the case of *The Machine That Changed the World* (Womack, Jones and Ross, 1990), but indeed, multi-focusedness still remains more pursued within the transport industry (score= 0.82 versus 0.78 in the whole sample), even though the difference was not statistically significant. The one-way ANOVA test revealed that the industry factor significantly affects only the process ownership (p=0.008) and delegation (p=0.001) and in this case the electrical and machinery industry far exceed the other assembly industry (see Table 26-5).

Company Size. Company size was strongly related to the adoption of the paradigm. Small companies had lower scores than large- and medium-sized ones (see Table 26-4). The differences were statistically significant for all the three basic principles: for multi-focusedness (t-test: p=0.015), the difference mainly depends on the business level (T-test: p=0.010); in the case of integration (t-test: p=0.002) the dominance of large companies can be traced back to the differential integration between business and manufacturing strategy (t-test: p=0.000); finally the superior orientation to process ownership (t-test: p=0.003) within the large companies mainly relies on their ability to develop the knowledge of the process of their workers (t-test: p=0.050).

Two-way ANOVA allows us to test the independent influence of the size factor. In fact size and industry can explain separately the adoption of the paradigm within the sample, while no significant interaction was detected for all the principles and their sub-principles. Much the same was found for size and country factors. In this case no interaction of the factors was detected between *multi-focusedness* and *process ownership,* whilst *integration* shows some joint effect of the two factors, from business and manufacturing strategy integration (two-way ANOVA: p=0.032) and production-engineering integration (two-way ANOVA: p=0.016). In fact, because the American and the Japanese units within the sample are larger than the others, it was hard to extract size or country as independent factors.

Market trend. The hypothesis about the influence of market trends on the adoption of the *multi-focused manufacturing paradigm* was not confirmed (see Table 26-4), since the paradigm was also adopted in growing markets, where the competition is expected to be less keen. A significant difference was revealed as to multi-focusedness (t-test: p=0.031) and to the integration of manufacturing and business strategy (t-test: p=0.015), even in favour of companies operating in growing markets.

Planning environment. Finally, the planning environment did not affect the orientation to the paradigm (see Table 26-4). No statistically significant difference was detected when comparing *make-to-order* companies with *make-to-stock* ones, except for the involvement of the workers. This was significantly higher in a make-to-order environment (t-test: p=0.019). Two-way ANOVA did not show any interactions of planning environment and market trends with other influencing factors—size, industry and country.

26.7. The Multi-Focused Manufacturing Paradigm and Performance Improvements

The effectiveness of the multi-focused manufacturing paradigm is linked to the capability it gives the adopters to improve the performances of the production systems more and faster than the non adopters (see Section 26-3). The ability to distinguish different degrees of belonging to the paradigm allows us to test the hypotheses about the effects of its adoption on performance improvements.

The average degree of improvement is based the following question from the IMSS:

> We ask you to mentally construct an index for each manufacturing performance indicator. We ask you to assume that the beginning of 1990 is the base with index 100. How large would you estimate that the percentage change in the index today (1992/1993) would be?

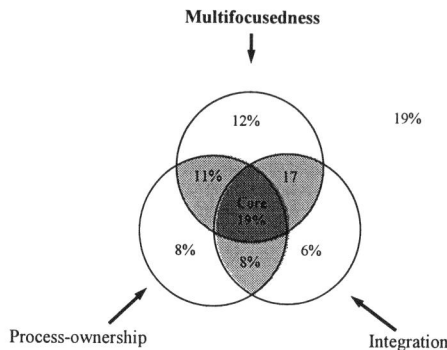

Figure 26-4. The star model: distribution of the sample in the different classes

Table 26-4.

	n.	A1 Multi-focusedness			A2 Integration					A3 Process Ownership				Belonging to the paradigm (final score)
		Business Level	Mfg. Level	Tot. A1	Suppliers	Prod. Engr.	Internal Prod.	Mfg. Strat.	Tot. A2	Delegation	Process Knowledge	Process Involvement	Tot. A3	
Industry														
381 (metal)	142	0.83	0.86	0.78	0.67	0.69	0.66	0.61	0.47	0.27	0.35	0.27	0.17	0.13
382 (machinery)	66	0.83	0.88	0.79	0.74	0.65	0.64	0.56	0.49	0.38	0.31	0.33	0.24	0.19
383 (electrical)	92	0.83	0.85	0.76	0.73	0.73	0.70	0.60	0.53	0.38	0.39	0.28	0.22	0.17
384 (transport)	55	0.84	0.91	0.82	0.72	0.66	0.73	0.62	0.52	0.28	0.38	0.22	0.19	0.15
385 (instruments)	40	0.78	0.88	0.76	0.70	0.65	0.76	0.49	0.47	0.29	0.43	0.26	0.22	0.16
missing or other	48	0.81	0.91	0.80	0.69	0.52	0.71	0.57	0.43	0.27	0.33	0.19	0.12	0.09
Size														
Small (1)	256	0.80	0.86	0.76	0.69	0.64	0.68	0.55	0.45	0.30	0.34	0.26	0.17	0.13
Medium-large (2)	177	0.86	0.90	0.82	0.72	0.69	0.70	0.63	0.53	0.33	0.39	0.28	0.21	0.18
Market trend														
Growing	117	0.86	0.90	0.83	0.72	0.64	0.70	0.64	0.49	0.31	0.37	0.27	0.27	0.15
Stable-declining	313	0.81	0.87	0.77	0.70	0.68	0.69	0.57	0.49	0.31	0.36	0.27	0.19	0.15
Planning environment														
Make-to-Order (3)	267	0.83	0.88	0.79	0.70	0.68	0.69	0.58	0.49	0.31	0.36	0.29	0.19	0.15
Make-to-Stock (4)	149	0.82	0.85	0.77	0.70	0.63	0.70	0.60	0.48	0.31	0.37	0.23	0.18	0.15
Total	**443**	**0.82**	**0.88**	**0.78**	**0.71**	**0.66**	**0.69**	**0.59**	**0.49**	**0.31**	**0.36**	**0.27**	**0.19**	**0.15**

To explore the performance improvements linked to the adoption of the paradigm, we used the framework described in paragraph 29.3.2—the "star model" -, which allows us to distinguish among different degrees of adoption. The 443 companies were distributed as follows (Figure 26-4): 83 companies (19%) of the sample could be classified as core adopters; approximately 36% of the companies resulted to be stars with on two principles out of three a score higher than the mean; the remaining 45% represented insufficient scores or no adoption at all.

It seems as if the multi-focused manufacturing paradigm generally provides its adopters with a higher improvement capability compared with non-adopters. Looking at the global sample in Table 26-5 and referring to the three hypotheses it appears that:

- Core adopters of the paradigm improved their performance more than non-adopters on almost all performance criteria. There is a general dominance of the adopters over the non-adopters. In fact, when comparing the adopters with the rest of the sample (stars and other) four differences in performance improvements are significantly better, namely (a) inventory turnover, (b) speed of product development, (c) customer service and (d) delivery lead time (see Table 26-5).
- Partial adoption of the paradigm also resulted in advantages in a subset of performances, i.e., a partial dominance over the non-adopters.
- As full adoption is a general dominance and partial adoption is a partial dominance as to performance improvement, the fuzzy-logic approach is enforced.

Table 26-5. Performance improvements within different classes of adoption of the multi-focused manufacturing paradigm

Performance criteria	Average improvement			Average improvement		
	Core adopters (%)	Partial adopters (stars) + Non-adopters	t-test significance (%)	Partial adopters (%)	Non-adopters (%)	t-test significance (%)
Conformance to specification	39.87	26.04		32.82	21.10	
Average unit manufacturing cost	16.80	12.61		16.14	9.45	
Inventory turnover	40.87	22.38	1.9	28.84	18.01	
Speed of product development	29.49	15.55	0.8	18.11	13.76	
On-time deliveries	46.35	21.60		27.52	16.63	1.9
Equipment changeover	25.82	16.21		20.47	13.32	2.7
Market share	12.56	11.16		18.97	5.66	
Profitability	8.12	10.27		16.16	7.09	
Customer service	26.99	17.83	4.4	22.46	13.99	0.6
Manufacturing lead time	45.95	23.07		31.10	16.35	0.4
Procurement lead time	36.03	15.12		18.37	12.33	
Delivery lead time	36.28	19.75	1.5	22.77	16.53	4.1
Product variety	19.03	13.03		13.06	12.91	

This also implies that co-presence of principles enforces improvement gains (the more you put together the more you gain).

These findings support the idea that the multi-focused manufacturing paradigm is effective since it allows companies to improve more. Indeed, some methodologica specification is needed. Given the cross-sectional nature of the IMSS data, no strict causality can be inferred in an absolute sense between the degree of the adoption of the paradigm and performance improvements. As Hamblin and Lettman (1994) have pointed out, the usual statistical tests do not allow us to state a causal link between techniques and performances. In fact, one may contend that the performance improvements, for example in inventory turnover and market share can create additional resources (cash-flows) to be invested in the multi-focused manufacturing, so that the causal link would be the reverse (more improvements: innovation towards the multi-focused manufacturing). To state strict causality we should employ two-way models based on time series on the two classes of variables (Granger causality (Granger, 1969)), which we cannot do. Anyway, when considering manufacturing performances(cost, delivery time, etc.) rather than business ones (profitability and market share) the causal link between the degree of adoption of the paradigm and the degree of performance improvements may be reasonably assumed.

26.8.   Managerial Implications

In this section we will discuss the consequences for management if they want to pursue the multi-focused manufacturing paradigm. We distinguished the managerial functions, the managerial roles and the management skills (Daft, 1991; Mintzberg, 1989) affected and required by the paradigm.

26.8.1. MANAGERIAL FUNCTIONS

The multi-focused manufacturing paradigm affects the content of the classical managerial functions of planning, organising, leading and controlling. Below the impact of the paradigm on the classical managerial functions will be discussed.

- *Planning* - Goals for the future organisation performance should be set on *multi-focusedness* and allow for strategic flexibility. The tasks need to be defined precisely, in order to prevent non-focusedness or a too narrow focus. The planning function should also foresee that resources are required that facilitate the intended multi-focusedness.
- *Organising* - By assigning tasks specific attention should be paid to integration with suppliers and customers, and integration between production and engineering. This also has consequences for grouping the tasks and allocating resources to

departments in order to facilitate the required integration and sufficient level of process-ownership (see Paashuis and Boer (1995)).

- *Leading* - When it comes to the leading function, management should motivate the workforce in such a way that the goals will be achieved. In a company oriented to the new paradigm this motivation can be achieved by improving the degree of process-ownership through a higher degree of delegation, process knowledge at hand, and involvement.

- *Controlling* - The classical control by direct supervision should be transited to a higher degree of self-management by the workforce. This can be achieved through job-enlargement and enrichment. So, instead of being monitored by the hierarchical boss it is advisable to create a working environment where the workforce is capable of assessing their own performance and extracting what actions are needed in the future to improve it.

The managerial functions are mainly geared towards multi-focusedness by means of a high degree of integration and a high degree of process ownership. This also affects the managerial roles.

26.8.2. MANAGERIAL ROLES

Managerial roles traditionally distinguished are the decisional, the informational and the interpersonal ones.

- *Decision aspects* - The entrepreneurial role will to a certain degree be shared by the workforce if they are not only capable but also allowed to proactively manage their processes. The disturbance handling role will for a large part being transferred to the process-owners and management becomes more a facilitator instead of a "verdict" teller. The traditional one way resource allocation by management will be partly overtaken by the process owners who are allowed to bring in their suggestions in order to improve the processes they are involved in. In general we can say that the bottom-up information flow will increase.

- *Information role* - Traditionally, the monitoring role is the sole domain of higher level managers. The multi-focused manufacturing paradigm requires that the monitoring role will also be done by lower level managers and by the workforce. This role not only involves keeping in pace with the technical development in the functional areas but also taking into account the capabilities of competitors and helping to formulate bottom-up the manufacturing and corporate strategy. The top-down approach to disseminate and communicate the intentions of the company as a whole becomes even more important due to the multi-focusedness and the required integration between the company, suppliers and customers.

- *Interpersonal role* - Due to the required integration and the high degree of delegation the co-ordination by means of mutual adjustment becomes more important. In this role the liaison function—next to being figurehead and leader—grows in importance. This forces far more management by walking around in order to get hold of both technical and organisational problems between different stages of the production process.

26.8.3. MANAGEMENT SKILLS

The changes in management functions and roles also affect the skills and capabilities management needs to have in order to create a company oriented to new paradigm or to work effectively in a firm that already adopted it.

- *Conceptual* - Management needs to see the organisation as a whole and must be capable of recognising the critical relationships with other organisations. This especially relates to the integration part of the multi-focused manufacturing paradigm.
- *Human* - Due to the growing importance of process-ownership management needs to have the ability to work with and through other people and work effectively as a group member.
- *Technical* - Through the required higher degree of flow and product layout management needs to have sufficient understanding of and proficiency in the performance of certain stages of the whole production chain.

26.9. CONCLUSIONS

This contribution has investigated the diffusion and the implications of the multi-focused manufacturing paradigm based on the simultaneous implementation of strategic multi-focusedness, integration of business processes and process ownership, and the relationship between the degree of belonging to this paradigm and the improvements in performances.

The paradigm rises as a coherent set of principles underpinning the wide range of techniques and approaches for the innovation of the manufacturing systems. The identification of the paradigm has been based on internal and external consistency, as implied by today's business environments. Such a post-Fordist paradigm, that partly embodies previous post-Fordist experiences and partly introduces new elements, has been operationalised through a fuzzy-logic and hierarchical methodology. Using data from a sub-sample of 443 companies from International Manufacturing Strategy Survey database, a wide adoption across industries and countries has been detected. Large orientation to the multi-focused manufacturing paradigm has been discovered mainly in the Scandinavian area. Also large cross-industrial transferability emerges, not only in the automotive, and mainly within the electrical and machinery industry. On the other hand, large companies show to be more oriented to the paradigm than small ones.

On the whole, process ownership is not very much implemented at the moment. It is expected to be the most difficult part of the paradigm to reach, given that the orientation to the multi-focused manufacturing paradigm is a step-by-step process and not a switching on. In addition, big differences across countries have been found about process ownership, which requires more interpretation on the basis of cultural and institutional differences.

The empirical evidence also suggest that a higher degree of belonging to the paradigm results in a higher performance improvement. This leads us to conclude that the three principles reinforce one another.

If companies want to pursue the multi-focused manufacturing they should consider the changing role of management, management functions, and the skills a manager must have. Main implications are that management should (1) regard trade-offs as shiftable and continously improvable, (2) educate the workforce in such a way that they are not only capable but also allowed to proactively manage their processes, and (3) group tasks and allocate resources to department in order to facilitate the required integration and sufficient level of process ownership.

On the basis of these results, current investigation is addressed to the hypothesis that the multi-focused manufacturing paradigm does not act as a new one best way to organise manufacturing activities, but actually preserves noticeable space for different manufacturing strategies, that is different mixes of goals and innovations.

26.9.1. ACKNOWLEDGEMENTS

Financial support of "Trasferimento delle tecnologie dei progetti finalizzati" by C.N.R. (National Research Council of Italy) is gratefully acknowledged. The contribution is due to the joint work of the authors. However, G. Spina has written section 1, 3.2, 5, 5.1 and 5.2; E. Bartezzaghi section 3.1, 3.3.1 and 8; R. Cagliano section 4, 5.3; 5.4 and 5.5; H. Boer, D. Draaijer and A. Bert jointly section 2, 3.3.2, 6 and 7.

26.9.2. REFERENCES

Abo T. (1990) 'Local production of Japanese automobile and electronics firms in the United States - the "application" and adaptation of Japanese style of management', Research Report No. 23, University of Tokyo, Institute of Social Science.

Bartezzaghi, E. (1992) 'I nuovi modelli del manufacturing', in R. Filippini, G. Pagliarani and G. Petroni (eds.), *Progettare e gestire l'impresa innovativa,* Etaslibri, Milano (in Italian)

Bartezzaghi, E., Turco, F., and Spina, G. (1992) 'The impact of JIT approach on production system performance: a survey of Italian industry', *International Journal of Operations and Production Mangement,* **12**, 1, 5-17.

Bolwijn, P.T., Boorsma, J., van Breukelen, Q.H., Brinkman, S. and Kumpe, T. (1986) *Flexible manufacturing: integrating technical and social innovation,* Elsevier Science Publishers Amsterdam.

Bonazzi, G. (1993) 'Il tubo di cristallo: Modello giapponese e fabbrica integrata alla FIAT Auto', *Il Mulino,* Bologna (in Italian).

Cagliano, R. and Spina, G. (1996) 'Assessing the Orientation of Factories to the Emerging Manufacturing Paradigm: A Fuzzy-Based Research Methodology', Proceedings, Third Conference of SIGEF, Buenos Aires, 10-13 November.

Daft, R.L. (1991) Management, 2nd edition, The Dryden Press, London.

Doshe, K., Jurgens, U. and Malsch, T. (1985) 'From Fordism to Toyotism? The social organization of the labour process in the Japanese automobile industry', Politics and Society, 14, 2, 115-146.

Draaijer, D. (1993) Market-Oriented Manufacturing Systems, Ph.D. dissertation, University of Twente, Enschede.

Draaijer, D.J., and Boer, H. (1995) 'Designing market-oriented production systems: theory and practice', Integrated Manufacturing Systems, 6, 5.

Drucker, P.F. (1990) 'The emerging theory of manufacturing', Harvard Business Review, May-June, 94-102.

Gilbert, J.P. (1990) 'The state of JIT implementation and development in the USA', International Journal of Production Research, 28, 6, 1099-1109.

Granger, W.W.J. (1969) 'Investigating causal Relations by econometric models and non-spectral methods', Econometrica, 37, 24-36.

Hamblin, D. and Lettman, A. (1994) 'Performance causality in manufacturing research', Proceedings of the 1st International EurOMA Conference: Operations Strategy and Performance 27-29 June, Cambridge University Press, 409-414.

Hayes, R.H., and Pisano, G.P. (1994) 'Beyond world-class: the new manufacturing strategy', Harvard Business Review, Jan-Feb, 77-86.

Hayes, R.H., Wheelwright, S.C., and Clark, K.B. (1988) Dynamic Manufacturing. Creating the Learning Organization, The Free Press, New York.

Hayes, R., and Wheelwright, S.C. (1984) Restoring Our Competitive Edge: Competing Through Manufacturing, John Wiley and Sons, New York.

Hill, T. (1985) Manufacturing Strategy, MacMillan Publishers, Basingstoke.

Kenney, M., and Florida, R. (1989), 'Japan's role in a post-Fordist age', Futures, April, 136-151.

Liepitz, A. (1987) Mirages and Miracles, Verso, London.

Miller, D., and Friesen, P.H. (1986) 'Porter's (1980) generic strategies and performance: an empirical examination with American data', Organization Studies, 7, 1, 37-55.

Mintzberg, H. (1989) Mintzberg on Management, The Free Press, New York.

Odaka, K. (1984) Japanese Style of Management, Chu o Koroshua, Tokyo..

Paashuis, V., and Boer, H. (1995) 'New product design: organising for integration', Proceedings of the 2nd EurOMA Conference on Management and New Production Systems, May 28-31, University of Twente, Enschede.

Pfeffer, J. (1977) 'On usefulness of the concept', in New Perspectives on Organizational Effectiveness, P.S. Goodman and J.M. Pennings, (eds.), Jossey-Bass, San Francisco.

Piore, M., and Sabel, C. (1984) The Second Industrial Divide: Possibilities for Prosperities, Basic Books, New York.

Porter, M.E. (1980) Competitive Strategy: Techniques for Analyzing Industries and Competitors, The Free Press, New York.

Schonberger, R.J. (1990) Building a Chain of Customers, The Free Press, New York.

Skinner, W. (1985) Manufacturing: The Formidable Competitive Weapon, John Wiley & Sons, New York.

Voss, C.A., and Robinson, S.J. (1987) 'Application of JIT Manufacturing Techniques in the UK', International Journal of Operations and Production Management, 7, 4, 46-52.

Womack, J.P., Jones, D.T., and Ross, D. (1990) The Machine That Changed the World, McMillan, London.

Zadeh, L.A. (1973) 'The concept of linguistic variable and its application to approximate reasoning', Memorandum ERL-M 411, Berkeley, October.

Zimmermann, H.J. (1993) Fuzzy Set Theory and Its Application, 6th ed, Kluwer Academic Publishers, London.

APPENDIX 26-1. THE MEMBERSHIP FUNCTIONS (Some examples)

MULTI-FOCUSEDNESS

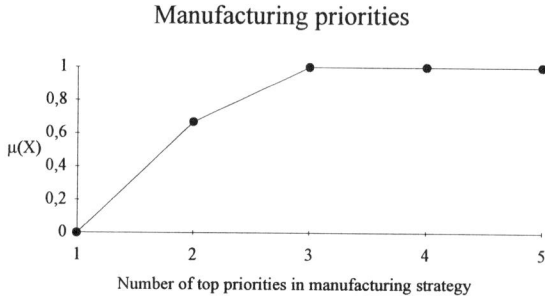

Manufacturing priorities

Number of top priorities in manufacturing strategy

INTEGRATION

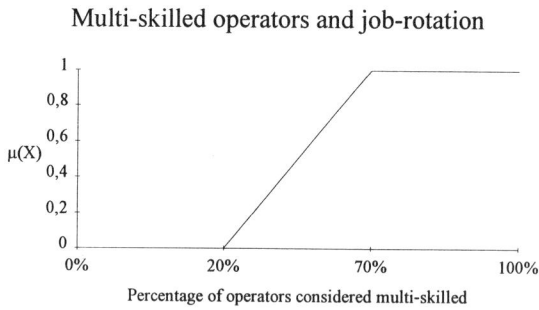

Multi-skilled operators and job-rotation

Percentage of operators considered multi-skilled

PROCESS OWNERSHIP

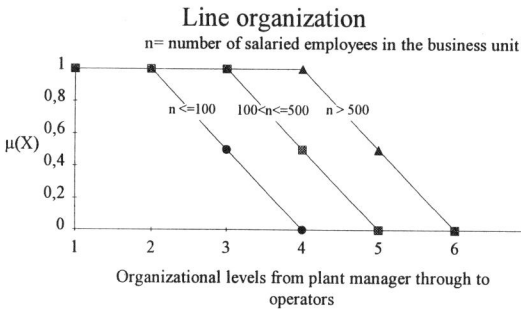

Line organization

n= number of salaried employees in the business unit

Organizational levels from plant manager through to operators

APPENDIX 26-2. THE AGGREGATION OPERATORS

AND $\qquad$ $\min \{\mu_a(X), \mu_b(X)\}$

FUZZY AND $\qquad$ $\alpha * \min \{\mu_a(X), \mu_b(X)\} + (1-\alpha) * \text{average} \{\mu_a(X), \mu_b(X)\}$

FUZZY OR $\qquad$ $\alpha * \max \{\mu_a(X), \mu_b(X)\} + (1-\alpha) * \text{average} \{\mu_a(X), \mu_b(X)\}$

OR $\qquad$ $\max \{\mu_a(X), \mu_b(X)\}$

FUZZY AND CONDITIONED $\qquad$ if $\quad \mu_{\text{fuzzy and}}(X) = 0 \quad$ then $\quad 0$

$\qquad$ else $\quad \mu_{\text{fuzzy and}}(X)$

APPENDIX 26-3   THE DROPPING PROCEDURE

The standard procedure used to select processable companies is aimed at determining which are the companies that can be assigned a correct score as adopters or non-adopters of the *multi-focused manufacturing paradigm*. The problem concerns missing answers to some of the questions used in the filter.

The following rules were used when processing a single company:

- a missing value prevails over a *zero* value (non orientation to the paradigm) if they are combined either through an *or* or a *fuzzy or*;
- a missing val`ue prevails over a generic *non-zero* value (some orientation to the paradigm) if they are combined either through an *and* or a *fuzzy and*;
- in the other cases it is possible to evaluate correctly the fuzzy score of the aggregation of a missing information with whatever data. In fact, *"missing" or "non-zero"* ⇒ *"non-zero"*, and *"missing" and "zero"* ⇒ *"zero"*;
- this algorithm is pushed from the leaves of the filter up to the three basic principles of the paradigm;
- a company is discarded if it is impossible to assign a fuzzy score to each of these principles.

APPENDIX - TABLES

APPENDIX - TABLES

Table 1 - Strategies, Goals and Finances

Question	A1 Description of the business unit								A2 Industry (SIC code) of the business unit												
	Company		Division		Plant		Other		381		382		383		384		385		Other	
Country		Ranking		Ranking		Ranking		Ranking		Ranking		Ranking		Ranking		Ranking		Ranking		Ranking
Argentina	56.1%	6	19.5%	12	24.4%	11	0.0%	5	71.1%	2	5.3%	17	13.2%	13	2.6%	17	5.3%	10	2.6%	2
Australia	41.4%	12	44.8%	2	13.8%	16	0.0%	5	40.7%	8	7.4%	16	22.2%	10	22.2%	5	7.4%	8	0.0%	3
Austria	51.9%	7	25.9%	9	18.5%	15	3.7%	3	29.6%	13	18.5%	5	37.0%	2	11.1%	11	3.7%	13	0.0%	3
Brazil	32.1%	17	17.9%	13	50.0%	1	0.0%	5	28.6%	14	14.3%	12	0.0%	17	28.6%	1	28.6%	1	0.0%	3
Canada	39.1%	14	34.8%	4	26.1%	9	0.0%	5	36.4%	9	31.8%	1	4.5%	16	9.1%	12	18.2%	3	0.0%	3
Chile	83.3%	1	0.0%	19	0.0%	18	16.7%	1	100.0%	1	0.0%	18	0.0%	17	0.0%	19	0.0%	15	0.0%	3
Denmark	82.4%	2	17.6%	14	0.0%	18	0.0%	5	16.7%	18	25.0%	3	41.7%	1	16.7%	8	0.0%	15	0.0%	3
Finland	47.1%	10	29.4%	7	23.5%	12	0.0%	5	43.8%	4	25.0%	3	6.3%	15	12.5%	10	12.5%	7	0.0%	3
Great Britain	22.2%	18	50.0%	1	27.8%	7	0.0%	5	42.4%	5	9.1%	14	27.3%	8	6.1%	15	15.2%	6	0.0%	3
Germany	60.9%	4	8.7%	17	26.1%	9	4.3%	2	40.9%	7	18.2%	7	31.8%	5	9.1%	12	0.0%	15	0.0%	3
Netherlands	63.0%	3	14.8%	15	22.2%	13	0.0%	5	34.6%	10	15.4%	11	11.5%	14	23.1%	4	15.4%	5	0.0%	3
Italy	36.6%	15	34.1%	5	26.8%	8	2.4%	4	19.5%	16	17.1%	8	29.3%	6	17.1%	7	17.1%	4	0.0%	3
Japan	22.2%	18	29.6%	6	48.1%	2	0.0%	5	23.8%	15	9.5%	13	33.3%	4	28.6%	1	4.8%	12	0.0%	3
Mexico	51.6%	9	29.0%	8	19.4%	14	0.0%	5	41.0%	6	7.7%	15	28.2%	7	17.9%	6	5.1%	11	0.0%	3
Norway	33.3%	16	22.2%	11	44.4%	3	0.0%	5	55.6%	3	16.7%	9	16.7%	12	5.6%	16	5.6%	9	0.0%	3
Portugal	60.0%	5	5.0%	18	35.0%	5	0.0%	5	17.1%	17	28.6%	2	22.9%	9	28.6%	1	2.9%	14	0.0%	3
Spain	51.7%	8	13.8%	16	34.5%	6	0.0%	5	33.3%	11	18.5%	5	18.5%	11	7.4%	14	22.2%	2	0.0%	3
Sweden	40.3%	13	24.2%	10	35.5%	4	0.0%	5	33.3%	11	16.7%	9	36.7%	3	13.3%	9	0.0%	15	0.0%	3
USA	46.3%	11	43.9%	3	9.8%	17	0.0%	5	0.0%	19	0.0%	18	0.0%	17	2.4%	18	0.0%	15	97.6%	1
Total	46.6%		26.1%		26.6%		0.7%		33.6%		14.6%		21.5%		14.2%		8.6%		7.6%	

P. Lindberg et al. (eds.), International Manufacturing Strategies, 465-496.
© 1998 Kluwer Academic Publishers. Printed in the Netherlands.

Table 1 - Strategies, Goals and Finances (continued)

Question	A3 Average proportion of business unit turnover spent on:			A4A The degree of importance of the following goals (1 = not important, 5 = very important)														
	R&D (%)		Proc. Equip (%)		Train & Educ (%)		A4A Lower Mfg. Costs		A4B Delivery Speed		A4C Customer Service		A4D Des & Mfg Quality		A4E Delivery Reliability		A4F Product Range	
Country		Ranking		Ranking		Ranking		Ranking		Ranking		Ranking		Ranking		Ranking		Ranking
Argentina	11.10	1	40.32	1	8.08	2	4.63	3	4.07	13	4.34	13	4.63	8	4.05	14	3.33	11
Australia	3.25	17	3.77	18	2.22	7	4.17	13	4.14	9	4.45	10	4.59	9	4.34	7	3.10	16
Austria	4.95	9	4.88	15	1.23	16	4.00	15	4.22	5	4.48	8	4.37	16	3.79	18	2.81	19
Brazil	3.94	12	9.59	7	2.46	6	4.75	2	4.54	1	4.89	1	4.86	2	4.68	3	3.68	3
Canada	5.58	7	8.33	8	2.78	4	4.26	11	4.43	2	4.87	2	4.74	5	4.70	2	3.35	10
Chile	10.50	2	18.14	2	10.33	1	4.00	15	4.00	16	4.33	14	4.33	17	4.33	8	3.17	14
Denmark	6.73	4	4.19	17	1.26	15	3.76	18	3.94	18	4.41	12	4.71	6	4.29	10	3.12	15
Finland	3.32	16	6.60	10	0.89	17	4.00	15	4.06	15	4.12	18	4.47	13	4.31	9	2.94	17
Great Britain	5.59	6	5.82	13	1.31	14	4.42	8	3.89	19	4.33	14	4.54	12	4.00	15	3.39	8
Germany	2.98	18	2.25	19	1.84	10	4.63	4	4.38	3	4.17	17	4.38	15	4.29	11	3.21	12
Netherlands	7.79	3	10.59	5	1.87	9	3.67	19	4.00	16	4.44	11	4.33	17	3.52	19	3.44	6
Italy	4.52	10	6.26	12	0.81	18	4.44	7	4.07	13	3.80	19	4.68	7	4.15	13	3.93	1
Japan	3.43	15	6.50	11	0.59	19	4.89	1	4.19	7	4.46	9	4.93	1	3.96	16	3.42	7
Mexico	2.01	19	10.99	4	2.95	3	4.27	10	4.26	4	4.19	16	4.55	11	4.37	6	3.79	2
Norway	3.51	14	4.95	14	1.89	8	4.25	12	4.10	11	4.70	4	4.26	19	4.80	1	2.85	18
Portugal	3.98	11	9.68	6	1.78	11	4.56	5	4.22	6	4.58	6	4.76	4	4.60	4	3.58	4
Spain	3.67	13	7.75	9	1.62	12	4.45	6	4.10	10	4.55	7	4.38	14	3.89	17	3.38	9
Sweden	5.73	5	4.77	16	1.61	13	4.12	14	4.15	8	4.63	5	4.55	10	4.43	5	3.19	13
USA	5.18	8	15.44	3	2.65	5	4.32	9	4.08	12	4.78	3	4.83	3	4.25	12	3.50	5
Total	4.85		9.95		2.27		4.33		4.16		4.44		4.60		4.26		3.39	

Table 1 - Strategies, Goals and Finances (continued)

Question	A5A The market aims of the business unit		A5B		A5C		A6A Dominant product line:		A6B		A6C		A7 Market share of leading competitor		A8 Return on investment (ROI) for the last fiscal year:				A9 Inventory turnover for the last year:			
Country	Market coverage (Ranking)		Customer Focus (Ranking)		Geographic Focus (Ranking)		Market Share (Ranking)		No. units /year (Ranking)		Market Devel. (Ranking)		Competitor Mkt. Share (Ranking)		Net Profit	Total Assets	ROI (Ranking)		Net Sales	Inventory	Inventory Turnover (Ranking)	
Argentina	3.00	17	3.37	11	2.83	18	47.27	2	5241	5	2.37	19	30.15	18	1267	11545	10.98	13	15775	8	6.38	15
Australia	3.29	10	3.79	4	3.55	15	44.81	3	4238	6	2.93	11	26.22	12	7	92	13.78	5	104	12	7.13	14
Austria	3.12	15	3.44	10	4.19	3	22.19	17	13586	4	3.00	8	22.45	8	53	593	6.21	16	1042	174	5.60	18
Brazil	3.43	6	3.25	14	3.14	17	40.06	5	52205	2	2.93	11	21.12	4	1519	14064	9.69	12	100	25	7.43	11
Canada	3.09	16	3.22	15	3.61	13	38.20	7	158	16	3.14	3	21.02	3	2	35	12.69	6	40	14	7.80	9
Chile	3.17	11	3.50	8	3.17	16	40.00	6	101	17	2.40	18	27.50	14	13	90	18.50	3	160	33	4.66	19
Denmark	3.12	14	2.59	19	4.06	4	19.82	18	723	13	3.06	6	21.97	3	38	614	3.31	19	335	70	5.97	16
Finland	2.59	19	2.94	18	3.82	7	26.36	16	1316	10	3.24	1	26.85	9	94	638	10.42	9	27	3	5.97	17
Great Britain	3.42	7	2.97	16	4.53	1	26.84	15	1014	11	2.82	13	22.27	9	5	25	21.64	2	30	4	9.33	2
Germany	3.50	4	3.26	13	3.74	10	19.68	19	261	15	3.09	4	30.53	11	31	239	15.24	4	1012	164	7.24	13
Netherlands	2.96	18	3.74	5	3.93	6	37.20	8	252759	1	2.77	14	23.05	4	6	29	112.10	1	49	10	8.75	4
Italy	3.68	2	3.59	7	4.41	2	27.37	14	748	12	2.95	10	22.08	3	0	0	5.81	18	0	0	8.31	7
Japan	3.42	1	3.84	1	3.65	12	29.62	12	614	14	2.74	15	26.95	6	-	-	9.89	11	-	-	16.81	1
Mexico	3.82	1	3.94	2	2.79	19	41.22	4	32	18	2.63	17	19.94	2	0	0	6.06	17	0	0	8.36	5
Norway	3.15	12	2.95	17	3.60	14	49.07	1	31	19	3.00	8	25.82	4	1	84	7.84	14	108	27	8.33	6
Portugal	3.12	13	3.33	12	3.66	11	33.36	11	34333	3	3.15	2	23.35	2	126	5060	6.71	15	922	194	7.25	12
Spain	3.37	9	3.48	9	3.78	8	35.74	9	2577	8	3.07	5	27.28	4	1212	75691	10.74	8	35129	3609	7.49	10
Sweden	3.54	3	3.61	6	4.00	5	28.09	13	1742	9	3.02	7	19.76	1	-190	1782	10.26	10	1868	2202	8.25	8
USA	3.45	5	4.00	1	3.78	9	35.42	10	3495	7	2.67	16	27.14	2	0	1	11.66	7	1	0	9.14	3
Total	3.35		3.49		3.68		33.92		18945		2.88		23.60		224	7031	13.62		3428	534	8.14	

Table 1 - Strategies, Goals and Finances (continued)

Question	A10A	A10B	A10C	A10D

Past and anticipated changes for company in strategic market and product activities:

Country	Production Volume (units) 1991	Ranking	last 5	Ranking	next 5	Ranking	Number of Different Products for 1991	Ranking	last 5	Ranking	next 5	Ranking	Percentage of Revenues from New Products 1991	Ranking	last 5	Ranking	next 5	Ranking	Number of Suppliers 1991	Ranking	last 5	Ranking	next 5	Ranking
Argentina	3628	8	42.7	7	46.4	7	102	16	20.5	9	33.5	5	19.6	9	35.5	5	27.0	11	125	18	13.1	7	18.9	4
Australia	6455	4	47.0	5	21.4	15	896	8	22.2	8	11.4	16	13.6	15	16.2	13	22.3	13	387	11	-4.3	15	-9.2	15
Austria	3733	7	23.0	13	5.9	19	1218	5	12.4	17	1.8	19	19.4	10	12.5	16	11.6	19	593	5	2.3	10	-7.1	11
Brazil	37671	1	17.5	15	46.2	8	165	15	14.1	15	17.6	13	11.2	17	12.0	17	27.7	9	996	3	-13.3	17	-8.5	12
Canada	1738	12	35.2	9	47.6	5	513	11	40.8	4	25.7	10	26.2	5	27.4	8	31.8	7	417	9	7.3	9	-6.6	10
Chile	3788	6	3.5	18	53.3	4	80	17	24.5	7	28.3	8	16.5	13	21.7	11	13.8	18	191	16	44.6	3	10.3	5
Denmark	475	16	68.9	3	70.1	2	300	12	72.1	1	93.4	2	31.6	2	82.7	2	24.1	12	1147	2	23.6	6	-9.0	14
Finland	55	18	24.2	11	22.9	14	909	7	19.5	11	28.0	9	21.5	6	30.0	7	35.0	6	130	17	0.0	11	4.0	6
Great Britain	2125	10	14.9	16	34.7	10	563	10	14.3	14	14.5	14	17.9	11	15.8	15	16.1	16	280	13	-30.2	19	-12.1	16
Germany	1018	14	8.3	17	12.2	18	1639	2	12.7	16	18.2	12	43.0	1	16.0	14	21.2	14	435	8	53.8	8	25.2	2
Netherlands	746	15	39.3	8	90.7	1	929	6	14.5	13	25.0	11	26.9	4	27.0	9	137.8	1	201	15	24.4	5	24.6	3
Italy	1496	13	33.3	10	27.7	13	684	9	39.6	5	29.8	7	21.4	8	45.4	3	75.1	2	709	4	-7.4	16	-8.6	13
Japan	415	17	62.1	4	58.9	3	3817	1	59.0	2	37.7	4	17.2	12	141.0	1	52.9	3	412	10	54.9	1	43.5	1
Mexico	30	19	20.1	14	32.0	12	53	19	10.0	19	13.2	15	11.7	16	7.5	18	17.7	15	60	19	-18.4	18	-37.3	19
Norway	30003	3	24.2	12	33.5	11	1300	3	12.1	18	7.1	17	9.2	19	40.7	4	35.7	5	476	7	-2.5	13	-20.0	18
Portugal	32384	2	72.8	2	41.9	9	68	18	42.5	3	33.3	6	9.7	18	26.9	10	28.2	8	363	12	43.8	4	3.0	7
Spain	2279	9	2.5	19	18.2	16	194	14	16.2	12	2.3	18	21.5	6	7.1	19	15.5	17	477	6	8.2	8	-12.7	17
Sweden	6267	5	43.3	6	18.0	17	265	13	20.1	10	290.8	1	15.7	14	19.6	12	27.5	10	219	14	-4.1	14	-4.5	9
USA	1912	11	95.5	1	47.1	6	1248	4	32.0	6	50.1	3	27.8	3	31.6	6	36.7	4	1718	1	-1.9	12	-4.0	8
Total	6655		37.8		36.6		706		25.8		50.8		18.9		30.3		31.6		469		4.2		-5.1	

Table 1 - Strategies, Goals and Finances (continued)

Question	A11 Present cost structure in manufacturing (estimated)					
Country	Direct Materials	Ranking	Direct Salaries	Ranking	Mfg. OH	Ranking
Argentina	48.08	16	26.17	4	25.72	7
Australia	60.89	3	17.52	14	21.59	16
Austria	49.05	15	28.40	2	22.55	12
Brazil	56.50	8	19.82	9	23.67	9
Canada	56.65	7	16.24	17	27.12	6
Chile	70.17	1	17.83	13	12.00	19
Denmark	58.34	6	19.01	10	22.66	11
Finland	56.14	9	23.14	6	20.71	18
Great Britain	51.39	14	17.41	15	31.20	1
Germany	44.24	19	25.76	5	30.00	3
Netherlands	47.56	17	29.19	1	23.23	10
Italy	60.05	4	18.24	11	22.37	13
Japan	64.45	2	14.56	19	20.99	17
Mexico	58.39	5	16.91	16	24.70	8
Norway	44.65	18	26.56	3	28.78	5
Portugal	55.49	11	21.72	8	22.37	14
Spain	56.13	10	21.74	7	22.06	15
Sweden	52.61	13	18.22	12	29.17	4
USA	53.60	12	15.91	18	30.49	2
Total	54.58		20.31		25.12	

Table 2 - Current Manufacturing Practice Facilities

Country	B1 Plant location				B2A a. Proportion of purchases by geographic location			B2B b. Proportion of sales by geographic location		
	The only plant in our company (Ranking)	The only plant in this economic area (Ranking)	Several in economic area, one in country (Ranking)	One of several plants in this country (Ranking)	Other parts of this country (Ranking)	Outside country, within economic area (Ranking)	Outside this economic area (Ranking)	Other parts of this country (Ranking)	Outside country, within economic area (Ranking)	Outside this economic area (Ranking)
Argentina	74% (3)	5% (12)	10% (14)	10% (11)	68.6 (4)	14.0 (15)	14.4 (8)	77.0 (4)	9.6 (16)	8.8 (17)
Australia	46% (15)	21% (2)	21% (7)	11% (10)	56.0 (9)	9.7 (16)	34.9 (1)	71.9 (5)	9.9 (15)	15.3 (6)
Austria	62% (7)	0% (5)	35% (2)	4% (17)	35.5 (16)	55.5 (1)	9.0 (17)	34.0 (18)	53.8 (1)	10.5 (13)
Brazil	54% (11)	18% (5)	11% (13)	18% (7)	71.3 (3)	5.5 (17)	23.8 (3)	71.6 (6)	15.7 (13)	13.6 (8)
Canada	55% (9)	5% (13)	36% (1)	5% (16)	39.5 (14)	51.6 (4)	8.9 (18)	38.7 (16)	49.2 (2)	12.1 (10)
Chile	83% (1)	0% (15)	0% (19)	17% (8)	39.6 (13)	16.8 (14)	21.6 (4)	59.2 (7)	5.0 (19)	1.3 (19)
Denmark	71% (4)	12% (9)	6% (18)	12% (9)	34.7 (17)	40.1 (6)	25.2 (2)	32.1 (19)	42.3 (5)	25.6 (1)
Finland	63% (6)	19% (4)	19% (9)	0% (19)	50.0 (12)	38.4 (7)	11.6 (12)	42.4 (14)	33.0 (12)	24.6 (2)
Great Britain	50% (14)	19% (3)	22% (6)	8% (12)	63.5 (8)	27.7 (10)	8.8 (19)	44.3 (12)	33.7 (10)	22.2 (3)
Germany	22% (19)	22% (1)	17% (11)	39% (3)	67.0 (5)	21.3 (13)	11.7 (11)	54.1 (9)	34.3 (9)	11.7 (12)
Netherlands	52% (13)	7% (11)	33% (4)	7% (14)	30.1 (19)	47.9 (5)	15.1 (7)	42.4 (15)	41.0 (6)	16.7 (5)
Italy	54% (10)	10% (10)	7% (17)	29% (4)	64.9 (7)	22.7 (12)	12.0 (10)	48.6 (11)	37.1 (8)	14.1 (7)
Japan	38% (17)	0% (15)	13% (12)	50% (1)	80.1 (1)	1.8 (19)	18.1 (5)	78.7 (2)	11.5 (14)	9.8 (14)
Mexico	66% (5)	2% (14)	8% (15)	25% (5)	66.5 (6)	23.6 (11)	9.9 (15)	84.1 (1)	9.0 (17)	6.9 (18)
Norway	76% (2)	0% (15)	18% (10)	6% (15)	38.1 (15)	51.9 (3)	10.0 (14)	42.9 (13)	48.2 (3)	8.9 (16)
Portugal	60% (8)	0% (15)	21% (7)	19% (6)	34.0 (18)	54.8 (2)	9.8 (16)	49.1 (10)	38.6 (7)	11.8 (11)
Spain	42% (16)	15% (7)	35% (2)	8% (13)	50.5 (11)	30.2 (9)	12.2 (9)	56.9 (8)	33.2 (11)	9.0 (15)
Sweden	52% (12)	13% (8)	31% (5)	3% (18)	51.7 (10)	37.0 (8)	10.7 (13)	36.6 (17)	47.7 (4)	16.9 (4)
USA	37% (18)	16% (6)	8% (16)	39% (2)	79.5 (2)	3.1 (18)	16.8 (6)	78.5 (3)	9.0 (18)	12.2 (9)
Total	54%	10%	19%	17%	56.7	28.3	13.8	57.5	28.9	13.3

Table 2 - Current Manufacturing Practice (continued)
Facilities (continued)

Country	B3 Relationship with principal parts/materials suppliers: 1 = Distant, 5 = Close		B4 Variation in customer/market demand? 1 = Large, 5 = Small		B5 Mix of order sizes on same equipment						B6 Different process routings for different products: 1 = Many routings, 5 = Single routing	
					Same equipment		Different equipment		No mix			
		Ranking		Ranking		Ranking		Ranking		Ranking		Ranking
Argentina	3.22	13	3.44	1	71.1%	17	23.7%	3	5.3%	11	3.50	5
Australia	3.52	3	3.38	3	85.2%	6	14.8%	8	0.0%	15	2.79	17
Austria	3.08	17	2.65	18	84.6%	7	15.4%	7	0.0%	15	2.88	15
Brazil	3.50	4	2.82	17	84.6%	7	3.8%	18	11.5%	4	3.54	2
Canada	3.14	15	3.27	4	85.7%	5	9.5%	13	4.8%	12	2.55	19
Chile	3.17	14	2.83	16	100.0%	1	0.0%	19	0.0%	15	3.00	12
Denmark	3.44	6	3.19	7	87.5%	4	6.3%	14	6.3%	9	3.76	1
Finland	3.06	18	3.18	8	88.2%	3	5.9%	15	5.9%	10	3.53	3
Great Britain	3.47	5	3.17	9	77.1%	13	20.0%	4	2.9%	14	3.40	6
Germany	2.63	19	3.08	10	68.0%	18	20.0%	4	12.0%	3	2.88	16
Netherlands	3.31	9	3.19	6	80.8%	11	11.5%	10	7.7%	6	3.08	10
Italy	3.59	2	3.41	2	75.6%	14	9.8%	12	14.6%	1	3.51	4
Japan	3.11	16	2.96	13	74.1%	16	25.9%	1	0.0%	15	3.19	8
Mexico	3.42	7	2.18	19	81.5%	10	14.8%	8	3.7%	13	2.89	14
Norway	3.28	11	3.22	5	94.4%	2	5.6%	16	0.0%	15	3.24	7
Portugal	3.24	12	2.90	15	82.1%	9	5.1%	17	12.8%	2	3.10	9
Spain	3.38	8	2.97	12	67.9%	19	25.0%	2	7.1%	7	2.93	13
Sweden	3.30	10	2.95	14	74.1%	15	19.0%	6	6.9%	8	3.07	11
USA	3.61	1	2.98	11	79.5%	12	10.3%	11	10.3%	5	2.68	18
Total	3.32		3.00		79.2%		14.3%		6.5%		3.11	

Table 2 - Current Manufacturing Practice (continued)
Capacity Planning

	B7 Overall manufacturing capacity policy is to keep capacity:						B8 Proportion of forecast vs. customer orders:				B9 Capacity utilisation of the main processes:				B10 Average days of inventory in:						B11 Average leadtime from customer order to delivery	
	Higher than demand	Ranking	Equal to demand	Ranking	Lower than demand	Ranking	Forecast orders	Ranking	Customer orders	Ranking	Hours/day	Ranking	% planned capacity	Ranking	Raw material/components (days)	Ranking	WIP (days)	Ranking	Finished goods (days)	Ranking	Leadtime (days)	Ranking
Country																						
Argentina	41%	8	44%	12	15%	4	37.1	6	64.1	14	13.9	10	76.8	16	46.4	18	22.3	7	15.0	5	39.7	7
Australia	52%	4	41%	14	7%	15	37.3	5	62.7	15	13.4	14	76.4	17	30.3	8	16.8	4	24.3	13	24.1	3
Austria	19%	19	65%	3	15%	4	35.6	7	66.7	13	14.0	8	78.0	14	40.1	15	25.8	14	19.1	9	66.7	13
Brazil	21%	18	71%	2	7%	16	23.2	17	76.8	3	13.9	9	82.0	9	41.5	16	23.2	8	11.4	2	99.2	17
Canada	41%	9	50%	6	9%	12	19.8	18	80.3	2	16.7	4	77.8	15	32.9	10	24.3	11	25.5	14	67.5	14
Chile	83%	1	17%	18	0%	19	47.0	2	53.0	18	11.7	17	79.0	13	46.7	19	21.8	6	26.3	15	10.7	1
Denmark	41%	7	47%	9	12%	9	29.3	12	70.7	9	17.4	1	81.0	11	29.3	5	24.7	13	17.6	7	101.5	18
Finland	35%	13	47%	9	18%	3	13.8	19	86.7	1	14.6	7	75.3	18	33.5	12	19.7	5	9.7	1	37.2	6
Great Britain	67%	3	25%	17	8%	13	33.3	9	66.8	12	13.6	12	82.4	7	17.9	2	15.3	3	11.8	3	75.7	15
Germany	39%	11	57%	4	4%	17	26.3	14	73.7	7	15.3	6	86.8	4	31.8	9	37.4	17	46.6	18	34.8	5
Netherlands	22%	17	44%	5	33%	1	29.4	11	68.3	11	13.5	13	88.3	2	39.4	13	42.7	18	54.5	19	63.1	12
Italy	34%	15	54%	5	12%	8	39.9	4	59.6	17	13.7	11	82.6	6	45.1	17	24.0	10	26.9	16	85.1	16
Japan	23%	16	73%	1	4%	18	30.7	10	69.3	10	17.4	2	88.7	1	10.9	1	12.1	1	20.5	10	32.5	4
Mexico	77%	2	13%	19	10%	11	59.3	1	40.7	19	9.1	19	71.3	19	33.4	11	34.2	16	16.8	6	21.1	2
Norway	35%	14	35%	16	30%	2	24.6	16	75.4	4	11.4	18	87.3	3	30.2	7	66.0	19	35.3	17	131.8	19
Portugal	49%	5	37%	15	15%	6	27.7	13	75.4	5	13.0	16	86.5	5	39.5	14	24.6	12	19.0	8	56.4	11
Spain	39%	10	50%	6	11%	10	25.9	15	73.9	6	15.7	5	82.2	8	30.2	6	23.3	9	22.8	12	40.2	8
Sweden	39%	12	47%	8	14%	7	34.2	8	70.9	8	13.1	15	81.3	10	29.3	4	14.3	2	13.6	4	44.9	10
USA	49%	6	44%	12	8%	14	41.7	3	61.2	16	16.7	3	80.7	12	24.8	3	26.3	15	21.5	11	41.8	9
Total	44%		44%		12%		34.7		66.3		13.9		80.7		33.1		24.4		20.9		54.9	

Table 2 - Current Manufacturing Practice (continued)
Manufacturing Process and Technology

Country	B12 Value added for manufacturing? % value added	Ranking	B13 Proportion of manufacturing by process type: One-off, unique	Ranking	Batch	Ranking	Line	Ranking	B14 Use of cellular layout in manufacturing: 1=0%, 5=100%	Ranking	B15 Value added for assembly % value added	Ranking	B16 Proportion of assembly by process type: One-off, unique	Ranking	Batch	Ranking	Line	Ranking	B17 Use of cellular layout in assembly: 1=0%, 5=100%	Ranking
Argentina	68.8	3	10.5%	13	60.5%	6	28.9%	8	2.5	15	31.2	17	14.8%	16	29.6%	8	55.6%	4	3.30	5
Australia	51.6	12	10.3%	14	51.7%	10	37.9%	5	2.6	12	48.4	8	26.9%	9	26.9%	13	46.2%	6	3.18	7
Austria	54.2	10	12.0%	12	68.0%	4	20.0%	13	2.8	11	45.8	10	33.3%	4	33.3%	6	33.3%	14	2.84	11
Brazil	54.0	11	38.1%	1	33.3%	17	28.6%	9	2.6	13	46.0	9	33.3%	4	20.8%	16	45.8%	7	2.76	13
Canada	48.3	18	24.0%	6	60.0%	7	16.0%	17	3.2	3	51.8	2	42.9%	3	23.8%	14	33.3%	14	3.50	2
Chile	67.5	4	16.7%	10	50.0%	12	33.3%	7			32.5	16	0.0%	19	66.7%	2	33.3%	14		
Denmark	49.1	16	25.0%	4	50.0%	12	25.0%	11	2.5	16	50.9	4	31.8%	6	27.3%	12	40.9%	10	3.12	9
Finland	65.7	6	15.4%	11	69.2%	3	15.4%	18	3.7	1	34.3	14	69.2%	1	23.1%	15	7.7%	19	3.13	8
Great Britain	46.1	19	24.1%	5	65.5%	5	10.3%	19	2.4	17	53.9	1	16.1%	15	48.4%	3	35.5%	13	2.72	14
Germany	58.2	9	18.8%	9	37.5%	16	43.8%	3	2.8	9	41.8	11	29.2%	7	29.2%	9	41.7%	9	2.76	12
Netherlands	49.0	17	6.7%	17	73.3%	1	20.0%	13	3.1	5	51.0	3	28.0%	8	32.0%	7	40.0%	11	2.71	15
Italy	49.6	15	8.6%	15	57.1%	9	34.3%	6	2.6	14	50.4	5	22.9%	11	17.1%	18	60.0%	3	2.03	18
Japan	72.5	1	6.5%	18	29.0%	19	64.5%	1	2.0	18	27.5	19	5.6%	17	5.6%	19	88.9%	1	2.50	17
Mexico	69.1	2	1.6%	19	51.6%	11	46.8%	2	2.9	7	30.9	18	3.2%	18	71.0%	1	25.8%	17	3.76	1
Norway	66.6	5	33.3%	2	46.7%	15	20.0%	13	3.2	4	33.4	15	44.4%	2	44.4%	4	11.1%	18	3.45	3
Portugal	61.1	7	29.4%	3	32.4%	18	38.2%	4	2.9	6	38.9	13	21.4%	12	17.9%	17	60.7%	2	2.52	16
Spain	51.0	13	7.7%	16	73.1%	2	19.2%	16	3.2	2	49.0	7	19.2%	13	42.3%	5	38.5%	12	3.25	6
Sweden	50.9	14	18.9%	8	58.5%	8	22.6%	12	2.8	8	49.1	6	26.9%	9	28.8%	10	44.2%	8	3.40	4
USA	61.0	8	23.3%	7	50.0%	12	26.7%	10	2.8	10	39.0	12	17.9%	14	28.6%	11	53.6%	5	3.06	10
Total	57.5		15.7%		53.4%		30.9%		2.8		42.5		23.2%		33.7%		43.0%		3.05	

Table 2 - Current Manufacturing Practice (continued)
Manufacturing Process and Technology (continued)

	B18				Machines/system use by type:										B19				B20	
	FMS/FMC		NC-machines		Conventional Machines		Machining centers		Robots for mfg.		Robots for assembly		Flexible Assembly Systems		Level of automation				Throughput efficiency	
															General level		Highest level			
Country		Ranking		Ranking		Ranking		Ranking		Ranking		Ranking		Ranking		Ranking		Ranking		Ranking
Argentina	0.00	18	4.45	17	98.6	6	4.07	8	0.47	16	0.00	17	1.84	10	2.04	19	3.83	16	49.2	6
Australia	1.29	9	7.82	14	59.0	8	2.14	11	0.77	13	3.27	5	2.09	8	3.25	10	5.31	12	33.4	9
Austria	3.08	3	13.38	6	47.0	11	1.83	12	2.50	8	0.42	13	2.21	7	3.88	4	6.07	6	31.2	12
Brazil	0.21	15	19.32	4	160.9	4	1.17	17	0.67	14	0.36	14	0.00	17	2.50	16	2.94	18	68.0	3
Canada	0.72	12	7.94	13	24.2	16	1.72	13	0.39	17	2.00	8	2.82	5	3.50	6	5.63	9	39.6	7
Chile	0.00	18	0.00	19	52.5	9	7.50	3	0.00	19	0.00	17	0.00	17	4.00	3			72.0	2
Denmark	0.71	13	27.06	3	362.6	2	5.35	5	2.53	7	2.56	7	0.50	14	2.08	18	4.75	14	28.8	13
Finland	1.86	7	9.80	11	39.4	13	4.20	7	3.67	5	0.07	15	0.00	17	3.07	12	5.60	10	17.5	15
Great Britain	0.15	16	11.46	8	32.9	15	1.00	18	1.32	10	1.17	11	0.31	15	2.81	14	4.00	15	15.0	17
Germany	1.41	8	6.38	15	52.1	10	2.91	10	1.55	9	1.35	9	2.00	9	2.86	13	5.75	7	33.3	10
Netherlands	1.22	10	6.25	16	22.4	17	1.55	16	1.15	11	0.48	12	2.57	6	3.12	11	6.23	4	32.9	11
Italy	2.21	4	18.12	5	104.3	5	6.65	4	4.35	4	3.43	4	1.11	12	3.35	8	6.26	3	26.3	14
Japan	8.89	1	69.05	1	651.3	1	8.57	2	50.80	1	27.81	1	4.27	2	4.09	2	6.61	2	74.2	1
Mexico	0.09	17	1.75	18	20.7	19	0.44	19	0.18	18	0.04	16	0.22	16	2.79	15	3.16	17	4.8	18
Norway	0.71	13	8.88	12	20.9	18	1.59	14	1.06	12	0.00	17	0.54	13	2.20	17	5.57	11	16.4	16
Portugal	1.00	11	10.93	9	63.6	7	1.56	15	0.59	15	2.57	6	15.04	1	3.85	5	5.21	13	66.2	4
Spain	2.00	5	10.64	10	33.6	14	3.96	9	3.15	6	1.35	10	1.18	11	4.73	1	6.62	1	53.3	5
Sweden	1.95	6	12.67	7	46.4	12	4.73	6	6.05	3	15.12	3	3.05	3	3.39	7	6.11	5	33.7	8
USA	3.43	2	35.70	2	348.6	3	20.14	1	6.71	2	21.03	2	3.03	4	3.32	9	5.63	8	1.9	19
Total	1.6		14.38		106.5		3.92		4.35		4.81		2.39		3.17		5.12		32.7	

Table 2 - Current Manufacturing Practice (continued)
Organisation

Country	B21 Employees during the last fiscal year				B22 Levels (plant manager to operator) Organisational levels		B23 Average span of control of a foreman			
	Total empl.	Ranking	Salaried empl.	Ranking		Ranking	Manufacturing	Ranking	Assembly	Ranking
Argentina	301	18	189	15	4.3	14	19.8	14	18.6	14
Australia	612	11	161	16	3.9	6	25.9	6	25.8	9
Austria	805	7	428	7	4.2	13	28.6	4	31.9	3
Brazil	1476	4	1301	2	4.0	10	39.9	1	40.0	1
Canada	303	17	131	17	3.6	3	18.4	15	17.8	15
Chile	163	19	111	18	4.3	15	15.5	18	13.6	18
Denmark	1783	2	689	4	3.3	1	24.6	9	29.0	5
Finland	472	14	229	13	4.0	10	22.3	12	24.3	10
Great Britain	506	13	194	14	3.9	6	16.9	16	14.8	17
Germany	642	9	319	9	4.0	8	32.0	2	28.8	6
Netherlands	545	12	339	8	4.4	17	29.1	3	19.7	13
Italy	1128	5	595	5	3.4	2	26.5	5	26.9	7
Japan	1674	3	1514	1	6.3	19	16.9	17	22.1	12
Mexico	375	15	256	12	4.0	12	25.0	8	17.1	16
Norway	317	16	110	19	3.8	5	23.4	10	31.9	4
Portugal	677	8	486	6	5.0	18	20.5	13	22.9	11
Spain	931	6	317	10	4.0	9	23.1	11	32.2	2
Sweden	634	10	276	11	3.8	4	25.8	7	26.4	8
USA	2879	1	748	3	4.3	16				
Total	869.1		445.3		4.1		23.7		25.5	

Table 2 - Current Manufacturing Practice (continued)
Organisation (continued)

	B24A Design of the payment system for direct employees:						B24B Incentive basis							
Country	Group incentive	Ranking	Individual incentive	Ranking	Fixed salary	Ranking	Quality	Ranking	Productivity	Ranking	Profit	Ranking	Output	Ranking
Argentina	16.7%	12	27.1%	7	56.3%	10	24.2%	9	57.6%	3	3.0%	15	15.2%	13
Australia	41.4%	5	6.9%	16	51.7%	11	17.4%	15	39.1%	11	13.0%	7	30.4%	7
Austria	25.0%	10	53.6%	3	21.4%	16	28.1%	6	37.5%	12	3.1%	14	31.3%	6
Brazil	10.3%	17	0.0%	18	89.7%	1	33.3%	5	25.0%	16	33.3%	2	8.3%	15
Canada	12.5%	14	8.3%	14	79.2%	3	22.2%	11	22.2%	17	33.3%	2	22.2%	10
Chile	42.9%	4	0.0%	18	57.1%	9	25.0%	8	50.0%	6	0.0%	17	25.0%	9
Denmark	77.8%	1	16.7%	12	5.6%	19	18.2%	14	50.0%	6	4.5%	11	27.3%	8
Finland	58.8%	2	35.3%	5	5.9%	18	13.6%	16	40.9%	10	4.5%	11	40.9%	1
Great Britain	27.8%	7	8.3%	14	63.9%	7	23.8%	10	33.3%	14	4.8%	10	38.1%	3
Germany	13.6%	13	54.5%	2	31.8%	13	20.0%	12	32.0%	15	8.0%	9	40.0%	2
Netherlands	0.0%	19	18.5%	11	81.5%	2	36.4%	2	36.4%	13	9.1%	8	18.2%	11
Italy	31.7%	6	19.5%	10	48.8%	12	36.4%	2	57.6%	3	0.0%	17	6.1%	16
Japan	11.5%	15	73.1%	1	15.4%	17	4.2%	19	50.0%	6	41.7%	1	4.2%	17
Mexico	3.2%	18	27.4%	6	69.4%	4	10.5%	18	21.1%	18	31.6%	4	36.8%	4
Norway	27.8%	7	5.6%	17	66.7%	6	12.5%	17	62.5%	2	25.0%	6	0.0%	18
Portugal	10.9%	16	21.7%	9	67.4%	5	35.3%	4	64.7%	1	0.0%	17	0.0%	18
Spain	20.7%	11	48.3%	4	31.0%	14	19.4%	13	45.2%	9	3.2%	13	32.3%	5
Sweden	45.5%	3	24.2%	8	30.3%	15	27.1%	7	54.3%	5	1.4%	16	17.1%	12
USA	27.0%	9	13.5%	13	59.5%	8	36.8%	1	21.1%	18	31.6%	4	10.5%	14
Total	24.1%		24.8%		51.1%		23.7%		44.4%		10.1%		21.8%	

Table 2 - Current Manufacturing Practice (continued)
Organisation (continued)

Country	B25 Different job classifications Ranking		B26 Suggestions per employee per year Per empl./year	Ranking	B27 Proportion of the work force working in teams (%)	Ranking	B28 Hours of training - new production workers hours	Ranking	B29 Hours of training - regular work force hours/year	Ranking	B30 Multi-skilled operators (%)	Ranking	B31 Frequency of job rotation (1 = Never, 5 = Frequently)	Ranking	jobs	Ranking	B32 Turnover for direct employees (%)	Ranking	B33 Short-term absenteeism for direct employees (%)	Ranking
Argentina	8.11	5	1.87	15	31.21	13	112.31	8	33.54	7	41.13	13	2.95	15	3.80	5	7.29	15	4.40	11
Australia	11.72	10	5.42	5	42.86	5	47.63	18	21.12	15	59.14	2	3.50	6	3.04	13	4.43	10	3.79	9
Austria	10.81	7	1.96	14	26.50	18	62.73	14	13.56	19	37.57	15	2.77	18	2.48	17	3.92	7	3.50	6
Brazil	40.85	19	1.58	17	34.04	9	60.63	15	43.35	5	43.40	11	3.57	5	3.62	7	3.92	8	2.35	1
Canada	17.82	15	2.27	12	31.23	12	50.93	17	38.60	6	52.73	5	3.45	7	3.16	12	3.64	5	3.02	4
Chile	10.83	8	3.40	7	60.83	2	35.00	19	15.00	18	16.60	19	2.83	17	3.80	6	9.33	16	2.75	2
Denmark	4.69	1	2.73	9	32.21	10	111.25	9	45.12	4	47.35	7	3.29	11	2.84	15	5.41	12	3.83	10
Finland	24.07	16	0.85	19	27.35	17	113.50	7	30.35	8	45.00	9	2.65	19	2.33	19	2.72	2	5.06	12
Great Britain	6.25	2	2.47	11	41.56	6	67.72	12	22.79	14	60.35	1	3.63	4	3.82	4	1.83	1	2.92	3
Germany	6.33	3	10.45	3	28.91	16	225.90	2	107.67	1	26.14	18	3.00	13	3.25	10			6.87	18
Netherlands	13.74	13	14.50	2	41.21	7	71.86	11	27.40	11	44.62	10	3.30	10	3.02	14	4.31	9	6.48	17
Italy	7.58	4	3.03	8	30.04	15	106.29	10	27.00	12	37.44	16	3.37	9	2.64	16	6.42	14	7.33	19
Japan	8.58	6	66.48	1	62.82	1	136.83	6	19.38	16	27.38	17	3.17	12	3.94	3	3.63	4	3.03	5
Mexico	13.56	9	2.72	10	12.22	19	54.75	16	60.88	3	42.87	12	4.08	1	3.38	9	30.68	18	5.66	15
Norway	11.00	9	2.25	13	30.75	14	144.60	5	19.27	17	45.44	8	3.00	13	3.53	8	3.19	3	3.66	8
Portugal	39.00	18	4.81	6	41.03	8	229.24	1	78.27	2	38.00	14	2.95	16	2.42	18	6.24	13	6.00	16
Spain	12.59	11	1.59	16	32.05	11	206.85	3	28.05	10	51.78	6	3.59	4	8.13	1	11.05	17	5.17	13
Sweden	17.56	14	1.41	18	50.57	3	145.96	4	25.93	13	54.75	4	3.70	2	3.20	11	3.66	6	5.27	14
USA	25.73	17	5.49	4	47.58	4	66.47	13	30.26	9	54.87	3	3.45	8	4.07	2	4.45	11	3.58	7
Total	16.06		7.06		36.50		109.85		35.56		45.73		3.38		3.51		8.08		4.73	

Table 2 - Current Manufacturing Practice (continued)
Planning and Control Systems

Country	B34 How far ahead production schedule is frozen	Ranking	B35 Planning dept. (%)	Ranking	Foreman or supervisor (%)	Ranking	Operators (%)	Ranking	B36 Raw materials and components delivered JIT (%)	Ranking	B37 Pct. orders delivered late (%)	Ranking	Machine capacity	Prod. bottle-necks	Qual. problems	Due date chges	Lab. shortage	Matl shortage	Des. chges	Other
Argentina	4.76	9	70%	3	25%	16	5%	10	22.0	15	12.1	13	13%	17%	13%	19%	6%	23%	2%	6%
Australia	2.98	16	48%	13	42%	8	10%	6	28.0	11	8.1	5	3%	19%	10%	0%	6%	48%	10%	3%
Austria	4.64	10	44%	15	52%	3	4%	12	18.1	18	11.5	11	8%	19%	11%	11%	5%	30%	16%	0%
Brazil	5.11	6	68%	4	29%	15	4%	13	19.6	16	12.0	12	0%	9%	12%	15%	0%	41%	9%	15%
Canada	4.93	8	58%	8	38%	12	4%	11	27.4	12	13.9	16	0%	19%	6%	25%	3%	16%	22%	9%
Chile	3.60	13	67%	5	33%	14	0%	16	67.0	1	19.2	18	13%	25%	0%	25%	13%	25%	0%	0%
Denmark	6.21	3	17%	19	67%	2	17%	3	35.6	5	6.1	2	0%	6%	0%	22%	6%	50%	11%	6%
Finland	3.20	15	24%	18	71%	1	6%	8	36.0	4	15.1	17	0%	6%	0%	6%	0%	41%	29%	18%
Great Britain	1.60	18	51%	12	46%	4	3%	15	25.7	13	13.3	15	2%	14%	20%	16%	0%	41%	6%	0%
Germany	7.57	2	56%	10	38%	11	6%	8	16.8	19	6.4	3	9%	18%	18%	23%	5%	41%	14%	5%
Netherlands	5.92	4	54%	11	39%	9	7%	7	30.6	8	11.0	9	7%	15%	15%	5%	2%	37%	17%	2%
Italy	5.08	7	85%	1	15%	19	0%	16	19.4	17	12.6	14	8%	11%	8%	11%	0%	39%	22%	0%
Japan	1.65	17	65%	6	35%	13	0%	16	52.7	3	4.8	1	8%	4%	25%	46%	0%	8%	0%	8%
Mexico	4.34	11	82%	2	15%	18	3%	14	24.4	14	21.2	19	0%	31%	32%	3%	5%	29%	0%	0%
Norway	14.73	1	32%	17	44%	5	24%	2	63.8	2	7.7	4	5%	36%	23%	0%	0%	23%	5%	9%
Portugal	0.79	19	45%	14	43%	7	12%	5	29.0	9	11.1	10	6%	25%	6%	15%	8%	17%	13%	10%
Spain	4.28	12	57%	9	43%	6	0%	16	33.0	6	10.6	8	5%	13%	13%	8%	3%	36%	10%	15%
Sweden	3.45	14	34%	16	39%	10	27%	1	31.2	7	9.6	6	1%	15%	19%	8%	3%	38%	14%	4%
USA	5.26	5	64%	7	23%	17	13%	4	29.0	10	9.8	7	0%	18%	26%	29%	0%	18%	3%	6%
Total	4.49		56%		36%		8%		29.4		11.8		4%	17%	15%	13%	3%	31%	11%	6%

Note under B37: Percentage of your orders are delivered late and general reason for lateness.

Table 2 - Current Manufacturing Practice (continued)

	B38 Proportion of maintenance spending is on:				B39 Proportion of quality spending on Quality							
Country	Preventative maintenance	Ranking	Rectifying maintenance	Ranking	Inspection/control	Ranking	Internal quality	Ranking	Prevention	Ranking	External quality	Ranking
Argentina	33.8	13	66.2	13	39.5	18	30.8	15	24.9	6	6.8	19
Australia	41.2	5	58.8	5	24.6	3	25.1	8	30.1	3	20.2	8
Austria	37.2	9	62.8	9	39.5	17	22.6	4	20.3	15	17.6	11
Brazil	32.8	15	67.2	15	35.5	15	22.2	3	26.9	4	16.4	13
Canada	23.5	19	76.5	19	33.8	14	30.6	14	20.9	14	14.6	14
Chile	34.3	12	65.7	12	29.5	9	45.0	19	15.5	17	9.8	18
Denmark	40.9	6	59.1	6	24.9	4	24.1	7	24.0	9	27.0	2
Finland	49.0	2	51.0	2	29.0	8	29.0	12	17.3	16	24.8	3
Great Britain	29.5	18	70.5	18	28.0	6	18.8	1	30.2	2	22.0	6
Germany	33.8	14	66.3	14	31.9	12	23.8	6	24.5	8	19.9	9
Netherlands	43.5	3	54.5	3	39.5	16	25.4	9	23.7	10	11.4	17
Italy	37.7	8	62.3	8	44.3	19	19.8	2	14.3	18	21.6	7
Japan	66.2	1	33.8	1	30.5	10	23.7	5	31.8	1	14.0	15
Mexico	34.7	11	65.3	11	28.8	7	40.1	18	12.1	19	19.1	10
Norway	30.0	17	70.0	17	17.2	1	31.8	16	22.7	11	28.2	1
Portugal	41.4	4	58.6	4	31.4	11	25.8	10	25.8	5	16.9	12
Spain	30.6	16	69.4	16	33.2	13	28.7	11	24.8	7	13.3	16
Sweden	39.5	7	60.5	7	25.6	5	29.8	13	21.4	12	23.7	4
USA	35.4	10	64.6	10	21.7	2	33.7	17	20.9	13	23.6	5
Total	37.2		62.7		31.0		28.2		22.1		18.9	

APPENDIX - TABLES

Table 2 - Current Manufacturing Practice (continued)
Product Development

Country	B40 Organisational coordination of design and manufacturing						B41 Information transfer from design to manufacturing (1 = One-way, 3 = Active)		B42 Job rotation between design and manufacturing (1 = Continuously, 3 = Never)		B43 Percentage of blueprints subject to ECOs (%)	
	Rules and standards	Formal meetings	Informal meetings	Cross-functional task forces	Personal contacts	Other		Ranking		Ranking		Ranking
Argentina	40.0%	17.8%	17.8%	6.7%	15.6%	2.2%	3.28	14	2.41	17	23.54	6
Australia	13.8%	27.6%	10.3%	48.3%	0.0%	0.0%	3.85	2	2.07	4	35.48	13
Austria	33.3%	16.7%	5.6%	19.4%	25.0%	0.0%	3.23	15	2.35	15	28.47	10
Brazil	36.1%	19.4%	8.3%	25.0%	5.6%	5.6%	3.67	6	2.26	10	23.06	5
Canada	21.4%	14.3%	17.9%	28.6%	17.9%	0.0%	3.45	11	2.37	16	32.81	11
Chile	16.7%	33.3%	0.0%	33.3%	16.7%	0.0%	3.50	9	2.60	19	100.00	19
Denmark	15.8%	15.8%	5.3%	52.6%	10.5%	0.0%	4.06	1	2.06	3	54.54	17
Finland	17.6%	5.9%	29.4%	23.5%	23.5%	0.0%	3.47	10	2.29	12	24.72	8
Great Britain	24.3%	27.0%	13.5%	35.1%	0.0%	0.0%	3.77	5	2.31	13	24.24	7
Germany	17.2%	20.7%	20.7%	27.6%	13.8%	0.0%	3.63	7	2.29	11	51.00	16
Netherlands	22.5%	25.0%	15.0%	22.5%	15.0%	0.0%	3.12	16	2.52	18	9.45	1
Italy	22.0%	7.3%	29.3%	36.6%	4.9%	0.0%	3.45	11	2.17	7	27.25	9
Japan	55.2%	34.5%	3.4%	3.4%	3.4%	0.0%	3.08	17	1.70	1	13.82	3
Mexico	11.3%	30.6%	29.0%	8.1%	17.7%	3.2%	2.27	19	2.21	9	12.23	2
Norway	29.2%	16.7%	16.7%	25.0%	12.5%	0.0%	3.56	8	2.19	8	42.25	15
Portugal	25.5%	18.2%	14.5%	27.3%	10.9%	3.6%	3.34	13	2.11	5	33.12	12
Spain	51.7%	13.8%	6.9%	17.2%	10.3%	0.0%	2.93	18	2.32	14	36.71	14
Sweden	20.8%	20.8%	9.7%	38.9%	8.3%	1.4%	3.80	3	2.15	6	22.46	4
USA	10.0%	22.5%	7.5%	60.0%	0.0%	0.0%	3.80	4	2.03	2	60.16	18
Total	25.2%	20.6%	14.7%	27.6%	10.7%	1.2%	3.38		2.21		29.51	

Table 3 - Past and Planned Activities in Manufacturing

Question / Country	C1A Improve conformance quality				C1B Reduce unit cost				C1C Reduce manufacturing lead time				C1D Reduce procurement lead time			
	Quantified goal (%)	Ranking	Degree of importance	Ranking	Quantified goal (%)	Ranking	Degree of importance	Ranking	Quantified goal (%)	Ranking	Degree of importance	Ranking	Quantified goal (%)	Ranking	Degree of importance	Ranking
Argentina	75.7%	14	4.30	8	78.9%	15	4.63	2	65.8%	16	4.00	8	38.9%	18	3.27	15
Australia	86.2%	6	4.21	12	82.8%	12	4.03	16	71.4%	14	3.41	19	65.5%	7	3.10	17
Austria	77.3%	13	3.85	15	79.2%	14	4.42	6	82.6%	6	3.71	15	60.9%	10	3.71	5
Brazil	70.4%	16	4.22	11	74.1%	16	4.26	9	82.1%	8	4.00	8	70.4%	4	3.89	2
Canada	85.7%	7	4.00	14	85.7%	10	4.05	15	80.0%	9	4.05	7	68.2%	5	3.62	8
Chile	83.3%	9	3.67	19	100.0%	1	4.00	18	100.0%	1	4.00	8	50.0%	15	2.75	19
Denmark	52.9%	19	3.81	16	82.4%	13	4.13	13	64.7%	17	3.50	18	47.1%	17	3.40	12
Finland	68.8%	17	3.75	18	70.6%	18	4.06	14	82.4%	7	4.06	6	56.3%	11	3.50	11
Great Britain	93.9%	1	4.38	5	97.0%	3	4.53	3	75.8%	12	3.94	12	48.5%	16	2.82	18
Germany	63.2%	18	3.81	16	95.7%	4	4.52	4	78.3%	11	4.30	2	36.4%	19	3.29	14
Netherlands	83.3%	9	4.25	10	88.5%	8	4.00	18	87.5%	3	3.87	14	61.9%	8	3.18	16
Italy	92.7%	2	4.20	13	97.6%	2	4.24	10	82.9%	5	3.54	17	80.5%	1	3.32	13
Japan	80.8%	12	4.38	4	92.3%	6	4.81	1	92.3%	2	4.48	1	70.8%	3	3.60	9
Mexico	70.5%	15	4.38	7	50.8%	19	4.02	17	50.8%	19	4.21	3	52.5%	14	4.11	1
Norway	90.0%	3	4.50	3	90.0%	7	4.32	8	80.0%	9	4.16	5	56.3%	11	3.60	9
Portugal	82.1%	11	4.65	1	73.0%	17	4.46	5	67.6%	15	3.88	13	67.6%	6	3.79	3
Spain	85.7%	7	4.28	9	93.1%	5	4.36	7	64.3%	18	3.63	16	74.1%	2	3.69	6
Sweden	89.7%	5	4.38	6	88.1%	9	4.14	12	84.2%	4	4.21	4	61.1%	9	3.78	4
USA	90.0%	3	4.63	2	82.9%	11	4.24	10	72.5%	13	3.95	11	56.1%	13	3.68	7
Total	81.5%		4.29		82.4%		4.28		74.6%		3.96		59.6%		3.56	

Table 3 - Past and Planned Activities in Manufacturing (continued)

Question	C1E Reduce new product development cycle		C1F Reduce materials cost		C1G Reduce overhead costs		C1H Improve direct labour productivity	
Country	Quantified goal (%) Ranking	Degree of importance Ranking	Quantified goal (%) Ranking	Degree of importance Ranking	Quantified goal (%) Ranking	Degree of importance Ranking	Quantified goal (%) Ranking	Degree of importance Ranking
Argentina	37.1% 17	3.38 16	54.1% 18	3.85 16	70.3% 14	4.00 11	77.1% 12	4.50 2
Australia	50.0% 12	3.25 19	74.1% 14	3.85 15	78.6% 7	3.81 16	85.7% 6	4.00 9
Austria	47.8% 13	3.65 7	79.2% 8	3.91 12	72.0% 12	3.96 13	69.6% 16	3.96 11
Brazil	42.9% 16	3.36 17	77.8% 10	4.33 2	74.1% 11	4.18 6	85.2% 7	4.29 3
Canada	57.9% 8	3.50 13	76.2% 11	3.90 13	81.0% 6	3.95 14	95.5% 2	4.09 6
Chile	66.7% 3	3.83 3	66.7% 15	4.00 10	66.7% 17	3.67 18	66.7% 17	2.50 19
Denmark	47.1% 14	3.63 8	58.8% 16	4.06 7	76.5% 9	4.00 11	70.6% 15	3.81 13
Finland	47.1% 14	3.33 18	82.4% 7	4.06 8	68.8% 15	3.56 19	76.5% 13	3.65 16
Great Britai	60.6% 6	3.71 5	90.9% 2	3.85 14	78.1% 8	4.32 2	75.8% 14	3.64 17
Germany	59.1% 7	3.68 6	75.0% 12	4.33 3	90.9% 2	4.14 8	61.9% 18	4.05 7
Netherlands	57.1% 9	3.58 11	75.0% 12	3.50 19	70.8% 13	3.74 17	80.8% 8	3.65 15
Italy	80.5% 1	3.71 4	85.4% 4	3.73 18	90.2% 3	4.02 10	80.5% 9	3.45 18
Japan	68.0% 2	4.04 1	92.3% 1	4.56 1	100.0% 1	4.48 1	100.0% 1	4.74 1
Mexico	29.5% 19	3.54 12	49.2% 19	4.28 5	47.5% 19	4.31 4	41.0% 19	4.10 5
Norway	56.3% 10	3.38 15	78.9% 9	4.06 9	84.2% 5	4.16 7	95.0% 3	4.05 7
Portugal	35.3% 18	3.42 14	57.9% 17	4.00 10	68.4% 16	4.08 9	86.8% 5	4.23 4
Spain	62.5% 4	3.63 8	86.2% 3	4.25 6	89.3% 4	4.32 3	92.6% 4	3.96 10
Sweden	53.6% 11	3.87 2	82.5% 6	4.31 4	74.1% 10	4.21 5	78.9% 11	3.79 14
USA	61.5% 5	3.61 10	85.4% 4	3.80 17	65.9% 18	3.90 15	80.5% 9	3.87 12
Total	52.2%	3.59	74.3%	4.04	74.6%	4.10	77.5%	3.98

Table 3 - Past and Planned Activities in Manufacturing (continued)

Question / Country	C1I Reduce number of suppliers				C1J Improve supplier quality				C1K Reduce inventories				C1L Increase delivery reliability			
	Quantified goal (%)	Ranking	Degree of importance	Ranking	Quantified goal (%)	Ranking	Degree of importance	Ranking	Quantified goal (%)	Ranking	Degree of importance	Ranking	Quantified goal (%)	Ranking	Degree of importance	Ranking
Argentina	16.2%	19	2.37	19	50.0%	16	3.64	13	62.2%	16	3.70	13	51.4%	18	3.97	9
Australia	44.8%	6	2.58	15	65.5%	10	3.55	15	82.8%	9	3.83	11	65.5%	12	3.34	19
Austria	39.1%	13	2.53	16	56.5%	15	3.36	17	84.0%	8	3.82	12	65.2%	13	3.45	17
Brazil	41.7%	9	3.08	5	59.3%	11	4.14	3	88.9%	6	4.46	1	64.0%	15	4.30	2
Canada	42.1%	7	2.94	9	70.0%	7	3.90	7	95.5%	2	3.86	10	80.0%	2	4.10	6
Chile	33.3%	15	2.50	18	33.3%	19	3.33	18	66.7%	14	3.33	18	66.7%	10	4.00	8
Denmark	29.4%	17	3.07	6	35.3%	18	3.38	16	58.8%	17	3.31	19	52.9%	16	3.60	16
Finland	41.2%	11	2.73	13	58.8%	12	3.19	19	76.5%	11	3.69	14	76.5%	4	4.29	3
Great Britain	45.5%	5	2.91	10	75.8%	3	3.62	14	90.6%	4	3.52	16	69.7%	8	3.85	11
Germany	23.8%	18	3.23	3	85.7%	2	3.85	9	45.0%	19	3.50	17	45.5%	19	3.76	13
Netherlands	40.0%	12	2.75	12	72.7%	6	3.71	11	76.2%	12	3.68	15	75.0%	6	4.14	5
Italy	68.3%	1	2.76	11	92.7%	1	3.90	6	92.7%	3	3.90	9	80.5%	1	3.44	18
Japan	30.8%	16	2.71	14	57.7%	14	3.92	5	88.5%	7	4.00	5	69.2%	9	3.69	15
Mexico	34.4%	14	3.79	1	45.9%	17	3.72	10	50.8%	18	4.23	3	52.5%	17	4.05	7
Norway	42.1%	7	3.19	4	57.9%	13	3.67	12	63.2%	15	3.95	7	78.9%	3	4.37	1
Portugal	41.7%	9	3.06	7	65.8%	9	4.11	4	81.1%	10	4.13	4	64.9%	14	3.76	14
Spain	48.1%	4	2.52	17	74.1%	4	3.85	8	96.6%	1	4.31	2	66.7%	10	3.92	10
Sweden	53.4%	3	3.38	2	66.1%	8	4.16	2	75.9%	13	3.95	8	75.9%	5	4.20	4
USA	62.5%	2	3.00	8	73.2%	5	4.20	1	90.0%	5	3.95	6	75.0%	6	3.82	12
Total	42.5%		2.99		64.4%		3.82		77.4%		3.93		67.0%		3.90	

Table 3 - Past and Planned Activities in Manufacturing (continued)

Question	C1M Increase delivery speed				C1N Improve white collar productivity				C1O Improve ability to make rapid design changes				C1P Improve ability to make rapid volume changes			
	Quantified goal (%)	Ranking	Degree of importance	Ranking	Quantified goal (%)	Ranking	Degree of importance	Ranking	Quantified goal (%)	Ranking	Degree of importance	Ranking	Quantified goal (%)	Ranking	Degree of importance	Ranking
Country																
Argentina	54.1%	11	3.71	6	45.5%	9	3.32	14	31.4%	15	3.16	16	48.6%	7	3.42	10
Australia	31.0%	18	2.81	16	42.9%	14	3.11	19	33.3%	11	3.25	14	42.9%	12	3.23	13
Austria	56.5%	10	3.35	14	33.3%	17	3.24	16	52.2%	3	3.25	14	30.4%	16	2.78	19
Brazil	63.0%	6	4.04	2	40.0%	15	3.65	9	30.8%	16	3.78	2	44.4%	11	3.61	7
Canada	70.0%	2	3.39	13	45.0%	11	3.53	11	57.9%	2	3.50	6	52.6%	3	3.65	5
Chile	83.3%	1	3.83	3	66.7%	3	3.33	13	50.0%	4	3.00	18	50.0%	5	3.40	11
Denmark	35.3%	17	3.53	11	58.8%	4	4.06	2	23.5%	18	3.87	1	41.2%	14	3.25	12
Finland	58.8%	7	3.50	12	29.4%	19	3.71	7	29.4%	17	3.29	12	52.9%	2	3.63	6
Great Britain	36.4%	16	2.76	17	45.5%	9	3.27	15	33.3%	11	3.26	13	27.3%	18	2.97	18
Germany	52.4%	12	3.75	5	52.4%	7	4.06	3	42.9%	7	3.53	5	27.3%	18	3.79	4
Netherlands	65.2%	5	3.80	4	43.5%	12	3.24	16	31.8%	13	3.29	11	33.3%	15	3.00	17
Italy	58.5%	8	2.63	18	78.0%	1	3.41	12	61.0%	1	3.03	17	70.7%	1	3.10	15
Japan	57.7%	9	3.65	9	52.0%	8	4.00	4	44.0%	6	3.64	4	50.0%	5	4.00	1
Mexico	65.6%	4	4.16	1	39.3%	16	3.79	6	36.1%	10	3.72	3	50.8%	4	3.93	2
Norway					68.4%	2	4.11	1	31.6%	14	3.00	18	42.1%	13	3.81	3
Portugal	50.0%	14	3.63	10	55.3%	6	3.68	8	40.0%	8	3.39	10	47.4%	8	3.54	8
Spain	67.9%	3	3.66	8	56.0%	5	3.65	9	37.5%	9	3.42	9	46.2%	10	3.22	14
Sweden	42.1%	15	3.71	7	43.4%	13	3.86	5	23.2%	19	3.45	7	30.4%	17	3.49	9
USA	51.3%	13	3.32	15	32.5%	18	3.20	18	50.0%	4	3.44	8	46.3%	9	3.05	16
Total	54.1%		3.52		48.1%		3.60		38.5%		3.41		44.4%		3.43	

Table 3 - Past and Planned Activities in Manufacturing (continued)

Question	C2A Total Quality Management Program						C2B Statistical Process Control						C2C ISO 9000					
Country	Degree of use last 2 years	Ranking	Relative Payoff	Ranking	Adopt within next 2 years?	Ranking	Degree of use last 2 years	Ranking	Relative Payoff	Ranking	Adopt within next 2 years?	Ranking	Degree of use last 2 years	Ranking	Relative Payoff	Ranking	Adopt within next 2 years?	Ranking
Argentina	2.78	13	3.23	10	43%	7	2.85	9	3.43	2	88%	8	2.04	18	3.36	8	44%	14
Australia	3.62	3	3.33	9	33%	10	2.71	11	2.83	16	75%	9	3.81	5	3.32	9	75%	9
Austria	2.74	14	2.92	16	67%	2	2.65	12	2.82	17	29%	15	3.52	6	3.10	14	80%	8
Brazil	3.17	7	3.65	5			3.04	5	3.11	9	100%	4	3.28	11	3.06	15	100%	1
Canada	3.45	5	3.22	12	20%	12	2.37	16	2.56	19	200%	1	2.53	16	2.53	18	50%	13
Chile	2.60	16	2.75	18	67%	2	2.25	18	3.75	1	20%	16	1.00	19	3.00	16	88%	5
Denmark	2.17	18	3.00	14	100%	1	2.38	15	3.00	12	100%	4	3.36	9	3.20	12	33%	15
Finland	2.80	11	3.38	6	43%	4	1.64	19	2.71	18	50%	13	4.21	2	3.77	4	60%	11
Great Britai	2.79	12	2.81	17	33%	6	3.09	3	3.13	7	0%	18	4.03	3	3.97	2	100%	1
Germany							3.00	7	3.12	8	100%	4	3.31	10	3.60	6	100%	1
Netherlands	3.11	8	2.94	15	44%	3	2.61	13	3.00	12	100%	4	3.82	4	3.29	11	100%	1
Italy	2.66	15	3.14	13	38%	4	3.05	4	3.29	4	200%	1	2.71	14	3.64	5	81%	7
Japan	3.85	2	4.04	1	0%	6	2.73	10	3.22	6	57%	11	2.62	15	3.40	7	63%	10
Mexico	4.16	1	3.85	2			3.67	1	3.02	11			2.90	13	2.92	17	25%	16
Norway	3.08	9	3.75	3	0%	6	2.46	14	2.89	15	67%	10	4.23	1	4.15	1	25%	16
Portugal	3.07	10	3.65	4	33%	4	2.87	8	3.27	5	14%	17	2.93	12	3.20	12		
Spain	3.27	6	3.35	8	29%	4	3.46	2	3.32	3	57%	11	3.38	8	3.88	3	20%	18
Sweden	2.53	17	3.23	11	50%	1	2.36	17	2.94	14	200%	1	3.50	7	3.31	10	57%	12
USA	3.54	4	3.35	7	50%	1	3.03	6	3.05	10	50%	13	2.25	17	2.46	19	83%	6
Total	3.18		3.38		41%		2.89		3.08		103%		3.14		3.28		61%	

APPENDIX - TABLES

Table 3 - Past and Planned Activities in Manufacturing (continued)

Country	C2D MRP Degree of use last 2 years	Ranking	C2D MRP Relative Payoff	Ranking	C2D MRP Adopt within next 2 years?	Ranking	C2E MRP II Degree of use last 2 years	Ranking	C2E MRP II Relative Payoff	Ranking	C2E MRP II Adopt within next 2 years?	Ranking	C2F Just-in-Time Manufacturing, Lean Production Degree of use last 2 years	Ranking	C2F Relative Payoff	Ranking	C2F Adopt within next 2 years?	Ranking
Argentina	2.30	17	3.67	7	10%	5	1.50	19	2.75	15	21%	7	2.16	19	3.44	12	27%	6
Australia	2.96	13	3.00	16	0%	8	2.77	10	3.00	12	38%	4	3.32	5	3.73	7	0%	11
Austria	3.20	10	3.33	13	0%	8	2.58	12	2.75	15	17%	9	2.55	16	3.24	17	40%	3
Brazil	3.00	12	3.44	10			2.67	11	3.50	5	100%	1	3.00	9	3.58	9		
Canada	3.71	4	3.43	11	0%	8	2.85	9	3.67	3	20%	8	3.05	8	3.59	8	33%	4
Chile	1.00	19	1.00	19	0%	8	3.00	6	2.50	18	50%	3	2.67	14	3.33	14	25%	7
Denmark	3.75	3	4.00	3	33%	2	3.50	1	3.57	4	0%	15	3.17	7	4.25	1		
Finland	2.09	18	2.43	18	0%	8	1.50	19	1.50	19	0%	15	2.67	14	3.91	4	25%	7
Great Britai	2.75	14	2.92	17	0%	8	3.03	5	3.84	1	13%	12	2.74	13	2.90	19	50%	2
Germany	2.33	16	3.20	14	0%	8	1.88	17	2.67	17	0%	15	2.95	11	3.31	15	33%	4
Netherlands	3.88	1	4.07	2	7%	7	2.94	7	3.27	10	25%	5	3.18	6	3.47	11	0%	11
Italy	3.24	9	3.53	9	27%	3	1.93	16	3.46	8	14%	11	2.32	18	3.42	13	59%	1
Japan	3.33	6	3.88	4	0%	8	2.26	13	3.50	5	10%	13	3.93	1	4.12	2	0%	11
Mexico	3.27	8	3.62	8	0%	8	3.47	2	3.78	2	100%	1	3.61	2	3.81	5		
Norway	3.33	6	4.11	1	0%	8	2.86	8	3.50	5	0%	15	2.36	17	3.75	6	0%	11
Portugal	2.70	15	3.71	6	0%	8	2.05	14	3.10	11	0%	15	2.83	12	3.28	16	0%	11
Spain	3.79	2	3.76	5	25%	4	3.24	3	3.43	9	17%	9	3.00	9	3.00	18	22%	9
Sweden	3.06	11	3.20	14	50%	1	1.96	15	3.00	12	25%	5	3.38	3	4.10	3	20%	10
USA	3.70	5	3.41	12	10%	5	3.17	4	2.96	14	7%	14	3.36	4	3.54	10	0%	11
Total	3.16		3.47		11%		2.66		3.35		17%		3.03		3.60		27%	

Table 3 - Past and Planned Activities in Manufacturing (continued)

Question	C2G Just-in-Time (frequent) Deliveries to Customers			C2H SMED (Single Minute Exchange of Dies)			C2I Pull Scheduling (Kanban)		
Country	Degree of use last 2 years (Ranking)	Relative Payoff (Ranking)	Adopt within next 2 years? (Ranking)	Degree of use last 2 years (Ranking)	Relative Payoff (Ranking)	Adopt within next 2 years? (Ranking)	Degree of use last 2 years (Ranking)	Relative Payoff (Ranking)	Adopt within next 2 years? (Ranking)
Argentina	2.38 (17)	3.40 (11)	31% (2)	1.82 (13)	3.57 (4)	14% (7)	1.70 (18)	3.70 (6)	14% (8)
Australia	2.96 (8)	3.10 (14)	14% (7)	2.00 (10)	2.53 (14)	22% (3)	2.80 (4)	3.39 (9)	20% (4)
Austria	2.57 (13)	2.67 (18)	29% (3)	2.08 (9)	2.56 (12)	0% (10)	2.12 (14)	2.50 (18)	14% (8)
Brazil	2.87 (10)	3.46 (10)	0% (9)	2.08 (8)	3.75 (2)		3.44 (2)	3.94 (1)	0% (12)
Canada	2.89 (9)	3.19 (13)	33% (1)	1.57 (17)	2.00 (17)	20% (4)	2.47 (9)	2.67 (17)	0% (12)
Chile	1.75 (19)	3.50 (9)	0% (9)	1.67 (15)	2.00 (17)	0% (10)	1.00 (19)		50% (1)
Denmark	1.82 (18)	3.00 (15)	0% (9)	2.23 (6)	4.00 (1)	0% (10)	2.18 (12)	3.80 (4)	0% (12)
Finland	3.36 (3)	3.78 (4)	0% (9)	1.60 (16)	2.80 (10)	0% (10)	2.55 (7)	3.86 (2)	0% (12)
Great Britai	2.48 (14)	2.09 (19)	0% (9)	1.97 (11)	2.79 (11)	0% (10)	2.06 (15)	2.90 (14)	20% (4)
Germany	2.47 (15)	2.80 (17)	17% (6)	1.00 (19)	1.00 (19)	0% (10)	2.47 (10)	3.10 (11)	17% (7)
Netherlands	3.29 (4)	3.64 (6)	0% (9)	1.83 (12)	2.38 (15)	7% (9)	2.53 (8)	3.08 (12)	6% (10)
Italy	2.44 (16)	3.80 (2)	24% (4)	2.78 (3)	3.55 (5)	33% (1)	2.17 (13)	3.42 (8)	6% (10)
Japan	3.96 (1)	4.04 (1)	0% (9)	3.42 (2)	3.59 (3)	0% (10)	3.31 (3)	3.76 (5)	0% (12)
Mexico	3.69 (2)	3.71 (5)		3.84 (1)	3.55 (5)	0% (10)	3.67 (1)	3.81 (3)	
Norway	2.86 (11)	3.80 (2)	0% (9)	1.55 (18)	2.20 (16)	13% (8)	1.83 (17)	2.86 (15)	0% (12)
Portugal	3.04 (5)	3.57 (8)	0% (9)	2.50 (4)	3.00 (4)	0% (10)	2.46 (11)	3.07 (13)	0% (12)
Spain	3.00 (6)	3.29 (12)	0% (9)	2.25 (5)	2.90 (9)	17% (5)	2.74 (5)	3.50 (7)	27% (3)
Sweden	3.00 (6)	3.61 (7)	20% (5)	1.69 (14)	3.05 (7)	17% (5)	1.98 (16)	2.70 (16)	38% (2)
USA	2.64 (12)	2.84 (16)	9% (8)	2.23 (7)	2.56 (12)	24% (2)	2.69 (6)	3.31 (10)	17% (6)
Total	2.91	3.35	12%	2.40	3.07	12%	2.55	3.34	16%

Table 3 - Past and Planned Activities in Manufacturing (continued)

Question	C2J Zero Defect Programs			C2K CAM			C2L CAD		
Country	Degree of use last 2 years (Ranking)	Relative Payoff (Ranking)	Adopt within next 2 years? (Ranking)	Degree of use last 2 years (Ranking)	Relative Payoff (Ranking)	Adopt within next 2 years? (Ranking)	Degree of use last 2 years (Ranking)	Relative Payoff (Ranking)	Adopt within next 2 years? (Ranking)
Argentina	1.76 (17)	3.38 (6)	21% (5)	1.50 (18)	3.20 (10)	22% (5)	2.25 (18)	3.62 (13)	33% (5)
Australia	2.55 (11)	3.06 (12)	14% (9)	2.48 (14)	3.05 (15)	20% (6)	3.54 (13)	3.48 (14)	25% (8)
Austria	2.00 (14)	2.38 (18)	20% (6)	3.15 (1)	3.06 (14)	40% (1)	4.04 (2)	3.68 (10)	0% (10)
Brazil	3.08 (4)	3.55 (4)		3.00 (5)	4.00 (1)		3.43 (14)	3.79 (8)	
Canada	2.78 (6)	3.20 (11)	33% (2)	2.67 (11)	3.33 (9)	0% (9)	3.95 (5)	4.22 (2)	50% (3)
Chile	1.50 (19)	2.00 (19)	0% (11)	1.00 (19)		0% (9)	1.00 (19)		100% (1)
Denmark	1.58 (18)	4.50 (1)	0% (11)	2.85 (7)	3.70 (3)	0% (9)	4.00 (4)	4.00 (5)	0% (10)
Finland	2.00 (14)	2.88 (16)	11% (10)	2.38 (16)	2.63 (18)	0% (9)	4.17 (1)	3.36 (17)	
Great Britai	2.63 (8)	3.04 (13)	0% (11)	2.77 (8)	3.11 (12)	40% (1)	3.89 (7)	3.67 (11)	0% (10)
Germany	2.75 (7)	2.92 (15)	20% (6)	2.57 (13)	3.00 (16)	0% (9)	3.64 (12)	3.68 (9)	50% (3)
Netherlands	2.50 (12)	2.67 (17)	0% (11)	2.67 (11)	3.00 (16)	0% (9)	3.35 (15)	3.32 (18)	0% (10)
Italy	2.00 (14)	3.38 (6)	32% (3)	2.24 (17)	3.64 (4)	16% (7)	3.85 (9)	4.11 (4)	67% (2)
Japan	3.64 (2)	3.80 (2)	0% (11)	2.75 (10)	3.47 (6)	0% (9)	3.92 (6)	4.17 (3)	0% (10)
Mexico	4.11 (1)	3.73 (3)		3.05 (4)	3.11 (11)	0% (9)	3.00 (17)	3.45 (15)	0% (10)
Norway	2.17 (13)	3.33 (9)	40% (1)	3.00 (5)	4.00 (1)	0% (9)	3.64 (11)	4.27 (1)	0% (10)
Portugal	3.00 (5)	3.38 (6)	0% (11)	2.40 (15)	3.07 (13)	0% (9)	3.29 (16)	3.41 (16)	0% (10)
Spain	3.20 (3)	3.50 (5)	0% (11)	3.15 (1)	3.35 (8)	27% (3)	3.76 (10)	3.86 (7)	33% (5)
Sweden	2.60 (9)	3.21 (10)	20% (6)	2.77 (9)	3.38 (7)	25% (4)	3.86 (8)	3.63 (12)	33% (5)
USA	2.59 (10)	2.92 (14)	29% (4)	3.09 (3)	3.59 (5)	7% (8)	4.03 (3)	3.91 (6)	25% (8)
Total	2.73	3.28	16%	2.69	3.30	14%	3.60	3.74	26%

Table 3 - Past and Planned Activities in Manufacturing (continued)

Country	C2M Design for Assembly/Manufacturability (DFA/DFM)						C2N Quality Function Deployment (QFD)						C2O Value analysis / Redesign of Products					
	Degree of use last 2 years	Ranking	Relative Payoff	Ranking	Adopt within next 2 years?	Ranking	Degree of use last 2 years	Ranking	Relative Payoff	Ranking	Adopt within next 2 years?	Ranking	Degree of use last 2 years	Ranking	Relative Payoff	Ranking	Adopt within next 2 years?	Ranking
Argentina	1.30	19	5.00	1	6%	8	2.05	15	3.50	5	25%	2	2.24	17	3.40	7	27%	4
Australia	2.35	13	2.75	17	0%	9	2.75	7	2.85	13	0%	11	2.16	19	2.78	17	0%	8
Austria	3.00	3	2.92	13	33%	3	2.63	8	3.00	11	17%	7	2.90	6	3.30	10	0%	8
Brazil	2.69	6	3.82	5			3.00	6	3.75	2			2.88	7	3.25	12		
Canada	2.60	8	3.70	6	0%	9	2.31	11	2.73	16	0%	11	2.50	11	3.33	9	0%	8
Chile	1.50	17	4.00	2	17%	5	3.33	3	2.67	17	20%	4	3.25	2	3.50	4	43%	3
Denmark	3.43	1	4.00	2	0%	9	1.69	17	3.33	8			2.91	5	3.71	2		
Finland	1.90	14	2.40	19	0%	9	1.80	16	3.29	9	10%	10	2.83	8	3.38	8	50%	1
Great Britain	2.39	11	2.76	16	17%	5	2.39	9	2.82	15	17%	7	2.97	4	3.09	15	0%	8
Germany	1.90	14	2.83	15	0%	9	1.33	19	2.25	19	13%	9	2.70	9	3.27	11	0%	8
Netherlands	2.42	10	2.89	14	0%	9	2.33	10	2.56	18	18%	6	2.67	10	2.75	19	0%	8
Italy	1.80	16	3.19	11	20%	4	1.41	18	3.22	10	28%	1	2.17	18	3.43	5	16%	7
Japan	3.39	2	3.57	7	0%	9	3.73	1	3.80	1	0%	11	3.69	1	3.88	1	50%	1
Mexico	2.88	5	3.45	8	75%	1	3.22	4	3.48	6	0%	11	2.98	3	3.66	3	0%	8
Norway	1.38	18	2.50	18	0%	9	2.09	13	3.60	4	20%	4	2.33	16	3.43	5	0%	8
Portugal	3.00	3	3.35	10	0%	9	3.71	2	3.64	3	0%	11	2.44	13	2.76	18	0%	8
Spain	2.53	9	3.15	12	40%	2	3.13	5	3.45	7	0%	11	2.47	12	3.21	13	17%	6
Sweden	2.37	12	4.00	2	0%	9	2.07	14	3.00	11	0%	11	2.35	15	3.10	14	20%	5
USA	2.63	7	3.41	9	6%	7	2.16	12	2.84	14	21%	3	2.41	14	3.04	16	0%	8
Total	2.49		3.34		14%		2.56		3.24		16%		2.64		3.29		13%	

Table 3 - Past and Planned Activities in Manufacturing (continued)

Question	C2P Quality Policy Deployment (QPD)						C2Q Reorganise to "plant within a plant"						C2R Defining a Manufacturing Strategy					
Country	Degree of use last 2 years	Ranking	Relative Payoff	Ranking	Adopt within next 2 years?	Ranking	Degree of use last 2 years	Ranking	Relative Payoff	Ranking	Adopt within next 2 years?	Ranking	Degree of use last 2 years	Ranking	Relative Payoff	Ranking	Adopt within next 2 years?	Ranking
Argentina	2.29	16	3.38	6	33%	1	2.04	19	3.25	15	29%	5	2.96	14	3.80	6	0%	11
Australia	3.00	7	3.17	9	14%	10	2.78	12	2.95	18	0%	12	2.92	15	2.90	18	40%	4
Austria	3.08	6	2.91	14	33%	1	2.44	16	3.10	16	50%	2	2.54	19	2.75	19	40%	4
Brazil	3.00	7	3.75	2			3.21	3	3.82	4	0%	12	3.89	1	4.16	1	0%	11
Canada	2.88	10	3.14	10	0%	11	2.72	13	3.33	13	0%	12	3.37	7	3.94	3	0%	11
Chile	3.33	4	2.33	18	25%	5	3.00	6	3.00	17	0%	12	3.00	13	3.50	12	33%	6
Denmark	2.45	13	2.67	16			2.33	18	4.00	1	100%	1	3.60	5	3.91	4		
Finland	2.75	12	3.00	12	0%	11	2.67	14	3.43	10	17%	7	2.91	16	3.25	14	0%	11
Great Britai	2.29	17	2.65	17	0%	11	3.13	5	3.33	13	0%	12	3.24	10	3.17	16	33%	6
Germany	3.00	7	3.36	7	0%	11	2.79	11	3.63	7	17%	7	3.19	12	3.58	10	0%	11
Netherlands	1.40	19	2.00	19	0%	11	2.42	17	2.90	19	0%	12	2.86	17	3.08	17	25%	9
Italy	1.76	18	3.36	7	19%	7	2.83	8	3.78	5	7%	11	2.85	18	3.55	11	17%	10
Japan	4.32	1	4.16	1			3.14	4	3.37	12	33%	3	3.80	3	3.76	8	100%	1
Mexico	2.77	11	2.74	15	17%	8	3.67	1	3.91	3			3.86	2	3.98	2		
Norway	3.27	5	3.56	4	25%	5	2.58	15	4.00	1	17%	7	3.46	6	3.83	5	75%	2
Portugal	3.83	2	3.75	2	0%	11	3.28	2	3.50	8	0%	12	3.30	8	3.20	15	0%	11
Spain	3.59	3	3.55	5	33%	1	2.79	10	3.69	6	33%	3	3.67	4	3.79	7	0%	11
Sweden	2.31	15	3.11	11	33%	1	2.80	9	3.46	9	20%	6	3.25	9	3.67	9	50%	3
USA	2.33	14	2.95	13	17%	8	2.94	7	3.42	11	9%	10	3.19	11	3.33	13	33%	6
Total	2.78		3.19		17%		2.92		3.51		16%		3.31		3.58		26%	

Table 3 - Past and Planned Activities in Manufacturing (continued)

Question	C2S Simultaneous Engineering						C2T Activity-Based Costing (ABC)						C2U Implementing Team Approach (Work Groups)					
	Degree of use last 2 years	Ranking	Relative Payoff	Ranking	Adopt within next 2 years?	Ranking	Degree of use last 2 years	Ranking	Relative Payoff	Ranking	Adopt within next 2 years?	Ranking	Degree of use last 2 years	Ranking	Relative Payoff	Ranking	Adopt within next 2 years?	Ranking
Country																		
Argentina	1.61	18	3.33	4	0%	8	2.39	9	3.63	3	20%	10	2.55	19	3.19	18	27%	12
Australia	2.48	10	2.63	17	0%	8	2.05	12	2.11	19	44%	2	3.37	12	3.46	14	33%	9
Austria	2.82	5	3.38	3	0%	8	3.08	4	3.00	14	33%	5	3.00	15	3.07	19	67%	4
Brazil	2.88	4	3.54	2	0%	8	2.69	6	4.14	1			3.91	1	4.00	1		
Canada	2.69	7	3.00	13	0%	8	1.81	18	2.75	16	40%	4	3.38	11	3.60	13		
Chile	2.00	17	3.90	1	50%	1	2.00	13	3.50	5	33%	5	3.00	15	3.33	15	100%	1
Denmark	3.57	1	2.60	18	0%	8	1.83	17	3.50	5	0%	13	3.58	4	3.75	7	100%	1
Finland	1.60	19	2.79	15	33%	5	2.33	10	3.44	7	14%	11	2.77	18	3.78	6	100%	1
Great Britain	2.27	11	3.10	11	0%	8	1.94	16	2.67	17	0%	13	3.58	5	3.75	7	0%	13
Germany	2.54	9	2.56	19	50%	1	2.53	8	3.13	11	0%	13	2.94	17	3.64	12	50%	6
Netherlands	2.09	16	3.26	7	36%	4	2.64	7	3.08	12	33%	5	3.41	9	3.93	2	33%	9
Italy	2.15	14	3.26	7	44%	3	1.78	19	3.36	9	33%	5	3.46	7	3.67	11	50%	6
Japan	2.68	8	3.20	8	14%	6	3.19	1	3.38	8	0%	13	3.77	2	3.92	3	0%	13
Mexico	3.07	3	3.14	10	0%	8	3.11	3	3.66	2	100%	1	3.63	3	3.68	9		
Norway	2.70	6	3.15	9	0%	8	2.80	5	3.57	4	25%	9	3.53	6	3.88	4	50%	6
Portugal	2.26	12	2.75	16	0%	8	3.18	2	3.28	10	0%	13	3.30	13	3.29	16	0%	13
Spain	2.13	15	3.07	12	0%	8	2.00	13	2.91	15	0%	13	3.25	14	3.27	17	60%	5
Sweden	2.21	13	3.29	6	0%	8	2.10	11	3.04	13	13%	12	3.38	10	3.83	5	33%	9
USA	3.53	2	3.29	6	6%	7	1.94	15	2.28	18	42%	3	3.41	8	3.68	9	0%	13
Total	2.56		3.13		16%		2.40		3.12		28%		3.38		3.65		41%	

Table 3 - Past and Planned Activities in Manufacturing (continued)

Question	C2V Benchmarking						C2W KAIZEN (Continuous Improvement)						C2X Total Productive Maintenance					
	Degree of use last 2 years	Ranking	Relative Payoff	Ranking	Adopt within next 2 years?	Ranking	Degree of use last 2 years	Ranking	Relative Payoff	Ranking	Adopt within next 2 years?	Ranking	Degree of use last 2 years	Ranking	Relative Payoff	Ranking	Adopt within next 2 years?	Ranking
Country																		
Argentina	2.10	14	3.13	9	9%	12	2.17	16	3.11	15	18%	8	2.04	12	2.70	12	11%	9
Australia	2.54	8	2.57	16	17%	8	3.88	2	3.83	7	0%	11	2.21	10	2.20	18	14%	7
Austria	2.20	10	3.17	8	0%	13	2.70	13	2.78	18	25%	6	2.00	13	2.29	17	17%	6
Brazil	2.65	7	3.18	7	0%	13	3.48	5	3.89	4			2.67	5	3.39	6		
Canada	2.13	13	3.00	10	20%	7	3.63	4	3.84	6	33%	4	1.82	15	2.40	16	40%	1
Chile	3.50	2	2.50	17	14%	10	2.33	14	1.50	19	100%	1	2.33	8	3.00	9	0%	12
Denmark	2.14	12	3.60	2			2.92	10	4.33	2	0%	11	2.64	6	4.60	1		
Finland	2.00	16	3.00	10	25%	2	1.92	18	3.63	8	40%	3	2.42	7	2.80	11	0%	12
Great Britai	1.94	18	2.24	19	0%	13	2.82	11	2.89	17	20%	7	1.64	18	2.47	15	25%	3
Germany	2.27	9	3.00	10	14%	10	1.78	19	3.00	16	33%	4	1.67	17	2.67	13	0%	12
Netherlands	1.90	19	2.67	15	25%	2	3.00	8	3.60	9	0%	11	2.29	9	3.00	9	10%	11
Italy	2.07	15	3.58	3	23%	5	2.17	17	3.41	13	44%	2	1.78	16	3.44	5	28%	2
Japan	4.04	1	4.09	1	0%	13	4.23	1	4.60	1	0%	11	4.59	1	4.41	2		
Mexico	2.80	3	3.44	4	100%	1	3.72	3	3.94	3			2.71	4	3.02	8	0%	12
Norway	2.71	5	3.27	5	17%	8	2.82	11	3.88	5	0%	11	1.45	19	2.00	19	20%	4
Portugal	2.78	4	3.18	6	0%	13	3.00	8	3.53	10	0%	11	3.04	2	3.37	7	13%	8
Spain	2.67	6	2.80	14	22%	6	3.06	7	3.40	14	17%	10	2.76	3	3.53	3	0%	12
Sweden	1.98	17	2.88	13	0%	13	2.29	15	3.50	12	0%	11	1.92	14	3.47	4	20%	4
USA	2.15	11	2.45	18	24%	4	3.32	6	3.53	11	18%	9	2.12	11	2.53	14	11%	9
Total	2.42		3.08		22%		3.00		3.60		23%		2.36		3.08		15%	

Table 3 - Past and Planned Activities in Manufacturing (continued)

Question	C2Y Energy Conservation Programs						C2Z Environmental Protection Programs						C2AA Health and Safety Programs					
Country	Degree of use last 2 years	Ranking	Relative Payoff	Ranking	Adopt within next 2 years?	Ranking	Degree of use last 2 years	Ranking	Relative Payoff	Ranking	Adopt within next 2 years?	Ranking	Degree of use last 2 years	Ranking	Relative Payoff	Ranking	Adopt within next 2 years?	Ranking
Argentina	1.92	17	3.25	6	0%	11	2.48	17	3.00	14	10%	9	3.12	14	3.50	10	17%	2
Australia	2.52	9	2.95	11	0%	11	3.04	12	3.24	12	14%	8	4.14	2	3.89	4		
Austria	2.33	15	2.55	17	14%	7	3.38	6	3.28	10	0%	10	3.21	13	3.00	19	0%	4
Brazil	3.19	2	3.42	3			3.50	4	3.75	3			4.11	4	4.15	2		
Canada	2.81	5	3.19	7	33%	3	3.53	3	3.17	13	0%	10	3.86	5	3.71	7		
Chile	1.50	19	2.00	19	0%	11	2.25	18	3.25	11	50%	2	2.75	19	3.75	6		
Denmark	2.54	8	3.86	1	100%	1	3.42	5	3.75	3	100%	1	3.38	12	3.44	11		
Finland	1.60	18	2.57	16	20%	4	1.73	19	2.25	19	0%	10	2.86	17	3.20	17	0%	4
Great Britai	3.03	3	2.90	12	14%	7	3.00	13	2.64	17	0%	10	3.60	9	3.26	16	0%	4
Germany	2.19	16	3.00	9	0%	11	3.12	9	3.69	6	33%	4	2.78	18	3.43	12	0%	4
Netherlands	2.38	13	2.64	15	0%	11	3.30	7	2.75	16	33%	4	3.61	8	3.17	18		
Italy	2.37	14	3.37	5	36%	2	3.20	8	3.71	5	0%	10	3.49	10	3.66	9	0%	4
Japan	3.36	1	3.48	2			3.92	1	4.04	2			4.12	3	4.17	1		
Mexico	2.46	11	2.49	18	13%	9	3.07	10	3.34	8			3.10	15	3.32	15	0%	4
Norway	2.75	6	3.38	4	0%	11	3.00	13	4.11	1	50%	2	3.63	7	3.87	5		
Portugal	2.52	10	2.67	14	0%	11	2.94	16	3.55	7	0%	10	3.42	11	3.70	8	0%	4
Spain	2.63	7	2.75	13	17%	6	2.95	15	2.50	18	0%	10	3.73	6	3.40	13	75%	1
Sweden	2.40	12	3.15	8	20%	4	3.06	11	2.95	15	17%	7	3.10	16	3.39	14	0%	4
USA	3.00	4	3.00	9	12%	10	3.72	2	3.33	9	25%	6	4.15	1	3.98	3	13%	3
Total	2.58		2.97		14%		3.16		3.28		15%		3.50		3.58		14%	

Table 3 - Past and Planned Activities in Manufacturing

Question	C3A		C3B		C3C	
	To what degree is there a process for translating corporate and marketing goals into a manufacturing strategy?		How much can manufacturing influence the development of corporate strategies and goals?		To what degree is manufacturing driven by the need to meet short-term financial, budget and output requirements?	
Country	1 = None, 5 = Formal	Ranking	1 = Not at all, 5 = A lot	Ranking	1 = not at all, 5 = A lot	Ranking
Argentina	3.26	16	3.61	5	3.45	10
Australia	3.55	10	3.62	4	3.48	9
Austria	3.38	13	2.88	19	3.20	13
Brazil	3.79	5	3.54	8	3.61	5
Canada	3.32	14	3.00	16	3.50	7
Chile	3.80	4	4.00	2	2.83	19
Denmark	3.75	6	3.31	11	3.00	17
Finland	2.82	19	2.94	18	3.18	14
Great Britai	3.03	17	3.03	15	4.17	2
Germany	3.00	18	2.96	17	2.96	18
Netherlands	3.27	15	3.04	14	3.15	15
Italy	3.85	3	3.24	12	3.37	11
Japan	4.12	1	4.15	1	4.00	3
Mexico	3.69	8	3.68	3	4.23	1
Norway	3.45	12	3.60	6	3.50	7
Portugal	3.55	11	3.54	7	3.10	16
Spain	3.86	2	3.24	13	3.60	6
Sweden	3.60	9	3.42	10	3.33	12
USA	3.73	7	3.48	9	3.78	4
Total	3.53		3.39		3.53	

Table 4 - Manufacturing Performance Improvements over Past Two Years

Question / Country	D1A Conformance to specification (manufacturing quality) 1992 = 100	Ranking	D2B Average unit manufacturing cost 1992 = 100	Ranking	D2C Inventory turnover 1992 = 100	Ranking	D2D Speed of product development 1992 = 100	Ranking	D2E On-time deliveries 1992 = 100	Ranking	D2F Equipment changeover 1992 = 100	Ranking	D2G Market share 1992 = 100	Ranking
Argentina	139.6	5	126.6	1	132.2	4	135.4	1	131.3	5	124.3	6	126.5	2
Australia	146.1	3	114.5	9	126.8	8	114.6	14	121.6	11	118.3	10	105.3	12
Austria	111.4	19	105.4	18	116.3	15	113.5	15	118.7	15	110.7	14	104.1	14
Brazil	150.9	2	117.3	5	148.8	2	124.6	6	130.6	6	110.2	15	110.6	7
Canada	125.3	11	112.5	14	119.9	11	117.6	10	146.1	2	120.6	7	102.3	16
Chile	113.3	16	115.0	8	115.3	16	115.8	11	111.2	19	118.0	11	103.4	15
Denmark	119.5	14	110.1	16	120.7	10	113.0	17	120.5	13	127.2	3	99.5	18
Finland	114.1	15	110.3	15	117.6	14	112.3	18	122.9	9	103.2	19	105.7	10
Great Britai	133.4	6	120.9	4	129.2	5	109.1	19	111.3	18	113.0	12	111.8	6
Germany	113.1	17	106.0	17	110.4	18	120.7	7	123.5	8	120.1	8	104.1	14
Italy	125.6	9	113.1	13	122.5	9	118.1	9	120.5	12	124.6	5	112.7	4
Japan	127.0	8	115.1	7	111.7	17	115.8	12	115.7	17	107.0	18	105.5	11
Mexico	111.5	18	104.4	19	119.7	12	115.1	13	117.7	16	118.6	9	109.9	8
Netherlands	123.3	13	121.1	3	118.9	13	113.3	16	121.7	10	107.8	17	163.8	1
Norway	130.0	7	113.8	11	108.8	19	125.0	5	142.6	3	108.3	16	101.7	17
Portugal	124.1	12	113.3	12	128.2	7	119.8	8	120.3	14	125.3	4	111.8	6
Spain	166.3	1	115.4	6	136.3	3	126.3	4	175.2	1	127.8	2	95.2	19
Sweden	125.3	10	114.0	10	128.8	6	130.7	2	126.2	7	111.5	13	106.4	9
USA	142.8	4	121.5	2	159.4	1	130.1	3	140.5	4	136.2	1	125.8	3
Total	128.8		114.1		126.5		120.2		126.8		118.3		111.3	

Table 4 - Manufacturing Performance Improvements over Past Two Years (continued)

Question / Country	D2H Profitability 1992 = 100	Ranking	D2I Customer service 1992 = 100	Ranking	D2J Manufacturing lead time 1992 = 100	Ranking	D2K Procurement lead time 1992 = 100	Ranking	D2L Delivery lead time 1992 = 100	Ranking	D2M Product variety 1992 = 100	Ranking
Argentina	114.1	6	133.4	2	130.8	5	126.6	4	124.8	8	121.8	5
Australia	111.8	9	126.3	4	129.8	7	113.6	10	124.8	9	111.8	11
Austria	98.9	16	109.8	17	122.4	11	111.2	14	122.3	10	104.0	18
Brazil	111.1	10	128.0	3	126.5	9	120.8	5	127.3	5	109.2	16
Canada	127.4	2	121.8	6	154.3	2	118.5	7	155.9	1	119.1	7
Chile	125.0	3	118.3	8	115.3	17	111.0	15	113.0	18	115.0	9
Denmark	113.2	8	109.7	18	141.6	3	106.1	19	131.7	4	132.9	3
Finland	108.2	12	112.2	16	128.4	8	110.0	17	122.2	12	118.1	8
Great Britai	96.0	18	116.5	12	117.9	15	115.8	8	122.2	11	124.2	4
Germany	84.9	19	115.1	14	118.3	14	108.7	18	118.3	15	111.7	13
Italy	114.0	7	117.1	11	124.4	10	113.8	9	120.0	13	120.4	6
Japan	104.7	13	118.7	7	113.4	19	112.1	13	109.9	19	112.9	10
Mexico	101.5	14	117.7	9	118.8	13	112.7	12	116.9	17	109.8	15
Netherlands	110.8	11	115.5	13	116.5	16	110.4	16	119.8	14	111.8	12
Norway	115.2	5	109.3	19	120.9	12	113.0	11	125.0	7	94.6	19
Portugal	124.8	4	114.7	15	114.9	18	127.4	3	117.4	16	110.5	14
Spain	96.8	17	122.0	5	130.3	6	132.5	2	151.8	2	108.4	17
Sweden	100.5	15	117.5	10	134.6	4	118.8	6	125.4	6	137.0	2
USA	166.8	1	135.8	1	174.8	1	156.5	1	132.1	3	139.1	1
Total	110.8		119.6		128.3		118.9		124.2		117.4	

INDEX